Studies in Systems, Decision and Control

Volume 372

The series "Studies in Systems, Decision and Control" (SSDC) covers both new developments and advances, as well as the state of the art, in the various areas of broadly perceived systems, decision making and control–quickly, up to date and with a high quality. The intent is to cover the theory, applications, and perspectives on the state of the art and future developments relevant to systems, decision making, control, complex processes and related areas, as embedded in the fields of engineering, computer science, physics, economics, social and life sciences, as well as the paradigms and methodologies behind them. The series contains monographs, textbooks, lecture notes and edited volumes in systems, decision making and control spanning the areas of Cyber-Physical Systems, Autonomous Systems, Sensor Networks, Control Systems, Energy Systems, Automotive Systems, Biological Systems, Vehicular Networking and Connected Vehicles, Aerospace Systems, Automation, Manufacturing, Smart Grids, Nonlinear Systems, Power Systems, Robotics, Social Systems, Economic Systems and other. Of particular value to both the contributors and the readership are the short publication timeframe and the world-wide distribution and exposure which enable both a wide and rapid dissemination of research output.

Indexed by SCOPUS, DBLP, WTI Frankfurt eG, zbMATH, SCImago.

All books published in the series are submitted for consideration in Web of Science.

More information about this series at http://www.springer.com/series/13304

Cengiz Kahraman · Serhat Aydın
Editors

Intelligent and Fuzzy Techniques in Aviation 4.0

Theory and Applications

Springer

Editors
Cengiz Kahraman
Department of Industrial Engineering
Istanbul Technical University
Istanbul, Turkey

Serhat Aydın
Air Force Academy
National Defence University
Istanbul, Turkey

ISSN 2198-4182 ISSN 2198-4190 (electronic)
Studies in Systems, Decision and Control
ISBN 978-3-030-75066-4 ISBN 978-3-030-75067-1 (eBook)
https://doi.org/10.1007/978-3-030-75067-1

This Springer imprint is published by the registered company Springer Nature Switzerland AG
The registered company address is: Gewerbestrasse 11, 6330 Cham, Switzerland

I dedicate this book to my children Nazlı Ece, Yunus Emre, and Onur.

Prof. Cengiz Kahraman

I have greatest thanks to my mother Serpil Aydın and my father Zeki Aydın for supporting me during my education life.

Dr. Serhat Aydın

Preface

Industry 4.0 is the Fourth Industrial Revolution that focuses on interconnectivity, automation, machine learning and real-time data, mostly in engineering terms. It is the trend of automation and data exchange in today's manufacturing technology including cyber-physical systems, the Internet of Things, cloud computing and cognitive computation. It is the concept of factories in which machines are augmented with wireless connectivity and sensors, connected to a system that can visualize the entire production line and make autonomous decisions. Aviation is the totality of activities surrounding mechanical flight and the aircraft industry including all non-scheduled civil flying, both private and commercial. Aviation includes business flights, air charter, private aviation, flight training, ballooning, paragliding, parachuting, gliding, hang gliding, aerial photography, air ambulance, crop dusting, charter flights, traffic reporting, police air patrols and forest fire fighting. The application of interconnectivity, automation, cyber-physical systems, the Internet of Things, cloud computing, and cognitive computation machine learning and real-time data in aviation is called Aviation 4.0. The aim of this book is to present the theory and applications of intelligent Aviation 4.0 technologies. This book involves 22 chapters, each written by the experts of the related research area.

Chapter " Aviation 4.0 Revolution" explains the history of Industrial Revolution, Industry 4.0, and its technologies. Chapter " Intelligent Systems in Aviation 4.0 Industry" presents a literature review on the intelligent systems used in Aviation 4.0 applications through tabular and graphical analyses.

Chapter " Fuzzy Logic Controller for Aviation Parking with 5G Communication Technology" proposes a fuzzy logic controller (FLC) for the parking of aircrafts, and all the calculations are done using the fuzzy logic controller toolbox of MATLAB. Chapter " An Integrated Fuzzy Decision Making and Integer Programming Model for Robot Selection for a Baggage Robot System" deals with the development of an automated baggage robot system for airports, which is being newly constructed. Chapter " Complex Spherical Fuzzy Sets and an Application to Catering Services in Aviation 4.0" presents complex spherical fuzzy sets (CSFSs) and their operations and properties. An application is given for aviation industry 4.0 in order to assess the performance of catering services. Chapter " Digitalization on Aviation 4.0:

Designing a Scikit-Fuzzy Control System for In-Flight Catering Customer Satisfaction" develops a catering service evaluation system for customers, and an interface is designed to be used in the Aviation 4.0 industry. Chapter " Analysis of Intelligent Software Implementations in Air Cargo Using Fermatean Fuzzy CODAS Method" focuses on the problem of analyzing and selecting intelligent software in Air Cargo in the concept of Aviation 4.0. Chapter " Intelligent and Fuzzy Applications in Aircraft Handling Services with Aviation 4.0" introduces and analyses the state of the art of aircraft ground handling automation at airports. Chapter " Novel Spherical Fuzzy Eco-holonic Concept in Sustainable Supply Chain of Aviation Fuel" extends and combines two concepts of Eco-Holonic structures and spherical fuzzy sets to introduce a novel concept of spherical fuzzy Eco-Holonic structures. Chapter " On-Board Cost Index Computation Through Fuzzy Logic" deals with automated systems providing safety and efficiency to the operators and the passengers of commercial flights.

Chapter " Toward Joint Application of Fuzzy Systems and Augmented Reality in Aircraft Disassembly" presents a system architecture based on augmented reality and fuzzy intelligent decision-making for the disassembly of complex products. Chapter " Some Novel Preference Relations for Picture Fuzzy Sets and Selection of 3-D Printers in Aviation 4.0" establishes novel picture preference relations under picture fuzzy environment. A few numerical examples on selection of 3D printers in aviation sector demonstrate the applicability and validity of the proposed approach. Chapter " A Conceptual Framework for Estimating the Remaining Operational Lifetime of the Recovered Components from End of Life Aircraft Using Fuzzy Simulation and Digital Twin" proposes a framework for using fuzzy simulation and digital twins to estimate the remaining useful life (RUL) of recovered parts. The application of the model related to aircraft engines is discussed. Chapter " Designing a System Architecture for the Management of the Recovered Parts from End-of-Life Aircraft Using Fuzzy Decision Making and Blockchain" presents the recovery of high-value parts from end-of-life aircraft as a business model and a main source of revenue.

Chapter " A Hybrid Model Based on FAHP and WASPAS for Evaluation of Explosive and Narcotics Trace Detection Devices" applies fuzzy analytic hierarchy process for assigning weights of the attributes, and the weighted aggregated sum product assessment (WASPAS) method is used to determine the most suitable alternative device for explosive and narcotics trace detection. Chapter " Selection of the Best Face Recognition System for Check in and Boarding Services discusses the problems, developments and challenges of face recognition in terms of Aviation 4.0 to determine the best face recognition system for check in and boarding systems. Chapter " Intelligent and Fuzzy Approaches in Aviation 4.0 Transportation and Cargo Applications" aims at defining the functional requirements (FRs) that a smart container design to be used in air transportation should meet and the design parameters (DPs) that will meet these requirements.

Chapter "Blockchain Applications in UAV-Towards Aviation 4.0" is on embedding UAVs with block-chain technology (BCT), which is an increasing phenomenon since the UAVs should be operated in an environment with many obstacles and possibilities of cyber-attacks. Chapter "Intelligent and Fuzzy UAV Transportation

Applications in Aviation 4.0" presents different categories of delivery operation problems by using UAVs such as facility location problems, vehicle routing problems, path-planning problems and their applications. Chapter "Spherical Fuzzy Inference Systems (S-FIS) to Control UAVs' Communication Technologies" considers the UAVs' communication problem and provides a trustworthy solution for selecting the appropriate communication technology and controlling it. Chapter " Technology Analysis for Logistics 4.0 Applications: Criteria Affecting UAV Performances" handles ten different performance indicators related to UAVs with two different fuzzy number styles, namely hesitant and Pythagorean fuzzy numbers, and AHP method. Chapter "A Novel Mathematical Model to Design UAV Trajectory for Search and Rescue Operations in Disaster" designs a mathematical model to get an optimal path planning to steer the UAVs based on the potential risk degree (PRD) of the candidate location in the affected area.

We would like to thank the authors for their efforts in writing their studies and the anonymous reviewers for their hard works in selecting high-quality chapters in this book. We would like to express our sincere thanks to Prof. Janusz Kacprzyk for his continuous supports and helps.

Istanbul, Turkey
August 2021

Cengiz Kahraman
Serhat Aydın

Contents

Contributors

Aylin Adem Industrial Engineering Department, Gazi University, Ankara, Turkey

D. Ajay Sacred Heart College (Autonomous), Tirupattur District, Tamilnadu, India

Ahmet Aktas Department of Industrial Engineering, Gazi University, Ankara, Turkey

J. Aldring Sacred Heart College (Autonomous), Tirupattur District, Tamilnadu, India

Diego Alonso Tabares Chair of ISO TC 20/SC 9 – Air Cargo and Ground Equipment, Geneva, Switzerland

Shahzaib Ashraf Department of Mathematics, Abdul Wali Khan University, Mardan, Pakistan

Serhat Aydın Industrial Engineering Department, National Defence University, Air Force Academy, Yeşilyurt, Istanbul, Turkey

Reza Vatankhah Barenji Industrial Engineering Department, Hacettepe University, Ankara, Turkey

Esra Çakır Department of Industrial Engineering, Galatasaray University, Istanbul, Turkey

Selcuk Cebi Department of Industrial Engineering, Yildiz Technical University, Besiktas, Istanbul, Turkey

Ecem Cem Department of Industrial Engineering, Aydin University, Buyukcekmece, Istanbul, Turkey

Cihan Cetinkaya Management Information Systems Department, Adana Alparslan Turkes Science and Technology University, Adana, Turkey

Metin Dağdeviren Industrial Engineering Department, Gazi University, Ankara, Turkey

Yaser Donyatalab Industrial Engineering Department, Istanbul Technical University, Macka, Istanbul, Turkey

Fariba Farid Industrial Engineering Department, Istanbul Technical University, Macka, Istanbul, Turkey

Elmira Farrokhizadeh Industrial Engineering Department, Istanbul Technical University, Macka, Istanbul, Turkey

Mahmoud Golabi Université de Haute-Alsace, Mulhouse, France

Mehmet Kabak Department of Industrial Engineering, Gazi University, Ankara, Turkey

Cengiz Kahraman Industrial Engineering Department, Istanbul Technical University, Macka, Istanbul, Turkey

Samira Keivanpour Department of Mathematical and Industrial Engineering, Polytechnique Montreal, Montreal, Quebec, Canada

Fabio Krykhtine Industrial Engineering Department at Polytechnic School, Federal University of Rio de Janeiro, Rio de Janeiro, Brazil;
Fuzzy Logic Laboratory, Federal University of Rio de Janeiro, Rio de Janeiro, Brazil;
Innovation Agency, Federal University of Rio de Janeiro, Rio de Janeiro, Brazil

Dorota Kuchta Wroclaw University of Science and Technology, Wroclaw, Poland

Fatma Kutlu Gündoğdu Department of Industrial Engineering, National Defence University, Turkish Air Force Academy, Istanbul, Turkey

Felix Mora-Camino DUT, South Africa, Neometsys and ENAC, Toulouse, France, Fluminense Federal University, Niterói, Brazil;
Durban University of Technology (DUT), Durban, South Africa;
Ecole Nationale de L'Aviation Civile, Toulouse, France;
Fluminense Federal University/PURO, Rio das Ostras, Brazil

Mazyar Ghadiri Nejad Industrial Engineering Department, Girne American University, Kyrenia, TRNC, Turkey;
Department of Industrial Engineering, Girne American University, Kyrenia, TRNC, Turkey

Eren Ozceylan Industrial Engineering Department, Gaziantep University, Gaziantep, Turkey

Baris Ozkan Industrial Engineering Department, Ondokuz Mayıs University, Samsun, Turkey

Muhammad Saeed Depertment of Mathematics, University of Management and Technology, Lahore, Pakistan

Muhammad Saqlain School of Mathematics, Northwest University Xi'an, Xi'an, China;
Department of Mathematics, Lahore Garrison University, Lahore, Pakistan

Duygu Sergi Department of Industrial Engineering, MEF University, Maslak, Sarıyer, Istanbul, Turkey

Seyed Amin Seyfi-Shishavan Industrial Engineering Department, Istanbul Technical University, Macka, Istanbul, Turkey

Irem Ucal Sari Industrial Engineering Department, Istanbul Technical University, Macka, Istanbul, Turkey

Ziya Ulukan Department of Industrial Engineering, Galatasaray University, Istanbul, Turkey

Burcu Yilmaz Kaya Industrial Engineering Department, Gazi University, Ankara, Turkey

Introduction to Intelligent and Fuzzy Techniques in Aviation 4.0: Theory and Applications

Aviation 4.0 Revolution

Serhat Aydın and Cengiz Kahraman

Abstract The concept of Industry 4.0 arose in Germany in 2011 and spread to other countries. Industry 4.0 refers to a strategic step that was introduced by the German government seeking ways to change industrial manufacturing through digitalization and the development of new technologies. It introduced the world with new technologies such as cybersecurity, big data, the cloud, simulations, augmented reality, additive manufacturing, etc. The technologies also spread to other fields of business sectors such as aviation, agriculture, health, etc. Thanks to these technologies aviation sector underwent a change named Aviation 4.0. In this chapter, we first discuss the Industry revolution, Industry 4.0, and its technologies. Then, we explain Aviation 4.0 and technologies used in Aviation 4.0. Aviation 4.0 technologies are divided into three categories as follows: Ground services application in Aviation 4.0., Maintenance and production in Aviation 4.0, Unmanned aerial vehicles technology in Aviation 4.0. Furthermore, we give a brief discussion about Aviation 5.0. Finally, the conclusion is given.

Keywords Industry 4.0 · Aviation · Aviation 4.0 · Aviation 5.0

1 Introduction

Industry means that the companies and activities involved in the process of producing goods for sale, especially in a factory or special area. Another definition of industry is that a group of productive enterprises or organizations that produce or supply goods, services, or sources of income [1].

S. Aydın (✉)
Industrial Engineering Department, National Defence University, Air Force Academy, Yeşilyurt, Istanbul 34149, Turkey
e-mail: saydin3@hho.msu.edu.tr

C. Kahraman
Industrial Engineering Department, Istanbul Technical University, Macka, Istanbul, Turkey
e-mail: kahramanc@itu.edu.tr

C. Kahraman and S. Aydın (eds.), *Intelligent and Fuzzy Techniques in Aviation 4.0*,
Studies in Systems, Decision and Control 372,
https://doi.org/10.1007/978-3-030-75067-1_1

Thanks to the widespread of globalization that is followed by developments in technology and interconnected systems, the industry sector is continually evolving, and numerous companies are facing new and different challenges [2]. Depending on this widespread globalization, industrialization has evolved in four phases and that is defined as the industrial revolution. The industrial revolution is characterized by a new technology finding broad application in industry and thereby fundamentally changing established practices [3]. They are called revolutions, as they redefine the levels of industrial performance by dramatically changing how products are produced and of the degree of productivity in value creation [4].

With mechanization and mechanical power generation in the 1800s, the first industrial revolution began. It brought about the transition from manual labor to the first processes of production, mainly in the textile industry. A big factor of the shift was an increased quality of life [5]. It was the transition to new manufacturing processes using water and steam. It was immensely advantageous in terms of providing a greater variety of different items and establishing a better quality of life for some. Power sources such as steam and coal made the use of machines more feasible, and the concept of machine development has spread rapidly. Machines made production quicker and simpler, and all sorts of new inventions and developments have been made possible as welded.

The second industrial revolution was caused by electrification, which allowed industrialization and mass processing [6]. New technological systems were introduced during this period, most notably superior electrical technology that allowed for even higher output and more advanced machines.

The third industrial revolution was characterized by the incorporation of microelectronics and automation by digitalization. The third industrial revolution took place around 1970 and involved the use of electronics and information technology to further automate development. This encouraged flexible production in manufacturing, where several goods were produced on flexible production lines with machines that are programmable [5]. It can be said that it began with the first computer era. Compared to the computing power they were able to provide, these early computers were often very basic, unwieldy, and extremely large. In this era, networking, renewable energies, manufacturing and automation progressed considerably thanks to the internet access.

The concept of Industry 4.0 arose in Germany in 2011 and spread to other countries [6]. Industrial revolution 4.0 refers to a strategic step that was introduced by the German government seeking ways to change industrial manufacturing through digitalization and the development of new technologies [7]. Namely, the combination of smart machines, storage systems, and production systems into rational networks, merging the real and virtual worlds in cyber-physical systems (CPS) characterize the industrial revolution [8]. Harmoniously working production units which are the digital data, software and information technologies are connected with the industrial revolution [9]. The fourth industrial revolution is the era of smart machines, storage systems and manufacturing facilities that can autonomously share information without human interference, activate actions and control each other. Key elements of industry 4.0 are summarized in Table 1 (Fig. 1).

Table 1 Key elements of industry 4.0 [10]

	Technologies of industry 4.0
Advanced manufacturing	Such as collaborative, autonomous, and programmable robots, which are able to interact each other and with people
Augmented reality	Which is a set of tools that allow you to add information to those really feel
Additive manufacturing	This refers to the creating objects through additive production processes, mainly through 3D printing. Through various basic technologies that differ mainly due to the ability to employ different materials, additive manufacturing allows companies to produce prototypes and finished products directly on the market or produce individual components that can also value products in terms of design
Simulations	Which are necessary to optimize products and processes, minimizing the number of errors
Vertical and horizontal integration of information	Throughout the entire value chain, from supplier to end-consumer
Cybersecurity	This represents the need to protect the computer systems, ensuring a network security
Big data	Which represents the collection and analysis of large amounts of data to improve products and production processes
The cloud	Which represent the need to share large amounts of data or IT resources, available by the Internet and accessible at all times
The industrial Internet of things	Which represents the set of technologies and sensors that enable communication between artificial world and people, including products and production processes

Due to its high added value, aircraft manufacturing is one of the most important branches of durable goods production. This industry contributes to the development of new technologies that require highly qualified professionals. Therefore, there is a progression of aviation revolution up to the industry revaluation.

The first successful flight in the history of a self-propelled, heavier-than-air aircraft was realized by Orville and Wilbur Wright. The Wrights constructed their own wind tunnel and created several sophisticated devices to measure lift and drag on the 200 wing designs they tested [12]. They successfully flew it hundreds of times in 1902, and it worked much better than previous versions. The Wrights not only produced a working aircraft the following year, the Wright Flyer, but also helped advance the science of aeronautical engineering by using a systematic method of research, including wind tunnel testing of airfoils and flight testing of full-size prototypes [13]. According to the Smithsonian Institution and Fédération Aéronautique Internationale (FAI), the Wrights made the first sustained, controlled, powered heavier-than-air

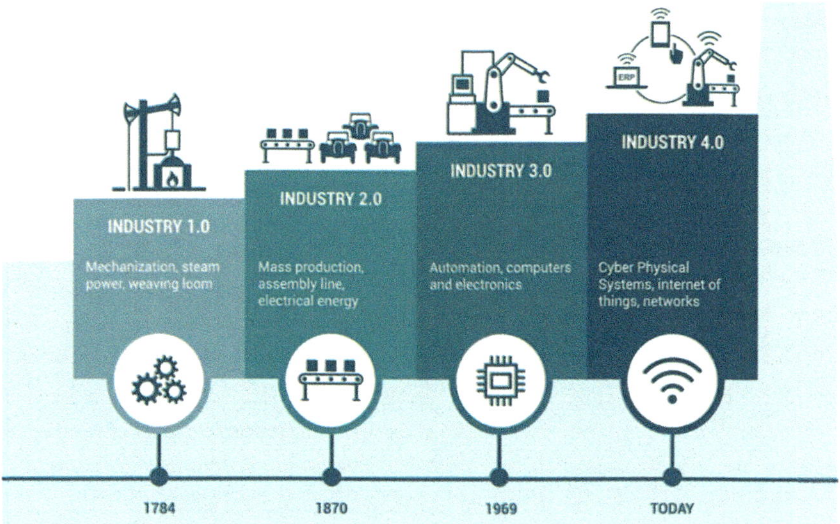

Fig. 1 Industry 4.0 revolution [11]

manned flight at Kill Devil Hills, North Carolina, 8 km south of Kitty Hawk, North Carolina on 17 December 1903 [14] (Fig. 2).

Nowadays, highly capable aircrafts that push human limits are produced. The aviation industry has gained highly capable of developing technology. This developing technology started with heavier-than-air manned flight by Wright brothers, today it is going on with new technologies by industry revolutions. Aviation 4.0 means using Industry 4.0 technologies in the aviation industry. After this explanation, the question

Fig. 2 First flight by Wright brothers [15]

arises why Aviation 4.0 is needed? To succeed in the global market, aviation companies need to remain present. They must produce goods quicker, cheaper, and better to accomplish that. Using the latest digital resources that Aviation 4.0 has to offer is the only way to ensure that this is achievable. Companies eliminate past problems and switch to a more productive way of operating that meets deadlines and satisfies consumers by embracing this digital approach. Aviation 4.0 instruments can allow businesses to reduce costs, improve efficiency, reduce inventory, increase the work rate, decrease waste, improve time to market and increase the ability to launch new products.

Aviation 1.0, the first evolutionary level, corresponding to the beginning of commercial aviation, where flight evolved under the rules of visual flight, following visual clues and signals, and hardly any instrumental assistance was available to help pilots fly. The technical challenges raised by how to create and fly an aircraft dominated this period. This period was named "*virtualization of the environment*". The technologies of this period could be classified as follows, altimeters, gyroscopes, speedometers, etc [16].

Aviation 2.0, the second evolutionary level, was dominated by using electrical devices to replace the old mechanism. In this stage, engineers improved airplanes to operate under extreme weather conditions and focused on air traffic control to orient airplanes in dense air traffic [16]. The technologies of this period could be classified as follows, instrument landing system, navigation instruments, electric autopilots.

Aviation 3.0, the third evolutionary level, involved the rapid introduction of electronics into the cockpit, powered by the availability of secure and accessible digital data processing and data communication technologies that penetrated the industry and society. In this stage, the complexity of instruments was greatly reduced by electronics. Integrated colored displays were developed for safety and more trusted flight. New liquid–crystal displays head up displays could be given as a new technology in this period.

Aviation 4.0 has started with the emergence of Industry 4.0. Supervisory control of production processes and big data collection and processing networks in Aviation 4.0 enables automation and integration of information technology systems to be possible. Systems for air traffic control was formulated as a cyber-physical system of systems (CPSS) [16]. Table 2 outlines the four phases of the revolution in commercial aviation, from Aviation 1.0 to Aviation 4.0 (Table 2).

2 Aviation 4.0 Technologies

With the emergence of Industry 4.0, the aviation industry has started to use Industry 4.0 technologies in many different areas of the aviation sector such as maintenance, baggage systems, eco-holonic systems, catering systems, etc. In many previous studies, researchers mainly focused on maintenance systems related to Aviation 4.0. This book focuses on different fields on Aviation 4.0 technologies. In the following

Table 2 Aviation 4.0 revolution [16]

Stage of aviation development	Characteristics	Characteristics of signal processing	Main challenges
Aviation 1.0: VFR	Airspace	Visual signals	How to build and fly an aircraft?
Aviation 2.0: IFR	Frequency space	Technical analog signals	How to fly an aircraft under adverse met conditions? How to control multiple aircraft flying in dense traffic in the same airspace?
Aviation 3.0: assistance systems; safety nets	Data space (Digitization; informatization)	Digital data processing; digital data communication	To support the people with the help of aggregated, visualized, understandable information to make informed decisions; SWIM
Aviation 4.0: AFR, RPAS, Decentralized decisions by systems	Cyber Space (Automation; Artificial Intelligence)	Cyber-physical systems	Cyber-physical systems to assist humans' physically strenuous, unpleasant or dangerous work. Cyber-physical systems to take decisions and to complete tasks autonomously

chapters, researchers discuss these different fields. In this chapter, we briefly mention some Aviation 4.0 technologies.

2.1 Ground Services Application in Aviation 4.0

Aircraft ground handling services describe servicing of an aircraft when aircraft is on the ground and (usually) parked at an airport terminal entrance. The performance of ground handling services is critical for a flight to take place on time. Aircraft ground handling services contain catering services, cabin services, passenger services, ramp services, etc. To provide effective services, ground handling companies are included in some applications in business processes. To give an example, the application of Augmented Reality (AR) is the best one. One of the last examples of this is ground handling services at Changi Airport in Singapore, personnel start using augmented reality glasses. When looking at the data matrix of cargo, all information regarding that cargo can be accessed with these glasses. Thus, ground services personnel can use this technology without contacting another channel for any information. They can carry out flight operations without wasting time. These glasses are made to the plane through cargo loading operations can be done faster and without errors [17].

Relevant AR technology applications have been also used in Airbus factories. Within the aircraft fuselage, tablets and sensors are mounted on the shop floor to monitor the location of parts and compare them to full-scale CAD models of the actual aircraft fuselage. Thus, this allows an image of the component of the region where they are operating to be downloaded to ensure that they have correctly fixed it. Location tags attached within the fuselage communicate with the sensors to enable tasks from any workplace to be visualized [18].

Another aircraft ground handling service is check-in, boarding and passport application. As it is known, passengers get on the plane from the point they enter the airport. Passengers have to perform many procedures as much as possible. They must own their documents and passport documents with them while traveling. Both passport control and boarding processes create a time pressure on passengers. To prevent this, new technology has developed named "hands in pocket" application. This application aims to be taken on the plane using their biometric features [19]. Another application of the biometric auto transition system has also been established in Sabiha Gökçen airport in Turkey. When passengers come to the first door, passengers have their electronic passport read by the reader. Then the system receives passengers data from the chip in the passport, the doors open, and passengers come in. When passengers come to the second door, they make the system scan their fingerprint and face. Thus, it matches the data it receives from passengers at that moment with the data it receives from passengers passport, identifies passengers and passes it through the second passport door (Figs. 3 and 4).

Baggage services application is another aircraft ground handling service application. The most common problem in baggage services is the loss of luggage. A related application has been developed to prevent this problem. With this application, by placing a card in the trunk, it can be monitored instantly at the airport where

Fig. 3 Augmented reality in aviation [20]

Fig. 4 Biometric auto transition system [21]

the baggage is. This system is called RFID (Radio Frequency Identification) technology, baggage tracking is carried out instantaneously and the baggage reaching the baggage delivery points is notified to passengers via SMS (Fig. 5).

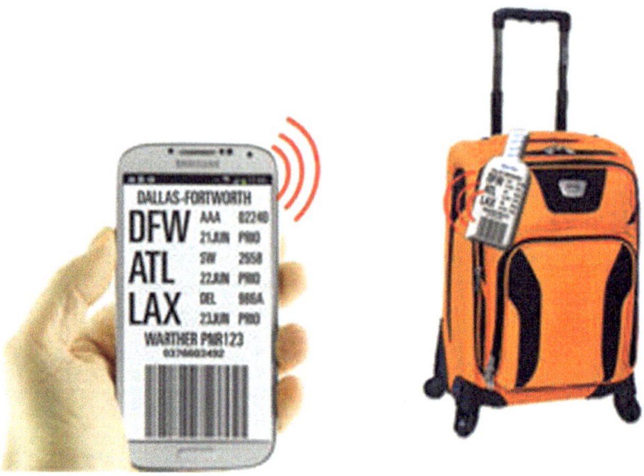

Fig. 5 RFID application in baggage services [22]

2.2 Maintenance and Production in Aviation 4.0

Maintenance is the process of ensuring that a system continually performs its intended function at its designed-in level of reliability and safety. Aviation production is a process of combining various material inputs and immaterial inputs to make something for aircraft. Engineers utilize the Industry 4.0 technologies for maintenance and product materials, some of these technologies are Collaborative robotic systems, Additive Manufacturing, Augmented reality, etc.

Robots increase productivity, favoring output savings and eliminating issues with lack of skilled labor. The high efficiency, low investment costs and, above all, adaptability of robots make them the ideal option for an efficient and simple solution for automation. Unique software are designed for robotics which aim to increase the repeatability and precision of the positioning of robots in this way. This feature helps to meet the requirements of aviation manufacturing.

Collaborative robots have been developed in Industry 4.0 era. Collaborative robots, referred to as cobots or co-robots, are the newest type of robots designed to interact with people in a shared work environment. The primary benefit of this form of robot is a cooperation between human robot in collaboration (Fig. 6).

Another Industry 4.0 technology used in aviation is Augmented Reality (AR). AR is defined as an integration of digital information with the user's environment in real time. Real-time information is used in the form of text, images, and audio enhancements integrated with actual objects with AR. AR is not only used in maintenance or production, also used in many areas such as AR creates a virtual checklist to help with pre-flight checks. Moreover, AR overlays a corridor view to show the

Fig. 6 Collaborative robot applications [19]

Fig. 7 Augmented reality in Aviation 4.0 [24]

appropriate path for taking off or landing. AR is also helpful with the cruise phase of airlines [23].

Airbus uses AR technology in the aircraft production process. The company produced a Mixed Reality Application. This application increases the productivity in production by scanning parts and detecting errors on structures Airbus has reduced the proportion of wrongly positioned or missing brackets in maintenance and production. Since 2015, Mixed Reality Application has been installed on about a hundred tablets and has been operated by more than 1000 users at production facilities across Europe. Moreover, Airbus factories use AR as follows: Tablets and sensors are mounted within the aircraft fuselage on the shop floor to monitor the location of components and compare them to full-scale real-world CAD models (Fig. 7).

Additive manufacturing, also called 3D printing process, is one of the Aviation 4.0 technologies. Additive manufacturing operates from the inside out, creating the end part layer by layer, as opposed to traditional manufacturing methods that obtain apart from a solid blank of raw material. The Additive manufacturing method prints very thin material layers over the other until the layers form a strong component that can be printed with high-quality materials such as plastics, aluminum, titanium, glass, and concrete [18].

3D printing enables more easier production process and easily produces very complex components. These components are smaller, easier to manufacture and, in the end, much less costly than traditional components. 3D printing provides the potential to increase the production of replacement parts by minimizing lead times and prices, which also leads to shorter maintenance times and less raw material usage, improving productivity. In September 2017, following thorough testing and EASA approval, the was installed on a serial production aircraft by Airbus. Furthermore,

Boeing applied additive manufacturing technology to the CST-100 program to reduce mass, cost, and cycle time.

2.3 *Unmanned Aerial Vehicle Technology in Aviation 4.0*

Unmanned Aerial Vehicles (UAVs) are flying robots that operating in conjunction with onboard sensors and GPS, can be remotely controlled or operate autonomously via software-controlled flight plans in its embedded systems. UAVs have a crucial role in Aviation 4.0. UAVs get new technologies with Aviation 4.0; therefore, they have undertaken many new missions. Industrial 4.0 technologies spread in many research fields such as agriculture, logistics, search and rescue operations, transportation, etc.

With the progression of Industry 4.0, UAVs start to use many technologies such as blockchain, 5G networks, smart sensors, internet of things, etc. UAVs find different application areas by using these technologies. Some of them can be summarized as follows.

UAVs bring a huge contribution to company safety. Detection of humans, vehicles, and objects by tracking sensors inside the UAVs allows for an increased level of security, particularly along the fence and property lines. UAVs can immediately recognize and follow any suspicious object or individual with precise visual recognition. The need and cost for traditional surveillance cameras and sensors are both minimized by these advanced approaches.

UAVs also bring a new concept named Agriculture 4.0. Farmers use agricultural UAVs to gather more data about their agriculture field. UAVs can collect data related to crop yield, the welfare of animals, soil quality, weather, nutrient measurements. consequently, farmers increase crop yields and increase overall profitability. Furthermore, farmers can observe whether there are any challenges regarding soil fertility, fertilizer conservation or soil dead areas by accessing 3D maps of actual soil by UAVs.

To sustain high yields, crops need coherent fertilization and spraying. This task has traditionally been done manually, by car, or even by airplane. Not only are these techniques inefficient and burdensome, but they can also be very expensive. Large reservoirs that can be filled with fertilizers, herbicides, or pesticides can be fitted with UAVs. It is far safer and cost-effective to use drones for crop spraying. Besides, UAVs can even be fully autonomous and programmed to operate on specific schedules and routes [25].

UAVs can survey and monitor terrain in harsh or unsafe conditions using integrated elevated map detection. In this way, UAVs can provide great convenience to protect people from dangers. Another use of UAVs with new technologies is search and rescue operations. UAVs can rapidly obtain disaster information with image analysis and a photogrammetric workflow. Also, UAVs gather the images of disaster and geographical perceptions of terrestrial targets for providing critical support to search and rescue operations.

UAVs are used to carry individual goods straight to the assembly line to ensure the supply of the right content at the right time. UAVs autonomously move parts up to 2.0 kg in weight directly to the appropriate area by moving by sensors. Also, UAVs are used by logistic companies to deliver goods to customers. Amazon Prime Air is a service that will deliver packages up to five pounds in 30 min or less using small drones. The aim of the service is to carry lightweight objects over short distances (Figs. 8, 9 and 10).

Fig. 8 Agriculture spreading with UAVs [26]

Fig. 9 Search and rescue with UAVs [27]

Fig. 10 Delivery with UAVs [28]

3 Aviation 4.0 in the Literature

In the literature, there are several research about Aviation 4.0 such as: Ghafar et al. [29] explored the employability prospective of aviation managers and examined the Industry 4.0-related management skills needed by aviation companies and analyzed the personal development intention of individual practitioners. They employed multistage-multilevel analysis on a single-case market study in a Malaysian Aviation Technical Training University. Findings from this study were segregated into three levels. At the industry level, it was concluded that there would be a brighter employability prospect of aviation managers at least until 2030. At the organizational level, there was an urgent need for the aviation professionals to upgrade themselves with managerial skills relevant to the Industry 4.0 such as data driven scenario planning, big data analysis, and collaborative decision making. At the individual practitioner's level, findings indicated that there was an increasing awareness among the individual aviation professional in particular the technical and engineering professional to engage in personal development in aviation management at postgraduate level. Ghafar et al. [30] aimed to conduct a preliminary investigation on the Industry 4.0's readiness level among small and medium aviation companies. Multiple case study had been conducted over 4 aviation companies (1 chartered airlines, and 3 maintenance, repair and overhaul (MRO) companies) operating in Malaysia. Data was collected using qualitative interview with managers of the respective companies and analysed using content analysis. Findings showed that the Industry 4.0's readiness status of small and medium aviation companies were still at the embryonic stage. The practical implication of this study contributes to the understanding of rising need for government agencies' supports to assist the small and medium aviation companies particularly in overcoming the Industry 4.0 implementation challenges. Büyüközkan et al. [31] analyzed the success factors of aviation 4.0. The factors were obtained via the literature survey and expert opinions. An integrated approach was used based on intuitionistic fuzzy multi-criteria decision making (IF MCDM) techniques. The weights of the success criteria were calculated by using the IF AHP (Analytic Hierarchy Process). IF VIKOR (Vise Kriterijumska Optimizacija I Kompromisno Resenje) method was used for selection of the most successful company among alternatives. Lehmann et al. [32] described the challenges of equipping a multilingual semantic application for Industry 4.0 with the right quality and the right quantity of semantics and a methodology to sustain such modeling effort. Mladkova [33] discussed human-technology relationship in the Industry 4.0 environment. She explored this relationship between human (tacit and explicit) and technical (explicit) factor in Industry 4.0 aviation environment. Arnaldo Valdes and Gómez Comendador [34] discussed the stages of aviation development from basic VFR flight rules at the Aviation 1.0, up to Aviation 4.0 stage where cyber-physical systems will be designed to assist humans physically tireless, unkind or hazardous work and to take decisions and to complete tasks autonomously. They illustrated current real cases of application of Aviation 4.0 concept to increase aviation safety and introduced future possible applications while outlines how they might significantly increase safety levels in aviation.

4 The Future of the Aviation Revolution

The new technologies of Industry 4.0 such as artificial intelligence (AI), machine learning (ML), robotics, Internet of Things (IoT), 3D printing, virtual and augmented reality (VR, AR), additive manufacturing (AM), etc. spread broadly in several fields of industry. This transition in cyber-physical infrastructure is summarized as the fourth industrial revolution, and the way we live, function, and connect to each other is radically shifting. Industry 4.0 applications are handled in many countries with new applications. Many businesses are still struggling to get into the concept and realization of Industry 4.0, but the next technological transition is already being discussed: Industry 5.0. Industry 5.0 is the future and an evolving trend: human–machine engagement and cooperation. The priority of industry 5.0 is to use machines and people effectively in a synergistic environment.

It is important to describe the next wave of technological revolution, how we communicate and how we define the laws of human–machine interaction. The aim of the Industry 4.0 is to provide the best quantity and mass production, on the other hand industry 5.0 will aim to be the new design of product with new technologies for more comfortable unprecedented life standard for human being. To achieve these goals industry 5.0 will bring unmanned technologies for society.

Humanoid robots are being built day by day to perform all human roles, especially with the awareness of Industry 5.0. Humanoid robots will be the focal point in industry 5.0. Humanoid robots will replace human muscle strength in the industry, in this way human being will begin to withdraw from the industrial workforce. It does not mean human being will lose their job; it means the new business fields will be created.

We can explain this situation as follows: Some professions are disappeared with emerging of industry 4.0 technologies, but on the other hand many professions are emerged such as social media experts, an increase in number of pilots for UAVs, etc. Furthermore, Industry 5.0 will improve both computer and human functions, leaving the mechanical with monotonous, boring activities and opening the imaginative side to the human being.

According to the development of industry 5.0, it would open a new era in aviation, named Aviation 5.0. This new name is so new that even a complete definition has not been made yet. Aviation 5.0 will use the new technologies developed in industry 5.0. It is inevitable that human and robot cooperate in aviation. Aviation is an industry with its own rules; therefore, there are various discussions on the risks of using humanoid robots in aviation by authorities.

Based on experience in industry 4.0, UAVs will get new roles and new technologies with Aviation 5.0. It is hard to predict where it will evolve from now, but UAVs will be more effective in agriculture, security, logistics applications. They will become smaller in size but will be more effective in use.

5 Conclusion

Industry 4.0 is the fourth wave of the industrial revolution that uses data-driven analytical approaches to handle business operations, allowed by technology such as artificial intelligence, machine learning, robotics, Internet of Things, 3D printing, virtual and augmented reality, additive manufacturing, etc. Industry 4.0 penetrate many fields of life such as aviation, agriculture, automobile sector, health sector, etc.

The aviation sector is one of the affected fields of Industry 4.0. Industry 4.0 technologies are transferred to the aviation sector, to support design, maintenance, in-flight structural health, flight management, etc. As a result, a new concept of the Aviation 4.0 has started up and has been developing day by day.

In this chapter, first we clarified the Industrial 4.0 revolution to understand the Aviation 4.0. We briefly discussed the technologies of Industrial 4.0 and explained the concept of Aviation 4.0. Furthermore, we defined the technologies of the Aviation 4.0 and stated the utilized areas of the technologies. We handled these technologies in three categories such as i) Ground services application in Aviation 4.0. ii)Maintenance and production in Aviation 4.0 and iii) Unmanned Aerial Vehicles technology in Aviation 4.0.

Furthermore, we discuss a new term: Aviation 5.0. It is a very new term in aviation, but the term inevitably will penetrate every corner of the aviation industry very soon. The concept will use humanoid robots and automation. Probably, many tasks will be realized only by UAVs such as good transportation, air ambulance services, search and rescue operations, firefighting activities, etc.

References

1. URL-1 (https://www.britannica.com/technology/industry)
2. Liao, Y., Deschamps, F., Loures, E.F.R., Ramos, L.F.P.: Past, present and future of industry 4.0: a systematic literature review research agenda proposal. Int. J. Prod. Res. **55**(12), 3609–3629 (2017)
3. Zhou, K., Liu, T., Zhou, L.: Industry 4.0: towards future industrial opportunities and challenges, 2015 12th International Conference on Fuzzy Systems and Knowledge Discovery (FSKD), IEEE, pp. 2147–2152 (2015)
4. Von Tunzelmann, N.: Historical coevolution of governance and technology in the industrial revolutions. Struct. Change Econ. Dyn. **14**(4), 365–384 (2003)
5. Puriwat, W., Tripopsakul, S.: Preparing for industry 4.0—will youths have enough essential skills?: An evidence from Thailand. Int. J. Instr. **13**(3), 89–104 (2020)
6. Magruk, A.: Uncertainty in the sphere of the industry 4.0—potential areas to research. Bus., Manage. Educ. **14**(2), 275–291 (2016)
7. Rojko, A.: Industry 4.0 concept: background and overview. Int. J. Interact. Mobile Technol. (IJIM), **11**(5), 77–90 (2017)
8. Petrillo, A., Felice, F. De, Cioffi, R., Zomparelli, F.: Fourth industrial revolution: current practices, challenges, and opportunities in digital transformation in smart manufacturing. InTech (2018)
9. Schuh, G., Potente, T., Wesch-Potente, C., Weber, A.R., Prote, J.P.: Collaboration mechanisms to increase productivity in the content of industrie 4.0. Paper presented at Robust Manufacturing Conference (RoMaC 2014). Bremen, Germany. Elsevier Science (2014) https://doi.org/10.1016/j.procir.2014.05.016
10. Baldassarre, F., Ricciardi, F., Campo, R.: The advent of industry 4.0 in manufacturing industry: literature review and growth opportunities. Dubrovnik Int. Econ. Meet. **3**(1), 632–643 (2017)
11. URL-2 https://optiware.com/page/2/?lang=it
12. Dodson, M.G.: "An Historical and Applied Aerodynamic Study of the Wright Brothers' Wind Tunnel Test Program and Application to Successful Manned Flight", US Naval Academy Technical Report, USNA-334, archived from the original on 5 September 2011, retrieved 11 March 2009 (2005)
13. URL-3. https://en.wikipedia.org/wiki/History_of_aviation#cite_note-Dodson-67
14. Telegram from Orville Wright in Kitty Hawk, North Carolina, to His Father Announcing Four Successful Flights, 1903 December 17". World Digital Library. 17 December 1903. Retrieved 21 July 2013
15. URL-4. https://www.wdl.org/en/item/9589/
16. Valdes, R.A., Comendador, V.F.G., Sanz, A.R., Castan, J.P. Aviation 4.0 more safety through automation and digitization. Aircr. Technol. **2**(4), 25–41 (2018)
17. Atalık, O, Akan, A.G.Ş., Bakır, A.G.M.: Havacilik 4.0: havayolu ve havaalani endüstrisinde güncel endüstri 4.0 uygulamaları. Sci. Committee 879
18. Barbosa, G.F., Aroca, R.V.: Advances of Industry 4.0 concepts on aircraft construction: an overview of trends. J. Steel Struct. Constr. 2472–0437 (2017)
19. URL-5. https://www.independent.co.uk/travel/news-and-advice/airportbiometrics-trial-face-recognition-klm-brisbane-sita-a7813271.html
20. URL-6. https://aviationvoice.com/ge-aviation-successfully-pilots-augmented-reality-in-maintenance-2-201709291015/
21. URL-7. https://www.aa.com.tr/tr/turkiye/sabiha-gokcen-havalimaninda-biyometrikotomatik-gecis/1474378
22. URL-8. https://makarfid.com/products/baggage-tracking/
23. URL-9. https://elearningindustry.com/augmented-reality-in-aviation-changing-face-sector-training-simulated-experience
24. URL-10. https://arpost.co/2019/02/01/augmented-reality-solution-to-shorten-aircraft-skin-inspection-time/

25. URL-11. https://uavcoach.com/agricultural-drones/
26. URL-12. https://www.rembeltech.com/yeni-nesil-tarim-dji-agras-mg-1/
27. URL-13. https://www.unmannedairspace.info/
28. URL-14. https://newatlas.com/amazon-prime-air-delivery-drone/29982/
29. Ghafar, N.H., Rahman, N.A.A., Mohammad, M.F.N., Shah, M.Z., Hassan, F.: Developing new aviation management postgraduate program in responding to industry 4.0: key findings from multistages-multilevels market study. IOP Conf. Ser.: Materials Sci. Eng. **645**(1), 28 October 2019. Int. Conf. Aerosp. Aviat. ICASA 2018; Institut Teknologi BandungBandung City; Indonesia; 23–25 October 2018
30. Ghafar, N.H., Harun, K., Lailawati, M.S., Rahman, N.A.A.: Preliminary investigation of industry 4.0 readiness among small and medium aviation companies in Malaysia. J. Phys.: Conf. Ser. **1573**, 012030. (2020) https://doi.org/10.1088/1742-6596/1573/1/012030
31. Büyüközkan, G., Feyzioğlu, O., Havle, C.A.: Analysis of success factors in aviation 4.0 using integrated intuitionistic fuzzy MCDM methods. International Conference on Intelligent and Fuzzy Systems, vol. 1029, pp. 598–606. INFUS 2019; Istanbul; Turkey; 23 July 2019 through 25 July 2019 (2020)
32. Lehmann, J., Heußner, A., Shamiyeh, M., Ziemer, S.: Extracting and modeling knowledge about aviation for multilingual semantic applications in industry 4.0, Proceedings—2018 6th International Conference on Enterprise Systems, ES 2018, 24 December 2018, Article number 8588258, pp. 56–606 International Conference on Enterprise Systems, ES 2018; Limassol; Cyprus; 1–2 October 2018
33. Mládková, L.: Industry 4.0: human-technology interaction: experience learned from the aviation industry. Proceedings of the European Conference on Knowledge Management, ECKM vol. 1, pp. 571–578 2018. 19th European Conference on Knowledge Management, ECKM 2018; Padua; Italy; 6–7 September 2018
34. Arnaldo Valdes, R., Gómez Comendador, V.F.: Aviation 4.0: more safety through automation and digitization. WIT Transactions on the Built Environment, vol. 174, pp. 225–236 (2018)

Intelligent Systems in Aviation 4.0 Industry

Cengiz Kahraman and Serhat Aydın

Abstract Aviation 4.0 is related to the design of cyber-physical systems which can support humans by aviation information systems and by helping them to make decisions and to complete tasks autonomously. Within aviation 4.0 context, robotics, augmented reality, radio frequency identification, and internet of things are some components of fundamental concepts and aviation 4.0 employs information technology systems and sensors. These connected systems can interact with each other using standard internet protocols and analyze data to predict failures, configure themselves, and adapt to changes. For this aim, intelligent systems are used to interact with human users in changing and dynamic physical and social environments. An intelligent system is a machine with an embedded, internet-connected computer that can gather and analyze data and communicate with other systems. This chapter presents a literature review on the intelligent systems used in aviation 4.0 through tabular and graphical analyses.

Keywords Intelligence systems · Aviation 4.0 · Particle swarm optimization · Ant colony optimization · Artificial bee colony · Fuzzy sets · Tabu search · Simulated annealing · Genetic algorithms · Differential evolution

1 Introduction

Intelligent systems exist all around us in terminals, digital televisions, home appliances, elevators, video games, cinema films, traffic lights, automobiles, digital signage and airplane controls, among a great number of other possibilities. Intelligence is an integral component of the developing internet of things, in which almost

C. Kahraman (✉)
Industrial Engineering Department, Istanbul Technical University, Macka, Istanbul, Turkey
e-mail: kahramanc@itu.edu.tr

S. Aydın
Industrial Engineering Department, National Defence University, Air Force Academy, Yeşilyurt, Istanbul 34149, Turkey
e-mail: saydin3@hho.msu.edu.tr

C. Kahraman and S. Aydın (eds.), *Intelligent and Fuzzy Techniques in Aviation 4.0*,
Studies in Systems, Decision and Control 372,
https://doi.org/10.1007/978-3-030-75067-1_2

everything imaginable can be provided with unique identifiers and the ability to automatically transfer data over a network without requiring human-to-human or human-to-computer interaction. These intelligent systems can use some heuristic or metaheuristic techniques to solve the complex problems for giving near optimal solutions.

Artificial intelligent can help to organize crew management including pilots, flight attendants, and engineers. The processes in an airline company require comprehensive planning and scheduling, artificial intelligence can help in all these processes. An artificial intelligence based maintenance system can help to predict the failures and can prevent flight delays and cancellations. Besides, artificial intelligence based ticketing systems can manage fair ticket prices and thus present competitive prices to the customers.

The rest of this chapter is organized as follows. Section 2 presents some of the intelligent optimization methods Sect. 3 presents intelligence in aviation 4.0 by using graphical analyses. Section 4 presents fuzzy systems in aviation 4.0. Section 5 presents different fuzzy MCDM methods. Section 6 gives the conclusions and future work directions.

2 Intelligent Systems

In this section we briefly introduce the intelligent optimization techniques. Particle swarm optimization, ant colony optimization, artificial bee colony, fuzzy sets, Tabu search, simulated annealing, genetic algorithms, differential evolution are among the intelligent optimization techniques. These techniques can be used to solve the complex problems of aviation 4.0 to obtain near optimal solutions.

Particle swarm optimization (PSO) method is inspired from the social behavior of biological swarm systems such as the movement of organisms in a bird flock or fish school. PSO method was developed originally by Kennedy and Eberhart [1]. It is a population-based computational method which achieves optimization by iteratively improving the candidate solutions. Particles, which are candidate solutions, form a population. The particles of the population are located in the search space according to the particle's position and velocity and the current optimum particles. The particles communicate either directly or indirectly with one another for search directions. As a result, the swarm is directed to the best solution. PSO has been used as an effective metaheuristic technique for various problem types of different applications.

Ant Colony Optimization (ACO) is a metaheuristic approach for solving hard combinatorial optimization problems. Ant colony optimization (ACO) algorithm based on the foraging behavior of ants has been first introduced by Dorigo and Gambardella [2]. The basic idea of ACO is to imitate the cooperative behavior of ant colonies. When searching for food, ants initially explore the area surrounding their nest in a random manner. As soon as an ant finds a food source, it evaluates it and carries some food back to the nest. During the return trip, the ant deposits a pheromone trail on the ground. The pheromone deposited, the amount of which may depend on the quantity and quality of the food, guides other ants to the food source [3]. Quantity of pheromone on the arc is decreased in time due to evaporating. Each ant decides to a path or way according to the quantity of pheromone which has been leaved by other ants. More pheromone trail consists in short path than long path. Because the ants drop pheromones every time they bring food, shorter paths are more likely to be stronger, hence optimizing the solution. The first developed ACO algorithm was the ant system (AS) [4], and since then several improvements of the AS have been devised [5–7].

Artificial bee colony (ABC) algorithm was proposed by Karaboğa [8]. Bee Colony Optimization (BCO) algorithm imitates the procedure of collective food search of honeybees. The initial search for the food is executed by a group of bees which inform their remaining bees in the hive about the location quantity and the quality of the food they have explored. A bee carrying out random search is called a scout. Moreover, the scout bees which will lead the followers also try to attract follower bees from the hive by a dance behavior named as waggle dance. During the waggle dance, the quantity of the food is also given to the followers. Besides, it is known that the quality food is an important factor for strong commitment among the bees. The foraging bees under the lead of the explorer bee leave the hive and collect the food in the explored area. The collected food is returned back to the hive. As the bees collect the food, they return back to the hive to store the food. Then, those bees may choose one of the following options to go through: (i) it may continue to collect food at the same location under its previous leader; (ii) it may choose to build up its own team and try to attract followers to join its team; or (iii) they may separate from the leader bee and become an uncommitted bee. The exploration of new areas and food collection processes continuously take place.

Genetic Algorithms (GAs) are heuristic procedures that use the principles of evolutionary algorithms. The methodology of Genetic algorithms has been developed by Holland [9] and applied extensively to various types of optimization problems. GAs are inspired from the biological process of natural selection and the survival of the fittest. A pool of solutions defined as a population of chromosomes and a search process is achieved by generations of crossovers. Improvement is aimed to be obtained by selecting the competitive chromosomes that weed out poor solutions and carry over the genetic material to the offspring. At each iteration, the competitive solutions are recombined with other solutions to obtain hopefully better solutions in terms of objective function value or the "fitness" value. The resulting better solutions are then used to replace inferior solutions in the population. For further details on Genetic Algorithms, the interested reader is referred to the study by Reeves [10].

Simulated annealing (SA) methods are the methods proposed for the problem of finding, numerically, a point of the global minimum of a function defined on a subset of a k-dimensional Euclidean space. The motivation of the methods lies in the physical process of annealing, in which a solid is heated to a liquid state and, when cooled sufficiently slowly, takes up the configuration with minimal inner energy. Metropolis et al. [11] described this process mathematically. SA uses this mathematical description for the minimization of other functions than the energy. SA algorithm is a technique to find a good solution of an optimization problem using a random variation of the current solution. A worse variation is accepted as the new solution with a probability that decreases as the computation proceeds. The slower the cooling schedule, or rate of decrease, the more likely the algorithm is to find an optimal or near-optimal solution [12].

The word Tabu (or taboo) comes from Tongan, a language of Polynesia, where it was used by the aborigines of Tonga island to indicate things that cannot be touched because they are sacred. According to Webster's Dictionary, the word now also means "a prohibition imposed by social custom as a protective measure" or of something "banned as constituting a risk." Difficulty in optimization problems encountered in practical settings such as telecommunications, logistics, financial planning, transportation and production has motivated in development of optimization techniques. Tabu search (TS) is a higher level heuristic algorithm for solving combinatorial optimization problems. It is an iterative improvement procedure that starts form an initial solution and attempts to determine a better solution.

Differential evolution (DE) is introduced by Storn and Price in 1996. DE is known as population-based optimization algorithm similar to GAs using similar operators; crossover, mutation and selection. According to Karaboğa and Ökdem [13], the main difference in constructing better solutions is that genetic algorithms rely on crossover while DE relies on mutation operation. This main operation is based on the differences of randomly sampled pairs of solutions in the population. DE algorithm uses mutation operation as a search mechanism and selection operation to direct the search toward the prospective regions in the search space. In addition to this, the DE algorithm uses a non-uniform crossover which can take child vector parameters from one parent more often than it does from others. By using the components of the existing population members to construct trial vectors, the recombination (crossover) operator efficiently shuffles information about successful combinations, enabling the search for a better solution space. An optimization task consisting of D parameters can be represented by a D-dimensional vector. In DE, a population of NP solution vectors is randomly created at the start. This population is successfully improved by applying mutation, crossover and selection operators.

Fuzzy sets capture the vagueness and impreciseness in the human thoughts and perceptions. The fuzzy set theory has been introduced by Zadeh in 1965 and it has been often employed in numerous areas from engineering sciences to social sciences. Fuzzy control theory has been used in many technologies such as air conditioners, automobile and such vehicle subsystems as automatic transmissions, ABS and cruise control, cameras, dishwashers, elevators, language filters on message boards and chat rooms for filtering out offensive text, the massive engine used in the Lord of the Rings films, which allowed large-scale armies to enact random yet orderly movements, pattern recognition in remote sensing, video game artificial intelligence, and home appliances (e.g., washing machine). Fuzzy control theory has a great chance to be applied to aviation 4.0 technologies such as automation, internet of things, artificial intelligence, cognitive computing, big data analytics, and digitization.

With the emergence of Industry 4.0, the aviation industry has started to use these technologies in many different areas of the aviation sector such as maintenance, baggage systems, eco-holonic systems, catering systems, etc.

3 Intelligence in Aviation 4.0

Scopus database has been searched for “aviation” and “fuzzy” or “intelligent” or “intelligence” or “heuristics”. This search gave 2990 publications whose illustrative analyses are presented below:

Figure 1 shows the frequencies of intelligent aviation publications with respect to years.

From Fig. 1, it is clearly seen that a significant increase in the slope of the curve after 2015.

Figure 2 presents the researchers making publications on intelligent aviation and their publication frequencies.

Figure 3 shows the affiliations of intelligent aviation publications. The leading institutes are Beihang University, NASA Ames Research Center, Nanjing University of Aeronautics and Astronautics, and Civil Aviation University of China.

Figure 4 presents the countries publishing on intelligent aviation. United states and China are by far leaders on intelligent aviation.

Figure 5 presents intelligent aviation documents by their types. Conference papers are by far leader among the other types.

Figure 6 presents the intelligent aviation publications by their sources. The leading sources for intelligence aviation are Aviation Week And Space Technology New York, Air Force Magazine, Lecture Notes in Computer Science Including Subseries Lecture Notes in Artificial Intelligence and Lecture Notes in Bioinformatics, Proceedings of SPIE The International Society for Optical Engineering, Advances in Intelligent Systems and Computing, and AIAA IEEE Digital Avionics Systems Conference Proceedings.

Figure 7 shows the subject areas of the intelligent aviation publications. Engineering and computer science are by far leaders among the other subjects.

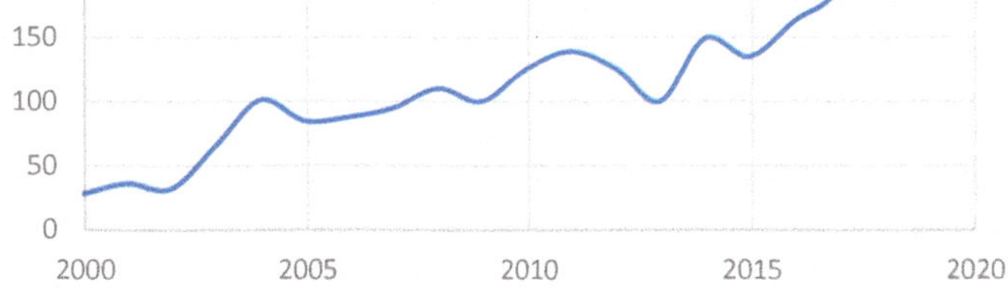

Fig. 1 Intelligent aviation publications by years

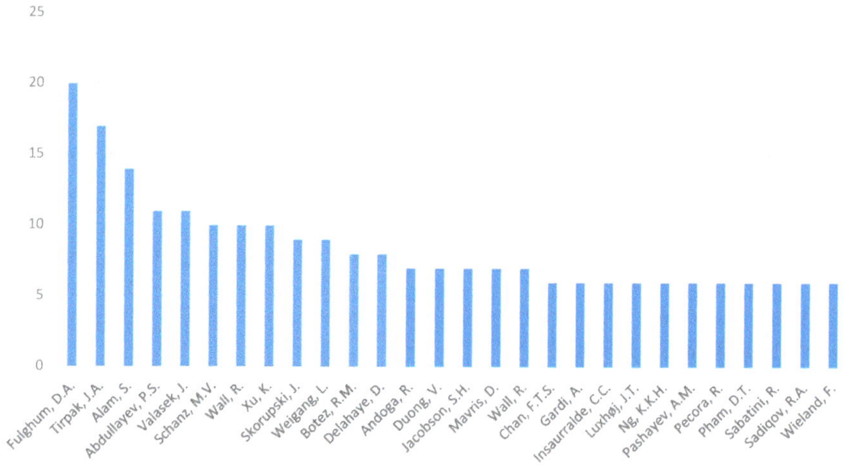

Fig. 2 Researchers publishing on intelligent aviation

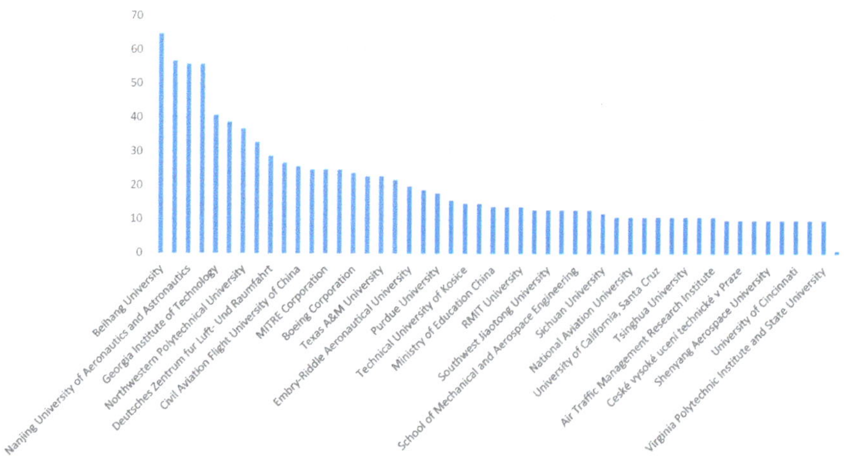

Fig. 3 Intelligent aviation publications by their affiliations

4 Fuzzy Systems in Aviation 4.0

In this section, we conducted a search on fuzzy articles on aviation since it is very popular to model aviation processes through fuzzy sets nowadays. Scopus database has been searched for the terms *aviation* and *fuzzy*. This search gave 270 publications whose illustrative analyses are presented below.

Figure 8 shows the frequencies of fuzzy aviation publications with respect to years. From Fig. 8, it is clearly seen that a significant increase in the slope of the curve after 2014 and decrease in the slope of the curve after 2018.

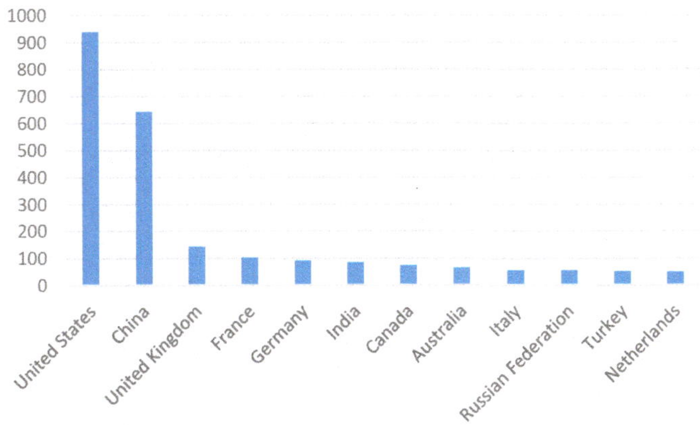

Fig. 4 Countries publishing on intelligent aviation

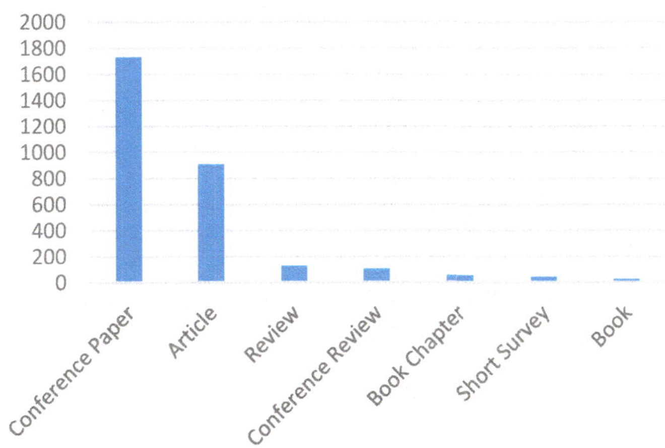

Fig. 5 Intelligent aviation documents by their types

Figure 9 shows the affiliations of fuzzy aviation publications. The leading institutes are Civil Aviation University of China, Civil Aviation Flight University of China.

Figure 10 presents the countries publishing on intelligent aviation. United states and China are by far leaders on fuzzy aviation.

Figure 11 presents fuzzy aviation documents by their subject areas.

Figure 12 presents intelligent aviation documents by their types. Conference papers are by far leader among the other types.

Fig. 6 Intelligent aviation publications by their sources

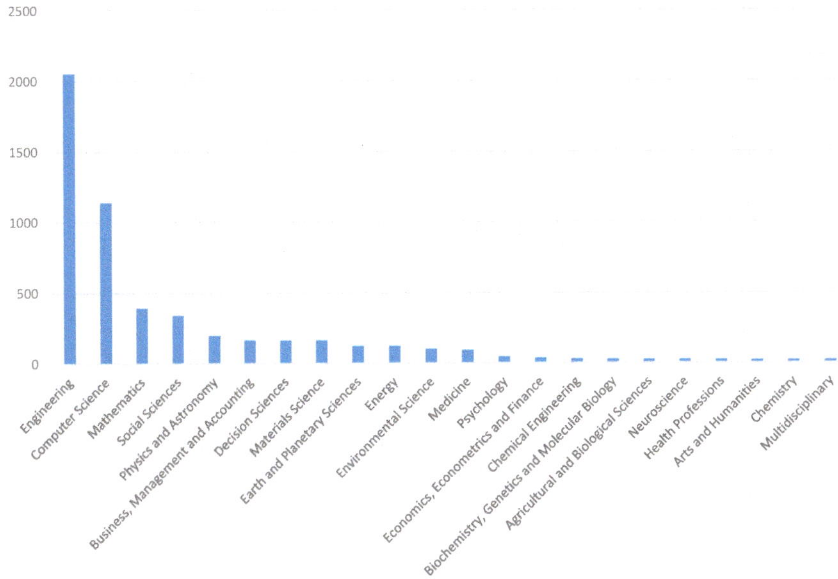

Fig. 7 Subject areas of intelligent aviation publications

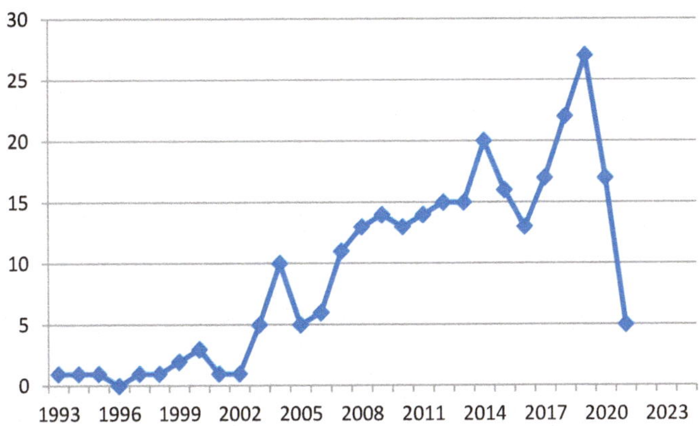

Fig. 8 Fuzzy aviation publications by years

5 Fuzzy MCDM Methods in Aviation 4.0

This book covers many fuzzy MCDM methods in Aviation 4.0. Therefore, we also conducted a research on fuzzy MCDM method articles in this section. In the following, these works are presented briefly.

Chang and Yeh [14] developed a new quantitative airline safety index based on comparative safety performance and efforts of airlines. They identified our safety

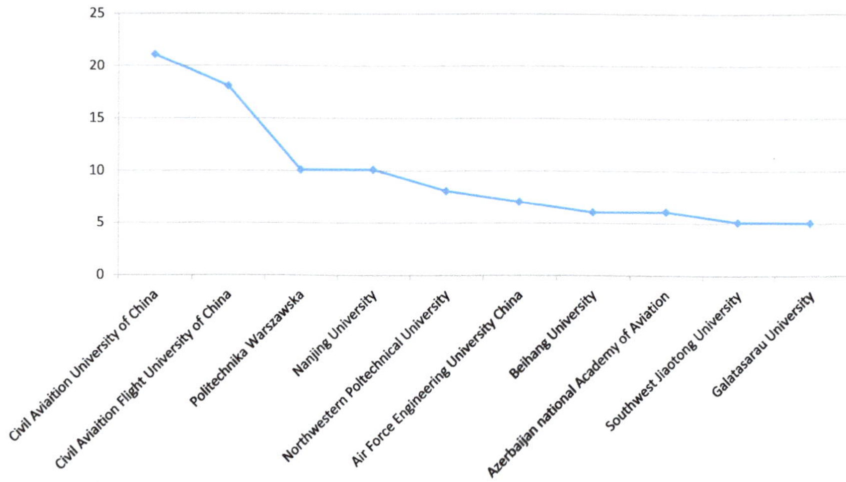

Fig. 9 Fuzzy aviation publications by their affiliations

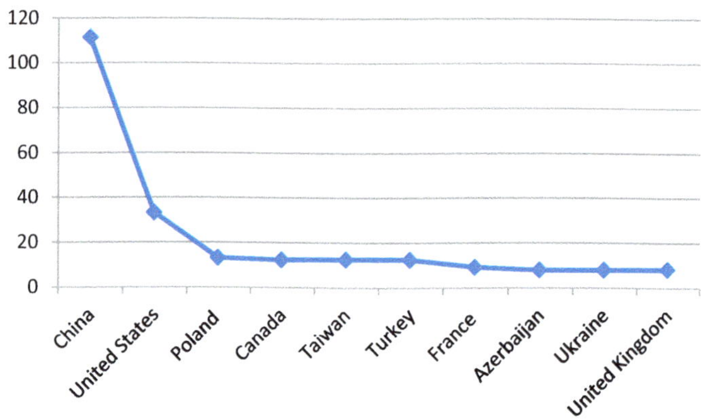

Fig. 10 Countries publishing on fuzzy aviation

dimensions corresponding to safety-related organizational divisions of airlines and their associated safety measures to facilitate the use of the index as a safety benchmarking and management tool. Based on the concept of optimality, a fuzzy multi-attribute decision making approach was developed to obtain a safety index for each airline. The airline safety index helped airlines understand their relative safety strengths and weaknesses in terms of manageable safety attributes, and identify functional areas for safety improvement.

Liou et al. [15] developed a method for building an effective safety management system for airlines is developed that incorporates organization and management factors. The method combines both fuzzy logic and Decision Making Trial and Evaluation Laboratory (DEMATEL). This method can map out the structural relations

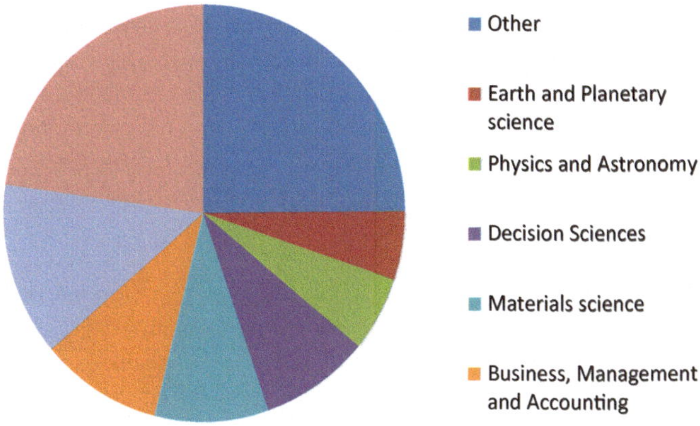

Fig. 11 Subject areas of fuzzy aviation publications

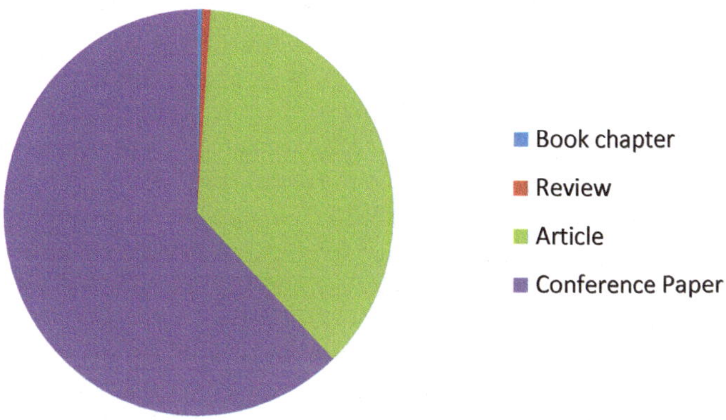

Fig. 12 Fuzzy aviation documents by their types

among diverse factors in a complex system and identify the key factors. They used data from the Taiwanese civil aviation industry to demonstrate the applicability of the model. Wang [16] utilized grey relation analysis to cluster financial ratios and find representative indicators. He also used fuzzy TOPSIS method to evaluate financial performance of airlines.

Hadjimichael [17] presented the Flight Operations Risk Assessment System (FORAS), a fuzzy expert system for proactive risk assessment. The aim of the model is to produce aviation risk models for air carriers, and includes software for the representation and application of those models. Nejati et al. [18] used a fuzzy TOPSIS method to review service quality factors of the airline industry. The research was conducted among graduate students of the University of Tehran, Iran. In order to meet the objectives of the study, Fuzzy TOPSIS approach was used. Required information was gathered through a questionnaire.

Chen and Chen [19] developed a hybrid fuzzy MCDM model based on a fuzzy analytic hierarchy process (FAHP) and VIKOR method was adopted to complete the construction of the aviate innovation system. The proposed model aims at providing Taiwanese airlines with innovation-oriented techniques for future strategy development. Liou and Chuang [20] proposed a fuzzy MCDM approach based on fuzzy AHP and SAW methods for evaluating corporate image and reputation in the airline market. A study of international airlines serving in Taiwan was conducted with the proposed method.

Kuo [21] presented an effective approach based on combining VIKOR, GRA, and interval-valued fuzzy sets. The approach was implemented using survey data of airline's service quality from the Chinese cross-strait passenger airlines. Torlak et al. [22] used fuzzy TOPSIS multi-methodological approach in the Turkish domestic airline industry. They made a comparative evaluation among air carriers in light of key success variables in the sector. Wang et al. [23] developed a hybrid model including the DEMATEL method and fuzzy linguistic method to accurately determine the characteristics of various situations in the contrived strategic process and appropriately describe the various evaluation criteria, so as to study the service quality connotations and the cause-effect relations and degrees of association among the various evaluation criteria of an ideal airline company.

Sevkli et al. [24] provided a quantitative basis to analytically determine the ranking of the factors in SWOT analysis via a conventional multi-criteria decision making method, Analytic Network Process (ANP). The proposed SWOT & fuzzy ANP methodology was implemented and tested for the Turkish airline industry.

Liao and Xu [25] proposed a method based on VIKOR and Hesitant fuzzy sets to evaluate the service quality among domestic airlines in Taiwan. Wu et al. [26] proposed FAHP method as an effective solution for resolving the uncertainty and imprecision in the evaluation of airlines' competitiveness. they reviewed the research of industrial international aviation competitiveness at both home and abroad, discuss a theoretical framework for the study of aviation competitiveness, establish an index system with five first-order indicators and 17 s-order indicators, set up a Chinese aviation competitiveness model based on simple fuzzy numbers from the fuzzy analytic hierarchy process, and evaluated the competitiveness of five major Chinese airlines.

Kilic et al. [27] suggested a fuzzy MCDM method based on fuzzy AHP and TOPSIS. They utilized fuzzy AHP method to obtain the relative importance/weights of the criteria. These weighted criteria were used as an input to the Technique for Order Preference by Similarity to Ideal Solution method to rank the decision alternatives. The proposed method was applied to the ERP selection problem at Turkish Airlines. Rezaei et al. [28] investigated supplier selection problems in the airline retail industry. They discussed a number of issues that make airline retail complex and distinguish it from conventional retail. They utilized a fuzzy AHP to evaluate the suppliers against the main criteria and sub-criteria. They applied the proposed methodology to one of the largest airlines in Europe, the Royal Dutch Airlines (KLM). Zhang and Xu [29] proposed a MCDM method including Pythagorean fuzzy sets and the TOPSIS method. They utilized the method to improve the service quality of domestic airline in Taiwan.

Ardakani et al. [30] proposed a hybrid model including SERVQUAL and fuzzy TOPSIS method. They aimed to identify service quality dimensions that could enhance the level of services provided by a variety of carriers and rank them. To do this aim, they focused on all Iranian passengers who traveled from Dubai International Airport to Imam Khomeini International Airport (Iran) using Iran Air, Mahan Air, and Emirates Airline within a one-month period, and the three air carriers were investigated by the proposed method. They proposed that "safety," "timeliness," and "variety and type of airplane" are the most important service quality criteria from the passengers' viewpoint. Chao and Kao [31] explored how an airline selects a strategic cargo alliance for membership. First, they collected dimensions and criteria for selecting a strategic cargo alliance through a review of the related literature. Then dimensions and criteria were screened using the Fuzzy Delphi Method (FDM). Finally, they calculated the weights of the selected dimensions and criteria using the fuzzy AHP.

Chiu and Hsieh [32] developed a latent human error analysis process, to explore the factors of latent human error in aviation maintenance tasks, and to provide an efficient improvement strategy for addressing those errors. They utilized the fuzzy TOPSIS method with four criteria to evaluate the error factors. They established a new analytic process for investigating latent human error and provided a strategy for analyzing human error using fuzzy TOPSIS. Garg [33] presented a model based approach of an AHP for evaluation of criteria and fuzzy TOPSIS for the selection of strategic alliance partner in Indian airline. Huang and Hsu [34] assessed the service requirements of combination air cargo carriers in Taiwan. They proposed a gap index based on fuzzy AHP to evaluate the perceived differences toward the service requirement attributes between combination air cargo carriers' users and combination air cargo carriers' operators. Lau et al. [35] developed a customer relationship management (CRM) business process management (BPM) model to identify airline customers with different degrees of relationship and profit potential, and select the highly profitable customers for developing retention strategy and processes, and convert the less profitable into profitable corporate accounts. They proposed a hybrid model including the FAHP and TOPSIS method. Ozdemir and Basligil [36] proposed a hybrid method for aircraft selection problem. They used fuzzy ANP and Choquet integral methods for evaluation, also they compared the results of both algorithms by using fuzzy AHP method.

Deveci et al. [37] focused on the process of new route selection, which has a high impact on the revenue and passenger numbers of airlines, and elected a new route at an airport in the North American region. They used interval type-2 fuzzy TOPSIS in order to find new aspects for new route feasibility. Dinçer et al. [38] evaluated the performance of the European airlines, using a balanced scorecard perspective. They used a hybrid multi-criteria approach including combining the fuzzy DEMATEL, fuzzy ANP, and MOORA methods. They demonstrated that customer dimensions and profit per customer are the most significant key factors in the balanced scorecard perspective. Görener et al. [39] proposed a three-phase hybrid approach to address the supplier performance evaluation problem in the aviation industry where the suppliers are obliged to follow tight tolerances regarding quality and delivery times. First, they selected the evaluation criteria, then, an interval Type-2 fuzzy AHP was employed to determine their relative importance. Finally, these weights were given as input to an Interval Type-2 fuzzy TOPSIS method. Li et al. [40] proposed a hybrid approach based on fuzzy AHP and 2-tuple fuzzy linguistic method to evaluate in-flight service quality. Their model consisted of three stages. In the first stage, the modified version of SERVQUAL instrument was used and a hierarchy of the evaluation index system for in-flight service quality was constructed. In the second stage, they used fuzzy AHP to analyze the structure of the in-flight service evaluation problem. Using linguistic variables, pairwise comparisons for evaluation criteria and sub-criteria were made to determine the weights of criteria and sub-criteria. In the third stage, the ratings of sub-criteria were assessed in linguistic values to express the qualitative evaluation of passengers' subjective opinions, and then the linguistic values are transformed into 2-tuples and the 2-tuple linguistic arithmetic mean operator is utilized to obtain the average ratings of 100 respondents. Liu et al. [41] proposed a hybrid model consisting of fuzzy DEMATEL and factor analysis to evaluate service quality in airlines. Zhang et al. [42] selected the best strategic freight forwarder by using a bi-objective intuitionistic fuzzy programming model.

Barak and Dahooei [43] proposed a novel hybrid method using fuzzy data envelopment analysis (DEA) and fuzzy multi-attribute decision making including fuzzy SAW, fuzzy TOPSIS, fuzzy VIKOR, ARAS-F, COPRAS-F and fuzzy MULTI-MOORA for ranking the airlines' safety. Chen and Ren [44] developed a multi-attribute sustainability evaluation model for assessing the sustainability of various alternative aviation fuels. They utilized fuzzy ANP to incorporate the interdependences and interactions among the criteria, and used fuzzy Grey Relational Analysis (FGRA) to determine the integrated priority of each alternative aviation fuel. Four alternative aviation fuels (petroleum refining, Fischer–Tropsch synthesis based on natural gas, algal-based fuel, and soybean-based fuel) were studied by the proposed model.

Karaman and Akman [45] suggested a model using fuzzy numbers and AHP method together to help disentangle the problem of selection decisions of Corporate Social Responsibility schemes in the airline sector. Kurtulmuşoğlu et al. [46] addressed critical service features prioritized by female and male passengers by ranking airlines on fuzzy COPRAS method. They determined that comfortable chair, seat space, leg room, and flexibility in reservation are the most impactful service features for females. On the other hand, availability of non-stop flight, frequency of flights, and airfare are the other impactful service features among males. Perçin [47] proposed a combined fuzzy multi-criteria decision making approach based on DEMATEL, ANP and VIKOR methodologies in a fuzzy environment for evaluating airline service quality. Sun et al. [48] suggested hesitant fuzzy TOPSIS method for airlines service quality evaluation.

Büyüközkan et al. [49] proposed a new and genuine DSQ model. The proposed model consists of digital tangibles, reliability, digital interaction, digital trust, and customer-centricity dimensions; and 35 criteria related to them. They determined importance degrees of criteria with the interval-valued intuitionistic fuzzy AHP method. Figure 13 represents different fuzzy MCDM method used in aviation 4.0.

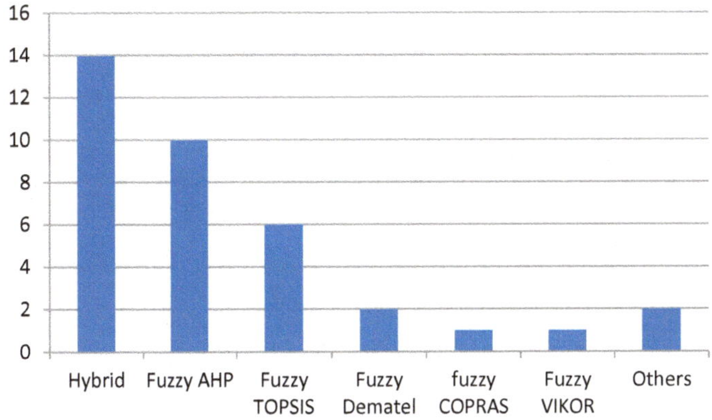

Fig. 13 Fuzzy MCDM publications

From Fig. 13, it is clearly seen that fuzzy hybrid methods are the most used ones in aviation.

Kandakoglu et al. [50] used fuzzy axiomatic design simulation for evaluating the alternative conceptual aircraft designs. Atli and Kahraman [51] tried to solve aircraft maintenance planning problems using fuzzy critical path analysis.

6 Conclusions

Aviation 4.0 describes a perfect service system characterized by better levels of controlling, organizing, and transforming all the processes, resulting in higher quality of services. Aviation 4.0 includes many different technologies such as; robotics, augmented reality, radio frequency identification, and internet of things etc.

As aviation 4.0 includes complex and various technologies, real life applications need utilize intelligent and fuzzy methods to solve the encountered problems. Therefore, this chapter proposes a literature survey on intelligent and fuzzy systems in aviation 4.0.

In this chapter, we have first presented a broad literature survey on intelligent methods in aviation including swarm optimization, ant colony optimization, artificial bee colony, fuzzy sets, etc. After graphical illustrations on intelligent applications in aviation 4.0, we have given a literature review including just fuzzy techniques since it is very often used in modeling aviation problems. Finally, we have presented a literature review on fuzzy MCDM methods since their extensive usage in aviation alternative selection or ranking problems.

In the future studies, new extensions of intelligent systems or newly developed hybrid intelligent techniques can be employed in modeling and solving aviation 4.0 problems.

References

1. Kennedy, J., Eberhart, R.: Particle swarm optimization. In: Proceedings of ICNN'95-International Conference on Neural Networks, vol. 4, pp. 1942–1948. IEEE (Nov 1995)
2. Dorigo, M., Gambardella, L.M.: Ant colony system: a cooperative learning approach to the traveling salesman problem. IEEE Trans. Evol. Comput. **1**(1), 53–66 (1997)
3. Socha, K., Dorigo, M.: Ant colony optimization for continuous domains. Eur. J. Oper. Res. **185**(3), 1155–1173 (2008)
4. Dorigo, M.: Optimization, Learning and Natural Algorithms. Ph. D. Thesis, Politecnico di Milano (1992)
5. Gambardella, L.M., Dorigo, M.: Ant-Q: a reinforcement learning approach to the traveling salesman problem. In: Machine Learning Proceedings, pp. 252–260. Morgan Kaufmann (1995)
6. Gambardella, L.M., Dorigo, M.: Solving symmetric and asymmetric TSPs by ant colonies. In: Proceedings of IEEE International Conference on Evolutionary Computation, pp. 622–627. IEEE (May 1996)
7. Stützle, T., Hoos, H.H.: MAX–MIN ant system. Future Gener. Comput. Syst. **16**(8), 889–914 (2000)
8. Karaboga, D.: An Idea Based on Honey Bee Swarm for Numerical Optimization, vol. 200, pp. 1–10. Technical report-tr06, Erciyes university, engineering faculty, computer engineering department (2005)
9. Holland, J.H.: Adaptation in natural and artificial systems: an introductory analysis with applications to biology, control, and artificial intelligence. MIT press (1992)
10. Reeves, C.R.: Genetic alorithms. In: Glover, F., Kochenberge, G.A. (eds.) Handbook of Metaheuristics, pp. 55–82. Kluwer Academic, Boston (2003)
11. Metropolis, N., Rosenbluth, A.W., Rosenbluth, M.N., Teller, A.H., Teller, E.: Equation of state calculations by fast computing machines. J. Chem. Phys. **21**(6), 1087–1092 (1953)
12. Xinchao, Z.: Simulated annealing algorithm with adaptive neighborhood. Appl. Soft Comput. **11**(2), 1827–1836 (2011)
13. Karaboğa, D., Ökdem, S.: A simple and global optimization algorithm for engineering problems: differential evolution algorithm. Turk. J. Electr. Eng. Comput. Sci. **12**(1), 53–60 (2004)
14. Chang, Y.H., Yeh, C.H.: A new airline safety index. Transp. Res. Part B: Methodological **38**(4), 369–383 (2004)
15. Liou, J.J., Yen, L., Tzeng, G.H.: Building an effective safety management system for airlines. J. Air Transport Manage. **14**(1), 20–26 (2008)
16. Wang, Y.J.: Applying FMCDM to evaluate financial performance of domestic airlines in Taiwan. Expert Syst. Appl. **34**(3), 1837–1845 (2008)
17. Hadjimichael, M.: A fuzzy expert system for aviation risk assessment. Expert Syst. Appl. **36**(3), 6512–6519 (2009)
18. Nejati, M., Nejati, M., Shafaei, A.: Ranking airlines' service quality factors using a fuzzy approach: study of the Iranian society. Int. J. Qual. Reliab. Manage. (2009)
19. Chen, J.K., Chen, I.S.: Aviatic innovation system construction using a hybrid fuzzy MCDM model. Expert Syst. Appl. **37**(12), 8387–8394 (2010)
20. Liou, J.J., Chuang, M.L.: Evaluating corporate image and reputation using fuzzy MCDM approach in airline market. Qual. Quant. **44**(6), 1079–1091 (2010)
21. Kuo, M.S.: A novel interval-valued fuzzy MCDM method for improving airlines' service quality in Chinese cross-strait airlines. Transp. Res. Part E: Logistics Transp. Rev. **47**(6), 1177–1193 (2011)
22. Torlak, G., Sevkli, M., Sanal, M., Zaim, S.: Analyzing business competition by using fuzzy TOPSIS method: an example of Turkish domestic airline industry. Expert Syst. Appl. **38**(4), 3396–3406 (2011)
23. Wang, R., Lin, Y.H., Tseng, M.L.: Evaluation of customer perceptions on airline service quality in uncertainty. Procedia-Soc. Behav. Sci. **25**, 419–437 (2011)

24. Sevkli, M., Oztekin, A., Uysal, O., Torlak, G., Turkyilmaz, A., Delen, D.: Development of a fuzzy ANP based SWOT analysis for the airline industry in Turkey. Expert Syst. Appl. **39**(1), 14–24 (2012)
25. Liao, H., Xu, Z.: A VIKOR-based method for hesitant fuzzy multi-criteria decision making. Fuzzy Optim. Decis. Making **12**(4), 373–392 (2013)
26. Wu, C., Zhang, X.Y., Yeh, I.C., Chen, F.Y., Bender, J., Wang, T.N.: Evaluating competitiveness using fuzzy analytic hierarchy process—a case study of Chinese airlines. J. advanced transportation **47**(7), 619–634 (2013)
27. Kilic, H.S., Zaim, S., Delen, D.: Development of a hybrid methodology for ERP system selection: the case of Turkish Airlines. Decis. Support Syst. **66**, 82–92 (2014)
28. Rezaei, J., Fahim, P.B., Tavasszy, L.: Supplier selection in the airline retail industry using a funnel methodology: conjunctive screening method and fuzzy AHP. Expert Syst. Appl. **41**(18), 8165–8179 (2014)
29. Zhang, X., Xu, Z.: Extension of TOPSIS to multiple criteria decision making with Pythagorean fuzzy sets. Int. J. Intell. Syst. **29**(12), 1061–1078 (2014)
30. Ardakani, S.S., Nejatian, M., Farhangnejad, M.A., Nejati, M.: A fuzzy approach to service quality diagnosis. Mark. Intell. Plann. (2015)
31. Chao, C.C., Kao, K.T.: Selection of strategic cargo alliance by airlines. J. Air Transport Manage. **43**, 29–36 (2015)
32. Chiu, M.C., Hsieh, M.C.: Latent human error analysis and efficient improvement strategies by fuzzy TOPSIS in aviation maintenance tasks. Appl. Ergon. **54**, 136–147 (2016)
33. Garg, C.P.: A robust hybrid decision model for evaluation and selection of the strategic alliance partner in the airline industry. J. Air Transport Manage. **52**, 55–66 (2016)
34. Huang, S.H.S., Hsu, W.K.K.: Evaluating the service requirements of combination air cargo carriers. Asia Pacific Manage. Rev. **21**(1), 1–8 (2016)
35. Lau, H., Nakandala, D., Samaranayake, P., Shum, P.K.: BPM for supporting customer relationship and profit decision. Bus. Process Manage. J. (2016)
36. Ozdemir, Y., Basligil, H.: Aircraft selection using fuzzy ANP and the generalized choquet integral method: the Turkish airlines case. J. Intell. Fuzzy Syst. **31**(1), 589–600 (2016)
37. Deveci, M., Demirel, N.Ç., Ahmetoğlu, E.: Airline new route selection based on interval type-2 fuzzy MCDM: a case study of new route between Turkey-North American region destinations. J. Air Transport Manage. **59**, 83–99 (2017)
38. Dinçer, H., Hacıoğlu, Ü., Yüksel, S.: Balanced scorecard based performance measurement of European airlines using a hybrid multicriteria decision making approach under the fuzzy environment. J. Air Transport Manage. **63**, 17–33 (2017)
39. Görener, A., Ayvaz, B., Kuşakcı, A.O., Altınok, E.: A hybrid type-2 fuzzy based supplier performance evaluation methodology: the Turkish airlines technic case. Appl. Soft Comput. **56**, 436–445 (2017)
40. Li, W., Yu, S., Pei, H., Zhao, C., Tian, B.: A hybrid approach based on fuzzy AHP and 2-tuple fuzzy linguistic method for evaluation in-flight service quality. J. Air Transport Manage. **60**, 49–64 (2017)
41. Liu, J., Yu, S., Chu, J., Gou, B.: Identifying and analyzing critical factors impacting on passenger comfort employing a hybrid model. Hum. Factors Ergon. Manuf. Serv. Ind. **27**(6), 289–305 (2017)
42. Zhang, W., Ju, Y., Liu, X., Giannakis, M.: A mathematical programming-based method for heterogeneous multicriteria group decision analysis with aspirations and incomplete preference information. Comput. Ind. Eng. **113**, 541–557 (2017)
43. Barak, S., Dahooei, J.H.: A novel hybrid fuzzy DEA-Fuzzy MADM method for airlines safety evaluation. J. Air Transport Manage. **73**, 134–149 (2018)
44. Chen, L., Ren, J.: Multi-attribute sustainability evaluation of alternative aviation fuels based on fuzzy ANP and fuzzy grey relational analysis. J. Air Transport Manage. **68**, 176–186 (2018)
45. Karaman, A.S., Akman, E.: Taking-off corporate social responsibility programs: an AHP application in airline industry. J. Air Transport Manage. **68**, 187–197 (2018)

46. Kurtulmuşoğlu, F.B., Can, G.F., Pakdil, F., Tolon, M.: Does gender matter? Considering gender of service in the airline industry. J. Air Transport Manage. **70**, 73–82 (2018)
47. Perçin, S.: Evaluating airline service quality using a combined fuzzy decision-making approach. J. Air Transport Manage. **68**, 48–60 (2018)
48. Sun, G., Guan, X., Yi, X., Zhou, Z.: An innovative TOPSIS approach based on hesitant fuzzy correlation coefficient and its applications. Appl. Soft Comput. **68**, 249–267 (2018)
49. Büyüközkan, G., Havle, C.A., Feyzioğlu, O.: A new digital service quality model and its strategic analysis in aviation industry using interval-valued intuitionistic fuzzy AHP. J. Air Transport Manage. **86**, 101817 (2020)
50. Kandakoglu, A., Kahraman, C., Topcu, I.: Fuzzy Axiomatic Design Simulation for Evaluating the Conceptual Aircraft Designs, Proceedings, 10th International FLINS Conference on Uncertainty Modeling in Knowledge Engineering and Decision Making (FLINS 2012) August 26–29, 2012, Istanbul, Turkey, (2012)
51. Atlı, Ö., Kahraman, C.: Aircraft maintenance planning using fuzzy critical path analysis. Int. J. Comput. Intell. Syst. **5**(3), 553–567 (2012)

Intelligent and Fuzzy Applications in Aviation 4.0 Ground Services

Fuzzy Logic Controller for Aviation Parking with 5G Communication Technology

Muhammad Saqlain and Muhammad Saeed

Abstract The fourth industrial revolution is the current trend of automation. Devices that are connected with multiple networks are still unable to connect fastly, and the world is looking for the application of the massive lot, whereas the 5G network is enabled to meet the demand. It is a dire need to automate the aviation industry with this technology. So, in this research, a Fuzzy Logic Controller (FLC) for the parking of aircrafts is proposed and all the calculations are done using the fuzzy logic controller toolbox of MATLAB. An aircraft can be parked without human interference by seeking the information of flights and the availability of parking space in the garage along with the clearance of the runway. Membership values of FLC are taken in the form of a neutrosophic number and then converted into a fuzzy number using the accuracy function. Results evaluate that 5G will automate the Aviation automation industry with precision and efficiency in operation. 5G also has an ability of flexibility support to the upcoming technologies which are unknown to us.

1 Introduction

Communication systems have come a long way from simple wired connections to the current fourth-generation (5G) wireless systems revolutionizing the communication technologies. Global information and communication technologies are single-handedly relying on these wireless communication systems. This journey started with communication to rich multimedia contents that include futuristic concepts like

M. Saqlain (✉)
School of Mathematics, Northwest University Xi'an, Xi'an, China
e-mail: msaqlain@lgu.edu.pk

Department of Mathematics, Lahore Garrison University, Lahore, Pakistan

M. Saeed
Depertment of Mathematics, University of Management and Technology, C-2 Johar Town, , Lahore 54000, Pakistan
e-mail: Muhammad.saeed@umt.edu.pk

C. Kahraman and S. Aydın (eds.), *Intelligent and Fuzzy Techniques in Aviation 4.0*, Studies in Systems, Decision and Control 372,
https://doi.org/10.1007/978-3-030-75067-1_3

3D gaming technologies, cloud computing, and high-speed wireless connectivity. The current network technologies that include Long-Term Evolution (LTE) hasn't been consistent in providing the great data rate to make a permanent move from mere telephony to new ways of communication including video conferencing, video streaming, and the emerging applications from the Internet of Things (IoT) where the devices are expected to communicate billions and trillions of bytes of data via machine to machine communication paradigm. This can all be done only if the stringent quality of service, data rates, energy efficiency, and reliability of the connection is better than that of LTE\LTE-A services that we have today [1].

This immense pressure put on cellular network companies by these day-to-day increasing technological advancements was met in the form of presentation of 5G technology. 5G technology has higher data rates, lower latency, higher spectral efficiency, greater network capacity, and more reliable connection as compared to LTE. In this case, researchers around the globe have started to investigate 5G solutions for enumerated problems [2–4]. The goal of the 5G networking is to achieve the capacity that is 10,000-fold the capacity of 4G networks with 100 times the data rates showing an end to end latency one millisecond with improvement in certain areas of network densification leading to 1000 times more connected devices [5, 6].

A report by Global Mobile Suppliers from April 2019 has identified a total of 224 operators in 88 countries which are testing or trailing, licensed by the Global System for Mobile Communications (GMSA), to conduct any type of trials regarding 5G technology (GMSA 2019). The first country to implement a 5G technological system on a large scale was South Korea in April 2019. According to another survey published by the Swedish telecommunication giant Ericsson, it is estimated that 5G internet will cover 65% of the world's population by 2025 [7].

The aviation industry has opted for these methods of wireless communication and has proven to be one of the fastest-growing technology sectors in the world in this context overtaking the aircraft manufacturing sector [8–10]. Using these high-speed wireless technologies in airlines and air transport facilities, 4G LTE and 5G technologies are now on the verge. They are used for in-flight broadband connectivity where they are planned for address passengers of the airbus personally, using augmented reality headsets for more safety [11–13]. But due to technological and logical barriers, the implementation of these technologies has proven to be a hurdle [14]. Some of the challenges that presented themselves include some aspects that are beyond the reach of 4G technologies such as high latency, low data rate, and most importantly the mobility requirement has proven to be the biggest hurdle in mixed content-centric telecommunication technological services [15, 16].

With the development of fuzzy sets [17], decision making becomes easier but later on, this theory was extended by [18] named as the Intuitionistic fuzzy number theory. To deal with more precision, accuracy, and indeterminacy this idea was extended by [19] called neutrosophy theory. To, discuss the applications of these theory number of developments were made but the most important one is the theory of soft set [20]. Later on, fuzzy, intuitionistic, and neutrosophy theories were extended to fuzzy soft set [21], intuitionistic soft set [22], and neutrosophic soft set [23]. In different fields, the applications of these theories are presented by many researchers [24–26] but with the development of multi-criteria decision making (MCDM) techniques, it becomes a more powerful tool to solve problems.

Now the question arises why we are using these techniques in this case study? To get the answer of this question, firstly you need to know about the linguistic variables [27], image processing, and fuzzy logic controllers (FLC) [28–31] along with MATLAB implementation, since FLC is mainly used in automation of different gadgets [32, 33]. Recently, researchers are also applying the MCDM technique in the automation industry to solve the problems [34–36]. Thus, the proposed case study consists of all the requirements, which makes it a perfect problem to apply the above-mentioned FLC techniques.

From an evolutionary perspective, aviation development Aviation 1.0 to Aviation 4.0 is connected with the manufacturing process of aircraft, digitalization of loT, and the safety levels in air transport [37, 38]. Currently, the modern engine of an aircraft includes more than 5000 sensors, which generates up to 10 GB/sec of data. The data generated by the aerospace industry alone could soon surpass the magnitude of the consumer internet. The challenges related to this data include safety assurance of hardware and software, decision making, and predefined situations [13].

In a very recent Chinese company, Huawei tested the 5G of telecommunication technology. With this development the data transfer touching the limits of speed 10 GB per second. It is very important now to advance the equipment and techniques we are using in our daily life with this development [43–47] . This is the motivation of the current study. This leads us to use this technology in the aircraft parking system. Because it will speed the procedure which is manually covered by many workers at the airport. Without the use of this technology, it takes too much time to park the aircraft and heavy amounts are paid in the form of salaries or landing fees per hour.

The layout of the chapter is presented in Fig. 1, Sect. 1 consists of the study background, literature review, and the motivation of the current study. In Sect. 2, some basic definitions are defined. In the next section, material and method are presented. Section 4 consists of the case study of aircraft parking. Finally, results, discussions, and conclusion are made in the last section along with future directions.

2 Preliminaries

In this section, some definitions are added which will help understand the rest of the chapter.

Definition 2.1: Crisp Set A set X is said to be a Crisp Set if an element belongs to the set or not, i.e., there exists a function f such that: $f : \mathrm{X} \rightarrow \{0, 1\}$.

Definition 2.2: Fuzzy Set A set Y is called a fuzzy set if $Y = \{y, \pi_Y(y); y \in Y, \pi_Y(y) \in [0, 1]\}$ where y is the crisp set of Y and $\pi_Y(y)$ belong to the interval [0,1] [17].

Fig. 1 The chapter presentation in pictorial form

Definition 2.3: Intuitionistic Set A set Z is an Intuitionistic set if $Z = \{\alpha, [T_Z(\alpha), F_Z(\alpha)]; \alpha \in X, T_Z(\alpha), F_Z(\alpha) \in [0, 1]\}$ where α is the crisp set of Z, and $F_Z(\alpha)$ is the falsity membership value, and $T_Z(\alpha)$ is the truth membership value [18].

Definition 2.4: Neutrosophic Set A set C is called neutrosophic set if $C = \{l, [T_C(l), I_C(l), F_C(l)]; l \in C, T_C(l), I_C(l), F_C(l) \in [0, 1]\}$ where $T_C(l)$ is the truth membership value,$I_C(l)$ is the indeterminacy membership value and $F_C(l)$ is the falsity membership value [19].

Definition 2.5: MCDM Multi-criterion decision making (MCDM) has major applications in all fields of science. MCDM concerns with structuring and formulating decision making and planning problems that involve multiple criteria or are influenced by numerous factors. MCDM techniques involve weighing the individual criteria based on their relative importance to find the best solution to the problem on the decision-making preferences. In most cases, there doesn't exist an optimal solution for a decision-making problem so the decision making has to make a preference among all the criteria that the decision-maker is experiencing in the problem [39].

Definition 2.6: MATLAB MATLAB is the acronym for "matrix laboratory". It is a software developed by MathWorks and is a proprietary programming language and is a multi-paradigm numerical computing environment. MATLAB allows many functions that include implementation of algorithms, creation of programs and user interfaces and interfacing with programs that are based on different languages, data manipulation, and plotting, etc. (mathworks.com).

Definition 2.7: MATLAB Fuzzy Logic Toolbox The simulating system based on fuzzy logic, MATLAB functions and apps, and designing is provided by the Fuzzy Logic Toolbox. Fuzzy Logic Toolbox offers a proper guide for designing a fuzzy inference system. Many functions that include fuzzy clustering adaptive neuro-fuzzy learning is provided by the software. (mathworks.com).

Definition 2.8: Fuzzy Logic Controller Fuzzy logic is the basis of a fuzzy control system. Its mathematical bases analyze processes analog input values as logical variables that take on continuous values between 0 and 1. On the contrary to digital or classical logic, these systems operate on discrete values that are either 0 or 1 [29].

Definition 2.9: Closed-Circuit Television (CCTV) CCTV technology is known as video surveillance and is used to record a specific place using video cameras and transmit the feed to a set of monitors. It is employed in point-to-point (P2P), point-to-multipoint (P2MP), mesh wired and\or wireless links. It differentiates itself from broadcast television in the way that the signal is not openly transmitted. The term CCTV can be used for all types of cameras, but the term is mostly used for monitoring purposes of paramount areas that include banks, stores, and any area where security is required. An important use of CCTV is videotelephony but is rarely categorized in that manner. It has great importance in the field of education due to the use of video in distance education [40].

Definition 2.10: Defuzzification It is the process of producing a quantifiable result in Crisp logic, given fuzzy sets and corresponding membership degrees. It is the process that maps a fuzzy set to a crisp set. It is mostly required in fuzzy control systems. These have a set of rules that allow us to transform several variables into a fuzzy result. The result is expressed in terms of membership in fuzzy sets [41, 42].

3 Material and Method

The method and the material used in this chapter are discussed in this section.

The method which is adopted in this chapter is the Fuzzy Logic Controller (FLC) using MATLAB fuzzy logic toolbox, this method is mainly used for the automation of the gadgets. FLC consists of three parts, firstly, fuzzifier; this takes linguistic inputs and a neutrosophic number is assigned to each linguistic variable with the help of decision-makers which are then converted into fuzzy numbers and will pass through the second part rule selector and in the last a useful information in the form of time will be obtained.

The important material used for this proposed case study; the following materials are considered.

- FLC
- MATLAB fuzzy logic toolbox
- CCTV
- Image Processing
- Radar
- 5G telecommunication technology
- Airport Model for the parking of aircraft in tarmac.

Note: These all materials are proposed; and no one is physically used. All the calculations are done using MATLAB.

4 Proposed Model

In this section, we have proposed a model according to the required calculations.

4.1 Case Study

To discuss the:

- Validity
- Applicability

Of the proposed FLC, parking of aircraft is considered.

4.2 Proposed Model

Parking of aircraft is no easy task, to make the parking of aircrafts efficient or less complex, in this chapter the system is proposed to automate the parking of aircraft by using the CCTVs footage and after the image processing, the neutrosophic values are assigned to each data, collected by the footages. This automation is based on both hardware and software. Any automation system has not been employed so far due to the limitation of communication technology but with the invention and implementation of 5G telecommunication technology, the proposed system can be applied. The pilot is intended to go there in the minimum time. So, the following linguistic inputs are considered:

- By considering three Linguistic inputs and an output.
- Clearance of Runways
- Clearance of Taxiways
- The availability of the space in Tarmac
- The time required to reach the Tarmac (Fig. 2).

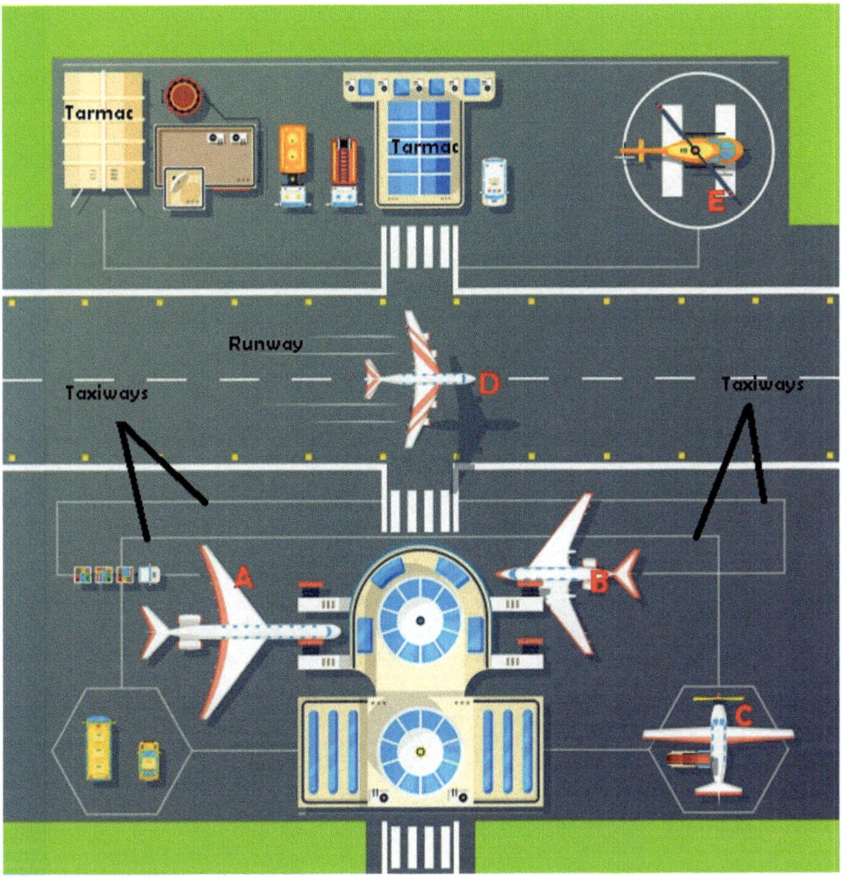

Fig. 2 The proposed model problem for the automatic system of aircraft parking with the three linguistic Inputs

4.3 *Parameters*

In this case study, the following linguistic variables are used which are further bifurcated. A short code is provided for the ease of table making or any mixing of information and all the possibilities are illustrated with the Fig. 3a–f. Since accuracy is required in this problem so that the neutrosophic numbers are being used for this purpose (Table 1).

4.4 Assumptions

- *Dependent attributes are considered.*
- *Hesitant Environment is not yet considered.*

The possible rules are listed here for the proposed model in Table 2.

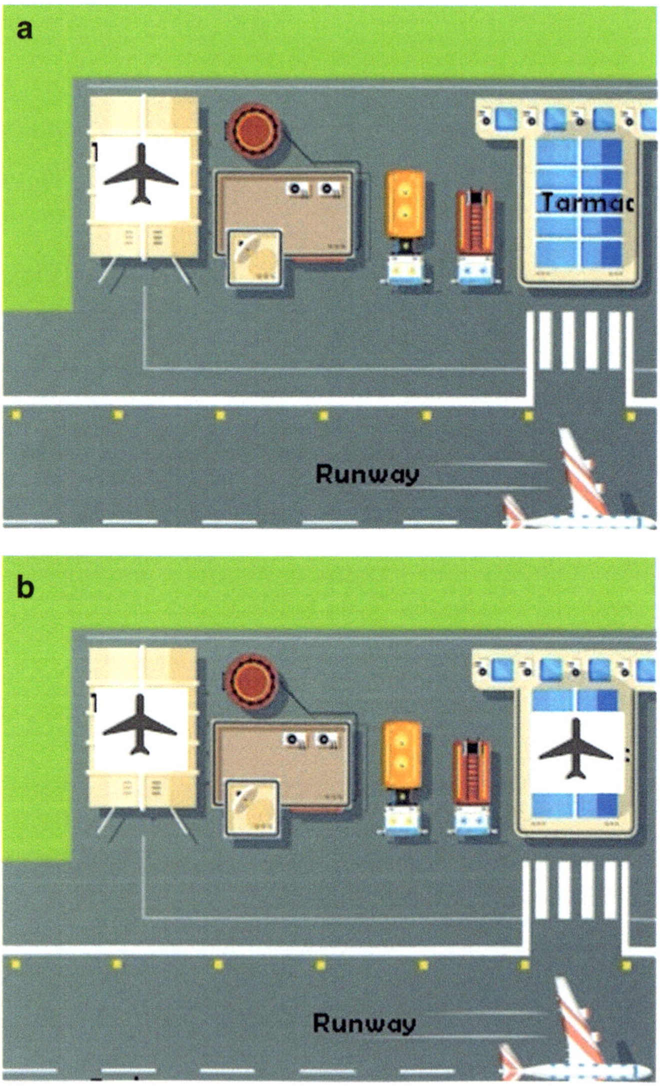

Fig. 3 **a** Tarmac with space (YES). **b** Tarmac with no space (No). **c** Runway not clear (NO). **d** Runway clear (YES). **e** Taxiway not clear. **f** Taxiway clear

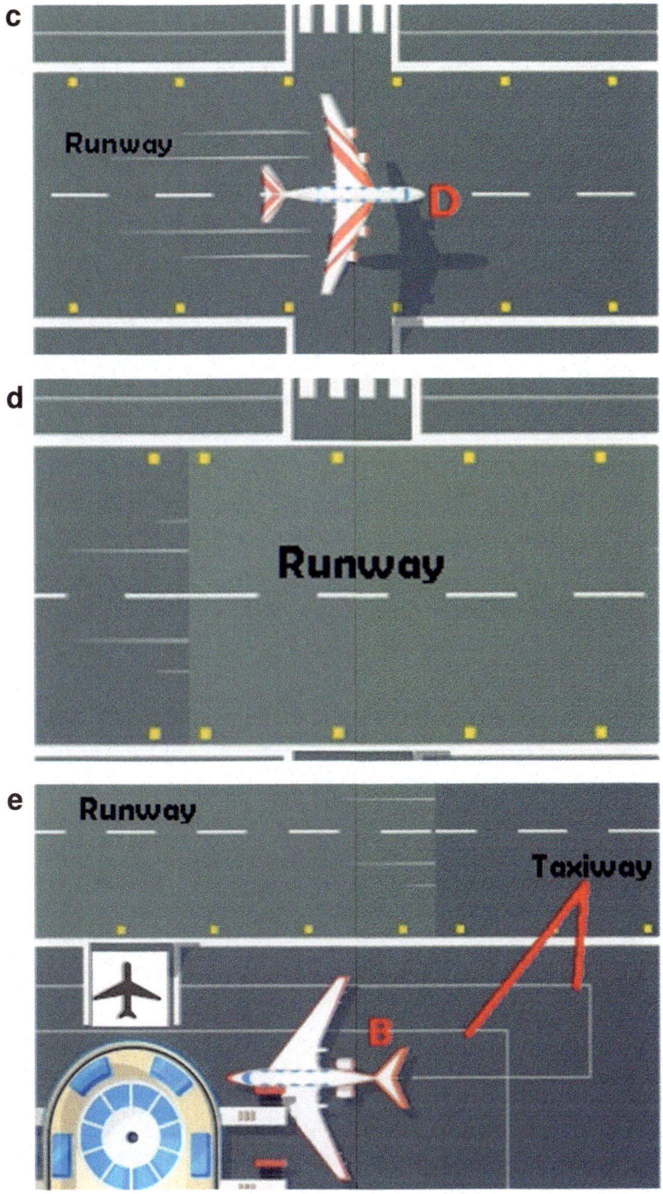

Fig. 3 (continued)

f

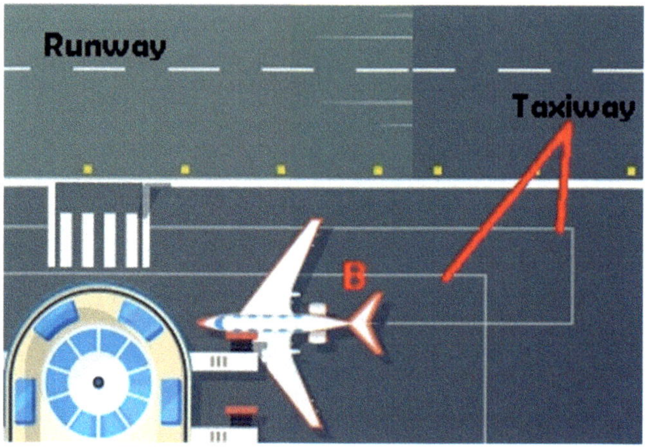

Fig. 3 (continued)

Table 1 Linguistic variables, codes and respective neutrosophic fuzzy numbers

Sr. No.	Linguistic variable		Code	Neutrosophic fuzzy number
1	Parking space in Tarmac	Yes	χ	(1.0, 0.1, 0.5)
2		No	Ϙ	(0.1, 0.3, 0.9)
3	Operational runway clearance	Yes	ɰ	(0.8, 0.2, 0.4)
4		No	þ	(0.2, 0.3, 0.7)
5	Clearance of taxiways	Yes	Ϋ	(0.7, 0.4, 0.6)
6		No	ꜧ	(0.3, 0.1, 0.8)

Table 2 Linguistic variables, codes and neutrosophic numbers

	Linguistic inputs			Linguistic output
Sr. No.	Parking space in Tarmac	Operational runway clearance	Clearance of taxiways	Time require to park aircraft
1	χ	ɰ	Ϋ	Time
2	χ	ɰ	ꜧ	Time
3	χ	þ	Ϋ	Time
4	χ	þ	ꜧ	Time
5	Ϙ	ɰ	Ϋ	Time
6	Ϙ	ɰ	ꜧ	Time
7	Ϙ	þ	Ϋ	Time
8	Ϙ	þ	ꜧ	Time

Table 3 Assigning of neutrosophic numbers to each rule from Table 2

	Linguistic inputs			Linguistic output
Sr. No.	Parking space in Tarmac	Operational runway clearance	Clearance of taxiways	Time require to park aircraft
1	(1.0, 0.1, 0.5)	(0.8, 0.2, 0.4)	(0.7 ,0.4, 0.6)	Time
2	(1.0, 0.1, 0.5)	(0.8, 0.2, 0.4)	(0.3, 0.1, 0.8)	Time
3	(1.0, 0.1, 0.5)	(0.2, 0.3, 0.7)	(0.7, 0.4, 0.6)	Time
4	(1.0, 0.1, 0.5)	(0.2, 0.3, 0.7)	(0.3, 0.1, 0.8)	Time
5	(0.1, 0.3, 0.9)	(0.8, 0.2, 0.4)	(0.7, 0.4, 0.6)	Time
6	(0.1, 0.3, 0.9)	(0.8, 0.2, 0.4)	(0.3, 0.1, 0.8)	Time
7	(0.1, 0.3, 0.9)	(0.2, 0.3, 0.7)	(0.7, 0.4, 0.6)	Time
8	(0.1, 0.3, 0.9)	(0.2, 0.3, 0.7)	(0.3, 0.1, 0.8)	Time

4.5 Assigning of Neutrosophic Numbers and De-Neutrosophication

The decision-maker will assign linguistic values from Table 1, according to his interest, knowledge, and experience, to the above-mentioned criteria, and the possible rules for alternatives are shown in Table 2, and assigned values are presented in below Table 3.

4.6 De-Neutrosophication

De-Neutrosophication is very imported to avoid complex solutions, so for this purpose, all the neutrosophic numbers are converted into fuzzy numbers by using accuracy function. This concept of conversion from neutrosophic to fuzzy was proposed.

The mathematical form of de-Neutrosophication is given below.

$$A(\chi) = \frac{\{T(x) + F(x) + I(x)\}}{3}$$

Table 4 Assigning of neutrosophic numbers to each rule from Table 2

Sr. No.	Parking space in Tarmac	Operational runway clearance	Clearance of taxiways
1	(1.0, 0.1, 0.5)	(0.8, 0.2, 0.4)	(0.7, 0.4, 0.6)

Example Now for the understanding purpose consider only one rule in Table 4, which is taken from Table 3, for the conversion of the neutrosophic number into fuzzy.

Let $X_1 = \{T(x_1) + F(x_1) + I(x_1)\}$ is $X_1 = (1.0, 0.1, 0.5)$

$$A(X_1) = \frac{\{T(x_1) + F(x_1) + I(x_1)\}}{3}$$
$$A(X_1) = \frac{(1.0, 0.1, 0.5)}{3}$$
$$= \frac{1.6}{3}$$
$$= 0.533$$

$X_2 = \{T(x_2) + F(x_2) + I(x_2)\}$ is $X_2 = (0.8, 0.2, 0.4)$

$$A(X_2) = \frac{\{T(x_2) + F(x_2) + I(x_2)\}}{3}$$
$$A(X_1) = \frac{(0.8, 0.2, 0.4)}{3}$$
$$= \frac{1.4}{3}$$
$$= 0.466$$

$X_3 = \{T(x_3) + F(x_3) + I(x_3)\}$ is $X_3 = (0.7, 0.4, 0.6)$

$$A(X_3) = \frac{\{T(x_3) + F(x_3) + I(x_3)\}}{3}.$$

$$A(X_3) = \frac{(0.7, 0.4, 0.6)}{3}$$
$$= \frac{1.7}{3}$$
$$= 0.566$$

Similarly, we convert all the eight rules listed in Table 3, into fuzzy numbers, and results are represented in Table 5.

4.7 FLC Approach for the Calculation of Time

A controller which converts linguistic values into logical values by considering continuous values between [0, 1]. This is a basic control system that relies on the degrees of the linguistic inputs and the output depends upon the degree of

Table 5 De-neutrosophication of each rule

	Linguistic inputs			Linguistic output
Sr. No.	Parking space in Tarmac	Operational runway clearance	Clearance of taxiways	Time require to park aircrafts
1	0.533	0.466	0.566	Time
2	0.533	0.466	0.400	Time
3	0.533	0.400	0.566	Time
4	0.533	0.400	0.400	Time
5	0.433	0.466	0.566	Time
6	0.433	0.466	0.400	Time
7	0.433	0.400	0.566	Time
8	0.433	0.400	0.400	Time

linguistic inputs. To get the output it is necessary to Defuzzify the values. In the model an FLC is based on three linguistic inputs that have further degrees and one linguistic output is required. Figure 4, shows the FLC for the model proposed here and its MATLAB approach is graphically shown in Fig. 5.

- The linguistic input ***parking space in tarmac*** is further bifurcated into the cases; if there is a parking space in tarmac then represented with "YES" otherwise "NO" respectively.

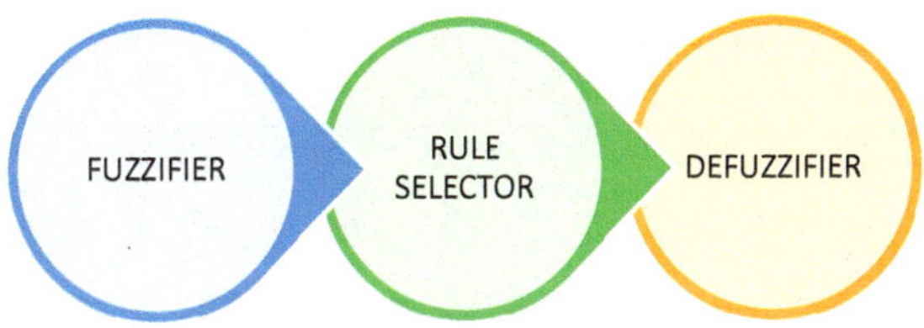

Fig. 4 FLC approach of the proposed model

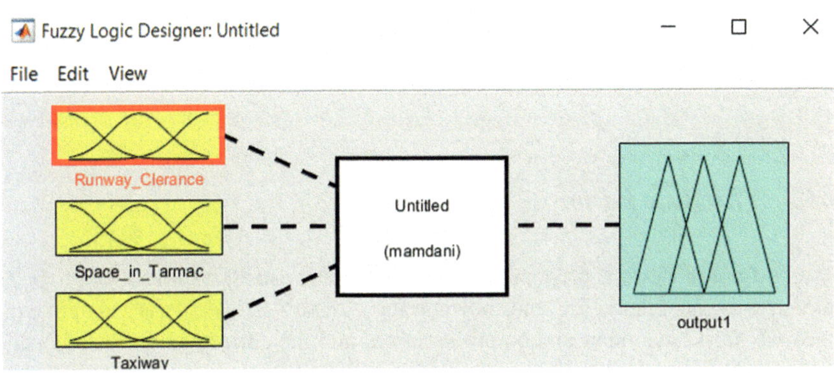

Fig. 5 MATLAB fuzzy logic controller approach of the proposed model

- Two possibilities are considered for ***operational runway clearance***. If the runway is clear no flight is landing or take off then it is represented with "YES" otherwise with "NO" respectively.
- Two possibilities are considered for the ***Clearance of taxiways***. If the taxiway is clear no aircraft is on the way then it is represented with "YES" otherwise with "NO" respectively.

Now before applying the fuzzy logic controller process it is necessary to allow them the membership values which we have been already calculated and are listed in Table 4. The process of the fuzzifier is represented in Table 3 whereas the rule selector is the possible input value so the only left is defuzzification.

4.8 Defuzzification

The procedure to convert the fuzzy output to a crisp value is said to be defuzzification, of the given fuzzy sets and the correlative membership degrees. It is an operation that maps a fuzzy set to a crisp set. The above-mentioned quantified results are obtained from the fuzzy interface technique. Defuzzification is done using the Centroid method.

$$\text{Time} \quad \overline{\omega} = \frac{\sum_{0}^{1} x.\mu(x)}{\sum_{0}^{1} .\mu(x)}$$

4.9 Membership Function Editor

This is the tool that allows us to display and change the rules of the inference engine. All the membership values, which are assigned to the linguistic variables are shown in Fig. 6.

4.10 The Rule Editor

To modify or to view the proposed rules Rule Editor is used. This is also said to be an inference engine of any FLC and shown in Figs. 7 and 8. It is necessary to define the rules we must have input and output variables and the corresponding membership values.

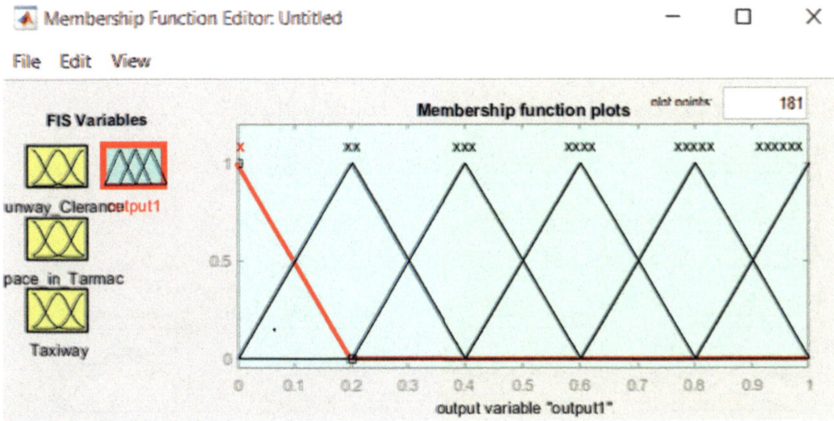

Fig. 6 Membership functions of inputs

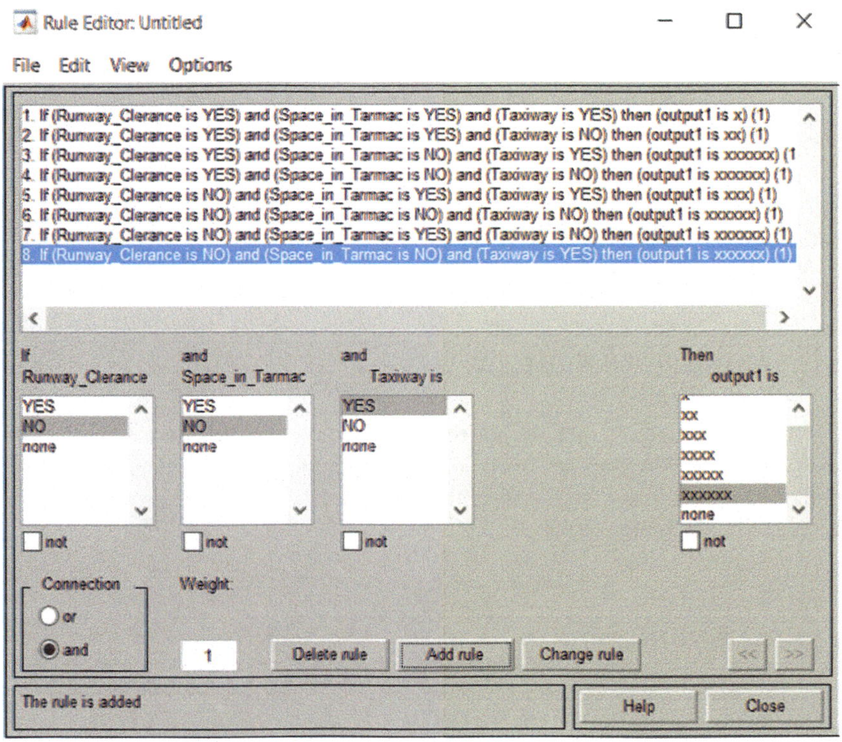

Fig. 7 MATLAB rule editor

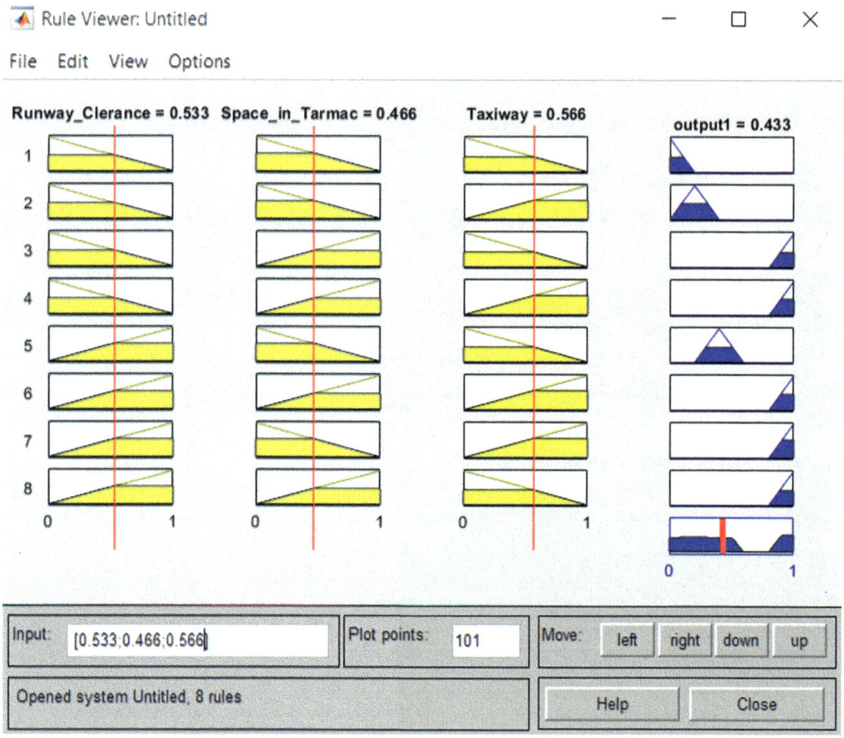

Fig. 8 MATLAB rule viewer

4.11 *Defuzzified Results*

Several built-in methods fuzzified the results of the proposed rules based on membership values. In this proposed method defuzzification is done using the centroid method. Table 6, and Fig. 9, shows the defuzzied results of the proposed model using MATLAB's Fuzzy toolbox.

4.12 *Mobile Usability*

One of the main purposes to propose FLC is its implementation. They can be implemented in versatile ways. To make a beneficial structure. In this proposed model three inputs are considered and it can be recognized by using the CCTV footages implementing image processing, which will send the information to the system, then by implementing the FLC the time required to park the aircraft will be calculated graphically shown in Fig. 10.

Table 6 Defuzzified results of proposed rules

	Linguistic inputs			Linguistic output
Sr. No.	Parking space in Tarmac	Operational runway clearance	Clearance of taxiways	Time require to park airCraft
1	0.533	0.466	0.566	0.433
2	0.533	0.466	0.400	0.475
3	0.533	0.400	0.566	0.593
4	0.533	0.400	0.400	0.612
5	0.433	0.466	0.566	0.813
6	0.433	0.466	0.400	0.816
7	0.433	0.400	0.566	0.847
8	0.433	0.400	0.400	0.902

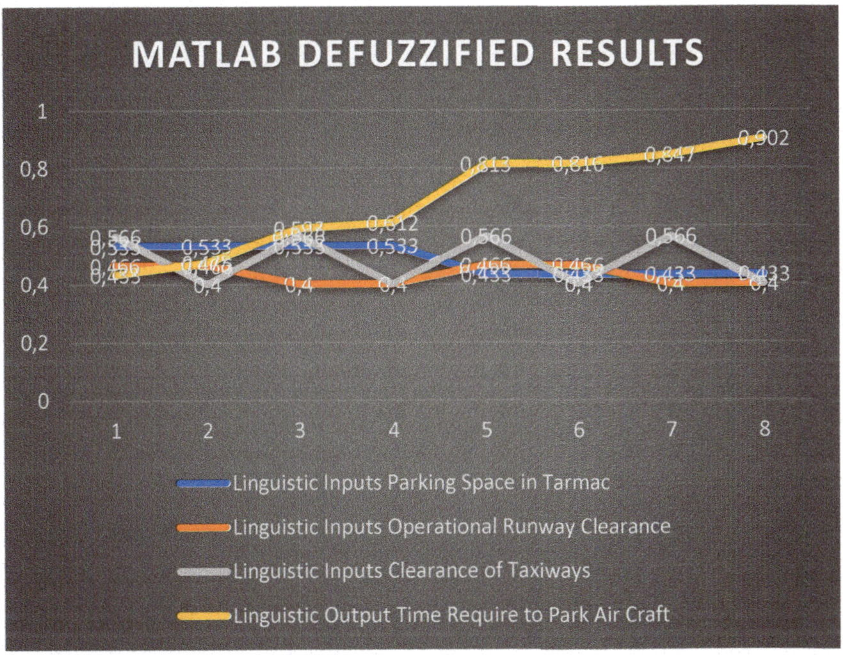

Fig. 9 Graphically defuzzified results

4.13 *How It Will Work? and the Role of 5G?*

It is a very important question so to find out the answer to this; firstly, we understand the concept of image processing which is the major component of this implementation.

Fig. 10 Mobile implementations

Image processing is the method containing some features, information, or characteristics using some software this can be done. So, collecting data from radar and CCTV footage will be processed and some useful information will be sent to the FLC. Here it is very important to transfer that data with high speed and accuracy so this task will be done using the 5G telecommunication technology.

FLC will read the data and it will send to the rule selector where it will be compared with all the possibilities already stored in the system after this it will select the optimal choice depending upon the inputs and information will be sent to the defuzzifier. At this point, all the information will be decoded, and instructions will be transferred to the pilot in the aircraft again by using the 5G means. Now the pilot will follow the instruction and will go the desired parking place within the calculated time. The pictorial form of this implementation is pretended in Fig. 11.

5 Result Discussion and Conclusion

In the current century, man is trying to automate gadgets. Also, the development and advancement of communication technologies are playing an important role in this arena. Due to this, artificial intelligence is also growing which is helping humanity by providing safe usage.

Artificial Intelligence (AI) is playing a vital role in the simulation of human intelligence using computers, laptops, and mobile phones.

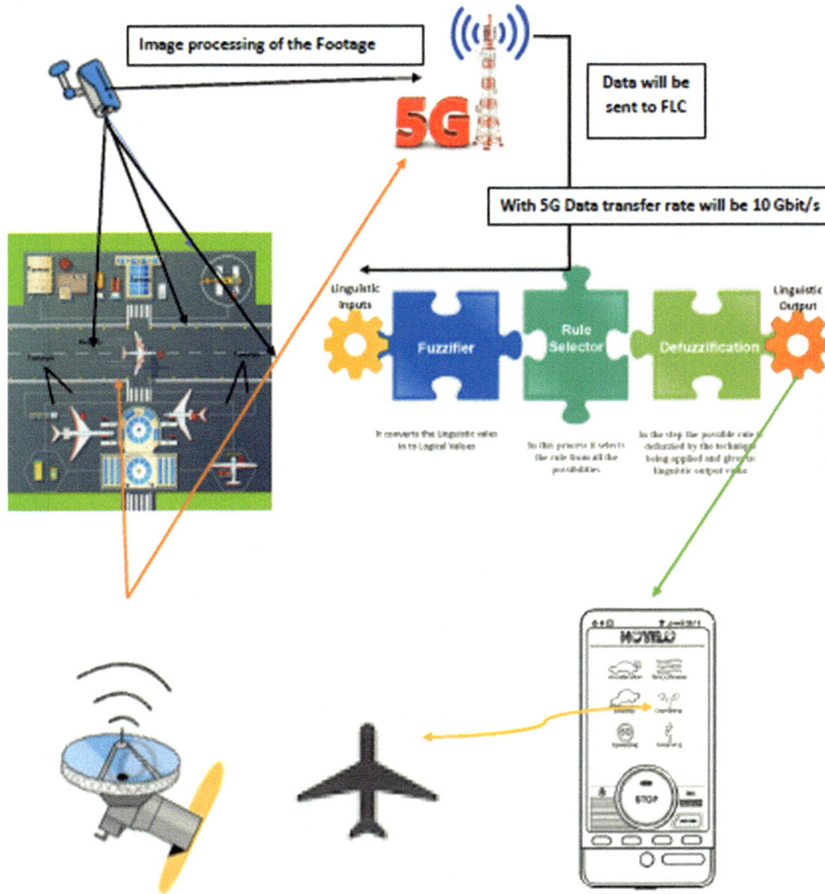

Fig. 11 The implementation of the proposed model

One of the key components of AI is FLC. Also, with the development of neutrosophic numbers, its implantation is more accurate, precise, and uncertain. In this chapter, a case study of the aircraft parking is considered.

This chapter aims to provide an idea to simulate or to automate the parking of aircraft at the airports and to reduce the parking time. Since time means money if we save the time it will automatically reduce human force and have a positive impact on the cost.

This is how it is implementing. Firstly, a dependent case study is considered in which each criterion, may depend on the other; which influences the output time. Secondly, With the help of image processing a linguistic number is assigned to each input data, after this a neutrosophic number is assigned to each input and then, these neutrosophic numbers are converted into fuzzy with the help of the de-neutrosophication technique. Now this fuzzy data is sent to FLC after defuzzification and the useful information in the form of time will be sent to the pilot.

Since this study has not yet been studied, a comparative study cannot be done with the existing methods.

Further, this proposed technique can be applied immensely at various parking places like, University parking, Shopping Mall parking, and many more.

In the future, the present work may be extended to other special types of parking problems like car parking in the alien city, etc.

Acknowledgements Authors are thankful to the Editor-In-Chief and Referees.

References

1. Le, L.B., Lau, V., Jorswieck, E., Dao, N.D., Haghighat, A., Kim, D.I., Le-Ngoc, T.: Enabling 5G mobile wireless technologies. J. Wireless Com. Network **218** (2015) https://doi.org/10.1186/s13638-015-0452-9
2. Agyapong, P.K., Iwamura, M., Staehle, D., Kiess, W., Benjebbour, A.: Design considerations for a 5G network architecture. IEEE Commun. Mag. **52**(11), 65–75 (2014)
3. Bergren, S. (2017). Design Considerations for a 5G Network Architecture. arXiv preprint arXiv:1705.02902
4. Thompson, J., Ge, X., Wu, H.C., Irmer, R., Jiang, H., Fettweis, G., Alamouti, S.: 5G wireless communication systems: prospects and challenges [Guest Editorial]. IEEE Commun. Mag. **52**(2), 62–64 (2014)
5. Hu, Y.C., Patel, M., Sabella, D., Sprecher, N., Young, V.: Mobile edge computing—a key technology towards 5G. ETSI White Paper **11**(11), 1–16 (2015)
6. Manasa, H.R., Pramila, S.: Architecture and technology components for 5G mobile and wireless communication. Int. J. Eng. Res. **4**(06) (2015)
7. CNBC Newsletters.: (2019) https://www.cnbc.com/2019/11/25/5g-will-span-two-thirds-of-global-population-in-6-years-ericsson-says.html
8. Atalık, O., Akan, Ş., Bakır, M.: Aviation 4.0: current practises of industry 4.0 in the airline and airport industries. (2019) https://doi.org/10.6084/m9.figshare.10316303
9. Kivits, R., Charles, M.B., Ryan, N.: A post-carbon aviation future: airports and the transition to a cleaner aviation sector. Futures **42**, 199–211 (2010)
10. Lee, J., Mo, J.: Analysis of technological innovation and environmental performance improvement in aviation sector. Int. J. Environ. Res. Public Health **8**, 3777–3795 (2011). https://doi.org/10.3390/ijerph8093777
11. Havle, C.A., Üçler, Ç.: Enablers for industry 4.0. In: 2018 2nd International Symposium on Multidisciplinary Studies and Innovative Technologies (ISMSIT), pp. 1–6. IEEE (2018)
12. Rüßmann, M., Lorenz, M., Gerbert, P., Waldner, M.: Industry 4.0: The Future of Productivity and Growth in Manufacturing Industries, pp. 1–14. (2015) https://www.zvw.de/media.media.72e472fb-1698-4a15-8858-344351c8902f.original.pdf
13. Valdés, R., Gomez, C., Victor, F., Sanz, Á., Javier, P.C.: Aviation 4.0: more safety through automation and digitization. pp. 25–41. (2018). https://doi.org/10.5772/intechopen.73688
14. Ustundag, A., Cevikcan, E.: Industry 4.0: managing the digital transformation. Springer, Cham, (2017)
15. Albreem, M.A.: 5G wireless communication systems: vision and challenges. In: 2015 International Conference on Computer, Communications, and Control Technology (I4CT). pp. 493–497. IEEE. (April 2015)

16. Idowu-Bismark, O., Okokpujie, K.O., Ryan, H., Adedokun, M.O.: 5G wireless communication network architecture and its key enabling technologies. Int. Rev. Aerosp. Eng. (I. RE. AS. E) **12**(2), 70–82 (2019)
17. Zadeh, L.A.: Fuzzy sets. Inf. Control **8**(3), 338–353 (1965)
18. Atanassov, K.T.: Intuitionistic fuzzy sets. Fuzzy Sets Syst. **20**(1), 87–96 (1986)
19. Smarandache, F.: Neutrosophy. Neutrosophic probability, set, and logic. ProQuest Information and Learning, Ann Arbor, Michigan, USA (1998)
20. Molodtsov, D.: Soft set theory-first results. Comput. Math. Appl. **37**, 19–31 (1999)
21 Maji, P., Biswas, R., Roy, A.R.: Fuzzy soft sets. Fuzzy Sets Syst. **9**, 589–602 (2001a)
22 Maji, P., Biswas, R., Roy, A.: Intuitionistic fuzzy soft sets. J. Fuzzy Math. 9 (2001b)
23. Maji, P.K.: Neutrosophic soft set. Ann. Fuzzy Math. Inform. **5**(1), 157–168 (2013)
24. Abdel-Baset, M., Chang, V., Gamal, A.: Evaluation of the green supply chain management practices: a novel neutrosophic approach. Comput. Ind. **108**, 210–220 (2019)
25. Riaz, M., Naeem, K., Ahmad, M.O.: Novel concepts of soft sets with applications. Ann. Fuzzy Mat. Inform. **13**(2), 239–251 (2017)
26. Wielki, J., Jurczyk, M., Madera, D.: Application of TOPSIS method for evaluation of IT application in the hospital. (2019). https://doi.org/10.34190/KM.19.134
27. Calvo-Flores, M., Verdegay, J., Vila, M.: Linguistic decision-making models. Int. J. Intell. Syst. **7**, 479–492 (1992). https://doi.org/10.1002/int.4550070507
28. Liu, J.: Fuzzy logic control. (2018) https://doi.org/10.1007/978-981-10-5263-7_4
29. Mamdani, E.H.: Application of fuzzy algorithms for control of simple dynamic plant. Proc. Inst. Electr. Eng. **121**(12), 1585–1588 (1974)
30. Shen, J., Jin, D., Li, Z.: Fuzzy logic control and fuzzy logic control chip. **25**, 61–66, 60 (1997)
31. Suwoyo, H., Tian, Y.Z., Hajar, M.: An flc-pso algorithm-controlled mobile robot. SINERGI. **24**, 177 (2020). https://doi.org/10.22441/sinergi.2020.3.002
32. Jafar, M.N., Saqlain, M., Mansoob, A., Riffat, A.: A best way to access gas station using fuzzy logic controller in neutrosophic environment. Sci. Inq. Rev. **4**(1), 30–45 (2020). Available at: https://doi.org/10.32350/sir.41.03
33. Saqlain, M., Saeed, M., Saeed, H.M.: Smart parking system using fuzzy logic controller for alien cities. Int. J. Math. Res. **9**(1), 62–71 (2020). https://doi.org/10.18488/journal.24.2020.91.62.71
34. Chatterjee, K., Kar, M.B., Kar, S.: Strategic decisions using intuitionistic fuzzy VIKOR method for information system (IS) outsourcing. In: 2013 International Symposium on Computational and Business Intelligence (ISCBI), pp. 123–126. IEEE, (August 2013)
35. Havle, C.A., Kılıç, B.: A hybrid approach based on the fuzzy AHP and HFACS framework for identifying and analyzing gross navigation errors during transatlantic flights. J. Air Transp. Manag. **76**, 21–30 (2019)
36. Şenel, M., Şenel, B., Havle, C.A.: Risk analysis of ports in maritime industry in Turkey using FMEA based intuitionistic fuzzy TOPSIS approach. In: ITM Web of Conferences, vol. 22, p 01018. EDP Sciences (2018)
37. Franco Barbosa G.: Aviation manufacturing towards to industry 4.0: a review. In: 4th International Conference and Exhibition on Mechanical & Aerospace Engineering, Florida, (2016)
38. Troiano, A., Pasero, E.: A runway surface monitor using internet of things. J. Electr. Eng.-Elektrotechnicky Casopis **65**(3), 169–173 (2014)
39. Hwang, C., Yoon, K.: Multiple Attribute Decision Making: Methods and Applications. Springer, A State-of-the-Art Survey (1981)
40. Dempsey, J.S.: Introduction to Private Security. Thomson Wadsworth, Belmont, CA, p. 78 (2008)
41. Ross, T.J.: Fuzzy Logic with Engineering Applications, pp. 90–91. Wiley India, (2010)
42. Van Leekwijck, W., Kerre, E.E.: Defuzzification: criteria and classification. Fuzzy Sets Syst. **108**(2), 159–178 (1999)
43. Ben-Ari, M., Mondada, F. (2018). Fuzzy logic control. https://doi.org/10.1007/978-3-319-62533-1_11

44. Büyüközkan, G., Feyzioglu, O., Havle, C. (2020). Analysis of Success Factors in Aviation 4.0 Using Integrated Intuitionistic Fuzzy MCDM Methods. https://doi.org/10.1007/978-3-030-23756-1_73
45. Global Mobile Supplier Association.: (2019). https://gsacom.com/paper/lte-5g-market-statistics-8-april-2019/
46. Gupta, A., Jha, R.K.: A survey of 5G network: architecture and emerging technologies. IEEE Access **3**, 1206–1232 (2015)
47. Notlagh, N.H., Taleb, T., Arouk, O.: Low-altitude unmanned aerial vehicles-based internet of things services: comprehensive survey and future perspectives. IEEE Int. Things J. **3**(6), 889–922 (2016)

An Integrated Fuzzy Decision Making and Integer Programming Model for Robot Selection for a Baggage Robot System

Ahmet Aktas and Mehmet Kabak

Abstract An efficient baggage-handling system is of critical importance to the aviation industry, especially for major hub airports that handle large volumes of passengers and connecting flights. Huge investments have been made over the years, and baggage handling has been long recognized as one of the most promising areas for airport automation. Based on the developments on Internet of Things, RFID, robotic and sensor technologies, modernization of airports are taken into consideration in the aviation industry, within the concept of Aviation 4.0. This chapter deals with the development of an automated baggage robot system for airports, which is being newly constructed. With this aim, hesitant fuzzy linguistic terms set based multi-criteria decision making model is used to evaluate alternative components of the baggage robot system. Then, 0–1 goal programming model is used to determine the optimal selection among alternative components under budget constraint. An example is presented to demonstrate the applicability of the proposed model. This model can be used by airport construction planners for the development of automated baggage robot systems.

1 Introduction

Due to the increase of air traffic volume, airport managements try coping with more flights and passengers. They are also looking for improvement ways to increase service quality, by reducing waiting time of passengers in the airport. In the last two decades, some technological developments about the airport processes, which are promising about obtaining more profit from airport operations, became useful. However, adaption of new technologies in the aviation industry is difficult, because of the conservative nature of the sector due to the strict regulations and safety standards [1].

A. Aktas (✉) · M. Kabak
Department of Industrial Engineering, Gazi University, 06570 Ankara, Turkey
e-mail: aaktas@gazi.edu.tr

C. Kahraman and S. Aydın (eds.), *Intelligent and Fuzzy Techniques in Aviation 4.0*,
Studies in Systems, Decision and Control 372,
https://doi.org/10.1007/978-3-030-75067-1_4

Baggage operations are one of the most important and vital processes among the many processes in the aviation sector. There are several actors of baggage operations from both airline companies and airport managements on management and operation sides. The process of baggage operations at the airport begins with the check-in of the baggage. A label is attached to each baggage at this step, which presents the information about the flight and the destination. Bar codes or RFID (radio-frequency identification) tags on the label provides more information such as the weight of the baggage and/or class of passenger like first class, business or economy.

Baggage handling processes are important for ground handling service companies, airlines and airports. According to the 2018 Baggage Report of SITA (The International Aeronautical Telecommunications Association) [2], the number of mishandled bags in 2017 were estimated to be around 22.7 million, which means US$ 2.3 billion extra cost for the aviation sector.

Bags must firstly be checked for security. Then they are delivered to the sorting and loading areas, which are the areas that they are loaded into containers called as unit loading devices (ULD). If an automated baggage handling system is in use in airport, bags are temporarily stored until a short time before departure. During the storage period, many operations are performed in the terminal buildings at several facilities which are leased or owned by airlines. So, the main decision maker for installation, upgrade or operation of the baggage system automation is generally airport management.

Based on the developments of Internet of Things, RFID, robotic and sensor technologies, automatization in industrial processes have rapidly been emerged. These changes are called as the Fourth Industrial Revolution or Industry 4.0. Aviation industry is also effected by the recent developments in technology and the changes related to Industry 4.0 in aviation industry are called as Aviation 4.0. Since the mishandled baggage operations cause significant costs and losses for airport managements, automation in baggage operations have taken attention of them, too. Self-check-in counters, RFID based baggage tracking systems, security checking by sensors, utilization of automated guided vehicles (AGVs) or baggage robots for transportation are some of the technologies have been implemented in airports so far.

This chapter proposes a decision-making model for determination of components of an automated baggage robot system for an airport which is being constructed. Within the proposed model, alternative robots for the baggage system were evaluated by using a hesitant fuzzy multi-criteria approach. Evaluation results of the multi-criteria analysis used as problem parameters in a binary integer programming model, which helps the decision maker on the selection of alternative robots.

The contributions of the study can be listed as follows:

- Conventional and automated baggage handling systems are described.
- An analytic decision model for baggage robot system design is proposed.
- Alternative baggage robots are evaluated by using a linguistic multi-criteria decision-making methodology.

- Determination of the number of baggage robots, which is an alternative element for baggage system automation, for a new airport is determined by using a binary integer programming formulation.

The rest of the chapter organized as follows: in the second part, a brief description of conventional baggage handling systems was presented. Next, the concept of baggage robot systems was given in the third part with the requirements and constraints taken into account for design of these systems. The fourth part explains the decision-making methodology presented in this chapter. Application steps and results of the proposed model for an airport was presented in the fifth part. This chapter was concluded in the sixth part with presenting managerial implications.

2 Literature Review

Baggage handling systems have taken attention of researchers in the last decade. Some studies on baggage handling operations can be summarized as follows.

A simulation study is conducted to analyze the impact of RFID reader number on baggage delivery performance [3]. A MILP model is proposed to solve route choice problem of destination coded vehicles [4]. Assignment of chutes to outgoing flights is modeled as a stochastic vector assignment problem under uncertainty assumption of flight delays and number of bags [5]. A baggage handling system of an international airport is analyzed by using a microscopic simulation model. All baggage-related subsystems including passenger arrival, baggage check-in, security screening, sorting, transportation to aircraft and loading are taken into account within the simulation model [6]. Discrete event simulation of a check-in area is performed to determine a balanced workload for a baggage handling system used by multiple airlines [7]. A column generation based solution approach is developed to determine the optimum assignment of outgoing baggage to handling facilities and scheduling [8]. A mathematical model in the form of colored timed Petri net is established in order to analyze hold baggage security screening system throughput [9]. By this way, detection of operating problems and determination of upgrading decisions can be possible [9]. Discrete event simulation model of an inbound baggage handling system at a regional airport in Italy is developed in order to determine bottlenecks and critical operations [10]. Service quality perception of passengers from a baggage handling is calculated by SERVQUAL method [11]. In this study, weights of quality dimensions are determined by the Best–Worst Method. A robust optimization model is proposed to determine flight assignments to unloading areas under arrival and departure times uncertainties [12].

Utilization of automated sortation conveyors on baggage handling is investigated in views of operations research perspective [13]. Identified problem areas are listed as layout planning, inbound problems (gate assignment, conveyor assignment, counter workload balancing), outbound problems (baggage claim assignment, carousel assignment) and short-term scheduling problems. In another study, colored

Petri Nets is used to model and simulate baggage handling systems in order to understand causes of failed bags [14]. A linguistic multi-criteria group decision making approach is used to determine the most appropriate modernization alternative for a hold baggage security screening system [15]. Coordination and sorting operations in an international airport is analyzed whether these operations can be conducted by multi-AGVs and stochastic model is used to determine the running priority and optimal path planning for AGVs [16].

As it seen from the studies reviewed, simulation and optimization techniques have been widely used for various problems of baggage handling operations. Furthermore, multi-criteria decision making techniques are also useful for determination of some strategic decisions related to baggage handling system design. In this study, an integrated model based on multi-criteria analysis and integer programming is used to obtain solution for determination of the number to be placed in a baggage robot system among a set of alternatives.

3 Conventional Baggage Handling Systems

Baggage handling process includes several activities based on the standards related to safety and security. A general representation of the process can be seen in Fig. 1 as follows.

According to the standards defined by IATA (The International Air Transport Association) in 2004, all baggage handling systems consist the following common functions:

1. Baggage check-in;
2. Hold Baggage Screening (HBS);

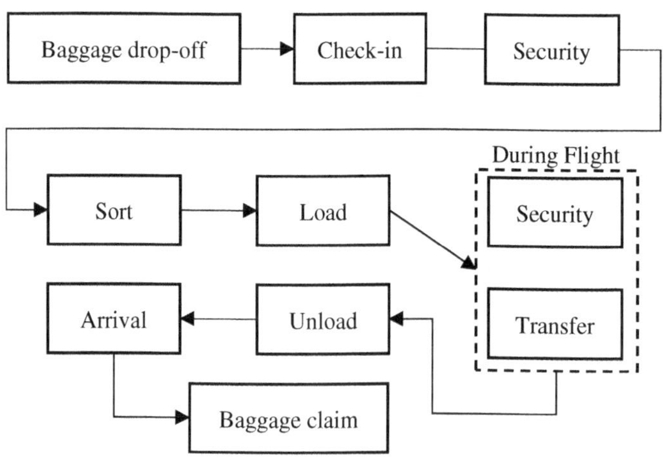

Fig. 1 Baggage flow process with tracking points [1]

3. Early baggage storage;
4. Early baggage flight build;
5. Manual coding;
6. Problem bag resolution,
7. Flight sortation, and
8. Flight build.

There are several ways to complete the check-in process. The most common types are the conventional way, which means checking-in by check-in desks, and the self check-in option, which refers to checking-in by self-service kiosks/bag drops. These check-in options constitute a common user terminal equipment (CUTE) [17]. By using CUTE technology, airport management increases the efficiency of check-in desks and other resources. This technology also gives the opportunity of desk allocation to different airlines at different times. So, the check-in desks get more flexible.

Baggage handling process begins after passenger's bag drop at the airport. Passengers can do the check-in for flight and bag-drop simultaneously. However, most of the passenger do their check-in for the flight online before arrival to the airport. Also, check-in kiosks, which provides the option of self-service check-in is another option to obtain boarding pass card. If a passenger with boarding pass want to drop their bags may apply to check-in desks or self-service bag drop point. At these points, a baggage label is printed after scanning the boarding pass and the printed label must be attached to the bag. This label contains some information about the flight like destination, flight number and airline company. Also, a barcode takes place on these labels, which makes it easier to read the information by hand held scanners and scanners of baggage conveyor systems. The dimensions and weight of the bags are also recorded at drop-off, but they are not seen on bag labels. So, this information is not considered in the following steps of baggage handling process. Passengers go through the security check processes and bags are move into the baggage handling area after attachment of bag labels.

To determine the correct aircraft for the correct bag, the following five activities performed in the baggage handling area once bags enter to the baggage handling area:

1. Security check of the bag contents
2. Sorting of bags
3. Loading the sorted baggage to the baggage carts at makeup stations.
4. Transportation of bags to the correct aircraft.
5. Loading bags into the aircraft.

In case of existence of more than one baggage handler at departure airport, the ground handler that handles the bag can be read on the bag label. After the bag reaches to the baggage handling area, security screening is firstly performed. Most of the airports use five screening levels, each screening level means a more detailed inspection.

Level 1 is an automatically examination of bag by using x-ray scanners whether it requires further inspection or not. Level 2 is a trained operator inspection on the x-ray picture in Level 2. The operator may request a new x-ray picture of the bag, then he/she may declare it is clear or sends to Level 3. At Level 3, a more detailed photo of the bag is taken by CTX-machine, which is a machine that is capable of making an acceptance of rejection decision on the bag. If CTX-machine rejects the bag, a trained operator analyzes the picture taken by the CTX-machine in Level 4. If the bag is still not cleared by the operator, then the bag owner is called and a manual investigation is performed in the presence of bag owner [18].

After clearance of baggage from security screening, it is proceeded by a conveyor belt through bag label reader. Bags are sorted by an automated system here by reading the information from the label. Then it is routed to the makeup area. The routing process is called as sorting process. Some of the passengers may drop their bags more than three hours before the flight departure. In these cases, their bags stored in early baggage storage. Most of the airports have early baggage storage. In airports with conventional baggage handling systems, these kind of bags are moved by workers manually to the early baggage storage and it increases labor costs and injury risks. Moving bags manually also makes it difficult to track and manage early dropped bags. The reason of this difficulty is that the workers work in shifts. So, mishandling of these bags becomes more possible and it also increases the costs of baggage handling [19].

Early bags are taken into sorting area at right time, if they are handled accurately. All bags are sorted in this area by using a conveyor belt, which moves the bags to correct sorting destinations. Bags, which passed security check, are assigned to different places corresponding the flights that they belong to by automated sorting machines. An infrared sensor reads the bag label and determines the place for each bag.

After the baggage handling system sorts bags, a worker checks the destinations of bags by using a manual barcode reader. Then, the worker loads the bags on a wheeled platform in order to send them into aircraft or storage area with trolleys. Unit load devices are another option for loading bags into the aircraft, especially for larger ones. After completion of loading bags and boarding of passengers, the aircraft can take off to the destination.

At the destination, the baggage process is reversed in a simpler shape. Firstly, the bags are transferred to the baggage handling system of the arrival airport from the aircraft. Bags are sorted here as transfer bags and final destination bags. Final destination bags sent to the baggage claim belt directly and they do not enter to baggage handling system. Transfer bags classified into regular transfer bags and bags will be transferred in a short time groups. Regular transfer bags enter into the baggage handling system from the security check process. If a bag is declared to be safe at any airport, further level security checks is not applied on this bag. Bags with a short transfer time are directly moved to the next aircraft from the arriving aircraft.

An exceptional case for baggage handling process is odd-sized items. Odd-sized items are baggage with heavier and/or larger items. Some examples for this type of items are bicycles, ski-boards, music instruments, etc. These items are generally about 32–80 kg and longer than 100 cm, taller than 65 cm and wider than 75 cm. Due to their dimensions and weight, handling of these items is not possible with conventional handling systems. So, they arc handled manually. That causes longer process times for these items, whereas their handling process is similar to other items. In the security check steps, odd-sized items directly go to the fifth level, because they cannot be checked by using x-ray or CTX machines. After security check of these items completed, they are directly loaded to the aircraft.

Unloading process of bags from aircraft is completed in less time. In general, the first bag is targeted to be placed on claim belt in less than 15 min' from the time that the aircraft door is opened at a medium scale airport. The time between the first bag and the last bag should be about ten minutes. Therefore, in 25 min, all bags must be loaded to the claim belt.

Several activities related to baggage handling process (sorting, makeup, etc.) can be performed by the same or different companies. In big hub airports, handling activities are made by different handlers. But, in small scaled airports all activities can be performed by a single company. Security process is commonly executed by a specialized security company in all airports.

4 Baggage Robot Systems

The aim of baggage handling systems is to transfer bags from baggage drop points to makeup stations, which is the place that bags loaded on carts to take them into the aircraft. As mentioned in the previous part, conventional baggage handling systems contain some jobs done by workers. So, some items can be mishandled in these systems.

Thanks to the development in technology, digitalization started to take place in most airport operations. By adoption of new technologies to the existing ones, passenger satisfaction in airports is increasing. The concept of improvement of current processes with the help of digitalization and new technologies can be defined as Aviation 4.0. Aviation 4.0 is the reflection of fourth industrial revolution to the aviation industry.

Industry 4.0 concept refers to the combination of industry and Internet of Things (IoT) technology. This concept was seen in different sectors so far, one of them is Airport 4.0. It is neither a new technology nor a business discipline. But it is a new approach as a result of technological developments over the last decade. Advancements in airport models from Airport 1.0 to Airport 4.0 is presented in Table 1.

Airports are generally leading places for technological innovations. Therefore, by using digital solutions airports have to improve their service quality and provide better customer experiences with operational improvements. Digital technologies are clustered into six groups as shown in Fig. 2 [20].

Table 1 Advancements in airport models

Airport model	Advancements
Airport 1.0	Manual and analog processes Long lag-times between resource demands and answer
Airport 2.0	Development of self-service stations for some processes
Airport 3.0	Optimization of flow monitoring by digitalization
Airport 4.0	Full-connection of stakeholders Perfect reaction to real-time demands

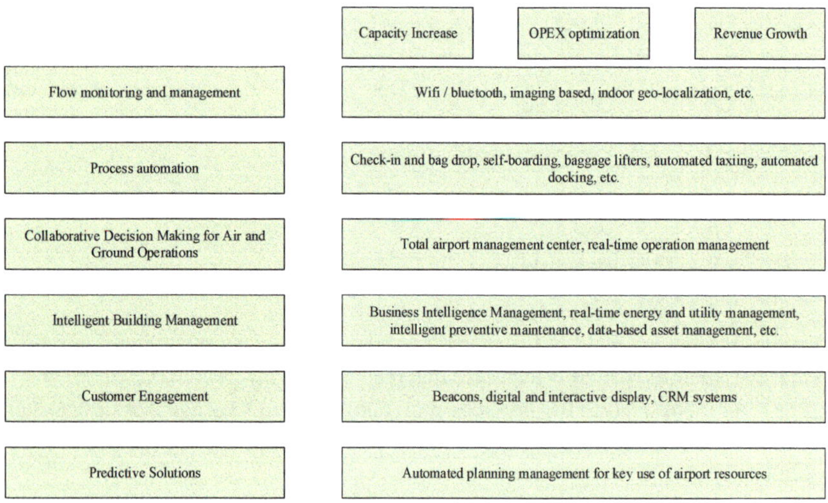

Fig. 2 Clustering of airport digital technologies

The potential benefit of Airport 4.0 technologies is expected to be 15–20% for capacity increase, reduction of operational expenses around 15% and revenue growth as a result of customer satisfaction improvements.

As a part of process automation automated baggage handling systems are extremely important for Airport 4.0 concept. Transition to automation in baggage handling would reduce costs by reduction of mishandled items and increase customer satisfaction by less service times. Baggage robot systems can be used for automated baggage handling.

Individual transport robots can be used to transfer bags between bag drop facilities and makeup stations instead of conveyors which takes part in conventional baggage handling systems. Transportation robots can decide their paths autonomously, so the bags can be moved between tracking points by using shortest paths.

Robots can deliver to or pick from security screening machines and take the cleared bags to the sorting area. The advantage of robots at this operation is that they can take the bag to the correct makeup station in a short time, instead of the current conveyor belts visit different makeup stations. Also, the robots can simultaneously complete transportation and sorting tasks.

Due to the charging requirement for robots, charging facilities are needed. Transportation robots are charged by inductive charging, so charging facilities can be placed anywhere in their movement area. The places that the robots are idle can be chosen as charging places. For example, after completion of any transportation job, robots are waiting for a new job at makeup stations and they are idle at that moment. This kind of places can be suitable for charging of robots.

To perform transportation and sorting jobs successfully, elements of the system must be carefully designed. The elements of a baggage robot system can be listed as follows:

1. Layout configuration and charging: This element is related to the location of charging facilities. Since the robots charged by inductive charging no cables needed. So, layout configuration is not a hard work. The only point is to determine the charging strategy in case of existence of many robots.
2. Control: Coordination of robot motions must be designed. System robustness and scalability is related to the control architecture [21].
3. Routing: The path that the bags follow in the handling system is the route. Robots must follow the shortest path.
4. Collision avoidance: In case of existence of more than one robot in the area, the risk of collision appears. To avoid collision of robots, a safe distance must be defined. While determining the control strategy, the actions must be defined in case of two robots get closer than the safe distance.
5. Deadlock resolution: Robots can wait for another robot's movement in order to avoid collision. If two or more robots wait for other robots' movement, the bags will not be delivered to makeup station in time, so mishandled items may occur because of the undelivered items to aircraft in time. This is called as deadlock. Deadlock situations must be considered as an important element of robot baggage systems.

Passengers leave their bags at a drop-off facility in the airport. A bag label is attached at that point and a small conveyor takes the bag into sorting area. Then, the first available robot is assigned to the bag and the robot carries the bag to the makeup station which is read by the robot from the label. If the robot is idle, it can be charged in the storage and charging area or it can wait for a new task when its charge is more than a threshold value.

So, the transportation robots are expected to pick up the bags from drop off stations, then sorting the bags that it picked up while transportation and drop them into the correct makeup station. The expectation from the robots can be called as functional requirements. Functional requirements of robots are given in Table 2.

In addition to the functional requirements, there are some non-functional requirements for baggage robot systems. The most important non-functional requirement for such a system is defined as being flexible in terms of capacity for baggage handling process. By this way, the system can adapt to variability of demand easily. Also,

Table 2 Functional requirements of robots

Function	Functional requirements
Transporting bags	Communication with conveyor to activate task and obtain the destination Picking the bag up Transport bag (handling the bag, prevent the bag from falling during transportation) Sorting bags Dropping the bag off at correct makeup station
While not transporting bags	Deactivate the robot if it is idle for a certain time Send the robot to charge, if its charge under the threshold value Transmit power from charging area to robot

the system should be flexible for capacity expansion for long term demand changes. Another requirement for baggage robot systems is the utilization rate of the system. That means an over dimensioned system may return lower utilization rates. But the capacity of the system, which is defined as sorted and transported bag numbers, should be enough for operational standards. Deficiencies in the system must be between an acceptable interval and the system should be able to continue to work. Also, the system should provide possibility of maintain working after maintenance in case of a deficiency. Lastly, if the system need human work for deficiencies, it should be safe for that workers. Non-functional requirements of baggage robot systems can be listed as:

- Capacity flexibility
- Expansion flexibility
- Utilization rate
- Capacity
- Deficiency resiliency
- Reliability
- Maintainability
- Safety.

Another important consideration for baggage robot systems is the limitations and operational conditions related to the system. Since the process starts with bag entrance, the first limitation is the input capacity of the system. After the entrance of the bag to the sorting area, a robot picks the bag up. The number of bags that a robot can carry at one time is defined as robot capacity. Next, the sorting system capacity is another limitation for the system, which is directly related to number of robots and robot capacity. The velocity of the robot must be defined in order to balance the sorting system and makeup area. Lastly, the size and shape of the area is an important limitation since they may cause collision and block some robot motions.

The basic conditions must also be taken into account. Firstly, the risk of collision for robots must be minimized. In case of collision occurrence between robots and robots or objects, robots or bags may get damaged. Also, when robots come closer to each other, deadlock situations can be faced. If the risk of collision is not minimized and collisions or deadlocks occur, sorting operations would be operated more slowly. It would also affect makeup operations and may cause mishandled items, which is not loaded on the correct aircraft on time. And the charge level of robots must be sufficient to complete the task that assigned to it.

The main characteristics of baggage robots are explained in this part. The next part, proposes an analytic decision model for robot selection for a baggage robot system.

5 Proposed Model

The robots used in baggage robot systems are directly effective on system success. For this reason, a robot set construction model based on hesitant fuzzy linguistic multi-criteria analysis and binary integer programming was proposed in this chapter. Hesitant fuzzy multi-criteria analysis is done because of the evaluation of different robots require consideration of different robots simultaneously. Binary integer programming is used in order to determine the number of required robots for the baggage robot system by considering system constraints and multi-criteria analysis results together.

In this part, the hesitant fuzzy linguistic multi-criteria analysis and linear programming methods was explained as follows.

Decision making is an activity that is done frequently in daily life. It is defined as selection of one or a subset among a set of alternatives. Decision maker can make a decision easily in case of existence of one criterion to consider. Increasing number of criteria to consider makes it difficult to decide among alternatives. Because, each alternative may be superior to the other in views of several criteria. Since the decision maker has to take all criteria into account simultaneously, the problem structure becomes very complex. Complex decision problems with a finite number of alternatives and a finite number of criteria are named as multi-criteria decision making (MCDM) problems. There are several approaches in the literature to solve MCDM problems.

Based on the structure of problem data, decision problems can be classified into three groups as (i) decision making under certainty, (ii) decision making under risk and (iii) decision making under uncertainty. The first group refers to existence of certain problem data. If the historical data for decision problem can be obtained and problem parameters can be expressed by probability distributions, this kind of problems defined as decision making under risk. The last group of decision making problems expresses the case of existence of uncertain problem data. Since fuzzy logic is a suitable tool for expressing uncertain or linguistic terms, it is commonly used for third group of decision making problems.

Since linguistic terms are not mathematically operable, fuzzy logic is a useful tool for modelling and solving problems with linguistic terms. One of the recent extensions of fuzzy set theory is Hesitant Fuzzy Sets [22]. Hesitant fuzzy sets are introduced to the literature to define uncertain situations that decision maker feel hesitant about several values of a parameter. Rodriguez et al. [23] proposed the concept of Hesitant Fuzzy Linguistic Terms Set (HFLTS), which helps expressing uncertain statements by using a context-free grammar. Yavuz et al. [24] extended HFLTS concept into group decision making problems. Their methodology needs hesitant fuzzy evaluations of decision makers to construct pairwise comparisons between them. By following the steps of their algorithm, criteria weights and alternative priority values can be obtained. Alternative priorities show the alternative to be selected.

The steps of the algorithm are given as follows:

Step 1: Definition of goal, criteria and alternatives related to the decision making problem.

A sample of hierarchical structure of a MCDM problem is given in Fig. 3.

Step 2: Definition of linguistic terms and context—free grammar.

Linguistic terms for importance degrees used in the study are presented in Table 3. Evaluations may be made by using a single term or a composite term which means a subset of linguistic terms, which are connected to the others by using an appropriate relation term from the set of "at most", "at least", "greater than", "lower than", "is" and "between".

Step 3: Collecting preferences of expert for criteria and alternatives.

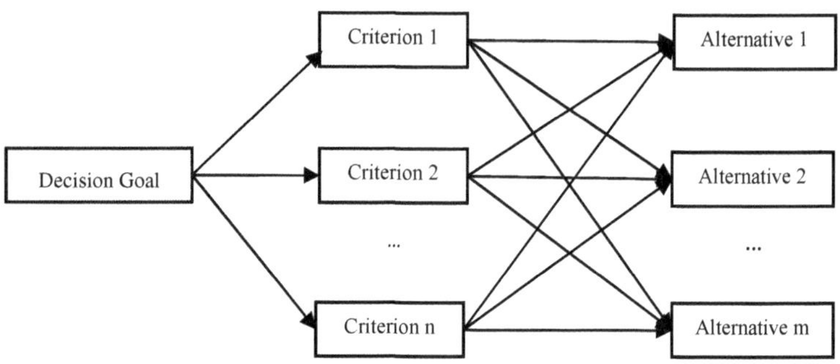

Fig. 3 Hierarchical structure of a MCDM problem with m alternatives and n criteria

Table 3 Importance degrees and related linguistic terms

Importance degree	Linguistic term
0	No importance (n)
1	Very low importance (vl)
2	Low importance (l)
3	Medium importance (m)
4	High importance (h)
5	Very high importance (vh)
6	Absolute importance (a)

Table 4 Importance degrees and related linguistic terms

Expert's preference	HFLTS equivalent
At most x	[n, x]
At least x	[x, a]
Lower than x	[n, x–1]
Greater than x	[x + 1, a]
Is x	[x, x]
Between x and y	[x, y]

Pairwise comparisons are made according to opinion of expert. Expert preferences are expressed as single or composite hesitant fuzzy linguistic terms.

Step 4: Transformation of experts' preferences into HFLTS.

By using grammar rules, experts' preferences are transformed into HFLTS. Grammar rules and HFLTS equivalents are presented in Table 4 as follows, where x and y represents the linguistic term in the preference of mentioned expert. x – 1, x + 1 represents the importance degree of the linguistic term for the lower and higher importance degree, respectively.

Step 6: Building the vector of intervals for collective preferences.

Vector of intervals are built by arithmetic mean of rows in pessimistic and optimistic preferences matrices.

Step 7: Obtaining priority values by normalization of interval values.

Interval midpoints are calculated and normalized interval midpoints show priority values.

Step 8: Selection of appropriate alternative.

The most appropriate alternative is determined by checking alternative priorities, which is defined by multiplication of alternative priorities in views of criteria and criteria weights. The alternative with the highest priority is determined as the best choice.

Linear programming is defined as a way to solve optimization problems [25]. Simplex algorithm was developed by George Dantzig in 1947 for obtaining solution to linear programming problems. Linear programming has been applied in a wide range of problems such as education, logistics, banking and forestry from the date the simplex algorithm developed.

Elements of a linear programming problem can be listed as: (i) decision variables, (ii) objective function, (iii) constraints, and (iv) sign restrictions. By using a linear programming, the following things are done:

1. Maximization or minimization of a linear function of decision variables. This element of a linear program called as objective function.
2. The linear program has a set of constraint, which must be satisfied. Constraints can be either equations or inequalities, but the values assigned to the decision variables must satisfy all constraints.
3. Each decision variable is restricted in terms of sign, which means that any variable must be defined as nonnegative or unrestricted in sign.

A linear programming model can be represented as follows:

$$z = \mathbf{cx}$$
$$\mathbf{Ax} = \mathbf{b}$$
$$\mathbf{x} \geq 0$$

The first equation in the representation shows the objective function of a linear program. z shows the objective value in objective function, where **c** is the objective coefficients and **x** expresses the decision variables. Constraints of the model are given by $\mathbf{Ax} = \mathbf{b}$. In this equation **A** is the technology coefficients matrix and **b** refers to the right hand side constants of constraints. The last element of the model is the sign restrictions of decision variables.

In some cases, values of decision variables of a linear program required to be integer. Personnel, equipment or vehicle assignment decisions can be counted as examples for such kind of problems. If the values of decision variables different from integers, the results may be insensible [26]. This type of linear programming problems called as integer programming problems.

The main difference between linear programming and integer programming is the additional sign restriction for decision variables which are required to be integers. If all decision variables are required to have integer values, the problem is named as pure integer programming. However, in some problems integer values are needed for only some of the decision variables and the problem is called mixed integer programming.

An important application area of integer programming containing decision variables to model yes or no decisions. Some decisions like assignment of a specific person to a job, selection of a location for a production plant, making an order for a material can be considered as examples of such situations. This type of decisions can be made by restriction of the decision variable as taking only one of 0 or 1 values. Decision variables to model such situations are named as binary variables. Integer programs with binary variables are called binary integer programming problems. Site selection, personnel scheduling, job scheduling, capital budgeting are some examples of binary integer programming problems.

6 Application of the Proposed Model

In this part, application steps and results of the proposed model was presented. With this aim, multi-criteria analysis of alternative baggage robots is firstly done. Then, elements of baggage robot system determined by binary integer programming model. The steps of the application are as follows:

Step 1: Definition of goal, criteria and alternatives related to the decision making problem.

The aim of the multi-criteria analysis is to determine the best alternative among baggage robots with respect to capacity (C1), maintainability (C2), reliability (C3), ease of control (C4) and safety (C5) criteria. Four alternative robots of different manufacturers (A1–A4) will be evaluated. Hierarchical structure of the decision problem is given in Fig. 4.

Step 2: Definition of linguistic terms and context—free grammar.

Linguistic terms defined in the previous part are used for constructing the single or composite evaluation statements of expert for decision elements in the model.

Step 3: Collecting preferences of expert for criteria and alternatives.

Pairwise comparisons are made according to opinion of an expert who is an airport management professional with fifteen years of experience. Expert preferences for criteria and alternatives are expressed as single or composite hesitant fuzzy linguistic terms and presented in Tables 5, 6, 7, 8, 9 and 10.

Step 4: Transformation of experts' preferences into HFLTS.

In this step, expert evaluations are expressed as HFLTS. According to the HFLTS equivalents presented in Table 4, pairwise comparison matrices are transformed into HFLTS and presented in Tables 11, 12, 13, 14, 15 and 16.

Step 5: Building the vector of intervals for collective preferences.

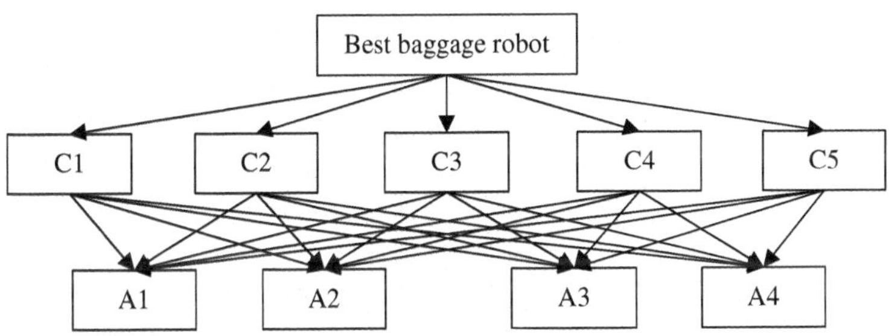

Fig. 4 Hierarchical structure of baggage robot alternatives evaluation

Table 5 Expert evaluation of criteria with respect to goal

	C1	C2	C3	C4	C5
C1	–	Between m and h	Between l and m	m	Between m and vh
C2	Between l and m	–	At most l	Between l and m	m
C3	Between m and h	At least h	–	m	Greater than m
C4	m	Between m and h	m	–	At least vh
C5	Between vl and m	m	Lower than m	At most vl	–

Table 6 Expert evaluation of alternatives with respect to capacity criterion

	A1	A2	A3	A4
A1	–	At least h	Between vl and m	At most m
A2	At most l	–	n	n
A3	Between m and vh	a	–	m
A4	At least m	a	m	–

Table 7 Expert evaluation of alternatives with respect to maintainability criterion

	A1	A2	A3	A4
A1	–	Between m and h	m	Between l and m
A2	Between l and m	–	Between l and m	Between vl and m
A3	m	Between m and h	–	m
A4	Between m and h	Between m and vh	m	–

Table 8 Expert evaluation of alternatives with respect to reliability criterion

	A1	A2	A3	A4
A1	–	m	m	Between l and m
A2	m	–	m	Between vl and m
A3	m	m	–	Between vl and m
A4	Between m and h	Between m and vh	Between m and vh	–

Table 9 Expert evaluation of alternatives with respect to ease of control criterion

	A1	A2	A3	A4
A1	–	Between l and m	At least h	Between h and vh
A2	Between m and h	–	At least vh	At least h
A3	At most l	At most vl	–	Between l and m
A4	Between vl and l	At most l	Between m and h	–

Table 10 Expert evaluation of alternatives with respect to safety criterion

	A1	A2	A3	A4
A1	–	Between m and vh	At least m	m
A2	Between vl and m	–	Between m and h	Between vl and m
A3	At most m	Between l and m	–	Between vl and m
A4	m	Between m and vh	Between h and vh	–

Table 11 HFLTS equivalent of pairwise comparison matrix of criteria

	C1	C2	C3	C4	C5
C1	–	[m, h]	[l, m]	[m, m]	[m, vh]
C2	[l, m]	–	[n, l]	[l, m]	[m, m]
C3	[m, h]	[h, a]	–	[m, m]	[h, a]
C4	[m, m]	[m, h]	[m, m]	–	[vh, a]
C5	[vl, m]	[m, m]	[n, l]	[n, vl]	–

Table 12 HFLTS equivalent of pairwise comparison matrix of alternatives with respect to C1

	A1	A2	A3	A4
A1	–	[h, a]	[vl, m]	[n, m]
A2	[n, l]	–	[n, n]	[n, n]
A3	[m, vh]	[a, a]	–	[m, m]
A4	[m, a]	[a, a]	[m, m]	–

Table 13 HFLTS equivalent of pairwise comparison matrix of alternatives with respect to C2

	A1	A2	A3	A4
A1	–	[m, h]	[m, m]	[l, m]
A2	[l, m]	–	[l, m]	[vl, m]
A3	[m, m]	[m, h]	–	[m, m]
A4	[m, h]	[m, vh]	[m, m]	–

Table 14 HFLTS equivalent of pairwise comparison matrix of alternatives with respect to C3

	A1	A2	A3	A4
A1	–	[m, m]	[m, m]	[l, m]
A2	[m, m]	–	[m, m]	[vl, m]
A3	[m, m]	[m, m]	–	[vl, m]
A4	[m, h]	[m, vh]	[m, vh]	–

Table 15 HFLTS equivalent of pairwise comparison matrix of alternatives with respect to C4

	A1	A2	A3	A4
A1	–	[l, m]	[h, a]	[h, vh]
A2	[m, h]	–	[vh, a]	[h, a]
A3	[n, l]	[n, vl]	–	[l, m]
A4	[vl, l]	[n, l]	[m, h]	–

Table 16 HFLTS equivalent of pairwise comparison matrix of alternatives with respect to C5

	A1	A2	A3	A4
A1	–	[m, vh]	[m, a]	[m, m]
A2	[vl, m]	–	[m, h]	[vl, m]
A3	[n, m]	[l, m]	–	[vl, l]
A4	[m, m]	[m, vh]	[h, vh]	–

Vector of intervals were built by arithmetic mean of rows in pessimistic and optimistic preferences, which refers to the lower and upper bound terms of each element in matrices presented in Step 4, respectively.

Step 6: Obtaining priority values by normalization of interval values.

Midpoint of each interval were calculated, then the midpoint values were normalized by vector normalization. The obtained normalized value gives the priority for each decision element. Interval utilities, midpoints and priority values are presented in Tables 17, 18, 19, 20, 21 and 22.

Table 17 Priority values of criteria

	Interval utilities		Midpoints	Weights
C1	2.750	3.750	3.250	0.217
C2	1.750	2.750	2.250	0.150
C3	3.500	4.750	4.125	0.275
C4	3.500	4.000	3.750	0.250
C5	1.000	2.250	1.625	0.108

Table 18 Priority values of alternatives in views of C1

	Interval utilities		Midpoints	Weights
A1	1.667	4.000	2.833	0.236
A2	0.000	0.667	0.333	0.028
A3	4.000	4.667	4.333	0.361
A4	4.000	5.000	4.500	0.375

Table 19 Priority values of alternatives in views of C2

	Interval utilities		Midpoints	Weights
A1	2.667	3.333	3.000	0.250
A2	1.667	3.000	2.333	0.194
A3	3.000	3.333	3.167	0.264
A4	3.000	4.000	3.500	0.292

Table 20 Priority values of alternatives in views of C3

	Interval utilities		Midpoints	Weights
A1	2.667	3.000	2.833	0.236
A2	2.333	3.000	2.667	0.222
A3	2.333	3.000	2.667	0.222
A4	3.000	4.667	3.833	0.319

Table 21 Priority values of alternatives in views of C4

	Interval utilities		Midpoints	Weights
A1	3.333	4.667	4.000	0.333
A2	4.000	5.333	4.667	0.389
A3	0.667	2.000	1.333	0.111
A4	1.333	2.667	2.000	0.167

Table 22 Priority values of alternatives in views of C5

	Interval utilities		Midpoints	Weights
A1	3.000	4.667	3.833	0.319
A2	1.667	3.333	2.500	0.208
A3	1.000	2.667	1.833	0.153
A4	3.333	4.333	3.833	0.319

As it can be clearly seen from the tables above, the most important aspect for selection of a baggage robot is reliability among the criteria considered in this study. Reliability is followed by ease of control and capacity. The priority order of rest of criteria goes as maintainability and safety. Moreover, none of the alternatives seen as dominant over the others. So, the application of multi-criteria evaluation is useful for the problem.

Step 7: Selection of appropriate alternative.

To be able to make decision among a number of alternatives, the priority values of alternatives must be aggregated. To do so, alternative priority matrix was constructed by collecting priority vectors of alternatives in views of criteria. Then, this matrix was multiplied by criteria priorities (weights) vector and final priority of alternatives were obtained.

$$\begin{bmatrix} 0.236 & 0.250 & 0.236 & 0.333 & 0.319 \\ 0.028 & 0.194 & 0.222 & 0.389 & 0.208 \\ 0.361 & 0.264 & 0.222 & 0.111 & 0.153 \\ 0.375 & 0.292 & 0.319 & 0.167 & 0.319 \end{bmatrix} \times \begin{bmatrix} 0.217 \\ 0.150 \\ 0.275 \\ 0.250 \\ 0.108 \end{bmatrix} = \begin{bmatrix} 0.2715 \\ 0.2161 \\ 0.2233 \\ 0.2891 \end{bmatrix}$$

According the final priority values of robot alternatives, the best alternative was considered to be A4. Moreover, the priority difference of A4 and A1 was very little (0.2891–0.2715 = 0.0176). The third alternative can be considered A3 and the worst option among the alternatives taken into account in this analysis was A2.

After the evaluation of alternative robots, a set of robots which will form the baggage robot system components will be determined by using an integer programming model. In this model, capacity of robots and budget limitations taken into account as constraints and maximization of system performance is considered as objective function. The decision to be determined in this model is the number of robots to be included to the system from each type. Priority values of alternative robots calculated in the multi-criteria analysis above are the objective coefficients of decision variables. Parameters, decision variables and mathematical formulation of the mathematical model of the problem are presented as follows:

Indices:

i: robot type ($i = 1, 2, 3, 4$)

Parameters:

p_i: priority value of robot type i
cap_i: transportation capacity of robot type i
c_i: purchasing cost of robot type i
V: number of bags needed to be transferred by robots
B: budget allocated for baggage robots

Decision variable:

x_i: number of robots of type i to be bought

Mathematical formulation:

$$\max z = \sum_{i=1}^{4} p_i x_i$$

subject to

$$\sum_{i=1}^{4} cap_i x_i \geq V$$

$$\sum_{i=1}^{4} c_i x_i \leq B$$

$x_i \geq 0$ *and integer.*, $i = 1, 2, 3, 4$

Problem data for robot alternatives are provided in Table 23.

Allocated budget by airport management is determined to be 1500 MU (Monetary Units) and the expected number of bags which will be transferred by baggage robot

Table 23 Problem data for robot alternatives

Alternative	p_i	cap_i (bags/hour)	c_i (MU)
A1	0.2715	26	18
A2	0.2161	5	10
A3	0.2233	38	27
A4	0.2891	40	35

Table 24 Solution results

Decision variable	x_1	x_2	x_3	x_4
Value	75	15	0	0

system is predicted to be 2000 bags/hour. According to these data, the mathematical model is solved by Microsoft Excel Solver. Optimal solution to the model is obtained in 0.2 s. The solution results are presented in Table 24.

The results of the integer programming model show that the optimal number of robots in baggage robot system is 90, 75 robots of A_1 and 15 robots of A_2. Based on this configuration, the capacity of system will be 2025 bags/hour and the required budget will be 1500 MU.

7 Conclusion

Baggage handling is one of the most important operations for airport management. Mishandled bags cause waste of workforce and extra costs to the management. So, reducing the number of mishandled items is vital.

Baggage robot systems are proposed as an automation option for baggage operations in airports. By using robots during the handling process, mishandled items caused by human errors are eliminated and the accuracy of baggage handling operations is increased.

According to the developments in Aviation 4.0 by time, automation of aviation processes is expected to be increased. By this way, errors occurred in airports and aviation management will be minimized.

In this chapter, a decision making model based on hesitant fuzzy multi-criteria analysis and integer programming is proposed. Within the methodology, priority values of four alternative robots are primarily determined via multi-criteria analysis, then the number of robots to satisfy functional requirements of system is determined by using an integer programming model. Because of the ease of application and flexibility of the model, the proposed model can be used as a decision support tool for airport managers.

The results of the study show that application results of a single method may lead to wrong decisions. In this study, if the decision maker takes only the multi-criteria analysis results into account, he/she would purchase only A4 type robots. However, according to the optimization model results by considering multi-criteria analysis, A4 type robots should not be purchased. The optimal result is a combination of A1 and A2 type robots. So, hybrid methods that combine advantages of different techniques can provide better results.

On the other hand, the application provided in this study contains a limited number of evaluation criteria, alternatives for multi-criteria evaluation and baggage volume and budget limitation in the optimization model. This can be considered as the main limitation of the study. In further studies, this study can be extended by considering different criteria and newly developed alternative robots. Also, consideration of different constraints and objective functions in the integer programming model can present more interesting results. Proposal of a multi-objective optimization model for the study would also bring a good research direction by comparison of several multi-objective optimization techniques.

References

1. World Maritime University: Transport 2040: Automation in Airports: Automatic Baggage Handling Systems—Technology and Transformation. Reports, 61 (2019)
2. SITA: Air Transport Industry Insights: The Baggage Report (2018)
3. Saygin, C., Natarajan, B.: RFID-based baggage handling system design. Sens. Rev. **30**(4), 324–335 (2010)
4. Tarau, A.N., de Schutter, B., Hellendoorn, J.: Predictive route control for automated baggage handling systems using mixed-integer linear programming. Transp. Res. Part C **19**(3), 424–439 (2011)
5. Huang, E., Mital, P., Goetschalckx, M., Wu, K.: Optimal assignment of airport baggage unloading zones to outgoing flights. Transp. Res. Part E **94**, 110–122 (2016)
6. Cavada, J.P., Cortés, C.E., Rey, P.A.: A simulation approach to modelling baggage handling systems at an international airport. Simul. Model. Pract. Theory **75**, 146–164 (2017)
7. Kim, G., Kim, J., Chae, J.: Balancing the baggage handling performance of a check-in area shared by multiple airlines. J. Air Transp. Manage. **58**, 31–49 (2017)
8. Frey, M., Kolisch, R., Artigues, C.: Column generation for outbound baggage handling at airports. Transp. Sci. **51**, 1226–1241 (2017)
9. Skorupski, J., Uchroński, P., Łach, A.: A method of hold baggage security screening system throughput analysis with an application for a medium-sized airport. Transp. Res. Part C **88**, 52–73 (2018)
10. Malandri, C., Briccoli, M., Mantecchini, L., Paganelli, F.: A discrete event simulation model for inbound baggage handling. Transp. Res. Proc. **35**, 295–304 (2018)
11. Rezaei, J., Kothadiya, O., Tavasszy, L., Kroesen, M.: Quality assessment of airline baggage handling systems using SERVQUAL and BWM. Tourism Manage. **66**, 85–93 (2018)
12. Huang, E., Liu, I., Lin, J.T.: Robust model for the assignment of outgoing flights on airport baggage unloading areas. Transp. Res. Part E **115**, 110–125 (2018)
13. Boysen, N., Briskorn, D., Fedtke, S., Schmickerath, M.: Automated sortation conveyors: asurvey from an operational research perspective. Eur. J. Oper. Res. **276**, 796–815 (2019)
14. Hafilah, D.L., Cakravastia, A., Lafdail, Y., Rakoto, N.: Modeling and simulation of air France baggage handling system with colored petri nets. IFAC PapersOnLine **52**(13), 2443–2448 (2019)

15. Skorupski, J., Uchronski, P.: Multi-criteria group decision-making approach to the modernization of hold baggage security screening system at an airport. J. Air Transp. Manage. **87**, 101841 (2020)
16. Shen, K., Li, C., Xu, D., Wu, W., Wan, H.: Sensor-network-based navigation of delivery robot for baggage handling in international airport. Int. J. Adv. Rob. Syst. **17**(4), 1729881420944734 (2020)
17. Bradley, A.: A comparison of whole life cycle costs of robotic, semi-automated, and manual build airport baggage handling systems. Ph.D. Thesis, Cranfield University, Cranfield (2013)
18. Grigoras, R., Hoede, C.: Design of a Baggage Handling System. University of Twente, Twente (2007)
19. Dadyala, G.: Short-term baggage storage systems lead to long-term success. Consultant Perspect. 6–7 (2013)
20. Little, A.D.: Airports 4.0: Impact of digital transformation on airport economics (2015). Retrieved from: https://www.adlittle.com/en/insights/viewpoints/airports-40. Access date 15 July 2020
21. Parker, L.E.: Springer handbook of robotics. Choice Rev. Online **46**(6), 46-3272 (2009)
22. Torra, V.: Hesitant fuzzy sets. Int. J. Intell. Syst. **25**, 529–539 (2010)
23. Rodríguez, R.M., Martínez, L., Herrera, F.: Hesitant fuzzy linguistic term sets for decision making. IEEE Trans. Fuzzy Syst. **20**, 109–119 (2012)
24. Yavuz, M., Oztaysi, B., Onar, S.C., Kahraman, C.: Multi-criteria evaluation of alternative-fuel vehicles via a hierarchical hesitant fuzzy linguistic model. Expert Syst. Appl. **42**(5), 2835–2848 (2015)
25. Winston, W.L., Goldberg, J.B.: Operations Research: Applications and Algorithms, vol. 3. Duxbury Press, Belmont (2004)
26. Hillier, F.S., Lieberman, G.J.: Introduction to Operations Research, 4th edn. Holden-Day, Oakland (1986)

Complex Spherical Fuzzy Sets and an Application to Catering Services in Aviation 4.0

D. Ajay and J. Aldring

Abstract In this research, the recent development of three dimensional fuzzy sets namely, spherical fuzzy sets is extended to complex spherical fuzzy sets (CSFSs) and the novelty of CSFSs with their operations and properties are investigated. The range of CSFSs membership functions lie in complex plane, not limited to [0, 1]. So, the CSFSs describe complex membership functions in a set in terms of complex numbers which leads to a new mathematical framework and comprehensive study of their properties. Some of the aggregation operators and distance measures are established with their mathematical proof. Further an intelligent system based on EDAS (Evaluation based on Distance from Average Solution) method has been developed using the proposed complex spherical fuzzy aggregation operators. An application is proposed for aviation industry 4.0 in order to assess the performance of catering services. Finally a numerical illustration has been demonstrated for the proposed innovative method.

Keywords Complex spherical fuzzy sets · Aggregation operators · MCDM · Aviation industry 4.0

1 Introduction

The assessment of a statement can be true or false in the context of conventional Boolean logic. In other words, an element can have two possibilities namely, to belong to a set that is under consideration or not to belong to it. But many real life statements or situations totally differ from this dichotomous argument. Such situations cannot be handled with possibilities yes or no, or true or false and the ability to make decision in such scenario is very much limited and the decision results may not be precise or may deviate from the reality. Such types of problems can be handled with multivalued logic or fuzzy logic based on fuzzy sets [1]. Researchers

D. Ajay (✉) · J. Aldring
Sacred Heart College (Autonomous), Tirupattur District, Tamilnadu 635601, India

C. Kahraman and S. Aydın (eds.), *Intelligent and Fuzzy Techniques in Aviation 4.0*, Studies in Systems, Decision and Control 372,
https://doi.org/10.1007/978-3-030-75067-1_5

have proposed several extensions of fuzzy sets based on the range and on the need to account for different types of uncertainty [1–6].

This research work is based on spherical fuzzy sets (SFSs) [6] which was introduced as one of the recent generalizations of fuzzy sets that deals with uncertain information in terms of membership, non-membership and hesitancy functions with the constraint that the sum of the squares of all these function values lies in the closed interval of zero and one. Recently spherical fuzzy QFD method has been introduced by Kutlu Gundogdu and Kahraman [7] and implemented to the linear delta robot technology. Further, Aydin and Kutlu Gundogdu [8] introduced fuzzy MULTIMOORA method to the application of industry 4.0 with interval valued SFSs. It is evident from the literature that many research works have been made under spherical fuzzy sets [9–15].

The notion of complex fuzzy sets was introduced by Ramot et al. [16] and it gave another important research direction to fuzzy sets. Later on, Bi et al. introduced complex fuzzy geometric [17] and arithmetic aggregation operators [18]. Recently, the power aggregation operator for complex fuzzy sets were presented in [19]. The idea of complex intuitionistic fuzzy sets was proposed by Alkouri and Salleh [20] from the idea of CFSs. Ullah et al. [21] developed complex Pythagorean fuzzy sets and their applications in pattern recognition. Also complex neutrosophic sets have been introduced by Ali and Smarandache [22]. More recently, Ali et al. [23] have utilized the operations which have been developed by Mahmood et al. [24] to develop complex T-Spherical fuzzy aggregation operators. Thus the fuzzy sets have evolved in complex scenario also with many applications in the field of MCDM. Many decision making approaches are available in the literature [25] and one of the most effective methods is EDAS. This approach was introduced by Keshavarz Ghorabaee et al. [26]. Then, Keshavarz Ghorabaee et al. [27] proposed an extended EDAS method for supplier selection. Kahraman et al. [28] established a new EDAS model under intuitionistic fuzzy information for solid waste disposal site selection. In this research work we have used the spherical fuzzy notion and basic operations to develop CSF-EDAS method for selecting the best catering services in aviation industry 4.0.

The main goal of industry 4.0 is for machine networking to operate on systematic processes or automatons, i.e. linking machines-machinery and machinery-human collaboration in order to predict future actions or to select feasible solutions from data or knowledge collection. These technologies have been developed for the growth of industry today with digital transformations. One of the most important digital industries is aviation sector, where all of these digital revolutions, such as robotics, automatons, Internet of Things, radio frequency recognition and digital manufacturing have all found immediate application. Literature informs us that research has already been conducted in this field with the help of fuzzy sets and also recently Buyukozkan et al. [29] analyzed success factors in aviation industry 4.0 using integrated intuitionistic fuzzy MCDM methods. A review of Industry 4.0 and related technologies are explained in detail by Oztemel and Gursev in [30] and Barbosa and Aroca [31] have discussed industry 4.0 concepts on aircraft construction with an application.

The advantage of the proposed complex spherical fuzzy numbers (CSFNs) lies in dealing with different intelligent systems for decision making and modelling real life problems when one deals with the information of periodicity/time and uncertainty at the same time. The theory of CSFs has become an evolutionary research area in recent years and stimulated applications in many realistic complex problems in the fields of medicine, engineering, biology, industry 4.0 and others. It is known that real-world knowledge is not obtained all at once or at the same time. In particular, the required knowledge also needs to be gathered and compiled from diverse levels at different times, for several concrete complicated situations, such as medical diagnosis problems, decision-making problems, and financial investment strategies. In such cases, complex spherical fuzzy theory can become an efficient framework for dealing with complex fuzzy information.

This chapter is organized as follows. Section 2 explains the details of methodology of the complex spherical fuzzy sets with EDAS model. Then the application of the proposed method is given in Sect. 3. The detailed comparison analysis has been presented in Sect. 4. Finally the conclusions have been listed in Sect. 5 with the indication of future scope for research in this direction.

2 Methodology

We discuss some of the basic concepts of fuzzy sets with their operations.

Definition 1 [1] If v is a particular element in the universe of discourse $\overset{*}{V}$, then a fuzzy set $\mathbb{S}$ is defined by a fuzzy membership function ($\mu_{\mathbb{S}}$) which associates to each v a membership value in the closed unit interval of zero and one. i.e. $\mu_{\mathbb{S}}(v) : X \rightarrow [0, 1]$

Definition 2 [2] An intuitionistic fuzzy sets $\widetilde{\mathcal{A}}$ is defined as a set of ordered pairs over a universal set U given by

$$\widetilde{\mathcal{A}} = \left\{ \left\langle x, \left(\xi_{\widetilde{\mathcal{A}}}(x), \psi_{\widetilde{\mathcal{A}}}(x), \right) \right\rangle | x \in U \right\}$$

where $\xi_{\widetilde{\mathcal{A}}}(x) : U \rightarrow [0, 1]$, $\psi_{\widetilde{\mathcal{A}}}(x) : U \rightarrow [0, 1]$ and with the condition $\xi_{\widetilde{\mathcal{A}}}(x) + \psi_{\widetilde{\mathcal{A}}}(x) \leq 1$ for each element $x \in U$. Here the membership and non-membership functions are denoted as $\xi_{\widetilde{\mathcal{A}}}(x)$ and $\psi_{\widetilde{\mathcal{A}}}(x)$ respectively.

Definition 3 [5] let X be a universe of discourse or a non empty set. Any object in the neutrosophic set A has the form $A_N = < x, \mu_A(x), \sigma_A(x), \gamma_A(x) > x \in X$, where $\mu_A(x), \sigma_A(x)$ and $\gamma_A(x)$ represent the degree of truth membership, the degree of indeterminacy and the degree of false membership respectively of each element $x \in X$ to the set A and it is defined as

$$A_N = \{x, \langle \mu_A(x), \sigma_A(x), \gamma_A(x) \rangle : x \in X\}$$

where $\mu_A(x), \sigma_A(x), \gamma_A(x) : X \rightarrow [0, 1]$ such that $0 \leq \mu_A(x) + \sigma_A(x) + \gamma_A(x) \leq 3$.

Definition 4 [6] Let $\widetilde{\mathcal{A}}_s$ be a *SFSs* in the universal discourse U defined as:

$$\widetilde{\mathcal{A}}_s = \left\{ \left\langle x, \left(\xi_{\widetilde{\mathcal{A}}_s}(x), \psi_{\widetilde{\mathcal{A}}_s}(x), \pi_{\widetilde{\mathcal{A}}_s}(x) \right) \right\rangle | x \in U \right\}$$

where $\xi_{\widetilde{\mathcal{A}}_s}(x) : U \to [0, 1], \psi_{\widetilde{\mathcal{A}}_s}(x) : U \to [0, 1], \pi_{\widetilde{\mathcal{A}}_s}(x) : U \to [0, 1]$ and $0 \le \xi^2_{\widetilde{\mathcal{A}}_s}(x) + \pi^2_{\widetilde{\mathcal{A}}_s}(x) + \pi^2_{\widetilde{\mathcal{A}}_s}(x) \le 1$ for each x. The values $\xi_{\widetilde{\mathcal{A}}_s}(x), \psi_{\widetilde{\mathcal{A}}_s}(x), \pi_{\widetilde{\mathcal{A}}_s}(x)$ are membership, non-membership and hesitancy functions of x in $\widetilde{\mathcal{A}}_s$ respectively.

Definition 5 [6] Score and accuracy functions of spherical fuzzy sets are defined respectively by $S\left(\widetilde{\mathcal{A}}_s(x)\right) = \left(\xi_{\widetilde{\mathcal{A}}_s}(x) - \pi_{\widetilde{\mathcal{A}}_s}(x)\right)^2 - \left(\psi_{\widetilde{\mathcal{A}}_s}(x) - \pi_{\widetilde{\mathcal{A}}_s}(x)\right)^2$, $A\left(\widetilde{\mathcal{A}}_s(x)\right) = \xi^2_{\widetilde{\mathcal{A}}_s}(x) + \psi^2_{\widetilde{\mathcal{A}}_s}(x) + \pi^2_{\widetilde{\mathcal{A}}_s}(x)$.

Definition 6 [6] Basic operations on spherical fuzzy numbers

(i) Algebraic sum

$$\widetilde{\mathcal{A}}_s \oplus \widetilde{\mathcal{B}}_s = \left\{ \sqrt{\xi^2_{\widetilde{\mathcal{A}}_s}(x) + \xi^2_{\widetilde{\mathcal{B}}_s}(x) - \xi^2_{\widetilde{\mathcal{A}}_s}(x)\xi^2_{\widetilde{\mathcal{A}}_s}(x)}, \quad \psi_{\widetilde{\mathcal{A}}_s}(x)\psi_{\widetilde{\mathcal{B}}_s}(x), \right.$$
$$\left. \sqrt{\left(1 - \xi^2_{\widetilde{\mathcal{B}}_s}(x)\right)\pi^2_{\widetilde{\mathcal{A}}_s}(x) + \left(1 - \xi^2_{\widetilde{\mathcal{A}}_s}(x)\right)\pi^2_{\widetilde{\mathcal{B}}_s}(x) - \pi^2_{\widetilde{\mathcal{A}}_s}(x)\pi^2_{\widetilde{\mathcal{B}}_s}(x)} \right\}$$

(ii) Algebraic product

$$\widetilde{\mathcal{A}}_s \otimes \widetilde{\mathcal{B}}_s = \left\{ \xi_{\widetilde{\mathcal{A}}_s}(x).\xi_{\widetilde{\mathcal{B}}_s}(x), \sqrt{\psi^2_{\widetilde{\mathcal{A}}_s}(x) + \psi^2_{\widetilde{\mathcal{B}}_s}(x) - \psi^2_{\widetilde{\mathcal{A}}_s}(x)\psi^2_{\widetilde{\mathcal{A}}_s}(x)}, \right.$$
$$\left. \sqrt{\left(1 - \psi^2_{\widetilde{\mathcal{B}}_s}(x)\right)\pi^2_{\widetilde{\mathcal{A}}_s}(x) + \left(1 - \psi^2_{\widetilde{\mathcal{A}}_s}(x)\right)\pi^2_{\widetilde{\mathcal{B}}_s}(x) - \pi^2_{\widetilde{\mathcal{A}}_s}(x)\pi^2_{\widetilde{\mathcal{B}}_s}(x)} \right\}$$

(iii) Scalar Multiplication: For $\lambda > 0$

$$\lambda.\widetilde{\mathcal{A}}_s = \left\{ \sqrt{1 - \left(1 - \xi^2_{\widetilde{\mathcal{A}}_s}(x)\right)^\lambda}, \psi^\lambda_{\widetilde{\mathcal{A}}_s}(x), \sqrt{\left(1 - \xi^2_{\widetilde{\mathcal{A}}_s}(x)\right)^\lambda - \left(1 - \xi^2_{\widetilde{\mathcal{A}}_s}(x) - \pi^2_{\widetilde{\mathcal{A}}_s}(x)\right)^\lambda} \right\}$$

(iv) Power : For $\lambda > 0$

$$\widetilde{\mathcal{A}}^\lambda_s = \left\{ \xi^\lambda_{\widetilde{\mathcal{A}}_s}(x), \quad \sqrt{1 - \left(1 - \psi^2_{\widetilde{\mathcal{A}}_s}(x)\right)^\lambda}, \quad \sqrt{\left(1 - \psi^2_{\widetilde{\mathcal{A}}_s}(x)\right)^\lambda - \left(1 - \psi^2_{\widetilde{\mathcal{A}}_s}(x) - \pi^2_{\widetilde{\mathcal{A}}_s}(x)\right)^\lambda} \right\}$$

2.1 *Complex Spherical Fuzzy Sets*

Definition 7 A CSFS $\breve{\mathcal{S}}$ defined on a universe of discourse $\overset{*}{V}$, is represented by degree of membership function $\widetilde{\Xi}_{\breve{\mathcal{S}}}(\dot{v})$, degree of non membership function $\widetilde{\Psi}_{\breve{\mathcal{S}}}(\dot{v})$ and a degree of hesitancy membership function $\widetilde{\Pi}_{\breve{\mathcal{S}}}(\dot{v})$ that takes complex valued grade of $\widetilde{\Xi}_{\breve{\mathcal{S}}}(\dot{v})$, $\widetilde{\Psi}_{\breve{\mathcal{S}}}(\dot{v})$ and $\widetilde{\Pi}_{\breve{\mathcal{S}}}(\dot{v})$ in $\breve{\mathcal{S}}$ for any $\dot{v} \in \overset{*}{V}$. The values $\widetilde{\Xi}_{\breve{\mathcal{S}}}(\dot{v})$, $\widetilde{\Psi}_{\breve{\mathcal{S}}}(\dot{v})$, $\widetilde{\Pi}_{\breve{\mathcal{S}}}(\dot{v})$ and their sum may all lie within the unit circle in the complex plane and are of the form,

$$\widetilde{\Xi}_{\breve{S}}(\dot{v}) = \xi_{\breve{S}}(\dot{v}).e^{j\mu_{\breve{S}}(\dot{v})}, \widetilde{\Psi}_{\breve{S}}(\dot{v}) = \psi_{\breve{S}}(\dot{v}).e^{j\nu_{\breve{S}}(\dot{v})}, \widetilde{\Pi}_{\breve{S}}(\dot{v}) = \pi_{\breve{S}}(\dot{v}).e^{j\omega_{\breve{S}}(\dot{v})}$$

where $j = \sqrt{-1}$, $\xi_{\breve{S}}(\dot{v}), \psi_{\breve{S}}(\dot{v}), \pi_{\breve{S}}(\dot{v})$ are amplitude terms whose values lie in [0, 1], and the phase terms are respectively $\mu_{\breve{S}}(\dot{v}), \nu_{\breve{S}}(\dot{v}), \omega_{\breve{S}}(\dot{v}) \in \Re$ or $[0, 2\pi]$. Since the phase terms will not affect amplitude terms, they are such that $0 \leq \xi^2_{\breve{S}}(\dot{v}) + \psi^2_{\breve{S}}(\dot{v}) + \pi^2_{\breve{S}}(\dot{v}) \leq 1$. The modulus of $\widetilde{\Xi}_{\breve{S}}$ is $\xi_{\breve{S}}$, denoted by $|\widetilde{\Xi}_{\breve{S}}|$. Similarly, the modulus of $\widetilde{\Psi}_{\breve{S}}, \widetilde{\Pi}_{\breve{S}}$ are $\psi_{\breve{S}}, \pi_{\breve{S}}$ respectively. The set builder form of the CSFS $\breve{S}$ is represented as

$$\breve{S} = \left\{ \dot{v}, \widetilde{\Xi}_{\breve{S}}(\dot{v}) = \breve{a}_{\Xi}, \widetilde{\Psi}_{\breve{S}}(\dot{v}) = \breve{a}_{\Psi}, \widetilde{\Pi}_{\breve{S}}(\dot{v}) = \breve{a}_{\Pi} : \dot{v} \in \overset{*}{V} \right\}$$

where, $\widetilde{\Xi}_{\breve{S}} : \overset{*}{V} \to \{\breve{a}_{\Xi} : \breve{a}_{\Xi} \in C, |\breve{a}_{\Xi}| \leq 1\}$, $\widetilde{\Psi}_{\breve{S}} : \overset{*}{V} \to \{\breve{a}_{\Psi} : \breve{a}_{\Psi} \in C, |\breve{a}_{\Psi}| \leq 1\}$, $\widetilde{\Pi}_{\breve{S}} : \overset{*}{V} \to \{\breve{a}_{\Pi} : \breve{a}_{\Pi} \in C, |\breve{a}_{\Pi}| \leq 1\}$ such that $\left|\widetilde{\Xi}^2_{\breve{S}}(\dot{v}) + \widetilde{\Psi}^2_{\breve{S}}(\dot{v}) + \widetilde{\Pi}^2_{\breve{S}}(\dot{v})\right| \leq 1$.

Definition 8 Let $\breve{S} = \langle \widetilde{\Xi}_{\breve{S}} = \xi_{\breve{S}}.e^{j\mu_{\breve{S}}}, \widetilde{\Psi}_{\breve{S}} = \psi_{\breve{S}}.e^{j\nu_{\breve{S}}}, \widetilde{\Pi}_{\breve{S}} = \pi_{\breve{S}}.e^{j\omega_{\breve{S}}} \rangle$ be complex spherical fuzzy value (CSFV), then three unary operators of CSFVs which include power, rotation, and reflection are defined as follows:

1. Power of $\breve{S}$

$$\breve{S}^k = \left\langle \widetilde{\Xi}^k_{\breve{S}} = \xi^k_{\breve{S}}.e^{jk\mu_{\breve{S}}}, \widetilde{\Psi}^k_{\breve{S}} = \psi^k_{\breve{S}}.e^{jk\nu_{\breve{S}}}, \widetilde{\Pi}^k_{\breve{S}} = \pi^k_{\breve{S}}.e^{jk\omega_{\breve{S}}} \right\rangle, k \neq 0. \quad (1)$$

2. Rotation of $\breve{S}$ with an angle θ

$$Rot_\theta(\breve{S}) = \langle \widetilde{\Xi}_{\breve{S}} = \xi_{\breve{S}}.e^{j(\mu_{\breve{S}}+\theta)}, \widetilde{\Psi}_{\breve{S}} = \psi_{\breve{S}}.e^{j(\nu_{\breve{S}}+\theta)}, \widetilde{\Pi}_{\breve{S}} = \pi_{\breve{S}}.e^{j(\omega_{\breve{S}}+\theta)} \rangle. \quad (2)$$

3. Reflection of $\breve{S}$

$$Ref(\breve{S}) = \langle \widetilde{\Xi}_{\breve{S}} = \xi_{\breve{S}}.e^{j(-\mu_{\breve{S}})}, \widetilde{\Psi}_{\breve{S}} = \psi_{\breve{S}}.e^{j(-\nu_{\breve{S}})}, \widetilde{\Pi}_{\breve{S}} = \pi_{\breve{S}}.e^{j(-\omega_{\breve{S}})} \rangle. \quad (3)$$

Definition 9 The complex spherical fuzzy operators (CSFOs) are defined on the complex plane, and the geometric properties such as rotational invariance and reflectional invariance are defined as follows:

(i) A CSFO $\Delta : \breve{S}^n \longrightarrow \breve{S}$ is rotationally invariant if and only if, for any θ,

$$\Delta\left(Rot_\theta(\breve{S}_1), Rot_\theta(\breve{S}_2), \ldots, Rot_\theta(\breve{S}_n)\right) = Rot_\theta\left(\Delta(\breve{S}_1, \breve{S}_2, \ldots, \breve{S}_n)\right) \quad (4)$$

(ii) A CSFO $\Delta : \breve{S}^n \longrightarrow \breve{S}$ is reflectionally invariant if and only if

$$\Delta\left(Ref(\breve{S}_1), Ref(\breve{S}_2), \ldots, Ref(\breve{S}_n)\right) = Ref\left(\Delta(\breve{S}_1, \breve{S}_2, \ldots, \breve{S}_n)\right) \quad (5)$$

Theorem 1 *The complex spherical fuzzy power operator is reflectionally invariant.*

Proof Let $\breve{\mathcal{S}} = \langle \widetilde{\Xi}_{\breve{S}} = \xi_{\breve{S}}.e^{j\mu_{\breve{S}}}, \widetilde{\Psi}_{\breve{S}} = \psi_{\breve{S}}.e^{j\nu_{\breve{S}}}, \widetilde{\Pi}_{\breve{S}} = \pi_{\breve{S}}.e^{j\omega_{\breve{S}}} \rangle$ be a CSFVs and $k > 0$. Then,

$$
\begin{aligned}
Ref\left(\breve{\mathcal{S}}^k\right) &= Ref\left(\left\langle \widetilde{\Xi}^k_{\breve{S}} = \xi^k_{\breve{S}}.e^{jk\mu_{\breve{S}}}, \widetilde{\Psi}^k_{\breve{S}} = \psi^k_{\breve{S}}.e^{jk\nu_{\breve{S}}}, \widetilde{\Pi}^k_{\breve{S}} = \pi^k_{\breve{S}}.e^{jk\omega_{\breve{S}}} \right\rangle\right), \\
&= \left\langle \xi^k_{\breve{S}}.e^{-jk\mu_{\breve{S}}}, \psi^k_{\breve{S}}.e^{-jk\nu_{\breve{S}}}, \pi^k_{\breve{S}}.e^{-jk\omega_{\breve{S}}} \right\rangle. \\
\left(Ref(\breve{\mathcal{S}})\right)^k &= \langle \widetilde{\Xi}_{\breve{S}} = \xi_{\breve{S}}.e^{-j\mu_{\breve{S}}}, \widetilde{\Psi}_{\breve{S}} = \psi_{\breve{S}}.e^{-j\nu_{\breve{S}}}, \widetilde{\Pi}_{\breve{S}} = \pi_{\breve{S}}.e^{-j\omega_{\breve{S}}} \rangle^k, \\
&= \left\langle \xi^k_{\breve{S}}.e^{-jk\mu_{\breve{S}}}, \psi^k_{\breve{S}}.e^{-jk\nu_{\breve{S}}}, \pi^k_{\breve{S}}.e^{-jk\omega_{\breve{S}}} \right\rangle.
\end{aligned}
$$

Thus, the power operator is reflectionaly invariant. □

Theorem 2 *If $k \neq 1$, the power operator of CSFVs is not rotationally invariant.*

Proof Let $\breve{\mathcal{S}} = \langle \widetilde{\Xi}_{\breve{S}} = \xi_{\breve{S}}.e^{j\mu_{\breve{S}}}, \widetilde{\Psi}_{\breve{S}} = \psi_{\breve{S}}.e^{j\nu_{\breve{S}}}, \widetilde{\Pi}_{\breve{S}} = \pi_{\breve{S}}.e^{j\omega_{\breve{S}}} \rangle$ be a CSFVs and $k > 0$. Then,

$$
\begin{aligned}
Rot_\theta\left(\breve{\mathcal{S}}^k\right) &= Rot_\theta\left(\left\langle \widetilde{\Xi}^k_{\breve{S}} = \xi^k_{\breve{S}}.e^{jk\mu_{\breve{S}}}, \widetilde{\Psi}^k_{\breve{S}} = \psi^k_{\breve{S}}.e^{jk\nu_{\breve{S}}}, \widetilde{\Pi}^k_{\breve{S}} = \pi^k_{\breve{S}}.e^{jk\omega_{\breve{S}}} \right\rangle\right), \\
&= \left\langle \xi^k_{\breve{S}}.e^{j(k\mu_{\breve{S}}+\theta)}, \psi^k_{\breve{S}}.e^{j(k\nu_{\breve{S}}+\theta)}, \pi^k_{\breve{S}}.e^{j(k\omega_{\breve{S}}+\theta)} \right\rangle. \\
\left(Rot_\theta(\breve{\mathcal{S}})\right)^k &= \langle \widetilde{\Xi}_{\breve{S}} = \xi_{\breve{S}}.e^{j(\mu_{\breve{S}}+\theta)}, \widetilde{\Psi}_{\breve{S}} = \psi_{\breve{S}}.e^{j(\nu_{\breve{S}}+\theta)}, \widetilde{\Pi}_{\breve{S}} = \pi_{\breve{S}}.e^{j(\omega_{\breve{S}}+\theta)} \rangle^k, \\
&= \left\langle \xi^k_{\breve{S}}.e^{jk(\mu_{\breve{S}}+\theta)}, \psi^k_{\breve{S}}.e^{jk(\nu_{\breve{S}}+\theta)}, \pi^k_{\breve{S}}.e^{jk(\omega_{\breve{S}}+\theta)} \right\rangle.
\end{aligned}
$$

Since $(k\mu_{\breve{S}} + \theta) \neq k(\mu_{\breve{S}} + \theta)$, $(k\nu_{\breve{S}} + \theta) \neq k(\nu_{\breve{S}} + \theta)$, and $(k\omega_{\breve{S}} + \theta) \neq k(\omega_{\breve{S}} + \theta)$.

The power operator of CSFVs is not rotationally invariant.

Let $\breve{\mathcal{S}} = \langle \xi_{\breve{S}}.e^{j\mu_{\breve{S}}}, \psi_{\breve{S}}.e^{j\nu_{\breve{S}}}, \pi_{\breve{S}}.e^{j\omega_{\breve{S}}} \rangle$ and $\breve{\mathcal{B}} = \langle \xi_{\breve{B}}.e^{j\mu_{\breve{B}}}, \psi_{\breve{B}}.e^{j\nu_{\breve{B}}}, \pi_{\breve{B}}.e^{j\omega_{\breve{B}}} \rangle$ be two CSFVs, then we have the following commonly used binary operations:

- Algebric sum;

$$
\begin{aligned}
\breve{\mathcal{S}} + \breve{\mathcal{B}} = \{&\left(\xi_{\breve{S}} \cos\mu_{\breve{S}} + \xi_{\breve{B}} \cos\mu_{\breve{B}}\right) + j\left(\xi_{\breve{S}} \sin\mu_{\breve{S}} + \xi_{\breve{B}} \sin\mu_{\breve{B}}\right), \\
&\left(\psi_{\breve{S}} \cos\nu_{\breve{S}} + \psi_{\breve{B}} \cos\nu_{\breve{B}}\right) + j\left(\psi_{\breve{S}} \sin\nu_{\breve{S}} + \psi_{\breve{B}} \sin\nu_{\breve{B}}\right), \\
&\left(\pi_{\breve{S}} \cos\omega_{\breve{S}} + \pi_{\breve{B}} \cos\omega_{\breve{B}}\right) + j\left(\pi_{\breve{S}} \sin\omega_{\breve{S}} + \pi_{\breve{B}} \sin\omega_{\breve{B}}\right)\}
\end{aligned}
$$

- Algebric product;

$$
\breve{\mathcal{S}} * \breve{\mathcal{B}} = \left\{\left(\xi_{\breve{S}}.\xi_{\breve{B}}.e^{j\mu_{\breve{S}}+\mu_{\breve{B}}}\right), \left(\psi_{\breve{S}}.\psi_{\breve{B}}.e^{j\nu_{\breve{S}}+\nu_{\breve{B}}}\right), \left(\pi_{\breve{S}}.\pi_{\breve{B}}.e^{j\omega_{\breve{S}}+\omega_{\breve{B}}}\right)\right\}
$$

The complex spherical fuzzy aggregation operation is given by a mapping

$$f : \breve{S}^n \longrightarrow \breve{S}$$

Then, the complex spherical fuzzy weighted arithmetic (CSFWA) operator

$$\begin{aligned}\text{(i)}\quad CSFWA\left(\breve{S}_1, \breve{S}_2, \ldots, \breve{S}_n\right) &= \sum_{i=1}^{n} w_i \breve{S}_i \\ &= \sum_{i=1}^{n} w_i \times \left\langle \widetilde{\Xi}_{\breve{S}_i} = \xi_{\breve{S}_i}.e^{j\mu_{\breve{S}_i}}, \widetilde{\Psi}_{\breve{S}_i} = \psi_{\breve{S}_i}.e^{j\nu_{\breve{S}_i}}, \widetilde{\Pi}_{\breve{S}_i} = \pi_{\breve{S}_i}.e^{j\omega_{\breve{S}_i}} \right\rangle\end{aligned} \tag{6}$$

where w_i are a complex valued weight vectors for all i and $\sum_{i=1}^{n} w_i = 1$.

Next, the complex spherical fuzzy weighted geometric (CSFWG) operator is given

$$\begin{aligned}\text{(ii)}\quad CSFWG\left(\breve{S}_1, \breve{S}_2, \ldots, \breve{S}_n\right) &= \prod_{i=1}^{n} \breve{S}_i^{w_i} \\ &= \prod_{i=1}^{n} \left\langle \widetilde{\Xi}_{\breve{S}_i} = \xi_{\breve{S}_i}.e^{j\mu_{\breve{S}_i}}, \widetilde{\Psi}_{\breve{S}_i} = \psi_{\breve{S}_i}.e^{j\nu_{\breve{S}_i}}, \widetilde{\Pi}_{\breve{S}_i} = \pi_{\breve{S}_i}.e^{j\omega_{\breve{S}_i}} \right\rangle^{w_i}\end{aligned} \tag{7}$$

where w_i are a complex valued weight vectors for all i and $\sum_{i=1}^{n} w_i = 1$.

Note: In this chapter, we consider only the real-valued weights i.e. $w_i \in [0, 1]$.

Example 1 Let $\breve{S}_1 = \left\langle 0.3e^{j\frac{3\pi}{4}}, 0.6e^{j\frac{2\pi}{3}}, 0.3e^{j0.6} \right\rangle$, $\breve{S}_2 = \left\langle 0.5e^{j\frac{2\pi}{4}}, 0.7e^{j\frac{4\pi}{3}}, 0.2e^{j0.35} \right\rangle$ be complex spherical fuzzy numbers (CSFNs) and assume that the weights are $w_1 = w_2 = 0.5$. Then, the CSFW operators

$$\begin{aligned}\text{(i)}\quad CSFWA\left(\breve{S}_1, \breve{S}_2\right) &= \sum_{i=1}^{2} w_i \breve{S}_i \\ &= w_1 \breve{S}_1 + w_2 \breve{S}_2 \\ &= 0.5\left(0.3e^{j\frac{3\pi}{4}}, 0.6e^{j\frac{2\pi}{3}}, 0.3e^{j0.6}\right) + 0.5\left(0.5e^{j\frac{2\pi}{4}}, 0.7e^{j\frac{4\pi}{3}}, 0.2e^{j0.35}\right) \\ &= \left\langle 0.15e^{j\frac{3\pi}{4}}, 0.3e^{j\frac{2\pi}{3}}, 0.15e^{j0.6} \right\rangle + \left\langle 0.25e^{j\frac{2\pi}{4}}, 0.35e^{j\frac{4\pi}{3}}, 0.1e^{j0.35} \right\rangle \\ &= \left\langle 0.3715e^{j1.8603}, 0.3279e^{j(-3.0091)}, 0.2481e^{j0.5001} \right\rangle\end{aligned}$$

Similarly, using CSFWG operator

(ii) $CSFWG\left(\breve{S}_1, \breve{S}_2\right) = \prod_{i=1}^{2} \breve{S}_i^{w_i} = \breve{S}_1^{w_1} \times \breve{S}_2^{w_2}$

$$= \left(0.3e^{j\frac{3\pi}{4}}, 0.6e^{j\frac{2\pi}{3}}, 0.3e^{j0.6}\right)^{0.5} \times \left(0.5e^{j\frac{2\pi}{4}}, 0.7e^{j\frac{4\pi}{3}}, 0.2e^{j0.35}\right)^{0.5}$$
$$= \left\langle 0.5477e^{j1.1781}, 0.7746e^{j1.0472}, 0.5477e^{j0.3000}\right\rangle$$
$$\times \left\langle 0.7071e^{j0.7854}, 0.8367e^{-j1.0472}, 0.4472e^{j0.1750}\right\rangle$$
$$= \left\langle 0.3873e^{j1.9635}, 0.6481e^{j0}, 0.2449e^{j0.4750}\right\rangle$$

The important properties of the aggregation operators are idempotency, boundedness and monotonicity.

Theorem 3 *Let $\breve{S}_1, \breve{S}_2, \ldots, \breve{S}_n$ be CSFVs and let w_i be the weight vectors. Then we have the following axioms*

(i) Idempotency: If all CSFVs are equal ($\breve{S}_1 = \breve{S}_2 = \ldots = \breve{S}_n$), then

$$CSFWA\left(\breve{S}_1, \breve{S}_2, \ldots, \breve{S}_n\right) = \breve{S}_1$$

(ii) Boundedness: $\left|CSFWA\left(\breve{S}_1, \breve{S}_2, \ldots, \breve{S}_n\right)\right| \leq max_i|\breve{S}_i|$

Proof (i). Since ($\breve{S}_1 = \breve{S}_2 = \ldots = \breve{S}_n$) and $\sum_{i=1}^{n} w_i = 1$, then

$$\begin{aligned} CSFWA\left(\breve{S}_1, \breve{S}_2, \ldots, \breve{S}_n\right) &= w_1\breve{S}_1 + w_2\breve{S}_2 + \cdots + w_n\breve{S}_n \\ &= w_1\breve{S}_1 + w_2\breve{S}_1 + \cdots + w_n\breve{S}_1 \\ &= (w_1 + w_2 + \cdots + w_n)\,\breve{S}_1 \\ &= \breve{S}_1 \end{aligned}$$

(ii) let $\breve{S} = max_i|\breve{S}_i|$. Since $\sum_{i=1}^{n} w_i = 1$, then

$$\begin{aligned} \left|CSFWA\left(\breve{S}_1, \breve{S}_2, \ldots, \breve{S}_n\right)\right| &= \left|w_1\breve{S}_1 + w_2\breve{S}_2 + \cdots + w_n\breve{S}_n\right| \\ &\leq |w_1\breve{S}_1| + |w_2\breve{S}_2| + \cdots + |w_n\breve{S}_n| \\ &= w_1|\breve{S}_1| + w_2|\breve{S}_2| + \cdots + w_n|\breve{S}_n| \\ &\leq w_1\breve{S} + w_2\breve{S} + \cdots + w_n\breve{S} \\ &= (w_1 + w_2 + \cdots + w_n)\,\breve{S} \\ &= \breve{S} \end{aligned}$$

Thus, the CSFWA operator satisfies idempotency and boundedness. It also satisfies the property of monotonicity.

Example 2 Let $\breve{\mathcal{S}}_1 = \left\langle 0.3e^{j\frac{3\pi}{4}}, 0.6e^{j\frac{2\pi}{3}}, 0.3e^{j0.6} \right\rangle$, $\breve{\mathcal{S}}_2 = \left\langle 0.5e^{j\frac{2\pi}{4}}, 0.7e^{j\frac{4\pi}{3}}, 0.2e^{j0.35} \right\rangle$ and $\breve{\mathcal{B}}_1 = \left\langle 0.5e^{j\frac{3\pi}{4}}, 0.3e^{j\frac{2\pi}{3}}, 0.2e^{j0.6} \right\rangle$, $\breve{\mathcal{B}}_2 = \left\langle 0.6e^{j\frac{2\pi}{4}}, 0.3e^{j\frac{4\pi}{3}}, 0.1e^{j0.35} \right\rangle$ be CSFVs and weights be $w_1 = w_2 = 0.5$. Then, we have

$$\begin{aligned} CSFWA\left(\breve{\mathcal{S}}_1, \breve{\mathcal{S}}_2\right) &= w_1\breve{\mathcal{S}}_1 + w_2\breve{\mathcal{S}}_2 \\ &= 0.5\left(0.3e^{j\frac{3\pi}{4}}, 0.6e^{j\frac{2\pi}{3}}, 0.3e^{j0.6}\right) + 0.5\left(0.5e^{j\frac{2\pi}{4}}, 0.7e^{j\frac{4\pi}{3}}, 0.2e^{j0.35}\right) \\ &= \left\langle 0.15e^{j\frac{3\pi}{4}}, 0.3e^{j\frac{2\pi}{3}}, 0.15e^{j0.6} \right\rangle + \left\langle 0.25e^{j\frac{2\pi}{4}}, 0.35e^{j\frac{4\pi}{3}}, 0.1e^{j0.35} \right\rangle \\ &= \left\langle 0.3715e^{j1.8603}, 0.3279e^{j(-3.0091)}, 0.2481e^{j0.5001} \right\rangle \end{aligned}$$

$$\begin{aligned} CSFWA\left(\breve{\mathcal{B}}_1, \breve{\mathcal{B}}_2\right) &= w_1\breve{\mathcal{B}}_1 + w_2\breve{\mathcal{B}}_2 \\ &= 0.5\left(0.5e^{j\frac{3\pi}{4}}, 0.3e^{j\frac{2\pi}{3}}, 0.2e^{j0.6}\right) + 0.5\left(0.6e^{j\frac{2\pi}{4}}, 0.3e^{j\frac{4\pi}{3}}, 0.1e^{j0.35}\right) \\ &= \left\langle 0.25e^{j\frac{3\pi}{4}}, 0.15e^{j\frac{2\pi}{3}}, 0.1e^{j0.6} \right\rangle + \left\langle 0.3e^{j\frac{2\pi}{4}}, 0.15e^{j\frac{4\pi}{3}}, 0.05e^{j0.35} \right\rangle \\ &= \left\langle 0.5085e^{j1.9259}, 0.1500e^{j3.1416}, 0.1490e^{j0.5169} \right\rangle \end{aligned}$$

Here $\left|CSFWA\left(\breve{\mathcal{S}}_1, \breve{\mathcal{S}}_2\right)\right| \leq \left|CSFWA\left(\breve{\mathcal{B}}_1, \breve{\mathcal{B}}_2\right)\right|$.

Therefore, $CSFWA$ operator satisfies the property of monotonicity.

Theorem 4 *Let* $\breve{\mathcal{S}}_j = \left\langle \widetilde{\Xi}_{\breve{\mathcal{S}}_j} = \xi_{\breve{\mathcal{S}}_j}.e^{j\mu_{\breve{\mathcal{S}}_j}}, \widetilde{\Psi}_{\breve{\mathcal{S}}_j} = \psi_{\breve{\mathcal{S}}_j}.e^{j\nu_{\breve{\mathcal{S}}_j}}, \widetilde{\Pi}_{\breve{\mathcal{S}}_j} = \pi_{\breve{\mathcal{S}}_j}.e^{j\omega_{\breve{\mathcal{S}}_j}} \right\rangle$ *and* $\breve{\mathcal{B}}_j = \left\langle \widetilde{\Xi}_{\breve{\mathcal{B}}_j} = \xi_{\breve{\mathcal{B}}_j}.e^{j\mu_{\breve{\mathcal{B}}_j}}, \widetilde{\Psi}_{\breve{\mathcal{B}}_j} = \psi_{\breve{\mathcal{B}}_j}.e^{j\nu_{\breve{\mathcal{B}}_j}}, \widetilde{\Pi}_{\breve{\mathcal{B}}_j} = \pi_{\breve{\mathcal{B}}_j}.e^{j\omega_{\breve{\mathcal{B}}_j}} \right\rangle$ *be two families of CSFVs. The weight vectors of* $CSFWG$ *operator be real valued i.e.,* $w_i \in [0, 1]\ \forall i$ *and* $\sum_{i=1}^{n} w_i = 1$. *Then we have the following axioms:*

(i) Idempotency: If all CSFVs are equal i.e.($\breve{\mathcal{S}}_1 = \breve{\mathcal{S}}_2 = \ldots = \breve{\mathcal{S}}_n$), *then*

$$CSFWG\left(\breve{\mathcal{S}}_1, \breve{\mathcal{S}}_2, \ldots, \breve{\mathcal{S}}_n\right) = \breve{\mathcal{S}}_1$$

(ii) Boundedness: Let $\breve{\mathcal{S}}_1, \breve{\mathcal{S}}_2, \ldots, \breve{\mathcal{S}}_n$ *be a CSFVs.*

$$\left|CSFWG\left(\breve{\mathcal{S}}_1, \breve{\mathcal{S}}_2, \ldots, \breve{\mathcal{S}}_n\right)\right| \leq max_i|\breve{\mathcal{S}}_i|$$

(iii) Monotonicity: If $\xi_{\breve{\mathcal{S}}_j} \geq \xi_{\breve{\mathcal{B}}_j}$, $\psi_{\breve{\mathcal{S}}_j} \leq \psi_{\breve{\mathcal{B}}_j}$, $\pi_{\breve{\mathcal{S}}_j} \leq \pi_{\breve{\mathcal{B}}_j}$ *then*

$$\left|CSFWG\left(\breve{\mathcal{S}}_1, \breve{\mathcal{S}}_2, \ldots, \breve{\mathcal{S}}_n\right)\right| \leq \left|CSFWG\left(\breve{\mathcal{B}}_1, \breve{\mathcal{B}}_2, \ldots, \breve{\mathcal{B}}_n\right)\right|$$

Proof The proof of the theorem follows from Theorem 3

Theorem 5 *The complex spherical fuzzy weighted arithmetic (CSFWA) operator is rotationally and reflectionally invariant.*

Proof Let $\breve{\mathcal{S}}_1, \breve{\mathcal{S}}_2, \ldots, \breve{\mathcal{S}}_n$ be CSFVs, then for any θ

$$\begin{aligned}
&CSFWA\left(Rot_\theta(\breve{\mathcal{S}}_1), Rot_\theta(\breve{\mathcal{S}}_2), \ldots, Rot_\theta(\breve{\mathcal{S}}_n)\right)\\
&\quad = CSFWA\left(\breve{\mathcal{S}}_1 e^{j\theta}, \breve{\mathcal{S}}_2 e^{j\theta}, \ldots, \breve{\mathcal{S}}_n e^{j\theta}\right)\\
&\quad = \left(w_1.\breve{\mathcal{S}}_1.e^{j\theta}, w_2.\breve{\mathcal{S}}_2.e^{j\theta}, \ldots, w_n.\breve{\mathcal{S}}_n.e^{j\theta}\right)\\
&\quad = \left(w_1.\breve{\mathcal{S}}_1, w_2.\breve{\mathcal{S}}_2, \ldots, w_n.\breve{\mathcal{S}}_n\right).e^{j\theta}\\
&\quad = CSFWA\left(\breve{\mathcal{S}}_1, \breve{\mathcal{S}}_2, \ldots, \breve{\mathcal{S}}_n\right).e^{j\theta}\\
&\quad = Rot_\theta\left(CSFWA\left(\breve{\mathcal{S}}_1, \breve{\mathcal{S}}_2, \ldots, \breve{\mathcal{S}}_n\right)\right)
\end{aligned}$$

Therefore, *CSFWA* operator is rotationally invariant for any θ.

Similarly we can prove that the *CSFWA* operator is reflectionally invariant

$$CSFWA\left(Ref(\breve{\mathcal{S}}_1), Ref(\breve{\mathcal{S}}_2), \ldots, Ref(\breve{\mathcal{S}}_n)\right) = Ref\left(CSFWA(\breve{\mathcal{S}}_1, \breve{\mathcal{S}}_2, \ldots, \breve{\mathcal{S}}_n)\right)$$

Example 3 Let $\breve{\mathcal{S}}_1 = \left\langle 0.3e^{j\frac{3\pi}{4}}, 0.6e^{j\frac{2\pi}{3}}, 0.3e^{j0.6}\right\rangle$, $\breve{\mathcal{S}}_2 = \left\langle 0.5e^{j\frac{2\pi}{4}}, 0.7e^{j\frac{4\pi}{3}}, 0.2e^{j0.35}\right\rangle$ be CSFNs and assume that the weights are $w_1 = w_2 = 0.5$. Then,

$$\begin{aligned}
CSFWA\left(\breve{\mathcal{S}}_1, \breve{\mathcal{S}}_2\right) &= w_1\breve{\mathcal{S}}_1 + w_2\breve{\mathcal{S}}_2\\
&= \langle 0.3715e^{j1.8603}, 0.3279e^{j(-3.0091)}, 0.2481e^{j0.5001}\rangle
\end{aligned}$$

(i) $\theta = \frac{\pi}{6}$, then

$$\begin{aligned}
Rot_{\frac{\pi}{6}}(\breve{\mathcal{S}}_1) &= \left\langle 0.3e^{j\left(\frac{3\pi}{4}+\frac{\pi}{6}\right)}, 0.6e^{j\left(\frac{2\pi}{3}+\frac{\pi}{6}\right)}, 0.3e^{j\left(0.6+\frac{\pi}{6}\right)}\right\rangle\\
&= \langle 0.3000e^{j2.8798}, 0.6000e^{j2.6180}, 0.3000e^{j1.1236}\rangle\\
Rot_{\frac{\pi}{6}}(\breve{\mathcal{S}}_2) &= \left\langle 0.5e^{j\left(\frac{2\pi}{4}+\frac{\pi}{6}\right)}, 0.7e^{\left(j\frac{4\pi}{3}+\frac{\pi}{6}\right)}, 0.2e^{\left(j0.35+\frac{\pi}{6}\right)}\right\rangle\\
&= \langle 0.5000e^{j2.0944}, 0.7000e^{j(-1.5708)}, 0.2000e^{j0.8736}\rangle
\end{aligned}$$

$$
\begin{aligned}
CSFWA\left(Rot_{\frac{\pi}{6}}(\breve{S}_1), Rot_{\frac{\pi}{6}}(\breve{S}_2)\right) &= w_1 Rot_{\frac{\pi}{6}}(\breve{S}_1) + w_2 Rot_{\frac{\pi}{6}}(\breve{S}_2) \\
&= 0.5\left(0.3e^{j2.8798}, 0.6e^{j2.6180}, 0.3e^{j1.1236}\right) + \\
&\quad 0.5\left(0.5e^{j2.0944}, 0.7e^{j(-1.5708)}, 0.2e^{j0.8736}\right) \\
&= \left\langle 0.15e^{j2.8798}, 0.3e^{j2.6180}, 0.15e^{j1.1236}\right\rangle + \\
&\quad \left\langle 0.25e^{j2.0944}, 0.35e^{j(-1.5708)}, 0.1e^{j0.8736}\right\rangle \\
&= \left\langle 0.3715e^{j2.3839}, 0.3279e^{j(-2.4855)}, 0.2481e^{j1.0237}\right\rangle
\end{aligned}
$$

$$
\begin{aligned}
Rot_{\frac{\pi}{6}}\left(CSFWA(\breve{S}_1, \breve{S}_2)\right) &= Rot_{\frac{\pi}{6}}\left(0.3715e^{j1.8603}, 0.3279e^{j(-3.0091)}, 0.2481e^{j0.5001}\right) \\
&= \left\langle 0.3715e^{j(1.8603+\frac{\pi}{6})}, 0.3279e^{j(-3.0091+\frac{\pi}{6})}, 0.2481e^{j(0.5001+\frac{\pi}{6})}\right\rangle \\
&= \left\langle 0.3715e^{j2.3839}, 0.3279e^{j(-2.4855)}, 0.2481e^{j1.0237}\right\rangle
\end{aligned}
$$

Thus we show that the CSFWA operator is rotationally invariant.

(ii) Reflectional invariance of CSFWA

$$
\begin{aligned}
Ref(\breve{S}_1) &= \left\langle 0.3e^{j(-\frac{3\pi}{4})}, 0.6e^{j(-\frac{2\pi}{3})}, 0.3e^{j(-0.6)}\right\rangle \\
&= \left\langle 0.3e^{j(-2.3562)}, 0.6e^{j(-2.0944)}, 0.3e^{j(-0.6)}\right\rangle
\end{aligned}
$$

$$
\begin{aligned}
Ref(\breve{S}_2) &= \left\langle 0.5e^{j(-\frac{2\pi}{4})}, 0.7e^{j(-\frac{4\pi}{3})}, 0.2e^{j(-0.35)}\right\rangle \\
&= \left\langle 0.5e^{j(-1.5708)}, 0.7e^{j(2.0944)}, 0.2e^{j(-0.35)}\right\rangle
\end{aligned}
$$

$$
\begin{aligned}
CSFWA\left(Rot_{\frac{\pi}{6}}(\breve{S}_1), Rot_{\frac{\pi}{6}}(\breve{S}_2)\right) &= w_1 Rot_{\frac{\pi}{6}}(\breve{S}_1) + w_2 Rot_{\frac{\pi}{6}}(\breve{S}_2) \\
&= 0.5\left(0.3e^{j2.8798}, 0.6e^{j2.6180}, 0.3e^{j1.1236}\right) + \\
&\quad 0.5\left(0.5e^{j2.0944}, 0.7e^{j(-1.5708)}, 0.2e^{j0.8736}\right) \\
&= \left\langle 0.15e^{j2.8798}, 0.3e^{j2.6180}, 0.15e^{j1.1236}\right\rangle + \\
&\quad \left\langle 0.25e^{j2.0944}, 0.35e^{j(-1.5708)}, 0.1e^{j0.8736}\right\rangle \\
&= \left\langle 0.3715e^{j2.3839}, 0.3279e^{j(-2.4855)}, 0.2481e^{j1.0237}\right\rangle
\end{aligned}
$$

$$
\begin{aligned}
Rot_{\frac{\pi}{6}}\left(CSFWA(\breve{S}_1, \breve{S}_2)\right) &= Rot_{\frac{\pi}{6}}\left(0.3715e^{j1.8603}, 0.3279e^{j(-3.0091)}, 0.2481e^{j0.5001}\right) \\
&= \left\langle 0.3715e^{j(1.8603+\frac{\pi}{6})}, 0.3279e^{j(-3.0091+\frac{\pi}{6})}, 0.2481e^{j(0.5001+\frac{\pi}{6})}\right\rangle \\
&= \left\langle 0.3715e^{j2.3839}, 0.3279e^{j(-2.4855)}, 0.2481e^{j1.0237}\right\rangle
\end{aligned}
$$

Theorem 6 *The complex spherical fuzzy weighted geometric (CSFWG) operator is rotationally and reflectionally invariant.*

Proof The proof is similar to that of Theorem 5 and hence is omitted.

Further we extend the basic spherical fuzzy operations [6] to *CSFSs* as follows:

Definition 10 Let $\breve{\mathcal{S}} = \langle \xi_{\breve{S}}.e^{j\mu_{\breve{S}}}, \psi_{\breve{S}}.e^{j\nu_{\breve{S}}}, \pi_{\breve{S}}.e^{j\omega_{\breve{S}}} \rangle$, $\breve{\mathcal{B}} = \langle \xi_{\breve{B}}.e^{j\mu_{\breve{B}}}, \psi_{\breve{B}}.e^{j\nu_{\breve{B}}}, \pi_{\breve{B}}.e^{j\omega_{\breve{B}}} \rangle$ be two CSFVs, then

[1] Union

$$\breve{\mathcal{S}} \cup \breve{\mathcal{B}} = \left\{ \begin{array}{c} \max\left\{\xi_{\breve{S}}, \xi_{\breve{B}}\right\}.e^{j\max\left\{\mu_{\breve{S}},\mu_{\breve{B}}\right\}}, \quad \min\left\{\psi_{\breve{S}}, \psi_{\breve{B}}\right\}.e^{j\min\left\{\nu_{\breve{S}},\nu_{\breve{B}}\right\}}, \\ \min\left\{\left(1 - \left(\max\left\{\xi_{\breve{S}}, \xi_{\breve{B}}\right\}\right)^2 + \left(\min\left\{\psi_{\breve{S}}, \psi_{\breve{B}}\right\}\right)^2\right)^{\frac{1}{2}}, \max\left\{\pi_{\breve{S}}, \pi_{\breve{S}}\right\}\right\}. \\ e^{j\min\left\{\left(1-\left(\max\left\{\mu_{\breve{S}},\mu_{\breve{B}}\right\}\right)^2+\left(\min\left\{\nu_{\breve{S}},\nu_{\breve{B}}\right\}\right)^2\right)^{\frac{1}{2}}, \max\left\{\omega_{\breve{S}},\omega_{\breve{S}}\right\}\right\}} \end{array} \right\}$$

[2] Intersection

$$\breve{\mathcal{S}} \cap \breve{\mathcal{B}} = \left\{ \begin{array}{c} \min\left\{\xi_{\breve{S}}, \xi_{\breve{B}}\right\}.e^{j\min\left\{\mu_{\breve{S}},\mu_{\breve{B}}\right\}}, \max\left\{\psi_{\breve{S}}, \psi_{\breve{B}}\right\}.e^{j\max\left\{\nu_{\breve{S}},\nu_{\breve{B}}\right\}}, \\ \max\left\{\left(1 - \left(\min\left\{\xi_{\breve{S}}, \xi_{\breve{B}}\right\}\right)^2 + \left(\max\left\{\psi_{\breve{S}}, \psi_{\breve{B}}\right\}\right)^2\right)^{\frac{1}{2}}, \min\left\{\pi_{\breve{S}}, \pi_{\breve{S}}\right\}\right\}. \\ e^{j\max\left\{\left(1-\left(\min\left\{\mu_{\breve{S}},\mu_{\breve{B}}\right\}\right)^2+\left(\max\left\{\nu_{\breve{S}},\nu_{\breve{B}}\right\}\right)^2\right)^{\frac{1}{2}}, \min\left\{\omega_{\breve{S}},\omega_{\breve{S}}\right\}\right\}} \end{array} \right\}$$

[3] Algebric Addition

$$\breve{\mathcal{S}} \oplus \breve{\mathcal{B}} = \left\{ \begin{array}{c} \left(\xi_{\breve{S}}^2 + \xi_{\breve{B}}^2 - \xi_{\breve{S}}^2.\xi_{\breve{B}}^2\right)^{\frac{1}{2}}.e^{j\left(\mu_{\breve{S}}^2+\mu_{\breve{B}}^2-\mu_{\breve{S}}^2.\mu_{\breve{B}}^2\right)^{\frac{1}{2}}}, \quad \psi_{\breve{S}}.\psi_{\breve{B}}.e^{j\nu_{\breve{S}}.\nu_{\breve{B}}}, \\ \left(\left(1-\xi_{\breve{B}}^2\right)\pi_{\breve{S}}^2 + \left(1-\xi_{\breve{S}}^2\right)\pi_{\breve{B}}^2 - \pi_{\breve{S}}^2.\pi_{\breve{B}}^2\right)^{\frac{1}{2}}.e^{j\left(\left(1-\mu_{\breve{B}}^2\right)\omega_{\breve{S}}^2+\left(1-\mu_{\breve{S}}^2\right)\omega_{\breve{B}}^2-\omega_{\breve{S}}^2.\omega_{\breve{B}}^2\right)^{\frac{1}{2}}} \end{array} \right\}$$

[4]. Algebric product

$$\breve{\mathcal{S}} \otimes \breve{\mathcal{B}} = \left\{ \begin{array}{c} \xi_{\breve{S}}.\xi_{\breve{B}}.e^{j\mu_{\breve{S}}.\mu_{\breve{B}}}, \quad \left(\psi_{\breve{S}}^2 + \psi_{\breve{B}}^2 - \psi_{\breve{S}}^2.\psi_{\breve{B}}^2\right)^{\frac{1}{2}}.e^{j\left(\nu_{\breve{S}}^2+\nu_{\breve{B}}^2-\nu_{\breve{S}}^2.\nu_{\breve{B}}^2\right)^{\frac{1}{2}}} \\ \left(\left(1-\psi_{\breve{B}}^2\right)\pi_{\breve{S}}^2 + \left(1-\psi_{\breve{S}}^2\right)\pi_{\breve{B}}^2 - \pi_{\breve{S}}^2.\pi_{\breve{B}}^2\right)^{\frac{1}{2}}.e^{j\left(\left(1-\nu_{\breve{B}}^2\right)\omega_{\breve{S}}^2+\left(1-\nu_{\breve{S}}^2\right)\omega_{\breve{B}}^2-\omega_{\breve{S}}^2.\omega_{\breve{B}}^2\right)^{\frac{1}{2}}} \end{array} \right\}$$

[5] Scalar Multiplication:

$$\text{For } \lambda > 0 \ \lambda.\breve{\mathcal{S}} = \left\{ \begin{array}{c} \left(1 - \left(1 - \xi_{\breve{S}}^2\right)^{\lambda}\right)^{\frac{1}{2}}.e^{j\left(1-\left(1-\mu_{\breve{S}}^2\right)^{\lambda}\right)^{\frac{1}{2}}}, \psi_{\breve{S}}^{\lambda}.e^{j\nu_{\breve{S}}^{\lambda}}, \\ \left(\left(1-\xi_{\breve{S}}^2\right)^{\lambda} - \left(1 - \xi_{\breve{S}}^2 - \pi_{\breve{S}}^2\right)^{\lambda}\right)^{\frac{1}{2}}.e^{j\left(\left(1-\mu_{\breve{S}}^2\right)^{\lambda}-\left(1-\mu_{\breve{S}}^2-\omega_{\breve{S}}^2\right)^{\lambda}\right)^{\frac{1}{2}}} \end{array} \right\}$$

[6] Power of $\breve{\mathcal{S}}$: For $\lambda > 0$

$$\breve{\mathcal{S}}^{\lambda} = \left\{ \begin{array}{c} \xi_{\breve{S}}^{\lambda}.e^{j\mu_{\breve{S}}^{\lambda}}, \left(1-\left(1-\psi_{\breve{S}}^{2}\right)^{\lambda}\right)^{\frac{1}{2}}.e^{j\left(1-\left(1-\nu_{\breve{S}}^{2}\right)^{\lambda}\right)^{\frac{1}{2}}}, \\ \left(\left(1-\psi_{\breve{S}}^{2}\right)^{\lambda}-\left(1-\psi_{\breve{S}}^{2}-\pi_{\breve{S}}^{2}\right)^{\lambda}\right)^{\frac{1}{2}}.e^{j\left(\left(1-\nu_{\breve{S}}^{2}\right)^{\lambda}-\left(1-\nu_{\breve{S}}^{2}-\omega_{\breve{S}}^{2}\right)^{\lambda}\right)^{\frac{1}{2}}} \end{array} \right\}$$

Definition 11 Let $\breve{\mathcal{S}} = \langle \xi_{\breve{S}}.e^{j\mu_{\breve{S}}}, \psi_{\breve{S}}.e^{j\nu_{\breve{S}}}, \pi_{\breve{S}}.e^{j\omega_{\breve{S}}} \rangle$ be a CSFS, then its score function S and accuracy function H are defined as:

$$S(\breve{\mathcal{S}}) = \frac{1}{2}\left(\left|(\xi_{\breve{S}}-\pi_{\breve{S}})^{2}-(\psi_{\breve{S}}-\pi_{\breve{S}})^{2}\right| + \left|\frac{1}{2\pi}\left[(\mu_{\breve{S}}-\omega_{\breve{S}})^{2}-(\nu_{\breve{S}}-\omega_{\breve{S}})^{2}\right]\right|\right) \tag{8}$$

$$H(\breve{\mathcal{S}}) = \frac{1}{2}\left|\left(\xi_{\breve{S}}^{2}+\psi_{\breve{S}}^{2}+\pi_{\breve{S}}^{2}\right)+\frac{1}{2\pi}\left(\mu_{\breve{S}}^{2}+\nu_{\breve{S}}^{2}+\omega_{\breve{S}}^{2}\right)\right| \tag{9}$$

For two *CSFNs* $\breve{\mathcal{S}}$ and $\breve{\mathcal{B}}$,

(i) If $S(\breve{\mathcal{S}}) > S(\breve{\mathcal{B}})$, then $\breve{\mathcal{S}} > \breve{\mathcal{B}}$; similarly $H(\breve{\mathcal{S}}) > H(\breve{\mathcal{B}})$, then $\breve{\mathcal{S}} > \breve{\mathcal{B}}$.

(ii) If $S(\breve{\mathcal{S}}) = S(\breve{\mathcal{B}})$, then $\breve{\mathcal{S}} = \breve{\mathcal{B}}$; similarly $H(\breve{\mathcal{S}}) = H(\breve{\mathcal{B}})$, then $\breve{\mathcal{S}} = \breve{\mathcal{B}}$.

(iii) If $S(\breve{\mathcal{S}}) = S(\breve{\mathcal{B}})$, $H(\breve{\mathcal{S}}) < H(\breve{\mathcal{B}})$, then $\breve{\mathcal{S}} < \breve{\mathcal{B}}$.

(iv) If $S(\breve{\mathcal{S}}) = S(\breve{\mathcal{B}})$, $H(\breve{\mathcal{S}}) > H(\breve{\mathcal{B}})$, then $\breve{\mathcal{S}} > \breve{\mathcal{B}}$.

Definition 12 Let $\breve{\mathcal{S}} = \langle \xi_{\breve{S}}.e^{j\mu_{\breve{S}}}, \psi_{\breve{S}}.e^{j\nu_{\breve{S}}}, \pi_{\breve{S}}.e^{j\omega_{\breve{S}}} \rangle$, $\breve{\mathcal{B}} = \langle \xi_{\breve{B}}.e^{j\mu_{\breve{B}}}, \psi_{\breve{B}}.e^{j\nu_{\breve{B}}}, \pi_{\breve{B}}.e^{j\omega_{\breve{B}}} \rangle$ be two CSFSs and $p_i, q_i, r_i, s_i \in [0, 1]$ such that $p_i + q_i + r_i + s_i = 1$. Then the distance measure (DM) of CSFSs is defined as follows:

$$\begin{aligned} DM^{1}_{CSFSs}\left(\breve{\mathcal{S}}, \breve{\mathcal{B}}\right) = \frac{1}{2}\sum_{i=1}^{n}\Big[\Big(p_1\left|\xi_{\breve{S}}^{2}(\dot{v}_i)-\xi_{\breve{B}}^{2}(\dot{v}_i)\right| + q_1\left|\psi_{\breve{S}}^{2}(\dot{v}_i)-\psi_{\breve{B}}^{2}(\dot{v}_i)\right| + \\ r_1\left|\pi_{\breve{S}}^{2}(\dot{v}_i)-\pi_{\breve{B}}^{2}(\dot{v}_i)\right| + s_1 \max\Big(\left|\xi_{\breve{S}}^{2}(\dot{v}_i)-\xi_{\breve{B}}^{2}(\dot{v}_i)\right|, \left|\psi_{\breve{S}}^{2}(\dot{v}_i)-\psi_{\breve{B}}^{2}(\dot{v}_i)\right|, \\ \left|\pi_{\breve{S}}^{2}(\dot{v}_i)-\pi_{\breve{B}}^{2}(\dot{v}_i)\right|\Big)\Big) + \frac{1}{2\pi}\Big(p_1\left|\mu_{\breve{S}}^{2}(\dot{v}_i)-\mu_{\breve{B}}^{2}(\dot{v}_i)\right| + \\ q_1\left|\nu_{\breve{S}}^{2}(\dot{v}_i)-\nu_{\breve{B}}^{2}(\dot{v}_i)\right| + r_1\left|\omega_{\breve{S}}^{2}(\dot{v}_i)-\omega_{\breve{B}}^{2}(\dot{v}_i)\right| + \\ s_1 \max\Big(\left|\mu_{\breve{S}}^{2}(\dot{v}_i)-\mu_{\breve{B}}^{2}(\dot{v}_i)\right|, \left|\nu_{\breve{S}}^{2}(\dot{v}_i)-\nu_{\breve{B}}^{2}(\dot{v}_i)\right|, \left|\omega_{\breve{S}}^{2}(\dot{v}_i)-\omega_{\breve{B}}^{2}(\dot{v}_i)\right|\Big)\Big)\Big] \end{aligned}$$

The above DM^{1}_{CSFSs} satisfies the following properties:

(i) $0 \le DM^{1}_{CSFSs}\left(\breve{\mathcal{S}}, \breve{\mathcal{B}}\right) \le 1$

(ii) $DM^{1}_{CSFSs}\left(\breve{\mathcal{S}}, \breve{\mathcal{B}}\right) = DM^{1}_{CSFSs}\left(\breve{\mathcal{B}}, \breve{\mathcal{S}}\right)$

(iii) $DM^{1}_{CSFSs}\left(\breve{\mathcal{S}}, \breve{\mathcal{B}}\right) = 1$ if $\breve{\mathcal{S}} = \breve{\mathcal{B}}$

(iv) If $\breve{\mathcal{S}}_1 \subseteq \breve{\mathcal{S}}_2 \subseteq \breve{\mathcal{S}}_3$ then $DM^{1}_{CSFSs}\left(\breve{\mathcal{S}}_1, \breve{\mathcal{S}}_3\right) > DM^{1}_{CSFSs}\left(\breve{\mathcal{S}}_1, \breve{\mathcal{S}}_2\right)$ and $DM^{1}_{CSFSs}\left(\breve{\mathcal{S}}_1, \breve{\mathcal{S}}_3\right) > DM^{1}_{CSFSs}$ $\left(\breve{\mathcal{S}}_2, \breve{\mathcal{S}}_3\right)$

Theorem 7 *The following axioms are valid for these complex spherical fuzzy sets (CSFSs)* $\breve{\mathcal{S}} = \left\langle \xi_{\breve{S}}.e^{j\mu_{\breve{S}}}, \psi_{\breve{S}}.e^{j\nu_{\breve{S}}}, \pi_{\breve{S}}.e^{j\omega_{\breve{S}}} \right\rangle$, $\breve{\mathcal{B}} = \left\langle \xi_{\breve{B}}.e^{j\mu_{\breve{B}}}, \psi_{\breve{B}}.e^{j\nu_{\breve{B}}}, \pi_{\breve{B}}.e^{j\omega_{\breve{B}}} \right\rangle$, *with the condition* $\lambda, \lambda_1, \lambda_2 > 0$

(i) $\breve{\mathcal{S}} \oplus \breve{\mathcal{B}} = \breve{\mathcal{B}} \oplus \breve{\mathcal{S}}$
(ii) $\breve{\mathcal{S}} \otimes \breve{\mathcal{B}} = \breve{\mathcal{B}} \otimes \breve{\mathcal{S}}$
(iii) $\lambda\left(\breve{\mathcal{S}} \oplus \breve{\mathcal{B}}\right) = \lambda\breve{\mathcal{S}} \oplus \lambda\breve{\mathcal{B}}$
(iv) $\lambda_1\breve{\mathcal{S}} \oplus \lambda_2\breve{\mathcal{S}} = (\lambda_1 + \lambda_2)\breve{\mathcal{S}}$
(v) $\left(\breve{\mathcal{S}} \oplus \breve{\mathcal{B}}\right)^{\lambda} = \breve{\mathcal{S}}^{\lambda} \oplus \breve{\mathcal{B}}^{\lambda}$
(vi) $\breve{\mathcal{S}}^{\lambda_1} \oplus \breve{\mathcal{S}}^{\lambda_2} = \breve{\mathcal{S}}^{\lambda_1+\lambda_2}$

Proof Using the above mentioned CSFSs operators, it is easy to prove. So the proof is omitted.

Theorem 8 *Complex spherical fuzzy weighted arithmetic mean (CSFWAM*) operator with respect to complex valued weight vectors w_i for all i and $w_i \in [0, 1]$ such that $\sum_{i=1}^{n} w_i = 1$, is defined as*

$$CSFWAM^{*}\left(\breve{\mathcal{S}}_1, \breve{\mathcal{S}}_2, \ldots, \breve{\mathcal{S}}_n\right) = \sum_{i=1}^{n} w_i\breve{\mathcal{S}}_i \tag{10}$$

$$= \left\{ \begin{array}{c} \left(1 - \prod_{i=1}^{n}\left(1 - \xi_{\breve{S}_i}^2\right)^{w_i}\right)^{\frac{1}{2}} .e^{j\left(1-\prod_{i=1}^{n}\left(1-\mu_{\breve{S}_i}^2\right)^{w_i}\right)^{\frac{1}{2}}}, \quad \prod_{i=1}^{n}\psi_{\breve{S}_i}^{w_i}.e^{j\prod_{i=1}^{n}\nu_{\breve{S}_i}^{w_i}}, \\ \left(\prod_{i=1}^{n}\left(1 - \xi_{\breve{S}_i}^2\right)^{w_i} - \prod_{i=1}^{n}\left(1 - \xi_{\breve{S}_i}^2 - \pi_{\breve{S}_i}^2\right)^{w_i}\right)^{\frac{1}{2}} .e^{j\left(\prod_{i=1}^{n}\left(1-\mu_{\breve{S}_i}^2\right)^{w_i}-\prod_{i=1}^{n}\left(1-\mu_{\breve{S}_i}^2-\omega_{\breve{S}_i}^2\right)^{w_i}\right)^{\frac{1}{2}}} \end{array} \right\}$$

Proof We use mathematical induction to prove the Eq. (10)
Step 1. For $n = 2$

$$w_1\breve{\mathcal{S}}_1 = \left\{ \begin{array}{c} \left(1 - \left(1 - \xi_{\breve{S}_1}^2\right)^{w_1}\right)^{\frac{1}{2}} .e^{j\left(1-\left(1-\mu_{\breve{S}_1}^2\right)^{w_1}\right)^{\frac{1}{2}}}, \quad \psi_{\breve{S}_1}^{w_1}.e^{j\nu_{\breve{S}_1}^{w_1}}, \\ \left(\left(1 - \xi_{\breve{S}_1}^2\right)^{w_1} - \left(1 - \xi_{\breve{S}_1}^2 - \pi_{\breve{S}_1}^2\right)^{w_1}\right)^{\frac{1}{2}} .e^{j\left(\left(1-\mu_{\breve{S}_1}^2\right)^{w_1}-\left(1-\mu_{\breve{S}_1}^2-\omega_{\breve{S}_1}^2\right)^{w_1}\right)^{\frac{1}{2}}} \end{array} \right\}$$

$$w_2\breve{\mathcal{S}}_2 = \left\{ \begin{array}{c} \left(1 - \left(1 - \xi_{\breve{S}_2}^2\right)^{w_2}\right)^{\frac{1}{2}} .e^{j\left(1-\left(1-\mu_{\breve{S}_2}^2\right)^{w_2}\right)^{\frac{1}{2}}}, \quad \psi_{\breve{S}_2}^{w_2}.e^{j\nu_{\breve{S}_2}^{w_2}}, \\ \left(\left(1 - \xi_{\breve{S}_2}^2\right)^{w_2} - \left(1 - \xi_{\breve{S}_2}^2 - \pi_{\breve{S}_2}^2\right)^{w_2}\right)^{\frac{1}{2}} .e^{j\left(\left(1-\mu_{\breve{S}_2}^2\right)^{w_2}-\left(1-\mu_{\breve{S}_2}^2-\omega_{\breve{S}_2}^2\right)^{w_2}\right)^{\frac{1}{2}}} \end{array} \right\}$$

$$CSFWAM^{*}\left(\breve{S}_1, \breve{S}_2\right) = w_1\breve{S}_1 \oplus w_2\breve{S}_2$$

$$= \left\{ \begin{array}{c} \left(1-\left(1-\xi_{\breve{S}_1}^2\right)^{w_1}\right)^{\frac{1}{2}}.e^{j\left(1-\left(1-\mu_{\breve{S}_1}^2\right)^{w_1}\right)^{\frac{1}{2}}}, \quad \psi_{\breve{S}_1}^{w_1}.e^{j\nu_{\breve{S}_1}^{w_1}}, \\ \left(\left(1-\xi_{\breve{S}_1}^2\right)^{w_1}-\left(1-\xi_{\breve{S}_1}^2-\pi_{\breve{S}_1}^2\right)^{w_1}\right)^{\frac{1}{2}}.e^{j\left(\left(1-\mu_{\breve{S}_1}^2\right)^{w_1}-\left(1-\mu_{\breve{S}_1}^2-\omega_{\breve{S}_1}^2\right)^{w_1}\right)^{\frac{1}{2}}} \end{array} \right\} \oplus$$

$$\left\{ \begin{array}{c} \left(1-\left(1-\xi_{\breve{S}_2}^2\right)^{w_2}\right)^{\frac{1}{2}}.e^{j\left(1-\left(1-\mu_{\breve{S}_2}^2\right)^{w_2}\right)^{\frac{1}{2}}}, \quad \psi_{\breve{S}_2}^{w_2}.e^{j\nu_{\breve{S}_2}^{w_2}}, \\ \left(\left(1-\xi_{\breve{S}_2}^2\right)^{w_2}-\left(1-\xi_{\breve{S}_2}^2-\pi_{\breve{S}_2}^2\right)^{w_2}\right)^{\frac{1}{2}}.e^{j\left(\left(1-\mu_{\breve{S}_2}^2\right)^{w_2}-\left(1-\mu_{\breve{S}_2}^2-\omega_{\breve{S}_2}^2\right)^{w_2}\right)^{\frac{1}{2}}} \end{array} \right\}$$

$$= \left\{ \begin{array}{c} \left[\left(\left(1-\left(1-\xi_{\breve{S}_1}^2\right)^{w_1}\right)^{\frac{1}{2}}\right)^2+\left(\left(1-\left(1-\xi_{\breve{S}_2}^2\right)^{w_2}\right)^{\frac{1}{2}}\right)^2-\left(\left(1-\left(1-\xi_{\breve{S}_1}^2\right)^{w_1}\right)^{\frac{1}{2}}\right)^2 \right. \\ \left. .\left(\left(1-\left(1-\xi_{\breve{S}_2}^2\right)^{w_2}\right)^{\frac{1}{2}}\right)^2\right]^{\frac{1}{2}} \\ .e^{j\left[\left(\left(1-\left(1-\mu_{\breve{S}_1}^2\right)^{w_1}\right)^{\frac{1}{2}}\right)^2+\left(\left(1-\left(1-\mu_{\breve{S}_2}^2\right)^{w_2}\right)^{\frac{1}{2}}\right)^2-\left(\left(1-\left(1-\mu_{\breve{S}_1}^2\right)^{w_1}\right)^{\frac{1}{2}}\right)^2.\left(\left(1-\left(1-\mu_{\breve{S}_2}^2\right)^{w_2}\right)^{\frac{1}{2}}\right)^2\right]^{\frac{1}{2}}}, \\ \left[\psi_{\breve{S}_1}^{w_1}\psi_{\breve{S}_2}^{w_2}\right].e^{j\left[\nu_{\breve{S}_1}^{w_1}\nu_{\breve{S}_2}^{w_2}\right]}, \\ \left[\left(1-\left(\left(1-\left(1-\xi_{\breve{S}_2}^2\right)^{w_2}\right)^{\frac{1}{2}}\right)^2\right)\left(\left(\left(1-\xi_{\breve{S}_1}^2\right)^{w_1}-\left(1-\xi_{\breve{S}_1}^2-\pi_{\breve{S}_1}^2\right)^{w_1}\right)^{\frac{1}{2}}\right)^2+\right. \\ \left(1-\left(\left(1-\left(1-\xi_{\breve{S}_1}^2\right)^{w_1}\right)^{\frac{1}{2}}\right)^2\right)\left(\left(\left(1-\xi_{\breve{S}_2}^2\right)^{w_2}-\left(1-\xi_{\breve{S}_2}^2-\pi_{\breve{S}_2}^2\right)^{w_2}\right)^{\frac{1}{2}}\right)^2- \\ \left.\left(\left(\left(1-\xi_{\breve{S}_1}^2\right)^{w_1}-\left(1-\xi_{\breve{S}_1}^2-\pi_{\breve{S}_1}^2\right)^{w_1}\right)^{\frac{1}{2}}\right)^2.\left(\left(\left(1-\xi_{\breve{S}_2}^2\right)^{w_2}-\left(1-\xi_{\breve{S}_2}^2-\pi_{\breve{S}_2}^2\right)^{w_2}\right)^{\frac{1}{2}}\right)^2\right]^{\frac{1}{2}} \\ .e^{j\left\{ \begin{array}{c} \left[\left(1-\left(\left(1-\left(1-\mu_{\breve{S}_2}^2\right)^{w_2}\right)^{\frac{1}{2}}\right)^2\right)\left(\left(\left(1-\mu_{\breve{S}_1}^2\right)^{w_1}-\left(1-\mu_{\breve{S}_1}^2-\omega_{\breve{S}_1}^2\right)^{w_1}\right)^{\frac{1}{2}}\right)^2+\right. \\ \left(1-\left(\left(1-\left(1-\mu_{\breve{S}_1}^2\right)^{w_1}\right)^{\frac{1}{2}}\right)^2\right)\left(\left(\left(1-\mu_{\breve{S}_2}^2\right)^{w_2}-\left(1-\mu_{\breve{S}_2}^2-\omega_{\breve{S}_2}^2\right)^{w_2}\right)^{\frac{1}{2}}\right)^2- \\ \left.\left(\left(\left(1-\mu_{\breve{S}_1}^2\right)^{w_1}-\left(1-\mu_{\breve{S}_1}^2-\omega_{\breve{S}_1}^2\right)^{w_1}\right)^{\frac{1}{2}}\right)^2.\left(\left(\left(1-\mu_{\breve{S}_2}^2\right)^{w_2}-\left(1-\mu_{\breve{S}_2}^2-\omega_{\breve{S}_2}^2\right)^{w_2}\right)^{\frac{1}{2}}\right)^2\right]^{\frac{1}{2}} \end{array} \right\}} \end{array} \right\}$$

$$= \left\{ \begin{array}{c} \left(1-\left(1-\xi_{\breve{S}_1}^2\right)^{w_1}\left(1-\xi_{\breve{S}_2}^2\right)^{w_2}\right)^{\frac{1}{2}}.e^{j\left(1-\left(1-\mu_{\breve{S}_1}^2\right)^{w_1}\left(1-\mu_{\breve{S}_2}^2\right)^{w_2}\right)^{\frac{1}{2}}}, \quad \psi_{\breve{S}_1}^{w_1}\psi_{\breve{S}_2}^{w_2}.e^{j\nu_{\breve{S}_1}^{w_1}\nu_{\breve{S}_2}^{w_2}}, \\ \left(\left(1-\xi_{\breve{S}_1}^2\right)^{w_1}\left(1-\xi_{\breve{S}_2}^2\right)^{w_2}-\left(1-\xi_{\breve{S}_1}^2-\pi_{\breve{S}_1}^2\right)^{w_1}\left(1-\xi_{\breve{S}_2}^2-\pi_{\breve{S}_2}^2\right)^{w_2}\right)^{\frac{1}{2}}. \\ e^{j\left(\left(1-\mu_{\breve{S}_1}^2\right)^{w_1}\left(1-\mu_{\breve{S}_2}^2\right)^{w_2}-\left(1-\mu_{\breve{S}_1}^2-\omega_{\breve{S}_1}^2\right)^{w_1}\left(1-\mu_{\breve{S}_2}^2-\omega_{\breve{S}_2}^2\right)^{w_2}\right)^{\frac{1}{2}}} \end{array} \right\}$$

$$= \left\{ \begin{array}{c} \left(1-\prod_{i=1}^{2}\left(1-\xi_{\breve{S}_i}^2\right)^{w_i}\right)^{\frac{1}{2}} .e^{j\left(1-\prod_{i=1}^{2}\left(1-\mu_{\breve{S}_i}^2\right)^{w_i}\right)^{\frac{1}{2}}}, \quad \prod_{i=1}^{2}\psi_{\breve{S}_i}^{w_i}.e^{j\prod_{i=1}^{2}\nu_{\breve{S}_i}^{w_i}}, \\ \left(\prod_{i=1}^{2}\left(1-\xi_{\breve{S}_i}^2\right)^{w_i}-\prod_{i=1}^{2}\left(1-\xi_{\breve{S}_i}^2-\pi_{\breve{S}_i}^2\right)^{w_i}\right)^{\frac{1}{2}} .e^{j\left(\prod_{i=1}^{2}\left(1-\mu_{\breve{S}_i}^2\right)^{w_i}-\prod_{i=1}^{2}\left(1-\mu_{\breve{S}_i}^2-\omega_{\breve{S}_i}^2\right)^{w_i}\right)^{\frac{1}{2}}} \end{array} \right\}$$

Obviously, $CSFWAM^*$ operator is true for $n = 2$

Step 2. Let us assume that Eq. (10) holds for n = k. That is,

$$CSFWAM^*\left(\breve{S}_1, \breve{S}_2, \ldots, \breve{S}_k\right)$$

$$= \left\{ \begin{array}{c} \left(1-\prod_{i=1}^{k}\left(1-\xi_{\breve{S}_i}^2\right)^{w_i}\right)^{\frac{1}{2}} .e^{j\left(1-\prod_{i=1}^{k}\left(1-\mu_{\breve{S}_i}^2\right)^{w_i}\right)^{\frac{1}{2}}}, \quad \prod_{i=1}^{k}\psi_{\breve{S}_i}^{w_i}.e^{j\prod_{i=1}^{k}\nu_{\breve{S}_i}^{w_i}}, \\ \left(\prod_{i=1}^{k}\left(1-\xi_{\breve{S}_i}^2\right)^{w_i}-\prod_{i=1}^{k}\left(1-\xi_{\breve{S}_i}^2-\pi_{\breve{S}_i}^2\right)^{w_i}\right)^{\frac{1}{2}} .e^{j\left(\prod_{i=1}^{k}\left(1-\mu_{\breve{S}_i}^2\right)^{w_i}-\prod_{i=1}^{k}\left(1-\mu_{\breve{S}_i}^2-\omega_{\breve{S}_i}^2\right)^{w_i}\right)^{\frac{1}{2}}} \end{array} \right\}$$

Step 3. For $n = k + 1$,

$$CSFWAM^*\left(\breve{S}_1, \breve{S}_2, \ldots, \breve{S}_{k+1}\right)$$

$$= \left\{ \begin{array}{c} \left(1-\prod_{i=1}^{k}\left(1-\xi_{\breve{S}_i}^2\right)^{w_i}\right)^{\frac{1}{2}} .e^{j\left(1-\prod_{i=1}^{k}\left(1-\mu_{\breve{S}_i}^2\right)^{w_i}\right)^{\frac{1}{2}}}, \quad \prod_{i=1}^{k}\psi_{\breve{S}_i}^{w_i}.e^{j\prod_{i=1}^{k}\nu_{\breve{S}_i}^{w_i}}, \\ \left(\prod_{i=1}^{k}\left(1-\xi_{\breve{S}_i}^2\right)^{w_i}-\prod_{i=1}^{k}\left(1-\xi_{\breve{S}_i}^2-\pi_{\breve{S}_i}^2\right)^{w_i}\right)^{\frac{1}{2}} .e^{j\left(\prod_{i=1}^{k}\left(1-\mu_{\breve{S}_i}^2\right)^{w_i}-\prod_{i=1}^{k}\left(1-\mu_{\breve{S}_i}^2-\omega_{\breve{S}_i}^2\right)^{w_i}\right)^{\frac{1}{2}}} \end{array} \right\}$$

$$\oplus\, w_{k+1}\breve{S}_{k+1}$$

$$= \left\{ \begin{array}{c} \left(1-\prod_{i=1}^{k}\left(1-\xi_{\breve{S}_i}^2\right)^{w_i}\right)^{\frac{1}{2}} .e^{j\left(1-\prod_{i=1}^{k}\left(1-\mu_{\breve{S}_i}^2\right)^{w_i}\right)^{\frac{1}{2}}}, \quad \prod_{i=1}^{k}\psi_{\breve{S}_i}^{w_i}.e^{j\prod_{i=1}^{k}\nu_{\breve{S}_i}^{w_i}}, \\ \left(\prod_{i=1}^{k}\left(1-\xi_{\breve{S}_i}^2\right)^{w_i}-\prod_{i=1}^{k}\left(1-\xi_{\breve{S}_i}^2-\pi_{\breve{S}_i}^2\right)^{w_i}\right)^{\frac{1}{2}} .e^{j\left(\prod_{i=1}^{k}\left(1-\mu_{\breve{S}_i}^2\right)^{w_i}-\prod_{i=1}^{k}\left(1-\mu_{\breve{S}_i}^2-\omega_{\breve{S}_i}^2\right)^{w_i}\right)^{\frac{1}{2}}} \end{array} \right\} \oplus$$

$$\left\{ \begin{array}{c} \left(1-\left(1-\xi_{\breve{S}_{k+1}}^2\right)^{w_{k+1}}\right)^{\frac{1}{2}} .e^{j\left(1-\left(1-\mu_{\breve{S}_{k+1}}^2\right)^{w_{k+1}}\right)^{\frac{1}{2}}}, \quad \psi_{\breve{S}_{k+1}}^{w_{k+1}}.e^{j\nu_{\breve{S}_{k+1}}^{w_{k+1}}}, \\ \left(\left(1-\xi_{\breve{S}_{k+1}}^2\right)^{w_{k+1}}-\left(1-\xi_{\breve{S}_{k+1}}^2-\pi_{\breve{S}_{k+1}}^2\right)^{w_{k+1}}\right)^{\frac{1}{2}} .e^{j\left(\left(1-\mu_{\breve{S}_{k+1}}^2\right)^{w_{k+1}}-\left(1-\mu_{\breve{S}_{k+1}}^2-\omega_{\breve{S}_{k+1}}^2\right)^{w_{k+1}}\right)^{\frac{1}{2}}} \end{array} \right\}$$

$$
= \left\{ \begin{array}{c}
\left[\left(\left(1 - \prod_{i=1}^{k} \left(1 - \xi_{\breve{S}_i}^2 \right)^{w_i} \right)^{\frac{1}{2}} \right)^2 + \left(\left(1 - \left(1 - \xi_{\breve{S}_{k+1}}^2 \right)^{w_2} \right)^{\frac{1}{2}} \right)^2 \right. \\
\left. - \left(\left(1 - \prod_{i=1}^{k} \left(1 - \xi_{\breve{S}_i}^2 \right)^{w_i} \right)^{\frac{1}{2}} \right)^2 . \left(\left(1 - \left(1 - \xi_{\breve{S}_{k+1}}^2 \right)^{w_2} \right)^{\frac{1}{2}} \right)^2 \right]^{\frac{1}{2}} \\
.e^{j \left\{ \begin{array}{c} \left[\left(\left(1 - \prod_{i=1}^{k} \left(1 - \mu_{\breve{S}_i}^2 \right)^{w_i} \right)^{\frac{1}{2}} \right)^2 + \left(\left(1 - \left(1 - \mu_{\breve{S}_{k+1}}^2 \right)^{w_2} \right)^{\frac{1}{2}} \right)^2 - \right. \\ \left. \left(\left(1 - \prod_{i=1}^{k} \left(1 - \mu_{\breve{S}_i}^2 \right)^{w_i} \right)^{\frac{1}{2}} \right)^2 . \left(\left(1 - \left(1 - \mu_{\breve{S}_{k+1}}^2 \right)^{w_2} \right)^{\frac{1}{2}} \right)^2 \right]^{\frac{1}{2}} \end{array} \right\}}, \\
\left[\prod_{i=1}^{k} \psi_{\breve{S}_i}^{w_i} \psi_{\breve{S}_{k+1}}^{w_2} \right] .e^{j \left[\prod_{i=1}^{k} \nu_{\breve{S}_i}^{w_i} \nu_{\breve{S}_{k+1}}^{w_2} \right]}, \\
\left[\left(1 - \left(\left(1 - \left(1 - \xi_{\breve{S}_{k+1}}^2 \right)^{w_2} \right)^{\frac{1}{2}} \right)^2 \right) \left(\left(\prod_{i=1}^{k} \left(1 - \xi_{\breve{S}_i}^2 \right)^{w_i} - \prod_{i=1}^{k} \prod_{i=1}^{k} \left(1 - \xi_{\breve{S}_i}^2 - \pi_{\breve{S}_i}^2 \right)^{w_i} \right)^{\frac{1}{2}} \right)^2 + \right. \\
\left(1 - \left(\left(1 - \prod_{i=1}^{k} \left(1 - \xi_{\breve{S}_i}^2 \right)^{w_i} \right)^{\frac{1}{2}} \right)^2 \right) \left(\left(\left(1 - \xi_{\breve{S}_{k+1}}^2 \right)^{w_2} - \left(1 - \xi_{\breve{S}_{k+1}}^2 - \pi_{\breve{S}_{k+1}}^2 \right)^{w_2} \right)^{\frac{1}{2}} \right)^2 - \\
\left(\left(\prod_{i=1}^{k} \left(1 - \xi_{\breve{S}_i}^2 \right)^{w_i} - \prod_{i=1}^{k} \left(1 - \xi_{\breve{S}_i}^2 - \pi_{\breve{S}_i}^2 \right)^{w_i} \right)^{\frac{1}{2}} \right)^2 . \\
\left. \left(\left(\left(1 - \xi_{\breve{S}_{k+1}}^2 \right)^{w_2} - \left(1 - \xi_{\breve{S}_{k+1}}^2 - \pi_{\breve{S}_{k+1}}^2 \right)^{w_2} \right)^{\frac{1}{2}} \right)^2 \right]^{\frac{1}{2}} \\
.e^{j \left\{ \begin{array}{c} \left[\left(1 - \left(\left(1 - \left(1 - \mu_{\breve{S}_{k+1}}^2 \right)^{w_2} \right)^{\frac{1}{2}} \right)^2 \right) \left(\left(\prod_{i=1}^{k} \left(1 - \mu_{\breve{S}_i}^2 \right)^{w_i} - \prod_{i=1}^{k} \left(1 - \mu_{\breve{S}_i}^2 - \omega_{\breve{S}_i}^2 \right)^{w_i} \right)^{\frac{1}{2}} \right)^2 + \right. \\ \left(1 - \left(\left(1 - \prod_{i=1}^{k} \left(1 - \mu_{\breve{S}_i}^2 \right)^{w_i} \right)^{\frac{1}{2}} \right)^2 \right) \left(\left(\left(1 - \mu_{\breve{S}_{k+1}}^2 \right)^{w_2} - \left(1 - \mu_{\breve{S}_{k+1}}^2 - \omega_{\breve{S}_{k+1}}^2 \right)^{w_2} \right)^{\frac{1}{2}} \right)^2 - \\ \left(\left(\prod_{i=1}^{k} \left(1 - \mu_{\breve{S}_i}^2 \right)^{w_i} - \prod_{i=1}^{k} \left(1 - \mu_{\breve{S}_i}^2 - \omega_{\breve{S}_i}^2 \right)^{w_i} \right)^{\frac{1}{2}} \right)^2 . \\ \left. \left(\left(\left(1 - \mu_{\breve{S}_{k+1}}^2 \right)^{w_2} - \left(1 - \mu_{\breve{S}_{k+1}}^2 - \omega_{\breve{S}_{k+1}}^2 \right)^{w_2} \right)^{\frac{1}{2}} \right)^2 \right]^{\frac{1}{2}} \end{array} \right\}}
\end{array} \right\}
$$

$$
= \left\{ \begin{array}{c}
\left(1 - \prod_{i=k+1}^{k} \left(1 - \xi_{\breve{S}_i}^2 \right)^{w_i} \right)^{\frac{1}{2}} .e^{j \left(1 - \prod_{i=k+1}^{k} \left(1 - \mu_{\breve{S}_i}^2 \right)^{w_i} \right)^{\frac{1}{2}}}, \quad \prod_{i=k+1}^{k} \psi_{\breve{S}_i}^{w_i} .e^{j \prod_{i=k+1}^{k} \nu_{\breve{S}_i}^{w_i}}, \\
\left(\prod_{i=k+1}^{k} \left(1 - \xi_{\breve{S}_i}^2 \right)^{w_i} - \prod_{i=k+1}^{k} \left(1 - \xi_{\breve{S}_i}^2 - \pi_{\breve{S}_i}^2 \right)^{w_i} \right)^{\frac{1}{2}} .e^{j \left(\prod_{i=k+1}^{k} \left(1 - \mu_{\breve{S}_i}^2 \right)^{w_i} - \prod_{i=k+1}^{k} \left(1 - \mu_{\breve{S}_i}^2 - \omega_{\breve{S}_i}^2 \right)^{w_i} \right)^{\frac{1}{2}}}
\end{array} \right\}
$$

Therefore, the operator is true for $n = k + 1$. Hence, the theorem.

Theorem 9 *Let $\breve{\mathcal{S}}_i = \left\langle \xi_{\breve{S}_i}.e^{j\mu_{\breve{S}_i}}, \psi_{\breve{S}}.e^{j\nu_{\breve{S}_i}}, \pi_{\breve{S}_i}.e^{j\omega_{\breve{S}_i}} \right\rangle$ be a family of CSFSs, the weight vectors of CSFWAM* operator be $w_i \in [0, 1]\ \forall i$ and $\sum_{i=1}^{n} w_i = 1$. Then we have the following axioms*

(i) Idempotency: If all CSFVs are equal i.e.($\breve{\mathcal{S}}_1 = \breve{\mathcal{S}}_2 = \ldots = \breve{\mathcal{S}}_n$), then

$$CSFWAM^*\left(\breve{\mathcal{S}}_1, \breve{\mathcal{S}}_2, \ldots, \breve{\mathcal{S}}_n\right) = \breve{\mathcal{S}}_1$$

(ii) Boundedness: Let $\breve{\mathcal{S}}_1, \breve{\mathcal{S}}_2, \ldots, \breve{\mathcal{S}}_n$ be a CSFSs.

$$\left|CSFWAM^*\left(\breve{\mathcal{S}}_1, \breve{\mathcal{S}}_2, \ldots, \breve{\mathcal{S}}_n\right)\right| \leq max_i|\breve{\mathcal{S}}_i|$$

(iii) Monotonicity:

$$\text{If } \xi_{\breve{S}_i} \geq \xi_{\breve{B}_i}, \psi_{\breve{S}_i} \leq \psi_{\breve{B}_i}, \pi_{\breve{S}_i} \leq \pi_{\breve{B}_i} \text{ then}$$
$$\left|CSFWAM^*\left(\breve{\mathcal{B}}_1, \breve{\mathcal{B}}_2, \ldots, \breve{\mathcal{B}}_n\right)\right| \leq \left|CSFWAM^*\left(\breve{\mathcal{S}}_1, \breve{\mathcal{S}}_2, \ldots, \breve{\mathcal{S}}_n\right)\right|.$$

Proof (i) Idempotency

$$CSFWAM^*\left(\breve{\mathcal{S}}_1, \breve{\mathcal{S}}_2, \ldots, \breve{\mathcal{S}}_n\right)$$
$$= \left\{ \begin{array}{c} \left(1 - \prod_{i=k}^{k}\left(1 - \xi^2_{\breve{S}_i}\right)^{w_i}\right)^{\frac{1}{2}} .e^{j\left(1-\prod_{i=k}^{k}\left(1-\mu^2_{\breve{S}_i}\right)^{w_i}\right)^{\frac{1}{2}}}, \quad \prod_{i=k}^{k} \psi^{w_i}_{\breve{S}_i}.e^{j\prod_{i=k}^{k} \nu^{w_i}_{\breve{S}_i}}, \\ \left(\prod_{i=k}^{k}\left(1 - \xi^2_{\breve{S}_i}\right)^{w_i} - \prod_{i=k}^{k}\left(1 - \xi^2_{\breve{S}_i} - \pi^2_{\breve{S}_i}\right)^{w_i}\right)^{\frac{1}{2}} .e^{j\left(\prod_{i=k}^{k}\left(1-\mu^2_{\breve{S}_i}\right)^{w_i} - \prod_{i=k}^{k}\left(1-\mu^2_{\breve{S}_i}-\omega^2_{\breve{S}_i}\right)^{w_i}\right)^{\frac{1}{2}}} \end{array} \right\}$$

Since all CSFVs are equal ($\breve{\mathcal{S}}_1 = \breve{\mathcal{S}}_2 = \ldots = \breve{\mathcal{S}}_n$), we have

$$CSFWAM^*\left(\breve{\mathcal{S}}_1, \breve{\mathcal{S}}_2, \ldots, \breve{\mathcal{S}}_n\right)$$
$$= \left\{ \begin{array}{c} \left(1 - \prod_{i=k}^{k}\left(1 - \xi^2_{\breve{S}}\right)^{w_i}\right)^{\frac{1}{2}} .e^{j\left(1-\prod_{i=k}^{k}\left(1-\mu^2_{\breve{S}}\right)^{w_i}\right)^{\frac{1}{2}}}, \quad \prod_{i=k}^{k} \psi^{w_i}_{\breve{S}}.e^{j\prod_{i=k}^{k} \nu^{w_i}_{\breve{S}}}, \\ \left(\prod_{i=k}^{k}\left(1 - \xi^2_{\breve{S}}\right)^{w_i} - \prod_{i=k}^{k}\left(1 - \xi^2_{\breve{S}} - \pi^2_{\breve{S}}\right)^{w_i}\right)^{\frac{1}{2}} .e^{j\left(\prod_{i=k}^{k}\left(1-\mu^2_{\breve{S}}\right)^{w_i} - \prod_{i=k}^{k}\left(1-\mu^2_{\breve{S}}-\omega^2_{\breve{S}}\right)^{w_i}\right)^{\frac{1}{2}}} \end{array} \right\}$$
$$= \left\{ \begin{array}{c} \left(1 - \left(1 - \xi^2_{\breve{S}_i}\right)^{\sum_{i=1}^{n} w_i}\right)^{\frac{1}{2}} .e^{j\left(1-\left(1-\mu^2_{\breve{S}}\right)^{\sum_{i=1}^{n} w_i}\right)^{\frac{1}{2}}}, \quad \psi^{\sum_{i=1}^{n} w_i}_{\breve{S}}.e^{j\nu^{\sum_{i=1}^{n} w_i}_{\breve{S}}}, \\ \left(\left(1 - \xi^2_{\breve{S}}\right)^{\sum_{i=1}^{n} w_i} - \left(1 - \xi^2_{\breve{S}} - \pi^2_{\breve{S}}\right)^{\sum_{i=1}^{n} w_i}\right)^{\frac{1}{2}} .e^{j\left(\left(1-\mu^2_{\breve{S}}\right)^{\sum_{i=1}^{n} w_i} - \left(1-\mu^2_{\breve{S}}-\omega^2_{\breve{S}}\right)^{\sum_{i=1}^{n} w_i}\right)^{\frac{1}{2}}} \end{array} \right\}$$
$$= \left\langle \xi_{\breve{S}_i}.e^{j\mu_{\breve{S}_i}}, \psi_{\breve{S}}.e^{j\nu_{\breve{S}_i}}, \pi_{\breve{S}_i}.e^{j\omega_{\breve{S}_i}} \right\rangle$$
$$= \breve{\mathcal{S}}$$

(ii) Boundedness: Let

$$\breve{\mathcal{S}}_i^+ = \left\langle \max_{1\le i\le n} \xi_{\breve{S}_i}.e^{j\mu_{\breve{S}_i}}, \max_{1\le i\le n} \psi_{\breve{S}}.e^{j\nu_{\breve{S}_i}}, \max_{1\le i\le n} \pi_{\breve{S}_i}.e^{j\omega_{\breve{S}_i}} \right\rangle,$$

$$\breve{\mathcal{S}}_i^- = \left\langle \min_{1\le i\le n} \xi_{\breve{S}_i}.e^{j\mu_{\breve{S}_i}}, \min_{1\le i\le n} \psi_{\breve{S}}.e^{j\nu_{\breve{S}_i}}, \min_{1\le i\le n} \pi_{\breve{S}_i}.e^{j\omega_{\breve{S}_i}} \right\rangle.$$

Then

$$\breve{\mathcal{S}}_i^- \le CSFWAM^* \le \breve{\mathcal{S}}_i^+$$

$$\left\{\begin{array}{c} \left(1-\prod_{i=1}^n\left(1-\min_{1\le i\le n}\xi_{\breve{S}_i}^2\right)^{w_i}\right)^{\frac{1}{2}}.e^{j\left(1-\prod_{i=1}^n\left(1-\mu_{\breve{S}_i}^2\right)^{w_i}\right)^{\frac{1}{2}}}, \quad \prod_{i=1}^n \min_{1\le i\le n}\psi_{\breve{S}_i}^{w_i}.e^{j\prod_{i=1}^n \nu_{\breve{S}_i}^{w_i}}, \\ \left(\prod_{i=1}^n\left(1-\min_{1\le i\le n}\xi_{\breve{S}_i}^2\right)^{w_i} - \prod_{i=1}^n\left(1-\min_{1\le i\le n}\xi_{\breve{S}_i}^2 - \right.\right. \\ \left.\left.\min_{1\le i\le n}\pi_{\breve{S}_i}^2\right)^{w_i}\right)^{\frac{1}{2}}.e^{j\left(\prod_{i=1}^n\left(1-\mu_{\breve{S}_i}^2\right)^{w_i}-\prod_{i=1}^n\left(1-\mu_{\breve{S}_i}^2-\omega_{\breve{S}_i}^2\right)^{w_i}\right)^{\frac{1}{2}}} \end{array}\right\} \le$$

$$\left\{\begin{array}{c} \left(1-\prod_{i=1}^n\left(1-\xi_{\breve{S}_i}^2\right)^{w_i}\right)^{\frac{1}{2}}.e^{j\left(1-\prod_{i=1}^n\left(1-\mu_{\breve{S}_i}^2\right)^{w_i}\right)^{\frac{1}{2}}}, \quad \prod_{i=1}^n \psi_{\breve{S}_i}^{w_i}.e^{j\prod_{i=1}^n \nu_{\breve{S}_i}^{w_i}}, \\ \left(\prod_{i=1}^n\left(1-\xi_{\breve{S}_i}^2\right)^{w_i} - \prod_{i=1}^n\left(1-\xi_{\breve{S}_i}^2-\pi_{\breve{S}_i}^2\right)^{w_i}\right)^{\frac{1}{2}}.e^{j\left(\prod_{i=1}^n\left(1-\mu_{\breve{S}_i}^2\right)^{w_i}-\prod_{i=1}^n\left(1-\mu_{\breve{S}_i}^2-\omega_{\breve{S}_i}^2\right)^{w_i}\right)^{\frac{1}{2}}} \end{array}\right\}$$

$$\le \left\{\begin{array}{c} \left(1-\prod_{i=1}^n\left(1-\max_{1\le i\le n}\xi_{\breve{S}_i}^2\right)^{w_i}\right)^{\frac{1}{2}}.e^{j\left(1-\prod_{i=1}^n\left(1-\mu_{\breve{S}_i}^2\right)^{w_i}\right)^{\frac{1}{2}}}, \quad \prod_{i=1}^n \max_{1\le i\le n}\psi_{\breve{S}_i}^{w_i}.e^{j\prod_{i=1}^n \nu_{\breve{S}_i}^{w_i}}, \\ \left(\prod_{i=1}^n\left(1-\max_{1\le i\le n}\xi_{\breve{S}_i}^2\right)^{w_i} - \prod_{i=1}^n\left(1-\max_{1\le i\le n}\xi_{\breve{S}_i}^2 - \right.\right. \\ \left.\left.\max_{1\le i\le n}\pi_{\breve{S}_i}^2\right)^{w_i}\right)^{\frac{1}{2}}.e^{j\left(\prod_{i=1}^n\left(1-\mu_{\breve{S}_i}^2\right)^{w_i}-\prod_{i=1}^n\left(1-\mu_{\breve{S}_i}^2-\omega_{\breve{S}_i}^2\right)^{w_i}\right)^{\frac{1}{2}}} \end{array}\right\},$$

$$\left\{\begin{array}{c} \left(1-\left(1-\min_{1\le i\le n}\xi_{\breve{S}_i}^2\right)^{\sum_{i=1}^n w_i}\right)^{\frac{1}{2}}.e^{j\left(1-\left(1-\mu_{\breve{S}_i}^2\right)^{\sum_{i=1}^n w_i}\right)^{\frac{1}{2}}}, \quad \min_{1\le i\le n}\psi_{\breve{S}_i}^{\sum_{i=1}^n w_i}.e^{j\nu_{\breve{S}_i}^{\sum_{i=1}^n w_i}}, \\ \left(\left(1-\min_{1\le i\le n}\xi_{\breve{S}_i}^2\right)^{\sum_{i=1}^n w_i} - \left(1-\min_{1\le i\le n}\xi_{\breve{S}_i}^2 - \right.\right. \\ \left.\left.\min_{1\le i\le n}\pi_{\breve{S}_i}^2\right)^{\sum_{i=1}^n w_i}\right)^{\frac{1}{2}}.e^{j\left(\left(1-\mu_{\breve{S}_i}^2\right)^{\sum_{i=1}^n w_i}-\left(1-\mu_{\breve{S}_i}^2-\omega_{\breve{S}_i}^2\right)^{\sum_{i=1}^n w_i}\right)^{\frac{1}{2}}} \end{array}\right\} \le$$

$$\left\{\begin{array}{c} \left(1-\left(1-\xi_{\breve{S}_i}^2\right)^{\sum_{i=1}^n w_i}\right)^{\frac{1}{2}}.e^{j\left(1-\left(1-\mu_{\breve{S}_i}^2\right)^{\sum_{i=1}^n w_i}\right)^{\frac{1}{2}}}, \quad \psi_{\breve{S}_i}^{\sum_{i=1}^n w_i}.e^{j\nu_{\breve{S}_i}^{\sum_{i=1}^n w_i}}, \\ \left(\left(1-\xi_{\breve{S}_i}^2\right)^{\sum_{i=1}^n w_i} - \left(1-\xi_{\breve{S}_i}^2-\pi_{\breve{S}_i}^2\right)^{\sum_{i=1}^n w_i}\right)^{\frac{1}{2}}.e^{j\left(\left(1-\mu_{\breve{S}_i}^2\right)^{\sum_{i=1}^n w_i}-\left(1-\mu_{\breve{S}_i}^2-\omega_{\breve{S}_i}^2\right)^{\sum_{i=1}^n w_i}\right)^{\frac{1}{2}}} \end{array}\right\}$$

$$\leq \left\{ \begin{array}{c} \left(1-\left(1-\max_{1\leq i\leq n}\xi_{\breve{S}_i}^2\right)^{\sum_{i=1}^n w_i}\right)^{\frac{1}{2}} .e^{j\left(1-\left(1-\mu_{\breve{S}_i}^2\right)^{\sum_{i=1}^n w_i}\right)^{\frac{1}{2}}}, \quad \max_{1\leq i\leq n}\psi_{\breve{S}_i}^{\sum_{i=1}^n w_i}.e^{j\nu_{\breve{S}_i}^{\sum_{i=1}^n w_i}}, \\ \left(\left(1-\max_{1\leq i\leq n}\xi_{\breve{S}_i}^2\right)^{\sum_{i=1}^n w_i}-\left(1-\max_{1\leq i\leq n}\xi_{\breve{S}_i}^2-\right.\right. \\ \left.\left.\max_{1\leq i\leq n}\pi_{\breve{S}_i}^2\right)^{\sum_{i=1}^n w_i}\right)^{\frac{1}{2}} .e^{j\left(\left(1-\mu_{\breve{S}_i}^2\right)^{\sum_{i=1}^n w_i}-\left(1-\mu_{\breve{S}_i}^2-\omega_{\breve{S}_i}^2\right)^{\sum_{i=1}^n w_i}\right)^{\frac{1}{2}}} \end{array} \right\},$$

$$\left\langle \min_{1\leq i\leq n}\xi_{\breve{S}_i}.e^{j\mu_{\breve{S}_i}}, \min_{1\leq i\leq n}\psi_{\breve{S}}.e^{j\nu_{\breve{S}_i}}, \min_{1\leq i\leq n}\pi_{\breve{S}_i}.e^{j\omega_{\breve{S}_i}}\right\rangle \leq$$

$$\left\{ \begin{array}{c} \left(1-\prod_{i=1}^n\left(1-\xi_{\breve{S}_i}^2\right)^{w_i}\right)^{\frac{1}{2}} .e^{j\left(1-\prod_{i=1}^n\left(1-\mu_{\breve{S}_i}^2\right)^{w_i}\right)^{\frac{1}{2}}}, \quad \prod_{i=1}^n\psi_{\breve{S}_i}^{w_i}.e^{j\prod_{i=1}^n\nu_{\breve{S}_i}^{w_i}}, \\ \left(\prod_{i=1}^n\left(1-\xi_{\breve{S}_i}^2\right)^{w_i}-\prod_{i=1}^n\left(1-\xi_{\breve{S}_i}^2-\pi_{\breve{S}_i}^2\right)^{w_i}\right)^{\frac{1}{2}} .e^{j\left(\prod_{i=1}^n\left(1-\mu_{\breve{S}_i}^2\right)^{w_i}-\prod_{i=1}^n\left(1-\mu_{\breve{S}_i}^2-\omega_{\breve{S}_i}^2\right)^{w_i}\right)^{\frac{1}{2}}} \end{array} \right\}$$

$$\leq \left\langle \max_{1\leq i\leq n}\xi_{\breve{S}_i}.e^{j\mu_{\breve{S}_i}}, \max_{1\leq i\leq n}\psi_{\breve{S}}.e^{j\nu_{\breve{S}_i}}, \max_{1\leq i\leq n}\pi_{\breve{S}_i}.e^{j\omega_{\breve{S}_i}}\right\rangle$$

Therefore,

$$\breve{\mathcal{S}}_i^- \leq CSFWAM^*\left(\breve{\mathcal{S}}_1, \breve{\mathcal{S}}_2, \ldots, \breve{\mathcal{S}}_n\right) \leq \breve{\mathcal{S}}_i^+$$

(iii) Let

$$\breve{\mathcal{S}}_i = \left\langle \xi_{\breve{S}_i}.e^{j\mu_{\breve{S}_i}}, \psi_{\breve{S}_i}.e^{j\nu_{\breve{S}_i}}, \pi_{\breve{S}_i}.e^{j\omega_{\breve{S}_i}}\right\rangle, \breve{\mathcal{B}}_i = \left\langle \xi_{\breve{B}_i}.e^{j\mu_{\breve{B}_i}}, \psi_{\breve{B}_i}.e^{j\nu_{\breve{B}_i}}, \pi_{\breve{B}_i}.e^{j\omega_{\breve{B}_i}}\right\rangle$$

be a two families of CSFSs, if $\xi_{\breve{S}_i} \geq \xi_{\breve{B}_i}$, $\psi_{\breve{S}_i} \leq \psi_{\breve{B}_i}$ and $\pi_{\breve{S}_i} \leq \pi_{\breve{B}_i}$, then

$$\left\{ \begin{array}{c} \left(1-\prod_{i=1}^n\left(1-\xi_{\breve{B}_i}^2\right)^{w_i}\right)^{\frac{1}{2}} .e^{j\left(1-\prod_{i=1}^n\left(1-\mu_{\breve{B}_i}^2\right)^{w_i}\right)^{\frac{1}{2}}}, \quad \prod_{i=1}^n\psi_{\breve{B}_i}^{w_i}.e^{j\prod_{i=1}^n\nu_{\breve{B}_i}^{w_i}}, \\ \left(\prod_{i=1}^n\left(1-\xi_{\breve{B}_i}^2\right)^{w_i}-\prod_{i=1}^n\left(1-\xi_{\breve{B}_i}^2-\pi_{\breve{B}_i}^2\right)^{w_i}\right)^{\frac{1}{2}} .e^{j\left(\prod_{i=1}^n\left(1-\mu_{\breve{B}_i}^2\right)^{w_i}-\prod_{i=1}^n\left(1-\mu_{\breve{B}_i}^2-\omega_{\breve{B}_i}^2\right)^{w_i}\right)^{\frac{1}{2}}} \end{array} \right\}$$

$$\leq \left\{ \begin{array}{c} \left(1-\prod_{i=1}^n\left(1-\xi_{\breve{S}_i}^2\right)^{w_i}\right)^{\frac{1}{2}} .e^{j\left(1-\prod_{i=1}^n\left(1-\mu_{\breve{S}_i}^2\right)^{w_i}\right)^{\frac{1}{2}}}, \quad \prod_{i=1}^n\psi_{\breve{S}_i}^{w_i}.e^{j\prod_{i=1}^n\nu_{\breve{S}_i}^{w_i}}, \\ \left(\prod_{i=1}^n\left(1-\xi_{\breve{S}_i}^2\right)^{w_i}-\prod_{i=1}^n\left(1-\xi_{\breve{S}_i}^2-\pi_{\breve{S}_i}^2\right)^{w_i}\right)^{\frac{1}{2}} .e^{j\left(\prod_{i=1}^n\left(1-\mu_{\breve{S}_i}^2\right)^{w_i}-\prod_{i=1}^n\left(1-\mu_{\breve{S}_i}^2-\omega_{\breve{S}_i}^2\right)^{w_i}\right)^{\frac{1}{2}}} \end{array} \right\}$$

$$\left|CSFWAM^*\left(\breve{\mathcal{B}}_1, \breve{\mathcal{B}}_2, \ldots, \breve{\mathcal{B}}_n\right)\right| \leq \left|CSFWAM^*\left(\breve{\mathcal{S}}_1, \breve{\mathcal{S}}_2, \ldots, \breve{\mathcal{S}}_n\right)\right|$$

Hence the theorem.

Theorem 10 *Complex spherical fuzzy weighted geometric mean* ($CSFWGM^*$) *operator with respect to complex valued weight vectors w_i for all i and $w_i \in [0, 1]$ such that $\sum_{i=1}^n w_i = 1$, is defined as*

$$CSFWGM^{*}\left(\breve{\mathcal{S}}_1, \breve{\mathcal{S}}_2, \ldots, \breve{\mathcal{S}}_n\right) = \sum_{i=1}^{n} \left(\breve{\mathcal{S}}_i\right)^{w_i} \tag{11}$$

$$= \left\{ \begin{array}{c} \prod_{i=1}^{n} \xi_{\breve{S}_i}^{w_i} . e^{j \prod_{i=1}^{n} \mu_{\breve{S}_i}^{w_i}}, \quad \left(1 - \prod_{i=1}^{n} \left(1 - \psi_{\breve{S}_i}^2\right)^{w_i}\right)^{\frac{1}{2}} . e^{j\left(1 - \prod_{i=1}^{n}\left(1 - \nu_{\breve{S}_i}^2\right)^{w_i}\right)^{\frac{1}{2}}}, \\ \left(\prod_{i=1}^{n} \left(1 - \psi_{\breve{S}_i}^2\right)^{w_i} - \prod_{i=1}^{n} \left(1 - \psi_{\breve{S}_i}^2 - \pi_{\breve{S}_i}^2\right)^{w_i}\right)^{\frac{1}{2}} . e^{j\left(\prod_{i=1}^{n}\left(1 - \nu_{\breve{S}_i}^2\right)^{w_i} - \prod_{i=1}^{n}\left(1 - \nu_{\breve{S}_i}^2 - \omega_{\breve{S}_i}^2\right)^{w_i}\right)^{\frac{1}{2}}} \end{array} \right\}$$

Proof The proof is similar to that of Theorem 8.

Theorem 11 *Let* $\breve{\mathcal{S}}_i = \left\langle \xi_{\breve{S}_i} . e^{j\mu_{\breve{S}_i}}, \psi_{\breve{S}} . e^{j\nu_{\breve{S}_i}}, \pi_{\breve{S}_i} . e^{j\omega_{\breve{S}_i}} \right\rangle$ *be a family of CSFSs, the weight vectors of CSFWGM* operator is real valued i.e.,* $w_i \in [0, 1]$ $\forall i$ *and* $\sum_{i=1}^{n} w_i = 1$. *Then we have the following axioms:*

(i) Idempotency: If all CSFVs are equal i.e.$(\breve{\mathcal{S}}_1 = \breve{\mathcal{S}}_2 = \ldots = \breve{\mathcal{S}}_n)$, *then*

$$CSFWGM^{*}\left(\breve{\mathcal{S}}_1, \breve{\mathcal{S}}_2, \ldots, \breve{\mathcal{S}}_n\right) = \breve{\mathcal{S}}_1$$

(ii) Boundedness: Let $\breve{S}_1, \breve{S}_2, \ldots, \breve{S}_n$ *be a CSFSs.*

$$\left|CSFWGM^{*}\left(\breve{\mathcal{S}}_1, \breve{\mathcal{S}}_2, \ldots, \breve{\mathcal{S}}_n\right)\right| \leq max_i|\breve{\mathcal{S}}_i|$$

(iii) Monotonicity: If $\xi_{\breve{S}_i} \geq \xi_{\breve{B}_i}, \psi_{\breve{S}_i} \leq \psi_{\breve{B}_i}, \pi_{\breve{S}_i} \leq \pi_{\breve{B}_i}$ *then*

$$\left|CSFWGM^{*}\left(\breve{\mathcal{B}}_1, \breve{\mathcal{B}}_2, \ldots, \breve{\mathcal{B}}_n\right)\right| \leq \left|CSFWGM^{*}\left(\breve{\mathcal{S}}_1, \breve{\mathcal{S}}_2, \ldots, \breve{\mathcal{S}}_n\right)\right|$$

Proof The proof is similar to that of Theorem 9.

2.2 The EDAS Model with Complex Spherical Fuzzy Information

Consider a decision making situation in which there are m alternatives $\{R_1, R_2, \ldots, R_m\}$ qualified by n attributes $\{C_1, C_2, \ldots, C_n\}$. Let l experts $\{E_1, E_2, \ldots, E_t\}$ be involved in the process of decision making and let $\{w_1, w_2, \ldots, w_n\}$ and $\{\varpi_1, \varpi_2, \ldots, \varpi_t\}$ be the attributes' weight vector and expert's weight vector respectively which statisfy the conditions $\sum_{i=1}^{n} w_i = 1$ and $\sum_{l=1}^{t} \varpi_i = 1$. The computing steps of the proposed EDAS model based on CSFSs can be established as follows:

Step 1. Construct the decision matrix $M^{\circ} = \left[e_{ij}^{l}\right]_{m\times n}$ $\forall i = 1, 2, 3, \ldots, m;\ j = 1, 2, 3, \ldots, n;\ l = 1, 2, 3, \ldots, t$ which can be depicted as follows:

$$M^{\circ} = \left[e_{ij}^{l}\right]_{m\times n} = \begin{matrix} \\ R_1 \\ R_2 \\ \vdots \\ R_m \end{matrix} \begin{matrix} \begin{matrix} C_1 & C_2 & \ldots & C_n \end{matrix} \\ \begin{pmatrix} e_{11}^{l} & e_{12}^{l} & \ldots & e_{1n}^{l} \\ e_{21}^{l} & e_{22}^{l} & \ldots & e_{2n}^{l} \\ \vdots & \vdots & \ddots & \vdots \\ e_{m1}^{l} & e_{m2}^{l} & \ldots & e_{mn}^{l} \end{pmatrix} \end{matrix}$$

where e_{ij}^{l} denotes the elements of CSFSs of alternative R_i on attributes C_j given by expert E^{l}.

Step 2. Normalize the decision matrix $M^{\circ} = \left[e_{ij}^{l}\right]_{m\times n}$ to $(M^{\circ})^{'} = \left[e_{ij}^{'l}\right]_{m\times n}$ For benefit attributes: $e_{ij}^{'l} = e_{ij}^{l} = \left\langle \xi_{\breve{S}_i}^{l}.e^{j\mu_{\breve{S}_i}^{l}}, \psi_{\breve{S}}^{l}.e^{j\nu_{\breve{S}_i}^{l}}, \pi_{\breve{S}_i}^{l}.e^{j\omega_{\breve{S}_i}^{l}} \right\rangle \forall i = 1, 2, 3, \ldots, m; j = 1, 2, 3, \ldots, n;\quad l = 1, 2, 3, \ldots, t$ For cost attributes: $e_{ij}^{'l} = \left(e_{ij}^{l}\right)^{c} = \left\langle \pi_{\breve{S}_i}^{l}.e^{j\omega_{\breve{S}_i}^{l}}, \psi_{\breve{S}}^{l}.e^{j\nu_{\breve{S}_i}^{l}}, \xi_{\breve{S}_i}^{l}.e^{j\mu_{\breve{S}_i}^{l}} \right\rangle$ $\forall i = 1, 2, 3, \ldots, m;\ \ j = 1, 2, 3, \ldots, n;\ \ l = 1, 2, 3, \ldots, t$

Step 3. According to the decision matrix $M^{\circ} = \left[e_{ij}^{l}\right]_{m\times n}$ and expert's weight vector $\{\varpi_1, \varpi_2, \varpi_3, \ldots, \varpi_t\}$, we can derive the overall e_{ij}^{l} to e_{ij} using *CSFSs* aggregation operators. The computing results can be represented by

$$M^{\circ} = \left[e_{ij}\right]_{m\times n} = \begin{matrix} \\ R_1 \\ R_2 \\ \vdots \\ R_m \end{matrix} \begin{matrix} \begin{matrix} C_1 & C_2 & \ldots & C_n \end{matrix} \\ \begin{pmatrix} e_{11} & e_{12} & \ldots & e_{1n} \\ e_{21} & e_{22} & \ldots & e_{2n} \\ \vdots & \vdots & \ddots & \vdots \\ e_{m1} & e_{m2} & \ldots & e_{mn} \end{pmatrix} \end{matrix}$$

Step 4. Compute the value of average solution $(\breve{\mathcal{A}}\mathcal{V})$ based on all proposed attributes

$$\breve{\mathcal{A}}\mathcal{V} = \left[\breve{\mathcal{A}}\mathcal{V}_j\right]_{1\times n} = \left[\frac{\sum_{i=1}^{m} e_{ij}}{m}\right]_{1\times n}$$

Based on the operations on CSFSs, we can get the following

$$\sum_{i=1}^{m} e_{ij} = \left\{ \begin{array}{c} \left(1 - \prod_{i=1}^{m}\left(1 - \xi_{\breve{S}_{ij}}^{2}\right)\right)^{\frac{1}{2}} .e^{j\left(1-\prod_{i=1}^{m}\left(1-\mu_{\breve{S}_{ij}}^{2}\right)\right)^{\frac{1}{2}}}, \prod_{i=1}^{m} \psi_{\breve{S}_{ij}}.e^{j\prod_{i=1}^{m}\nu_{\breve{S}_{ij}}}, \\ \left(\prod_{i=1}^{m}\left(1 - \xi_{\breve{S}_{ij}}^{2}\right) - \prod_{i=1}^{m}\left(1 - \xi_{\breve{S}_{ij}}^{2} - \pi_{\breve{S}_{ij}}^{2}\right)\right)^{\frac{1}{2}} .e^{j\left(\prod_{i=1}^{m}\left(1-\mu_{\breve{S}_{ij}}^{2}\right)-\prod_{i=1}^{m}\left(1-\mu_{\breve{S}_{ij}}^{2}-\omega_{\breve{S}_{ij}}^{2}\right)\right)^{\frac{1}{2}}} \end{array} \right\}$$

$$\breve{\mathcal{A}}\mathcal{V} = [\breve{\mathcal{A}}\mathcal{V}_j]_{1\times n} = \left[\frac{\sum_{i=1}^{m} e_{ij}}{m}\right]_{1\times n}$$

$$= \left\{ \begin{array}{c} \left(1-\prod_{i=1}^{n}\left(1-\xi_{\breve{S}_{ij}}^{2}\right)^{\frac{1}{m}}\right)^{\frac{1}{2}} . e^{j\left(1-\prod_{i=1}^{n}\left(1-\mu_{\breve{S}_{ij}}^{2}\right)^{\frac{1}{m}}\right)^{\frac{1}{2}}}, \prod_{i=1}^{n}\psi_{\breve{S}_{ij}}^{\frac{1}{m}} . e^{j\prod_{i=1}^{n}\nu_{\breve{S}_{ij}}^{\frac{1}{m}}}, \\ \left(\prod_{i=1}^{n}\left(1-\xi_{\breve{S}_{ij}}^{2}\right)^{\frac{1}{m}} - \prod_{i=1}^{n}\left(1-\xi_{\breve{S}_{ij}}^{2}-\pi_{\breve{S}_{ij}}^{2}\right)^{\frac{1}{m}}\right)^{\frac{1}{2}} . e^{j\left(\prod_{i=1}^{n}\left(1-\mu_{\breve{S}_{ij}}^{2}\right)^{\frac{1}{m}} - \prod_{i=1}^{n}\left(1-\mu_{\breve{S}_{ij}}^{2}-\omega_{\breve{S}_{ij}}^{2}\right)^{\frac{1}{m}}\right)^{\frac{1}{2}}} \end{array} \right\}$$

Step 5. Find the positive distance from average ($\mathcal{P}\breve{\mathcal{D}}\mathcal{A}$) and negative distance form average ($\mathcal{N}\breve{\mathcal{D}}\mathcal{A}$) using the formulae

$$\mathcal{P}\breve{\mathcal{D}}\mathcal{A}_{ij} = \left[\mathcal{P}\breve{\mathcal{D}}\mathcal{A}_{ij}\right]_{m\times n} = \frac{\max\left(0, \left(e_{ij} - \breve{\mathcal{A}}\mathcal{V}_j\right)\right)}{\breve{\mathcal{A}}\mathcal{V}_j}$$

$$\mathcal{N}\breve{\mathcal{D}}\mathcal{A}_{ij} = \left[\mathcal{N}\breve{\mathcal{D}}\mathcal{A}_{ij}\right]_{m\times n} = \frac{\max\left(0, \left(\breve{\mathcal{A}}\mathcal{V}_j - e_{ij}\right)\right)}{\breve{\mathcal{A}}\mathcal{V}_j}$$

For convenience, we can use the score function of CSFSs to determine the values of $\mathcal{P}\breve{\mathcal{D}}\mathcal{A}$ and $\mathcal{N}\breve{\mathcal{D}}\mathcal{A}$ as follows

$$\begin{aligned} \mathcal{P}\breve{\mathcal{D}}\mathcal{A}_{ij} &= \left[\mathcal{P}\breve{\mathcal{D}}\mathcal{A}_{ij}\right]_{m\times n} = \frac{\max\left(0, \left(S(e_{ij}) - S(\breve{\mathcal{A}}\mathcal{V}_j)\right)\right)}{S(\breve{\mathcal{A}}\mathcal{V}_j)} \\ \mathcal{N}\breve{\mathcal{D}}\mathcal{A}_{ij} &= \left[\mathcal{N}\breve{\mathcal{D}}\mathcal{A}_{ij}\right]_{m\times n} = \frac{\max\left(0, \left(S(\breve{\mathcal{A}}\mathcal{V}_j) - S(e_{ij})\right)\right)}{S(\breve{\mathcal{A}}\mathcal{V}_j)} \end{aligned} \qquad (12)$$

Step 6. Calculate the weighted sum of $\mathcal{P}\breve{\mathcal{D}}\mathcal{A}$ and $\mathcal{N}\breve{\mathcal{D}}\mathcal{A}$ and denote them as $\mathcal{S}\breve{\mathcal{P}}_i$ and $\mathcal{S}\breve{\mathcal{N}}_i$ respectively. The formula for the calculation is as follows

$$\mathcal{S}\breve{\mathcal{P}}_i = \sum_{j=1}^{n} w_j.\mathcal{P}\breve{\mathcal{D}}\mathcal{A}_{ij}, \quad \mathcal{S}\breve{\mathcal{N}}_i = \sum_{j=1}^{n} w_j.\mathcal{N}\breve{\mathcal{D}}\mathcal{A}_{ij} \qquad (13)$$

Step 7. The normalized values of $\mathcal{S}\breve{\mathcal{P}}_i$ and $\mathcal{S}\breve{\mathcal{N}}_i$ are calculated by

$$\mathcal{S}\breve{\mathcal{P}}_i^{n*} = \frac{\mathcal{S}\breve{\mathcal{P}}_i}{\max_i(\mathcal{S}\breve{\mathcal{P}}_i)}, \quad \mathcal{S}\breve{\mathcal{N}}_i^{n*} = 1 - \frac{\mathcal{S}\breve{\mathcal{N}}_i}{\max_i(\mathcal{S}\breve{\mathcal{N}}_i)} \qquad (14)$$

Step 8. Calculate the appraisal score ($\breve{\mathcal{A}}\mathcal{S}_i$) for all alternatives using

$$\breve{\mathcal{A}}\mathcal{S}_i = \frac{1}{2}\left(\mathcal{S}\breve{\mathcal{P}}_i^{n*} + \mathcal{S}\breve{\mathcal{N}}_i^{n*}\right)$$

Step 9. Rank the alternatives according to the values of $\breve{\mathcal{A}S}_i$ and the highest value of $\breve{\mathcal{A}S}_i$ is the best alternative.

3 Application of CSFSs for MCDM Problems

Recent developments in Aviation sector demand more mechanization and automation in every aspect of its operation. Catering services is one such area where the industrialists look for implementation of recent technology which will help improve the services as it involves complex decision making process from the preparation of menu to the disposal of food wastes. Hence the aviation industry catering services are more inclined to adopt the application of digital transformation proposed as part of industry 4.0. It is expected that the performance of aviation catering services will be very adequate and more comfortable to the customer needs in such adoption. So, in this illustrative problem, we have taken five aviation catering services which have inclination to adopt and have adopted these transformations as the alternatives. Further evaluation of these aviation catering services is to be concerned with customer satisfaction as well as the performance of the management. In this regard, the recent technologies in industry 4.0 have been chosen as the attributes namely, 1. Digital Manufacturing, 2. IoT (Internet of Things), 3. Artificial Intelligence / Automation Robotics, 4. RFID (Radio Frequency Identification).

These are very essential technologies that every aviation catering service should adopt to stay in competition with other players. Every stage in catering service involves human-machine interaction. The attributes/technologies are related to each other but are unique in their performance and different in their digital transformation applications. Therefore, the decision makers have assigned weighted values for each criterion respectively as follows 0.25, 0.18, 0.35, and 0.22. In this assessment, three experts belonging to the field of aviation industry were asked to evaluate the alternatives on criteria in terms of complex spherical fuzzy numbers. More briefly, these CSFNs have two terms namely, amplitude and phase term. The experts have measured the performance of the catering services under CSFNs. Usually the amplitude values lie between zero and one that represent the range of the performance of catering service and the phase terms represent the periodicity or time taken by the catering services. Through evaluation by experts, the complex spherical fuzzy decision matrices are formed and are represented in Tables 1, 2 and 3. Also according to the experience level, the weights of three experts are taken as 0.35, 0.2, and 0.45 respectively.

Step 1. The decision matrix $M^\circ = \left[e_{ij}^l\right]_{m\times n}$ $\forall i = 1, 2, 3, \ldots, m; j = 1, 2, 3, \ldots, n$

Step 2. Normalize the decision matrix $M^\circ = \left[e_{ij}^l\right]_{5\times 4}$ to $(M^\circ)' = \left[e_{ij}'^l\right]_{5\times 4}$. If all the criteria are benefit, the normalization is not needed.

Table 1 CSF evaluation matrix by E^1

M°	C_1	C_2	C_3	C_4
R_1	$\left(0.56e^{j\frac{2\pi}{3}}, 0.12e^{j\frac{2\pi}{7}}, 0.20e^{j\frac{2\pi}{4}}\right)$	$\left(0.62e^{j\frac{2\pi}{2}}, 0.15e^{j\frac{2\pi}{7}}, 0.23e^{j\frac{2\pi}{4}}\right)$	$\left(0.47e^{j\frac{2\pi}{3}}, 0.33e^{j\frac{2\pi}{3}}, 0.10e^{j\frac{2\pi}{7}}\right)$	$\left(0.51e^{j\frac{2\pi}{2}}, 0.34e^{j\frac{2\pi}{4}}, 0.15e^{j\frac{2\pi}{6}}\right)$
R_2	$\left(0.43e^{j\frac{2\pi}{4}}, 0.25e^{j\frac{2\pi}{5}}, 0.18e^{j\frac{2\pi}{6}}\right)$	$\left(0.50e^{j\frac{2\pi}{3}}, 0.28e^{j\frac{2\pi}{4}}, 0.22e^{j\frac{2\pi}{6}}\right)$	$\left(0.54e^{j\frac{2\pi}{3}}, 0.29e^{j\frac{2\pi}{4}}, 0.17e^{j\frac{2\pi}{6}}\right)$	$\left(0.64e^{j\frac{2\pi}{3}}, 0.17e^{j\frac{2\pi}{5}}, 0.19e^{j\frac{2\pi}{5}}\right)$
R_3	$\left(0.60e^{j\frac{2\pi}{2}}, 0.32e^{j\frac{2\pi}{5}}, 0.08e^{j\frac{2\pi}{8}}\right)$	$\left(0.58e^{j\frac{2\pi}{3}}, 0.12e^{j\frac{2\pi}{6}}, 0.30e^{j\frac{2\pi}{5}}\right)$	$\left(0.62e^{j\frac{2\pi}{2}}, 0.11e^{j\frac{2\pi}{6}}, 0.28e^{j\frac{2\pi}{5}}\right)$	$\left(0.80e^{j\frac{2\pi}{2}}, 0.15e^{j\frac{2\pi}{8}}, 0.05e^{j\frac{2\pi}{9}}\right)$
R_4	$\left(0.58e^{j\frac{2\pi}{2}}, 0.22e^{j\frac{2\pi}{7}}, 0.10e^{j\frac{2\pi}{8}}\right)$	$\left(0.61e^{j\frac{2\pi}{2}}, 0.13e^{j\frac{2\pi}{6}}, 0.26e^{j\frac{2\pi}{5}}\right)$	$\left(0.55e^{j\frac{2\pi}{3}}, 0.27e^{j\frac{2\pi}{7}}, 0.18e^{j\frac{2\pi}{6}}\right)$	$\left(0.67e^{j\frac{2\pi}{2}}, 0.26e^{j\frac{2\pi}{7}}, 0.07e^{j\frac{2\pi}{9}}\right)$
R_5	$\left(0.50e^{j\frac{2\pi}{3}}, 0.13e^{j\frac{2\pi}{6}}, 0.37e^{j\frac{2\pi}{4}}\right)$	$\left(0.65e^{j\frac{2\pi}{2}}, 0.10e^{j\frac{2\pi}{6}}, 0.25e^{j\frac{2\pi}{5}}\right)$	$\left(0.76e^{j\frac{2\pi}{6}}, 0.09e^{j\frac{2\pi}{9}}, 0.24e^{j\frac{2\pi}{7}}\right)$	$\left(0.47e^{j\frac{2\pi}{6}}, 0.35e^{j\frac{2\pi}{5}}, 0.18e^{j\frac{2\pi}{6}}\right)$

Table 2 CSF evaluation matrix by E^2

M°	C_1	C_2	C_3	C_4
R_1	$\left(0.48e^{j\frac{2\pi}{3}}, 0.35e^{j\frac{2\pi}{5}}, 0.17e^{j\frac{2\pi}{7}}\right)$	$\left(0.53e^{j\frac{2\pi}{4}}, 0.27e^{j\frac{2\pi}{7}}, 0.10e^{j\frac{2\pi}{8}}\right)$	$\left(0.61e^{j\frac{2\pi}{2}}, 0.28e^{j\frac{2\pi}{7}}, 0.19e^{j\frac{2\pi}{8}}\right)$	$\left(0.80e^{j\frac{2\pi}{2}}, 0.15e^{j\frac{2\pi}{7}}, 0.05e^{j\frac{2\pi}{9}}\right)$
R_2	$\left(0.53e^{j\frac{2\pi}{3}}, 0.27e^{j\frac{2\pi}{7}}, 0.10e^{j\frac{2\pi}{8}}\right)$	$\left(0.64e^{j\frac{2\pi}{3}}, 0.17e^{j\frac{2\pi}{8}}, 0.19e^{j\frac{2\pi}{6}}\right)$	$\left(0.43e^{j\frac{2\pi}{6}}, 0.37e^{j\frac{2\pi}{6}}, 0.20e^{j\frac{2\pi}{6}}\right)$	$\left(0.23e^{j\frac{2\pi}{7}}, 0.22e^{j\frac{2\pi}{8}}, 0.65e^{j\frac{2\pi}{3}}\right)$
R_3	$\left(0.66e^{j\frac{2\pi}{3}}, 0.20e^{j\frac{2\pi}{8}}, 0.14e^{j\frac{2\pi}{7}}\right)$	$\left(0.59e^{j\frac{2\pi}{4}}, 0.21e^{j\frac{2\pi}{6}}, 0.20e^{j\frac{2\pi}{6}}\right)$	$\left(0.18e^{j\frac{2\pi}{7}}, 0.11e^{j\frac{2\pi}{8}}, 0.77e^{j\frac{2\pi}{2}}\right)$	$\left(0.73e^{j\frac{2\pi}{2}}, 0.17e^{j\frac{2\pi}{8}}, 0.10e^{j\frac{2\pi}{9}}\right)$
R_4	$\left(0.41e^{j\frac{2\pi}{4}}, 0.28e^{j\frac{2\pi}{7}}, 0.31e^{j\frac{2\pi}{6}}\right)$	$\left(0.18e^{j\frac{2\pi}{8}}, 0.32e^{j\frac{2\pi}{6}}, 0.50e^{j\frac{2\pi}{3}}\right)$	$\left(0.29e^{j\frac{2\pi}{5}}, 0.32e^{j\frac{2\pi}{5}}, 0.39e^{j\frac{2\pi}{6}}\right)$	$\left(0.49e^{j\frac{2\pi}{5}}, 0.34e^{j\frac{2\pi}{7}}, 0.17e^{j\frac{2\pi}{8}}\right)$
R_5	$\left(0.07e^{j\frac{2\pi}{9}}, 0.39e^{j\frac{2\pi}{4}}, 0.64e^{j\frac{2\pi}{2}}\right)$	$\left(0.27e^{j\frac{2\pi}{8}}, 0.28e^{j\frac{2\pi}{3}}, 0.45e^{j\frac{2\pi}{7}}\right)$	$\left(0.55e^{j\frac{2\pi}{6}}, 0.27e^{j\frac{2\pi}{6}}, 0.08e^{j\frac{2\pi}{9}}\right)$	$\left(0.68e^{j\frac{2\pi}{2}}, 0.14e^{j\frac{2\pi}{8}}, 0.18e^{j\frac{2\pi}{8}}\right)$

Table 3 CSF evaluation matrix by E^3

M°	C_1	C_2	C_3	C_4
R_1	$\left(0.51e^{j\frac{2\pi}{3}}, 0.24e^{j\frac{2\pi}{6}}, 0.25e^{j\frac{2\pi}{6}}\right)$	$\left(0.70e^{j\frac{2\pi}{2}}, 0.12e^{j\frac{2\pi}{8}}, 0.18e^{j\frac{2\pi}{8}}\right)$	$\left(0.52e^{j\frac{2\pi}{3}}, 0.25e^{j\frac{2\pi}{6}}, 0.23e^{j\frac{2\pi}{6}}\right)$	$\left(0.56e^{j\frac{2\pi}{3}}, 0.12e^{j\frac{2\pi}{8}}, 0.20e^{j\frac{2\pi}{7}}\right)$
R_2	$\left(0.69e^{j\frac{2\pi}{2}}, 0.08e^{j\frac{2\pi}{9}}, 0.23e^{j\frac{2\pi}{7}}\right)$	$\left(0.59e^{j\frac{2\pi}{3}}, 0.21e^{j\frac{2\pi}{7}}, 0.10e^{j\frac{2\pi}{8}}\right)$	$\left(0.36e^{j\frac{2\pi}{6}}, 0.09e^{j\frac{2\pi}{9}}, 0.64e^{j\frac{2\pi}{2}}\right)$	$\left(0.76e^{j\frac{2\pi}{2}}, 0.09e^{j\frac{2\pi}{9}}, 0.15e^{j\frac{2\pi}{8}}\right)$
R_3	$\left(0.37e^{j\frac{2\pi}{5}}, 0.21e^{j\frac{2\pi}{7}}, 0.42e^{j\frac{2\pi}{5}}\right)$	$\left(0.45e^{j\frac{2\pi}{4}}, 0.23e^{j\frac{2\pi}{7}}, 0.32e^{j\frac{2\pi}{6}}\right)$	$\left(0.73e^{j\frac{2\pi}{2}}, 0.14e^{j\frac{2\pi}{7}}, 0.13e^{j\frac{2\pi}{8}}\right)$	$\left(0.55e^{j\frac{2\pi}{4}}, 0.27e^{j\frac{2\pi}{6}}, 0.08e^{j\frac{2\pi}{9}}\right)$
R_4	$\left(0.42e^{j\frac{2\pi}{5}}, 0.25e^{j\frac{2\pi}{6}}, 0.33e^{j\frac{2\pi}{6}}\right)$	$\left(0.66e^{j\frac{2\pi}{2}}, 0.19e^{j\frac{2\pi}{7}}, 0.15e^{j\frac{2\pi}{8}}\right)$	$\left(0.57e^{j\frac{2\pi}{3}}, 0.26e^{j\frac{2\pi}{7}}, 0.17e^{j\frac{2\pi}{6}}\right)$	$\left(0.18e^{j\frac{2\pi}{7}}, 0.32e^{j\frac{2\pi}{6}}, 0.50e^{j\frac{2\pi}{3}}\right)$
R_5	$\left(0.26e^{j\frac{2\pi}{7}}, 0.18e^{j\frac{2\pi}{8}}, 0.56e^{j\frac{2\pi}{3}}\right)$	$\left(0.47e^{j\frac{2\pi}{5}}, 0.34e^{j\frac{2\pi}{6}}, 0.19e^{j\frac{2\pi}{7}}\right)$	$\left(0.64e^{j\frac{2\pi}{2}}, 0.17e^{j\frac{2\pi}{8}}, 0.19e^{j\frac{2\pi}{8}}\right)$	$\left(0.82e^{j\frac{2\pi}{2}}, 0.10e^{j\frac{2\pi}{8}}, 0.08e^{j\frac{2\pi}{9}}\right)$

Table 4 The score value of e_{ij} and average solution $\breve{\mathcal{AV}}$

M°	C_1	C_2	C_3	C_4
R_1	0.1053	0.2893	0.1730	0.2875
R_2	0.1707	0.1864	0.0651	0.1738
R_3	0.0370	0.0492	0.2278	0.3432
R_4	0.0339	0.1837	0.0792	0.0533
R_5	0.0506	0.0376	0.1274	0.2113
$S(\breve{\mathcal{AV}})$	0.1085	0.0767	0.0761	0.1564

Table 5 The values of $\mathcal{P}\breve{\mathcal{D}}\mathcal{A}$ and $\mathcal{N}\breve{\mathcal{D}}\mathcal{A}$

$\mathcal{P}\breve{\mathcal{D}}\mathcal{A}$	C_1	C_2	C_3	C_4
R_1	0.0000	2.7726	1.2733	0.8379
R_2	0.5732	1.4306	0.0000	0.1115
R_3	0.0000	0.0000	1.9927	1.1942
R_4	0.0000	1.3962	0.0404	0.0000
R_5	0.0000	0.0000	0.6733	0.3508
$\mathcal{N}\breve{\mathcal{D}}\mathcal{A}$	C_1	C_2	C_3	C_4
R_1	0.0294	0.0000	0.0000	0.0000
R_2	0.0000	0.0000	0.1446	0.0000
R_3	0.6589	0.3580	0.0000	0.0000
R_4	0.6872	0.0000	0.0000	0.6591
R_5	0.5339	0.5092	0.0000	0.0000

Step 3. According to the decision matrix $M^\circ = \left[e_{ij}^l\right]_{5\times4}$ and expert's weight vector $\{\varpi_1 = 0.35, \varpi_2 = 0.2, \varpi_3 = 0.45\}$, we can derive the overall e_{ij}^l to e_{ij} using *CSFWAM** aggregation operators i.e. Eq. (10). The computing results are represented in Table 6.

Step 4. The value of average solution ($\breve{\mathcal{AV}}$) for all attributes are given in the Table 7.

Step 5. The positive distance from average ($\mathcal{P}\breve{\mathcal{D}}\mathcal{A}$) and negative distance from average ($\mathcal{N}\breve{\mathcal{D}}\mathcal{A}$) using Eq. (12) are listed in Tables 4 and 5

Step 6. The values of $\mathcal{S}\breve{\mathcal{P}}_i$ and $\mathcal{S}\breve{\mathcal{N}}_i$ that represent the weighted sum of $\mathcal{P}\breve{\mathcal{D}}\mathcal{A}$ and $\mathcal{N}\breve{\mathcal{D}}\mathcal{A}$, are calculated by Eq. (13) and the values are given as follows

$$\mathcal{S}\breve{\mathcal{P}}_1 = 1.1291, \mathcal{S}\breve{\mathcal{P}}_2 = 0.4254, \mathcal{S}\breve{\mathcal{P}}_3 = 0.9602, \mathcal{S}\breve{\mathcal{P}}_4 = 0.2655, \mathcal{S}\breve{\mathcal{P}}_5 = 0.3128$$

$$\mathcal{S}\breve{\mathcal{N}}_1 = 0.0073, \mathcal{S}\breve{\mathcal{N}}_2 = 0.0506, \mathcal{S}\breve{\mathcal{N}}_3 = 0.2292, \mathcal{S}\breve{\mathcal{N}}_4 = 0.3168, \mathcal{S}\breve{\mathcal{N}}_5 = 0.2251$$

Table 6 The aggregated CSF evaluation matrix

M°	C_1	C_2	C_3	C_4
R_1	$\left(0.5231e^{j2.0944}, 0.2031e^{j1.0290}, 0.2191e^{j1.2124}\right)$	$\left(0.6454e^{j2.6813}, 0.1526e^{j0.8453}, 0.1899e^{j1.1260}\right)$	$\left(0.5249e^{j2.2596}, 0.2818e^{j1.2942}, 0.1884e^{j0.9908}\right)$	$\left(0.6170e^{j2.5982}, 0.1807e^{j1.0281}, 0.1543e^{j1.0897}\right)$
R_2	$\left(0.5903e^{j2.2131}, 0.1520e^{j0.9018}, 0.2023e^{j1.0943}\right)$	$\left(0.5734e^{j2.0944}, 0.2226e^{j1.0630}, 0.1679e^{j0.9349}\right)$	$\left(0.4480e^{j1.1556}, 0.1798e^{j1.0056}, 0.4621e^{j2.1432}\right)$	$\left(0.6647e^{j1.9503}, 0.1344e^{j0.8780}, 0.2913e^{j2.0962}\right)$
R_3	$\left(0.5348e^{j1.7728}, 0.2410e^{j0.9832}, 0.2768e^{j1.4126}\right)$	$\left(0.5308e^{j1.7223}, 0.1799e^{j0.9770}, 0.2911e^{j1.1247}\right)$	$\left(0.6363e^{j2.2304}, 0.1226e^{j0.9224}, 0.3654e^{j2.6359}\right)$	$\left(0.6995e^{j2.2242}, 0.2004e^{j0.8939}, 0.0745e^{j0.8233}\right)$
R_4	$\left(0.4846e^{j1.6771}, 0.2445e^{j0.9621}, 0.2616e^{j1.2387}\right)$	$\left(0.5894e^{j2.3604}, 0.1846e^{j0.9770}, 0.2695e^{j2.2870}\right)$	$\left(0.5236e^{j1.8381}, 0.2746e^{j0.9601}, 0.2255e^{j1.1417}\right)$	$\left(0.4929e^{j1.2089}, 0.3012e^{j0.9621}, 0.3201e^{j2.0675}\right)$
R_5	$\left(0.3531e^{j0.9542}, 0.1875e^{j0.9978}, 0.5151e^{j2.4191}\right)$	$\left(0.5231e^{j1.5074}, 0.2131e^{j1.2029}, 0.2759e^{j1.1680}\right)$	$\left(0.6761e^{j1.3185}, 0.1493e^{j0.7983}, 0.2062e^{j1.2829}\right)$	$\left(0.7143e^{j1.6803}, 0.1658e^{j0.9258}, 0.1319e^{j1.6552}\right)$

Table 7 The value of average solution ($\breve{\mathcal{A}}\mathcal{V}$)

M°	C_1	C_2	C_3	C_4
$\breve{\mathcal{A}}\mathcal{V}$	$\begin{pmatrix} 0.5070e^{j0.5969}, 0.2026e^{j0.9738}, \\ 0.1931e^{j2.3756} \end{pmatrix}$	$\begin{pmatrix} 0.5759e^{j1.9980}, 0.1889e^{j1.0063}, \\ 0.2426e^{j1.3591} \end{pmatrix}$	$\begin{pmatrix} 0.5738e^{j1.6039}, 0.1911e^{j0.9835}, \\ 0.3039e^{j1.7811} \end{pmatrix}$	$\begin{pmatrix} 0.6483e^{j1.7936}, 0.1893e^{j0.9361}, \\ 0.2102e^{j1.8094} \end{pmatrix}$

Step 7. The normalized values of $\breve{\mathcal{SP}}_i$ and $\breve{\mathcal{SN}}_i$ are calculated by Eq. (14)

$$\breve{\mathcal{SP}}_1^{n*} = 1.0000, \breve{\mathcal{SP}}_2^{n*} = 0.3767, \breve{\mathcal{SP}}_3^{n*} = 0.8504, \breve{\mathcal{SP}}_4^{n*} = 0.2351, \breve{\mathcal{SP}}_5^{n*} = 0.2771$$

$$\breve{\mathcal{SN}}_1^{n*} = 0.9768, \breve{\mathcal{SN}}_2^{n*} = 0.8402, \breve{\mathcal{SN}}_3^{n*} = 0.2766, \breve{\mathcal{SN}}_4^{n*} = 0.0000, \breve{\mathcal{SN}}_5^{n*} = 0.2893$$

. **Step 8**. The appraisal score $\breve{\mathcal{SP}}_i$ for all alternatives are

$$\breve{\mathcal{AS}}_1 = 0.9884, \breve{\mathcal{AS}}_2 = 0.6085, \breve{\mathcal{AS}}_3 = 0.5635, \breve{\mathcal{AS}}_4 = 0.1176, \breve{\mathcal{AS}}_5 = 0.2832.$$

Step 9. The rank of alternatives according to the values of appraisal score $\breve{\mathcal{AS}}_i$ is listed as follows

$$\breve{\mathcal{AS}}_1 > \breve{\mathcal{AS}}_2 > \breve{\mathcal{AS}}_3 > \breve{\mathcal{AS}}_5 > \breve{\mathcal{AS}}_4; R_1 > R_2 > R_3 > R_5 > R_4$$

Similarly, we apply *CSFWGM** aggregation operator i.e. Eq. (11) for EDAS method. After executing all the steps involved in it, we calculate the appraisal score $\breve{\mathcal{AS}}_i$ for each alternatives as follows

$$\breve{\mathcal{AS}}_1 = 1.0000, \breve{\mathcal{AS}}_2 = 0.6068, \breve{\mathcal{AS}}_3 = 0.4752, \breve{\mathcal{AS}}_4 = 0.3166, \breve{\mathcal{AS}}_5 = 0.0856$$

Then, the rank of alternatives according to the values of $\breve{\mathcal{AS}}_i$ is

$$\breve{\mathcal{AS}}_1 > \breve{\mathcal{AS}}_2 > \breve{\mathcal{AS}}_3 > \breve{\mathcal{AS}}_4 > \breve{\mathcal{AS}}_5; R_1 > R_2 > R_3 > R_4 > R_5$$

4 Discussion

Further for comparative analysis and discussion, we have computed the EDAS method using ordinary spherical fuzzy aggregation operators such as spherical fuzzy weighted arithmetic mean aggregation operator (SFWAM) [6], spherical fuzzy weighted geometric mean aggregation operator (SFWGM) [6] and the appraisal scores have been calculated after executing all the steps involved in EDAS method for the same problem under analysis without considering phase terms. The detailed results are given in the Table 8 and also illustrated with graphical representation Fig. 1. According to the appraisal score values the alternatives are ranked.

From Table 9, we can see that the best alternative is R_1 and slight changes do exist in the ranking order. The CSF-EDAS method's results are consistent only for complex decision matrix. It would be very effective when the uncertainty that we account for is complex in nature having both the combination amplitude and phase terms, and we should not handle the terms separately.

Table 8 The appraisal score values of EDAS method

$\breve{AS}$	$CSFWAM^*$	$CSFWGM^*$	$SFWAM$	$SFWGM$
R_1	0.9884	1.0000	0.9476	1.0000
R_2	0.6085	0.6068	0.5790	0.4987
R_3	0.5635	0.4752	0.5692	0.2905
R_4	0.1176	0.3166	0.5370	0.5029
R_5	0.2832	0.0856	0.4700	0.4729

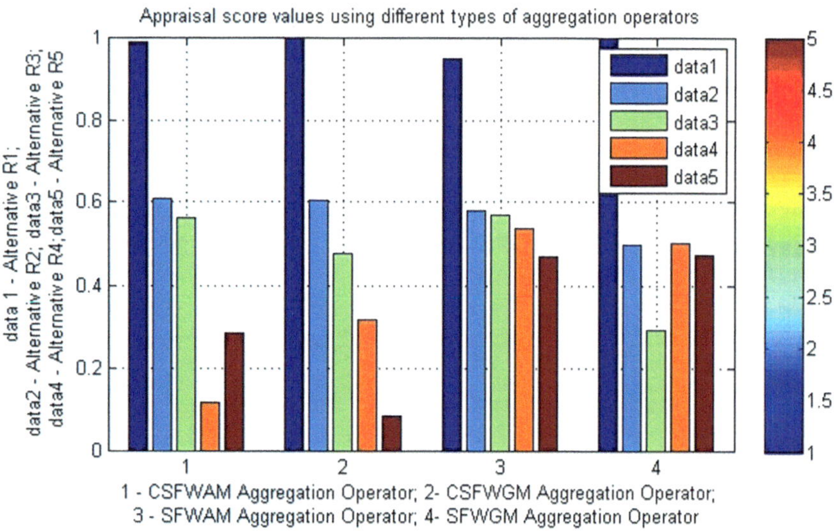

Fig. 1 Graphical Representation of appraisal score values

Table 9 The ranking order using EDAS method

Method	Ranking order
$CSFWAM^*$	$R_1 > R_2 > R_3 > R_5 > R_4$
$CSFWGM^*$	$R_1 > R_2 > R_3 > R_4 > R_5$
$SFWAM$ [6]	$R_1 > R_2 > R_3 > R_4 > R_5$
$SFWGM$[6]	$R_1 > R_4 > R_2 > R_5 > R_3$

5 Conclusions

In this chapter, the novel concept of complex spherical fuzzy sets has been introduced and also the limitations and structure of CSFSs are examined numerically. Particularly the geometrical properties of *CSFSs* are studied. The basic operations and some distance measures have been extended to CSFSs in order to develop intelligent systems for complex scenario. The main contributions of the chapter are:

1. The proposed novel concept of CSFSs is extended to basic addition and multiplication operations.
2. Based on the CSFSs, the aggregation operators are established to arithmetic and geometric mean.
3. With the help of some CSFSs operations an intelligent system is developed based on EDAS method.
4. Then the proposed intelligent system is applied to select the best catering services with adaption to aviation industry 4.0.
5. Comparative study has been made with existing operators and advantages and limitations have been discussed.

The proposed CSF-EDAS provides successful application in an intelligent system for selecting better catering service in aviation industry 4.0. Also it will be very efficient model in signal processing as well as in medical diagnosis due to its nature. The main advantage of the proposed model is that it is more applicable where there is complexity and periodicity.

More study needs to be done in the theory of CSFSs mainly in aggregation operators, various similarity, distance and, entropy measures. This research deliberates application in aviation industry 4.0 but it is not limited to it.

References

1. Zadeh, L.A.: Fuzzy sets. Inf. Control **8**, 338–353 (1965)
2. Atanassov, K.T.: Intuitionistic fuzzy sets. Fuzzy Sets Syst. **20**(1), 87–96 (1986)
3. Yager, R.R.: Pythagorean fuzzy subsets. In: Joint IFSA World Congress and NAFIPS Annual Meeting, Edmonton, Canada (2013)
4. Yager, R.R., Abbasov, A.M.: Pythagorean membership grades, complex numbers, and decision making. Int. J. Intell. Syst. **28**(5), 436–452 (2013)
5. Smarandache, F.: A Unifying Field in Logics, Neutrosophic Logic, Neutrosophy, Neutrosophic Set and Neutrosophic Probabilty, 4th edn. American Research Press, Rehoboth (1999)
6. Kutlu Gundogdu, F., Kahraman, C.: Spherical fuzzy sets and spherical fuzzy TOPSIS method. J. Intell. Fuzzy Syst. **36**(1), 337–352 (2019)
7. Kutlu Gundogdu, F., Kahraman, C.: A novel spherical fuzzy QFD method and its application to the linear delta robot technology development. Eng. Appl. Artif. Intell. **87**, (2020). https://doi.org/10.1016/j.engappai.2019.103348

8. Aydin, S., Kutlu Gundogdu, F.: Interval-valued spherical fuzzy MULTIMOORA method and its application to Industry 4.0. In: Kahraman, C., Kutlu Gündodu, F. (eds.) Decision Making with Spherical Fuzzy Sets. Studies in Fuzziness and Soft Computing, vol. 392. Springer, Cham (2021) . https://doi.org/10.1007/978-3-030-45461-6-13
9. Ashraf, S., Abdullah, S., Mahmood, T.: GRA method based on spherical linguistic fuzzy Choquet integral environment and its application in multi-attribute decision-making problems. Math. Sci. **12**(4), 263–275 (2018)
10. Ashraf, S., Abdullah, S., Mahmood, T., Ghani, F., Mahmood, T.: Spherical fuzzy sets and their applications in multi-attribute decision making problems. J. Intell. Fuzzy Syst. **36**, 2829–2844 (2019)
11. Ashraf, S., Abdullah, S.: Spherical aggregation operators and their application in multi-attribute group decision-making. Int. J. Intell. Syst. **34**(3), 493–523 (2019)
12. Jin, H., Ashraf, S., Abdullah, S.: Spherical fuzzy logarithmic aggregation operators based on entropy and their application in decision support systems. Entropy **21**(7), 628 (2019)
13. Jin, H., Ashraf, S., Abdullah, S., Qiyas, M., Bano, M., Zeng, S.: Linguistic spherical fuzzy aggregation operators and their applications in multi-attribute decision making problems. Mathematics **7**(5), 413 (2019)
14. Raq, M., Ashraf, S., Abdullah, S., Mahmood, T., Muhammad, S.: The cosine similarity measures of spherical fuzzy sets and their applications in decision making. J. Intell. Fuzzy Syst. **36**, 6059–6073 (2019)
15. Ashraf, S., Abdullah, S., Mahmood, T.: Spherical fuzzy Dombi aggregation operators and their application in group decision making problems. J. Ambient Intell. Human. Comput. 1–19 (Preprint) (2019)
16. Ramot, D., Milo, R., Friedman, M., Kandel, A.: Complex fuzzy sets. IEEE Trans. Fuzzy Syst. **10**, 2 (2002). https://doi.org/10.1109/91.995119
17. Bi, L., Dai, S., Hu, B.: Complex fuzzy geometric aggregation operators. Symmetry **10**(7), 251 (2018)
18. Bi, L., Dai, S., Hu, B.: Complex fuzzy arithmetic aggregation operators. J. Intell. Fuzzy Syst. **36**(3), 2765–2771 (2019)
19. Hu, B., Bi, L., Dai, S.: Complex fuzzy power aggregation operators. Math. Prob. Eng. (2019). https://doi.org/10.1155/2019/9064385
20. Alkouri, A.S., Salleh, A.R.: Complex intuitionistic fuzzy sets. AIP Conf. Proc. **1482**, 464 (2012). https://doi.org/10.1063/1.4757515
21. Ullah, K., Mahmood, T., Ali, Z.: On some distance measures of complex Pythagorean fuzzy sets and their applications in pattern recognition. Complex Intell. Syst. **6**(1) (2020)
22. Ali, M., Smarandache, F.: Complex neutrosophic set. Neural Comput. Appl. **28**, 1817–1834 (2017). https://doi.org/10.1007/s00521-015-2154-y
23. Ali, Z., Mahmood, T., Yang, M.S.: Complex T-spherical fuzzy aggregation operators with application to multi-attribute decision making. Symmetry **12**, 1311 (2020). https://doi.org/10.3390/sym10060193
24. Mahmood, T., Ullah, K., Khan, Q.: An approach toward decision-making and medical diagnosis problems using the concept of spherical fuzzy sets. Neural Comput. Appl. **31**, 7041–7053 (2019). https://doi.org/10.1007/s00521-018-3521-2
25. Kutlu Gundodu, F., Kahraman, C., Karaan, A.: Spherical fuzzy VIKOR method and its application to waste management. In: Kahraman, C., Cebi, S., Cevik Onar, S., Oztaysi, B., Tolga, A., Sari, I. (eds.) Intelligent and Fuzzy Techniques in Big Data Analytics and Decision Making. INFUS 2019. Advances in Intelligent Systems and Computing, vol. 1029. Springer, Cham (2020). https://doi.org/10.1007/978-3-030-23756-1-118
26. Keshavarz Ghorabaee, M., Zavadskas, E.K., Olfat, L., Turskis, Z.: Multi-criteria inventory classification using a new method of evaluation based on distance from average solution (EDAS). Informatica **26**, 435–451 (2015)
27. Keshavarz Ghorabaee, M., Zavadskas, E.K., Amiri, M., Turskis, Z.: Extended EDAS method for fuzzy multi-criteria decision-making: an application to supplier selection. Int. J. Comput. Commun. Control **11**, 358–371 (2016). https://doi.org/10.15837/ijccc.2016.3.2557

28. Kahraman, C., Keshavarz Ghorabaee, M., Zavadskas, E.K., Onar, S.C., Yazdani, M., Oztaysi, B.: Intuitionistic fuzzy EDAS method: an application to solid waste disposal site selection. J. Environ. Eng. Landscape Manage. **25**, 1–12 (2017). https://doi.org/10.3846/16486897.2017.1281139
29. Buyukozkan, G., Feyzioglu, O., Havle, C.A.: Analysis of success factors in aviation 4.0 using integrated intuitionistic fuzzy MCDM methods. In: Kahraman, C., Cebi, S., Cevik Onar, S., Oztaysi, B., Tolga, A., Sari, I. (eds.) Intelligent and Fuzzy Techniques in Big Data Analytics and Decision Making. INFUS 2019. Advances in Intelligent Systems and Computing, vol. 1029. Springer, Cham (2020). https://doi.org/10.1007/978-3-030-23756-1-73
30. Oztemel, E., Gursev, S.: Literature review of Industry 4.0 and related technologies. J. Intell. Manuf. **31**, 127–182 (2020). https://doi.org/10.1007/s10845-018-1433-8
31. Barbosa, G.F., Aroca, R.V.: Advances of Industry 4.0 concepts on aircraft construction: an overview of trends. J. Steel Struct. Constr. **3**, 125 (2017). https://doi.org/10.4172/2472-0437.1000125

Digitalization on Aviation 4.0: Designing a Scikit-Fuzzy Control System for In-Flight Catering Customer Satisfaction

Esra Çakır and Ziya Ulukan

Abstract The fourth industrial revolution is a vision that deeply affects the aviation industry in terms of cloud computing, artificial intelligence, digital transformation, and the internet of things. Although aviation 4.0 is at the beginning of its development, great competition begins to design suitable systems. Airline companies need well-designed software to use these digital transformation processes for data management and to receive customers' feedback to interpret them in the most suitable way. They should also develop programs that can be adapted to theirs pre-flight and in-flight systems to measure customer satisfaction and interpret crisp, linguistic, or fuzzy data. In this study, the catering service evaluation system for customers is designed and an interface is developed to be used in the aviation 4.0 industry. The study contributed to the literature by proposing a new fuzzy control system for catering customer evaluation and by designing an interface and coding system with a special SciKit fuzzy logic toolbox on Python environment. The implementation and the design of the system have been simulated several times and the results are evaluated by graphical and visual means.

Keywords Aviation 4.0 · Customer satisfaction measurement · Digitalization · Fuzzy control system design · In-flight catering evaluation · Python programming · SciKit fuzzy

1 Introduction

Three industrial revolutions have taken place in history. In the first industrial revolution called Industry 1.0, steam engines were used and production was increased. In the second industrial revolution (Industry 2.0) electrical energy was used. In the third

E. Çakır (✉) · Z. Ulukan
Department of Industrial Engineering, Galatasaray University, Istanbul, Turkey
e-mail: ecakir@gsu.edu.tr

Z. Ulukan
e-mail: zulukan@gsu.edu.tr

C. Kahraman and S. Aydın (eds.), *Intelligent and Fuzzy Techniques in Aviation 4.0*, Studies in Systems, Decision and Control 372,
https://doi.org/10.1007/978-3-030-75067-1_6

industrial revolution (Industry 3.0), production systems turned from analog to digital systems. Finally, with Industry 4.0, the world has come to the brink of a new industrial revolution. It is predicted that high technology and computerization tendency will increase and high efficiency in production will be achieved [1]. Industry 4.0 is the process of integrating the strengths of traditional methods with the internet, information and communication technologies. It is aimed to realize self-managed smart production systems. However, in order to establish this production system, some components must come together and operate in a holistic way [2]. The key components of Industry 4.0 can be listed as 3D printers, big data management, cloud computing systems, artificial intelligence and machine learning, internet of things, augmented reality, RFID, industrial robots and human machine interfaces [3]. With these features, Industry 4.0 allows companies with productivity, standardization, ergonomics, cleanliness, maintenance in order to connect digital and real life [4].

All sectors are affected by the technology revolution of industry 4.0. The aviation industry, which has shown the greatest development in the last 100 years, is undoubtedly one of them. Industry 4.0 components are used in the business processes of the airline industry and provide benefits to businesses in many ways. Industry 4.0 applications that find their place in the aviation sector (called Aviation 4.0) in various ways can be listed as: direction or way finding application at the airport [5], cargo arrangement with square code in ground handling services [6], applications that analyze the mental state of the customer in cabin services [7], thanks to facial recognition, check-in and boarding procedures without showing passport or ID [8], flight training applications and simulations with artificial intelligence [9] etc.

With the rapid increase in technological opportunities, airlines must compete not only in terms of technology but also in terms of customer satisfaction. Aviation firms make great efforts to determine to what extent customers are satisfied by using various tools. Therefore, many methods are used, from focus groups to quantitative surveys. Companies must keep up with the Aviation 4.0 era by adapting their all procedures using artificial intelligence, IoT, big data, cloud computing and other software.

In addition to the price, the experience during the flight is also an effective factor for the customer to choose the airline company. Reliable and high-quality service naturally attracts the customer. In this sense, customer feedback has an important place in the competition between companies. Catering services are an important selection criterion, especially on long journeys. King [10] states that some passengers are willing to change airlines, change travel patterns, and even pay more for high-quality in-flight meals/food served. Customer satisfaction is generally seen as a multi-dimensional structure of the same dimensions that constitute the service quality [11]. The studies [12] and [13] show that the service quality offered by airlines has a significant effect on customer satisfaction and this significantly affects customer loyalty.

Airline service quality in a Taiwanese airline are evaluated and researchers found that reliability and assurance were the first important dimensions, responsiveness was the second, empathy was the third, concrete things and flight model [14]. Other aspects of airline services include safety, handling customer complaints, courtesy of the crew, timely departure and arrival, comfort and cleanliness of seats, flexibility, friendliness, and integrity [15] In addition, nine evaluation criteria of service quality in airline companies are suggested and it is found that passengers are more interested in the comfort, interior decoration and services of airline companies [16]. Some other related research on in-flight catering and customer satisfaction can be listed as [17–21].

Studies in the literature show that measuring customer satisfaction is a multi-criteria decision-making problem. In recent years, a new approach has been developed in measuring customer satisfaction with the successful integration of fuzzy logic in multi-criteria decision-making systems. Fuzzy sets are useful to handle imprecision. Therefore, analysts may be interested in analyzing customer satisfaction surveys using a fuzzy rule-based decision support system to implement customer's vague statements [22]. One of first usage of fuzzy rule-based customer satisfaction is system developed for evaluating the financial management of state governments [23]. This success was followed by the financial management and capital management of major US cities, systems for generating bond notes for US cities and US public airports, and for assessing the financial health of public schools [24]. Knowledgebase fuzzy models have also been applied in some business and management disciplines [25].

These fuzzy rule-based control systems that are used successfully in the context of performance evaluation are also suitable for evaluating customer satisfaction.

In 2011, Pedregosa et al. [26] developed a Python module to handle fuzzy systems in software environment. Scikit-learn is a Python module which implements a wide range of state-of-the-art machine learning algorithms for supervised and unsupervised medium-scale problems. This toolkit aims to introduce machine learning to non-specialists using a high-level, general-purpose language [26]. Many fuzzy rule-based models can be coded in Python environment and these applications can be easily implemented into existing systems.

Thanks to the wide functioning of fuzzy logic and Scikit-Fuzzy module, models can be differentiated by adding features according to the subject. Considering the applicability of the methods described above, it is necessary to develop a new fuzzy rules based model and its graphical interface which is more effective and user-friendly while enabling the customer to convey information in a funny way for the evaluation of in-flight catering services in accordance with the age of Aviation 4.0.

The contribution of this paper to existing literature is to represent a fuzzy control system and an interface to measure customer satisfaction level on in-flight catering to understand the customer's expectation and measure its prestige. Proposed system is designed with Scikit Fuzzy toolkit which is a module that enables fuzzy logic operations and Tkinter [27] module that allows to create widgets and add features in Python environment. An application has been proposed for the aviation industry using these modules for the first time in the literature.

The rest of the paper is organized as follows. Section 2 gives preliminaries of fuzzy sets and introduce fuzzy control system design module. Proposed methodology is applied on in-flight catering customer level measurement in Sect. 3. Section 4 concludes the paper with further research.

2 Proposed Methodology

Scikit-Fuzzy is a collection of fuzzy logic algorithms intended for use in the SciPy Stack, written in the Python computing language. This SciKit is developed by the SciPy community [26]. Scikit-learn exposes a wide variety of machine learning algorithms, both supervised and unsupervised, using a consistent, task-oriented interface, thus enabling easy comparison of methods for a given application. Since it relies on the scientific Python ecosystem, it can easily be integrated into applications outside the traditional range of statistical data analysis. This package implements many useful tools and functions for computation and projects involving fuzzy logic.

In 1965, fuzzy logic is introduced by Zadeh [28]. It is a methodology predicated on the idea that the "truthiness" of something can be expressed over a continuum. This is a way to say that something is not exactly *true* or *false* but instead *partially true* or *partially false*. A fuzzy variable has a crisp value which takes on some number over a pre-defined domain (or universe). Zadeh classified a membership of an object between "0" and "1". In other words, an object contains that property with a membership between "0" and "1".

Definition 1 [28] A fuzzy set is characterized by a membership function mapping the elements of a domain, space or universe of discourse X to the unit interval [0, 1]. A fuzzy set $\tilde{A}$ in a universe of discourse X is defined as the following set of pairs: $\tilde{A}$ {(x, μ(x); x ∈ X}. μ: X → [0, 1] is a mapping called the degree of membership function of the fuzzy set $\tilde{A}$ and μ(x) is called the membership value of x ∈ X in the fuzzy set $\tilde{A}$. These membership grades are often represented by real numbers ranging from [0, 1].

Definition 2 [28] A fuzzy number $\tilde{A}$ is a fuzzy set on the real line R, must satisfy the following conditions:

- μ(x) is piecewise continuous.
- There exists at least one x ∈ R with μ(x) = 1.
- $\tilde{A}$ must be normal and convex.

Definition 3 [28] Triangular Fuzzy Number is defined as $\tilde{A}$ = {a, b, c}, where all a, b, c are real numbers and its membership function is given below.

$$\mu_{\tilde{A}}(x) = \begin{cases} \frac{(x-a)}{(b-a)} & for\ a \le x \le b \\ \frac{(c-x)}{(c-b)} & for\ b \le x \le c \\ 0 & otherwise \end{cases}$$

Definition 4 [28] A fuzzy set $\tilde{A}$ = (a, b, c, d) is said to trapezoidal fuzzy number if its membership function is given by where a ≤ b ≤ c ≤ d.

$$\mu_{\tilde{A}}(x) = \begin{cases} 0 & for\ x \leq a \\ \frac{(x-a)}{(b-a)} & for\ a \leq x \leq b \\ 1 & for\ b \leq x \leq c \\ \frac{(d-x)}{(d-c)} & for\ c \leq x \leq d \\ 0 & for\ x \geq d \end{cases}$$

A fuzzy variable also has several terms that are used to describe the variable. The terms taken together are the fuzzy set which can be used to describe the "fuzzy value/or fuzzy linguistic" of a variable. These terms are usually adjectives like "low," "medium," and "high." In essence, it describes "how connected to being high" something is [29].

A fuzzy control system links fuzzy variable using a set of rules. These rules are simply mappings that describe how one or more fuzzy variables relates to another. These are expressed in terms of an IF–THEN statement; the IF part is called the antecedent and the THEN part is the consequent [29].

Below is explained with a simple example how to set up a fuzzy control system using SciKit (a.k.a. skfuzzy) module. In the example, firstly, two parameters (i.e. technology used in the computer and memory capacity) are selected to measure the performance of a computer. Thus, ("technology used in the computer" and "memory capacity" are antecedents (Inputs) of fuzzy control system. The consequent (Output) is "performance". Note that the output does not have to be a single parameter. The inputs and outputs have two features, one of them is their universe and the other one is their fuzzy linguistic variables. Note that the fuzzy set used in this study consists of triangular and trapezoidal fuzzy numbers. Additionally, the skfuzzy module provides a lot of fuzzy set option such as Gaussian, sigmoid, S-function, Z-function, interval type 2 etc.

- **Antecedents (Inputs)**
  - **Technology used in computer**

    Universe: How high is the technology used in computer, on a scale of 1 to 10?
    Fuzzy set/linguistic variables: *low [0, 0, 5], medium [0, 5, 10], high [5, 10, 10]*

  - **Memory capacity**

    Universe: How much is the memory capacity of the computer, on a scale of 1 to 10?
    Fuzzy set/linguistic variables: *alittle [0, 0, 5], average [0, 5, 10], alot [0, 10, 10]*

- **Consequents (Outputs)**
  - **Performance**

    Universe: How great is the performance of the computer? on a scale of 1 to 100?
    Fuzzy set/linguistic variables: *bad [0, 20, 55], good [30, 60, 75], great [60, 90, 100].*

Secondly, fuzzy rules are set for these inputs and outputs. In this control system, one rule might be "IF the technology used in computer is "*high*" or memory capacity is "*a lot*" THEN the performance will be "*great*". The exact math related to how a rule is used to calculate the value of the consequent based on the value of the antecedent is outside the scope of this research. Three rules constructed for this example are listed as follows.

- **Rules**
  - IF the technology used in computer is "*high*" or memory capacity is "*a lot*" THEN the performance will be "*great*".
  - IF the technology used in computer is "*average*" THEN the performance will be "good".
  - IF the technology used in computer is "*low*" or memory capacity is "*a little*" THEN the performance will be "*bad*".

After all rules are defined, this information is given as a parameter to the simulation methods in the skyfuzz module as a list. Thus, the system is simulated for each new input received from the user. For example, if a user wants to get the rate of the performance of a computer, when s/he enters technology property as 7.1 and memory property 5.9, the program calculates a ratio of 64.807 for performance according to the determined rules.

To view a relational digraphs of fuzzy rules, "skyfuzz" module [29] provides an option with "view()" method and "matplotlib.pyplot" module [30] is useful to make graphs visible. The illustrative fuzzy membership functions also can be represented by this way. Additionally, the "numpy" module [31] is useful for mathematical operations, matrices and complex numbers. Finally, "Tkinter" [27] is an important module for interface design on Python. Widgets can be prepared according to the concept and customization can be made on the text, picture, color, windows etc.

3 Application

In this section, a software prepared to measure customer satisfaction on in-flight catering service is introduced. The content of the proposed fuzzy control system is explained, several simulations are made, and the results are visualized.

Table 1 Criteria and their fuzzy linguistic variables

Criteria	Linguistic variables		
Food quality	**bad** [0, 10, 30, 40]	**standard** [25, 45, 60, 70]	**exceptional** [40, 65, 90, 100]
Service quality	**insufficient** [0, 10, 20, 30]	**sufficient** [20, 40, 60]	**suitable** [40, 60, 80, 100]
Hygiene quality	**less** [0, 10, 30, 50]	**average** [30, 60, 60, 80]	**very** [70, 80, 90, 100]
Service personnel behavior	**fair** [0, 20, 50]	**good** [40, 60, 80]	**excellent** [60, 80, 100]
Product variety	**a little** [0, 10, 15, 25]	**well** [15, 25, 50]	**a lot** [25, 50, 75, 100]
Service scheduling	**disordered** [0, 10, 15, 25]	**regular** [15, 25, 50]	**adequate** [25, 50, 75, 100]
Timing	**late** [0, 10, 20, 40]	**fine** [20, 50, 70]	**punctual** [40, 70, 90, 100]
Accessibility	**few** [0, 10, 20, 50]	**enough** [20, 60, 80]	**much** [50, 60, 80, 100]

Aviation 4.0 aims at high technology and customization in technical equipment. In this sense, new software, algorithms, fast processing machines are necessary for the rapidly growing sector. At the same time, airline companies should consider customer feedback while keeping up with sectoral growth. At this point it is very important to collect and process information. Therefore, customers' feedback is valuable at every stage of airline usage.

In the application, a software in Python environment has been developed to evaluate the in-flight catering services that may affect the customer's preferences, especially on long distance flight. The most important feature of this software is to be able to receive and process imprecise opinions from the customer. For this reason, SciKit fuzzy control toolbox has been imported into the software. An interface is created by looking at the rules of the module and the content of the catering services and a fuzzy control system is prepared. This program is also a useful software that can be easily integrated into aircraft in-flight entertainment systems.

To develop this fuzzy control system, firstly, five experts contributes to determine criteria and their fuzzy membership functions on development fuzzy control system for in-flight catering customer satisfaction measurement to use aviation 4.0 industry. They have argued and decided eight criteria as in Table 1 (Figs. 1, 2, 3, 4, 5, 6, 7 and 8).

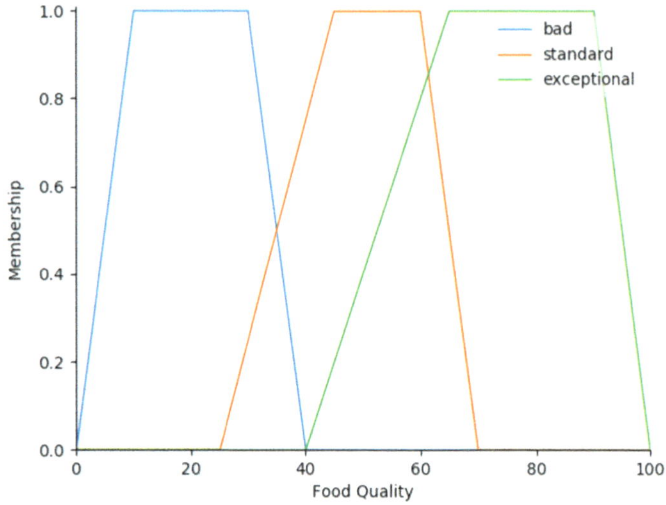

Fig. 1 Illustration of "Food Quality" criterion fuzzy membership function

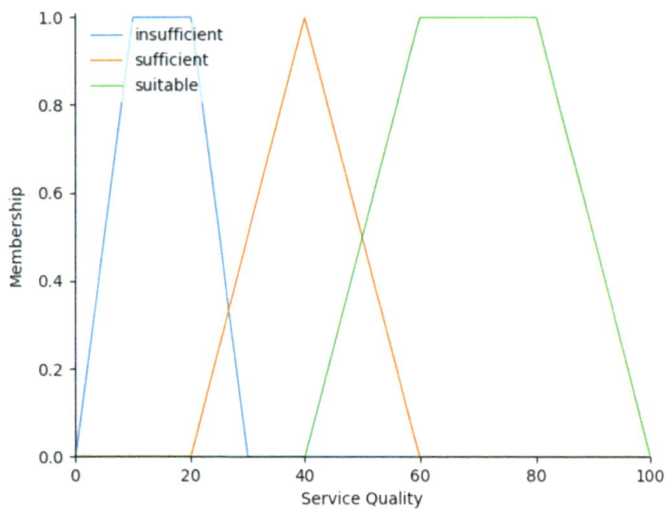

Fig. 2 Illustration of "Service Quality" criterion fuzzy membership function

The calculation satisfaction level is the goal of the system. This result is also expressed by fuzzy linguistic. Therefore, at the beginning of the program, this output should be determined, and its fuzzy membership function is proposed by experts as in Fig. 9.

According to determined criteria and their features, the fuzzy rules have been created by experts for fuzzy control system to measure in-flight catering customer satisfaction level. Five rules are listed as follows:

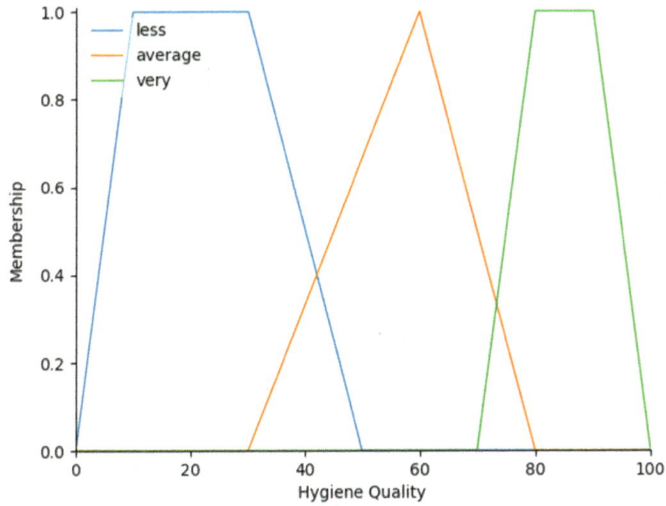

Fig. 3 Illustration of "Hygienne Quality" criterion fuzzy membership function

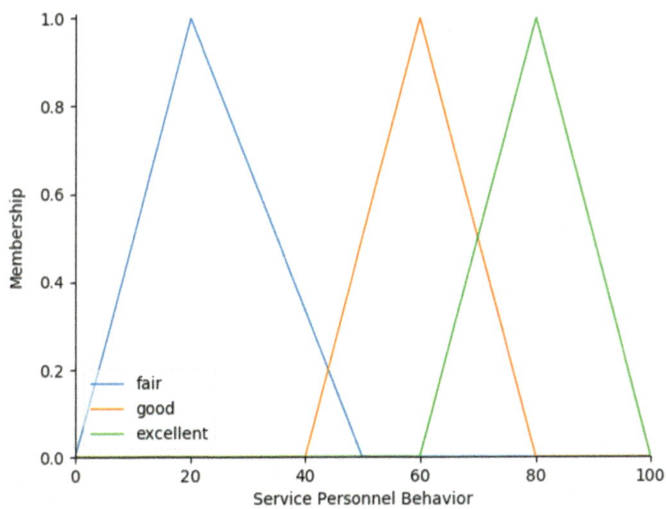

Fig. 4 Illustration of "Service Personnel Behavior" criterion fuzzy membership function

- Rule 1: (food['exceptional'] & service['suitable'] & hygiene['very'] & behavior['excellent'] & variety['alot'] & scheduling['adequate'] & timing['punctual'] & accessibility['much'], level['veryhigh'])
- Rule 2: (food['exceptional'] & service['suitable'] & hygiene['very'] & behavior['good'] & variety['alot'] & scheduling['disordered'] & timing['fine'] & accessibility['few'], level['high'])

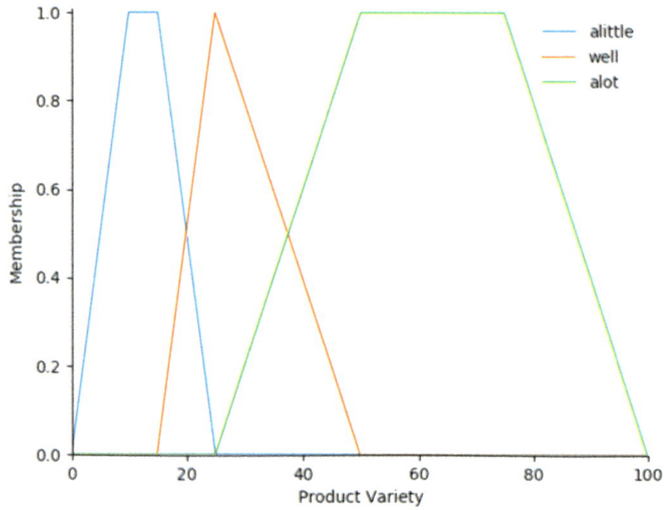

Fig. 5 Illustration of "Product Variety" criterion fuzzy membership function

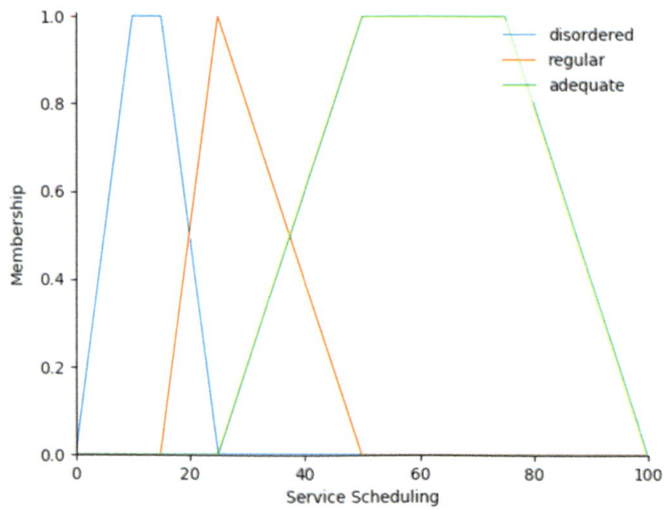

Fig. 6 Illustration of "Service Scheduling" criterion fuzzy membership function

- Rule 3: (food['standard'] & service['sufficient'] & hygiene['average'] & behavior['good'] & variety['well'] & scheduling['regular'] & timing['fine'] & accessibility['enough'], level['medium'])
- Rule 4: (food['bad'] & service['sufficient'] & hygiene['average'] & behavior['fair'] & variety['alittle'] & scheduling['regular'] & timing['punctual'] & accessibility['enough'], level['low'])

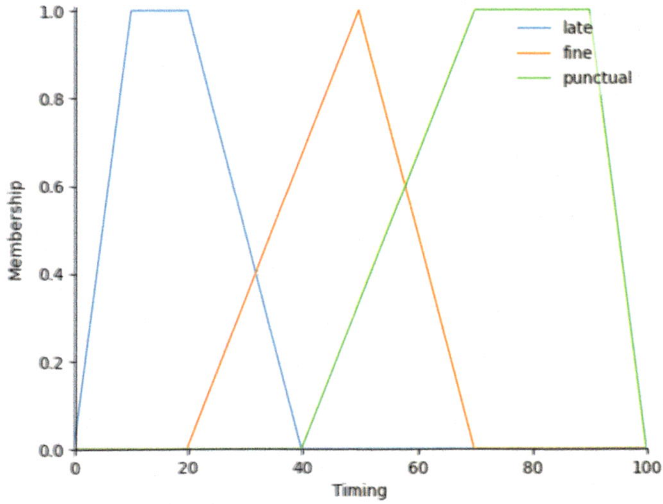

Fig. 7 Illustration of "Timing" criterion fuzzy membership function

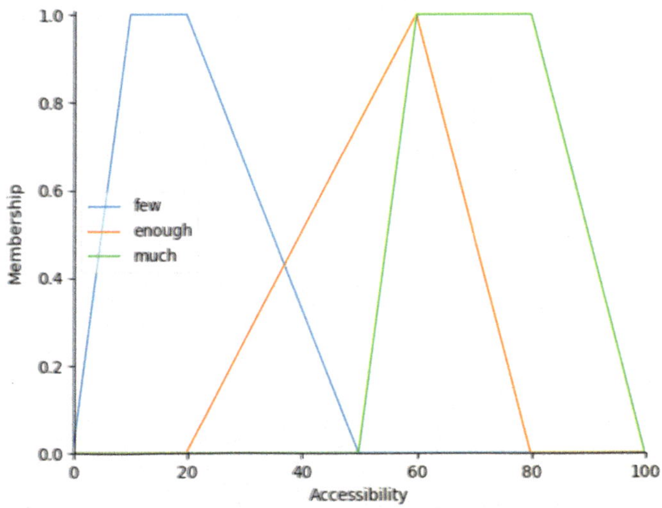

Fig. 8 Illustration of "Accessibility" criterion fuzzy membership function

- Rule 5: (food['exceptional'] & service['suitable'] & hygiene['very'] & behavior['good'] & variety['alot'] & scheduling['disordered'] & timing['fine'] & accessibility['few'], level['high']).

Note that by using eight criteria with three linguistic variables as in Table 1, one result with five linguistic variables as in Fig. 9 and three different operators such as "AND", "OR", "NOT", more than fifty million rules can be written for this fuzzy

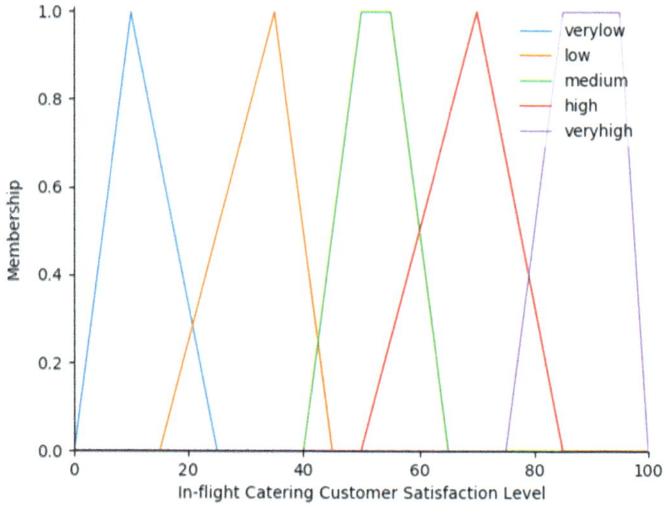

Fig. 9 Illustration of in-flight catering customer satisfaction level fuzzy membership function

control system. Above are the rules of evaluations used only in simulations for this paper. In fact, about one hundred rules have been written for this program. These rules can be increased depending on the computational sensitivity of the company. The relational digraph of five rules are illustrated as in Fig. 10, 11, 12, 13 and 14. Nodes represent eight criteria with their fuzzy variables and customer satisfaction level with its fuzzy variable.

An interface is designed using the Python programming language to gather opinions from customers. This interface can be easily implemented in the in-flight entertainment system. Interface features can be customized according to the logo, color and other visuals of the airline company to be selected. Depending on the airline's

Fig. 10 The relational digraph of rule 1

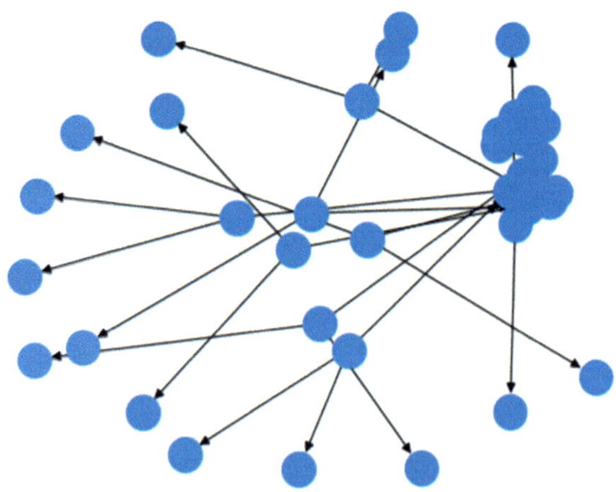

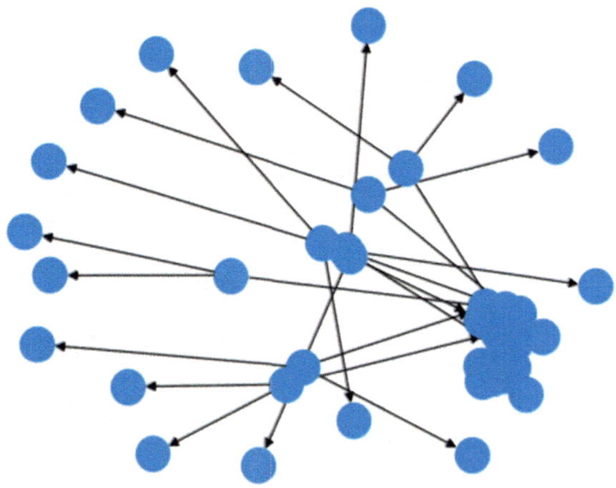

Fig. 11 The relational digraph of rule 2

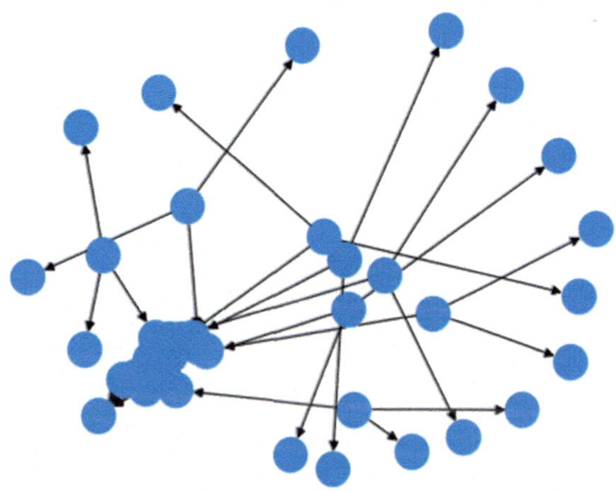

Fig. 12 The relational digraph of rule 3

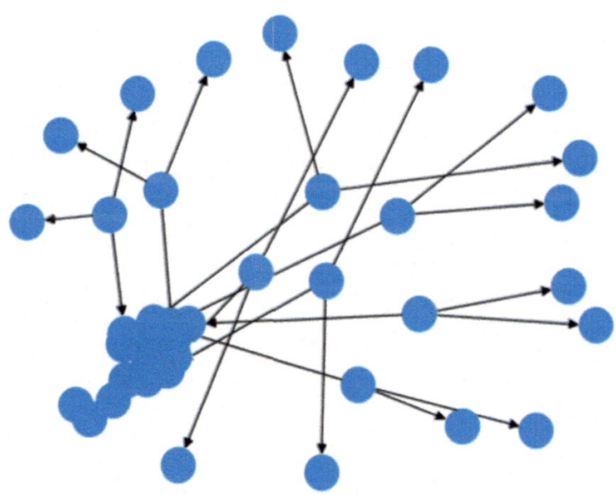

Fig. 13 The relational digraph of rule 4

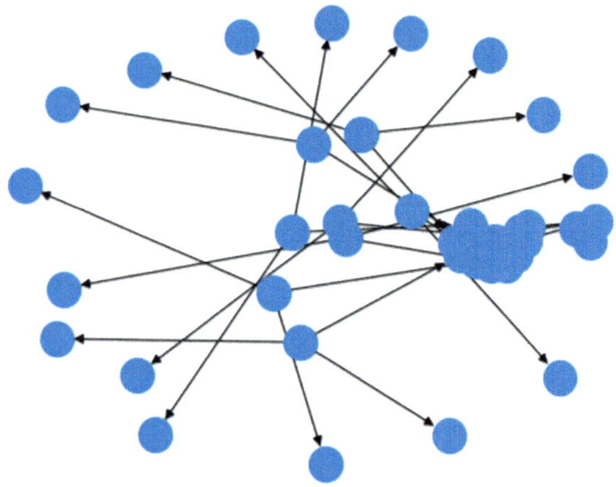

Fig. 14 The relational digraph of rule 5

system policy, a pop-up can be opened on the screen or the customer can find this section at his own request. In Fig. 15, an example of a designed interface is shown.

Using this interface, the customer can evaluate the in-flight catering service without any instruction. As in Fig. 15, there are eight criteria and their evaluation scale bar with fuzzy linguistic expressions. For example, in order to evaluate "Food Quality" of the in-flight catering service, customer express his/her opinion by moving the bar. The scale bar contains numbers from 0.0 to 100.0. The user can also see the fuzzy expressions and decide how close s/he feels to which situation. The customer may think the food quality is "standard" and decide which side is the closer "standard" by moving the bar.

The point to be noted is that the customer is not informed of the intervals in which fuzzy variables are defined. This is the system's own knowledge and system does the calculation accordingly. However, the customer can act as if s/he gives a score over 100, or s/he can express own closeness to linguistic expressions with the help of a score bar. Membership degrees are definitions made only for the back-end design of the constructed fuzzy control system.

After making the decision for eight criteria, the customer presses the "Submit" button and the satisfaction level is calculated according to the determined fuzzy rules. The overall score appears at the bottom of the interface.

The score is also printed for the developer in the interactive shell. If the customer has made an evaluation that return to be zero during the defuzzification phase, this situation is captured by try-except block used in the program and printed to both the interface for customer knowledge as "*Your overall score as 0.0. Thank you.*" and the interactive shell for the developer knowledge as "*Customer Satisfaction Level can not be computed. It is 0.0!*" with error explanation. This situation can occur in three ways.

1. In the first scenario, the user scored all criteria with a score of 0.0. Therefore, 0.0 is obtained in the defuzzification section. This is the opinion of the user and does not require intervention.

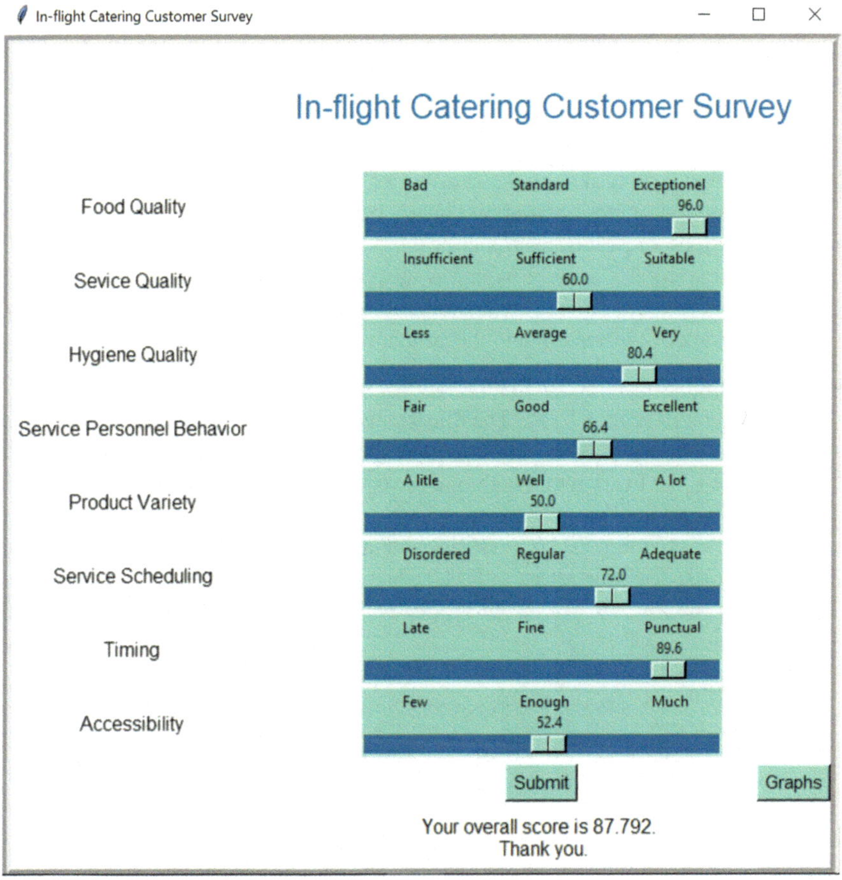

Fig. 15 In-flight customer survey interface

2. In the second scenario, an evaluation is made that does not comply with the fuzzy rules defined for the system on the back-end. In fact, this situation occurs when the program returns 0.0 in the defuzzification section as in the first scenario. But in this case the definition of the rule is missing. Then, this exception is a vulnerability and requires developer intervention. The fuzzy rule definition should be made by the system designer as not to validate any wrong evaluation received from the customer.
3. In the third scenario, two or more rules intersect. For this reason, the system does not decide according to which rule to calculate and prints the result as 0.0. As in second scenario, it is seen that a good system design is important. The developer of the system should fix these errors. It was previously tagged that over fifty million rules could be written for this program which means that missing rules can exist. The number of rules can be increased according to the company's sensitivity in calculations.

The "Graphs" button has been added to provide the researchers and developers with visuals of the transactions. For customer use only, this button can be removed from the front-end design. It calls the "graphs()" function defined on the back-end and displays the visuals of the definition of fuzzy membership degrees of the eight criteria and satisfaction level, the digraphs showing the relations of fuzzy rules and the visual form of the fuzzy score obtained as a result of customer evaluation, respectively. If the two exceptions mentioned above have occurred, the visual form of the fuzzy score obtained as a result of customer evaluation is not appear.

In this fuzzy control system, which is introduced for in-flight catering customer satisfaction measurement, different evaluations are requested from five experts as decision makers of the system and the simulation results are as follows:

- In first simulation, decision maker 1 (DM 1) made the evaluation as in the Fig. 16 with the overall score 87.792. The illustration of DM 1 customer satisfaction level fuzzy membership is in the Fig. 17. As in the figure, according to the determined fuzzy rule 1, the fuzzy membership of the overall level belongs to fuzzy linguistic variable "very high".
- In second simulation, DM 2 made the evaluation as in the Fig. 18 with the overall score 67.647. The illustration of DM 2 customer satisfaction level fuzzy membership is in the Fig. 19. As in the figure, according to the determined fuzzy rule

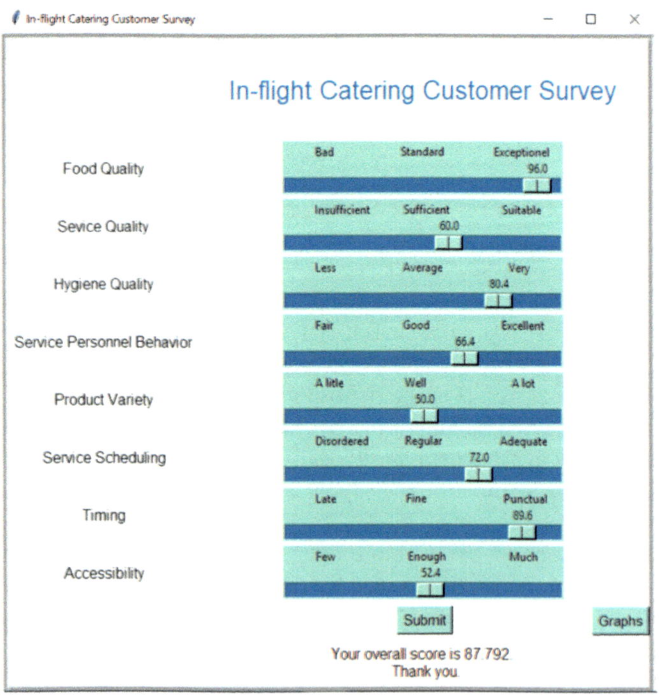

Fig. 16 In-flight customer survey of DM 1 with score 87.792

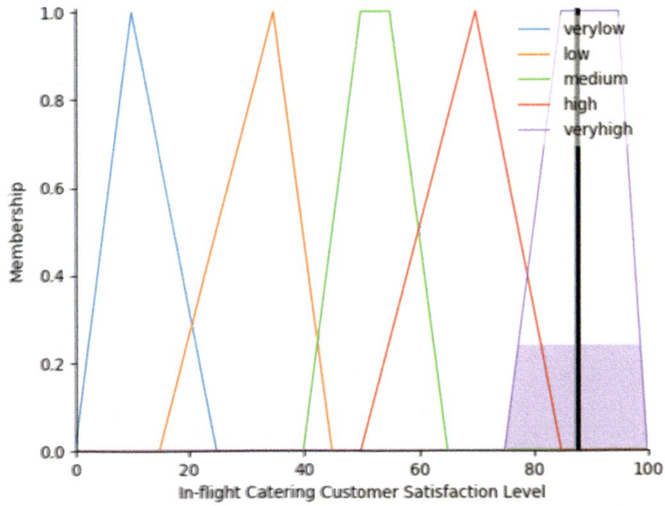

Fig. 17 Illustration of DM 1 customer satisfaction level fuzzy membership

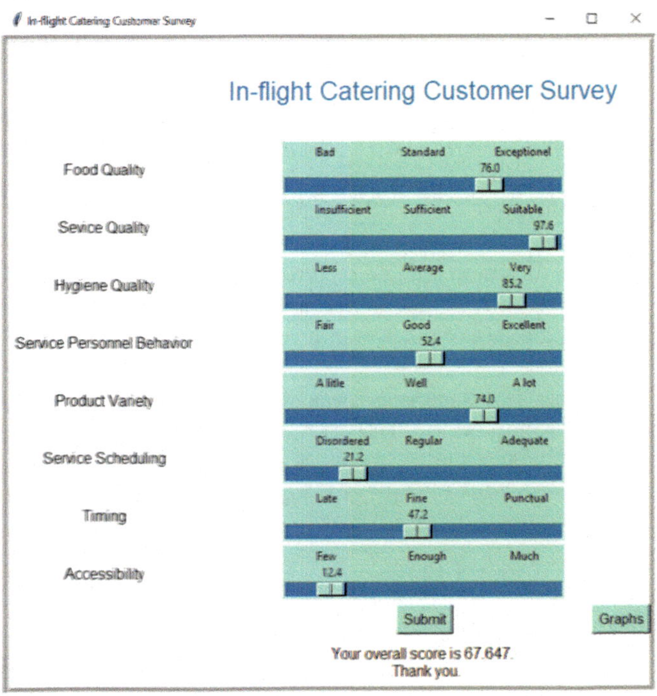

Fig. 18 In-flight customer survey of DM 2 with score 67.647

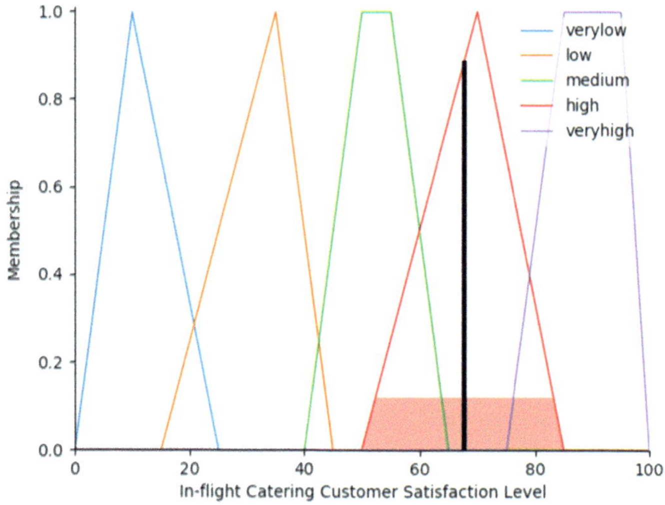

Fig. 19 Illustration of DM 2 customer satisfaction level fuzzy membership

2, the fuzzy membership of the overall level belongs to fuzzy linguistic variable "high".

- In third simulation, DM 3 made the evaluation as in the Fig. 20 with the overall score 52.5. The illustration of DM 3 customer satisfaction level fuzzy membership is in the Fig. 21. As in the figure, according to the determined fuzzy rule 3, the fuzzy membership of the overall level belongs to fuzzy linguistic variable "medium".
- In forth simulation, DM 4 made the evaluation as in the Fig. 22 with the overall score 30.05. The illustration of DM 4 customer satisfaction level fuzzy membership is in the Fig. 23. As in the figure, according to the determined fuzzy rule 4, the fuzzy membership of the overall level belongs to fuzzy linguistic variable "low".
- In fifth simulation, DM 5 made the evaluation as in the Fig. 24 with the overall score 12.214. The illustration of DM 5 customer satisfaction level fuzzy membership is in the Fig. 25. As in the figure, according to the determined fuzzy rule 5, the fuzzy membership of the overall level belongs to fuzzy linguistic variable "very few".

4 Conclusion and Discussion

With industry 4.0, the level of digitalization and personalization increased in many sectors. As in aviation, big data, artificial intelligence, cloud computing issues are involved in every stage of the industry. Airlines must receive and process a large amount of data in order to satisfy their customers and remain viable in the market.

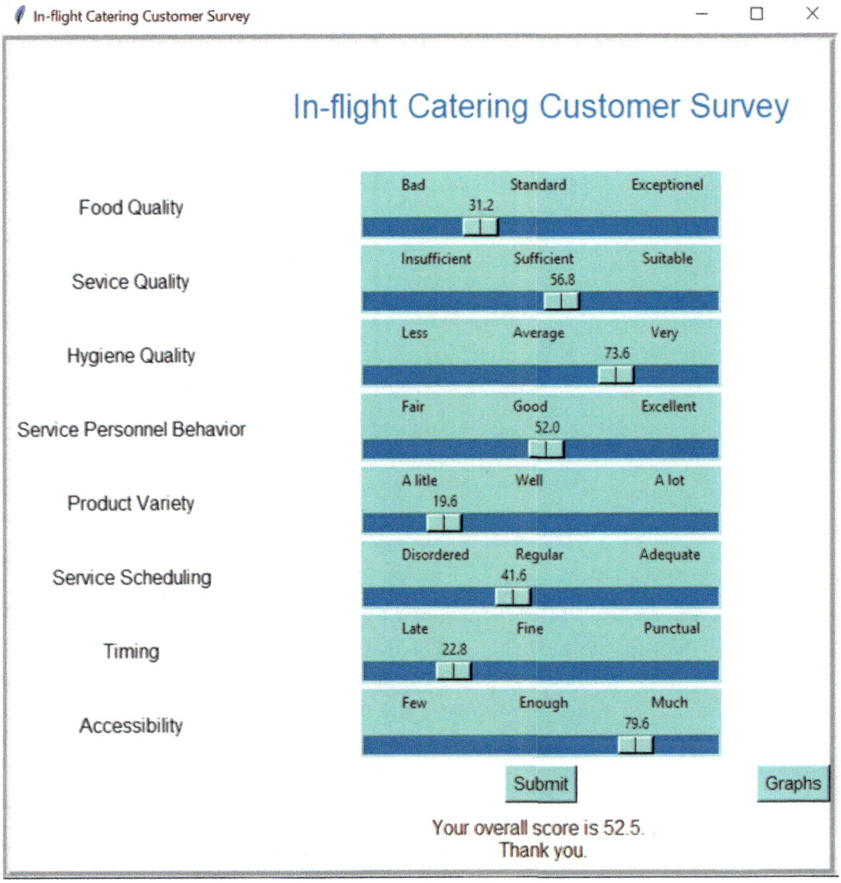

Fig. 20 In-flight customer survey of DM 3 with score 52.5

Therefore, up-to-date programs should be prepared for the customer's crisp, verbal or fuzzy feedback and they should be integrated into the systems quickly.

In this study, an interface design that includes a new fuzzy control system written in Python language is proposed. The program is prepared with Ski-Fuzzy toolkit developed to handle fuzzy sets. This software is based on a system of scoring in-flight catering services according to customers' linguistic, fuzzy or precise evaluations. Eight evaluation criteria with their fuzzy sets and fuzzy linguistic variables are determined by the experts. A set of fuzzy rules has been established to measure the level of customer satisfaction from these criteria. Since it can be easily adapted and programmed to the in-flight entertainment system, an interface has been developed in the Python environment. Thus, customers can make imprecise evaluations using scale bars for the eight criteria that appear on the screen. Customer satisfaction

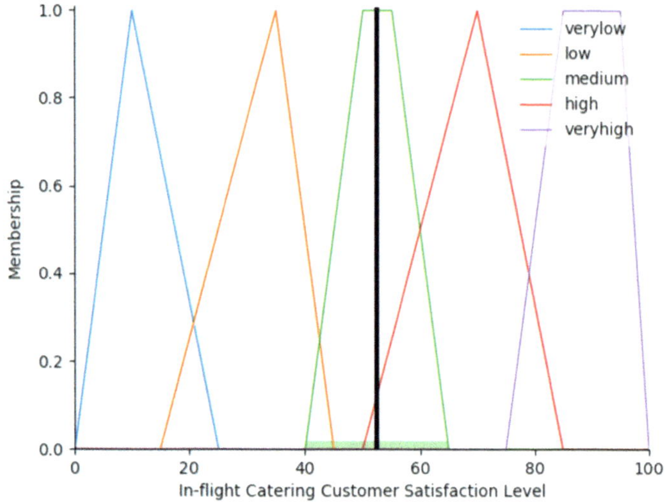

Fig. 21 Illustration of DM 3 customer satisfaction level fuzzy membership

is calculated according to rules determined by experts. The program can be easily customized according to the wishes of the airline companies.

Five simulations are made on this program. Experts are asked to respond like customers and demonstrate the validity of the rules they suggested. Not all rules can be shown due to the page constraint. However, evaluations are made following to the five rules and the results are visualized with graphics. The overall aim is to ensure that customer evaluations give an appropriate output in line with the established rules. Trials including the best and worst customer evaluations have been made in simulations. For each simulation, the software chooses the appropriate rule and calculates the result according to the customer evaluations. For example, in simulation 1, the customer rated the criteria with the highest expressions. Rule 1 states that the evaluation in the highest categories groups is evaluated with the highest degree of satisfaction. Thus, customer satisfaction is "very high" with 87,792. In simulation 2, the customer rated some criteria with the highest expressions and some with the lowest expressions. In this scenario, categories follow rule 2. Therefore, customer satisfaction is "high" with 67,648. In case the system cannot calculate, errors appear on the screen. These errors can be resolved by developer intervention. It is possible to write a feasible program with a good system design and clean coding.

As a result, an interface that measures the customer satisfaction level on in-flight catering service and includes an easily adaptable and customizable fuzzy control system has been designed using crisp, fuzzy or linguistic data requested from the customer. Scikit-Fuzzy toolkit designed to use fuzzy ecosystem in programming environment is used to design fuzzy control system. Also, Tkinter module is used in interface design to create widget and its features. Thus, in the age of aviation 4.0, a simple and easily adaptable software has been provided to enable airline companies

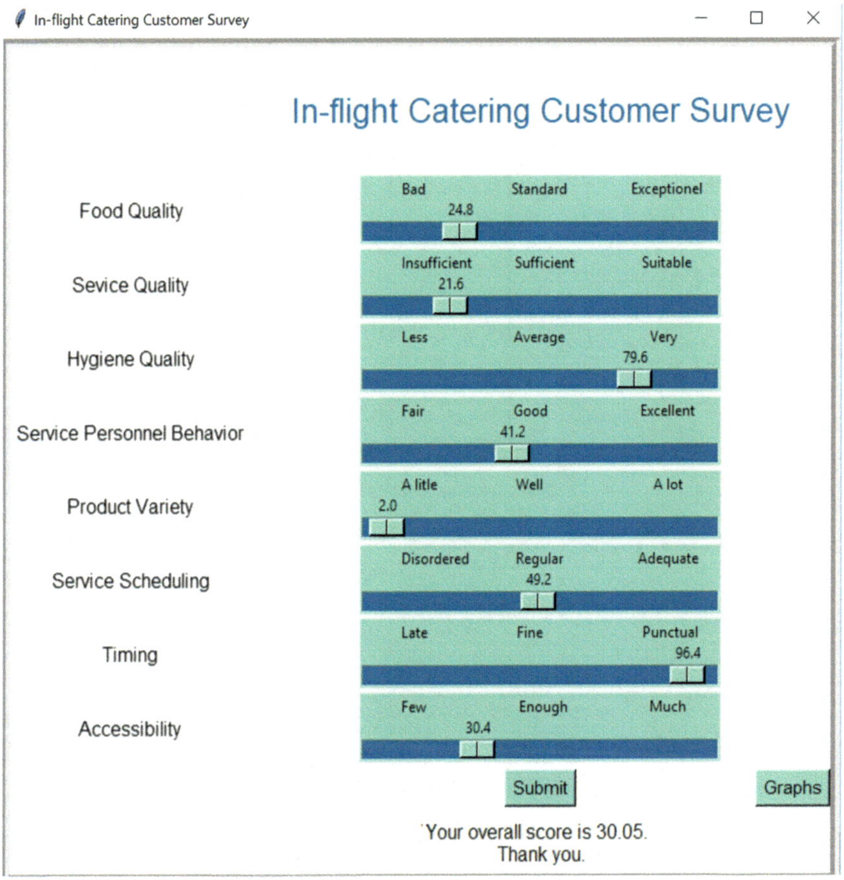

Fig. 22 In-flight customer survey of DM 4 with score 30.05

to instantly monitor the market situation by receiving feedback from their customers. Airlines can easily process this fuzzy data and stand out from the competition.

In future studies, this fuzzy control system can be adapted to measure customer satisfaction for different services of airlines in terms of Aviation 4.0. It can be supported with artificial intelligence algorithms and fuzzy control system can provide efficiency in cloud computing. In addition, rules for data evaluation on the Internet of Things can be developed and product categorization can be performed.

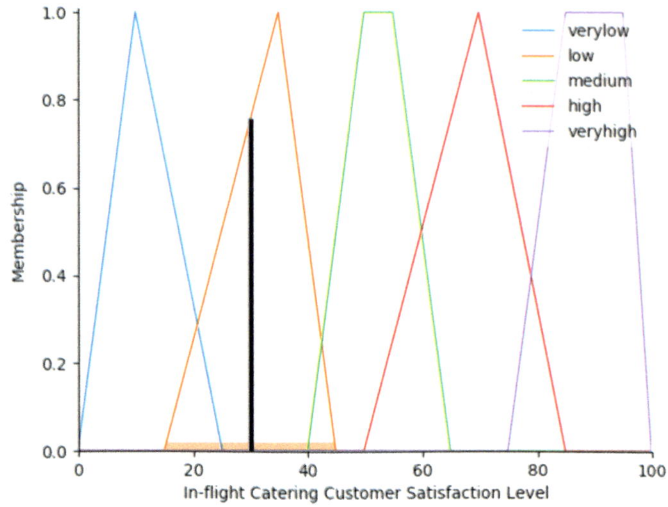

Fig. 23 Illustration of DM 4 customer satisfaction level fuzzy membership

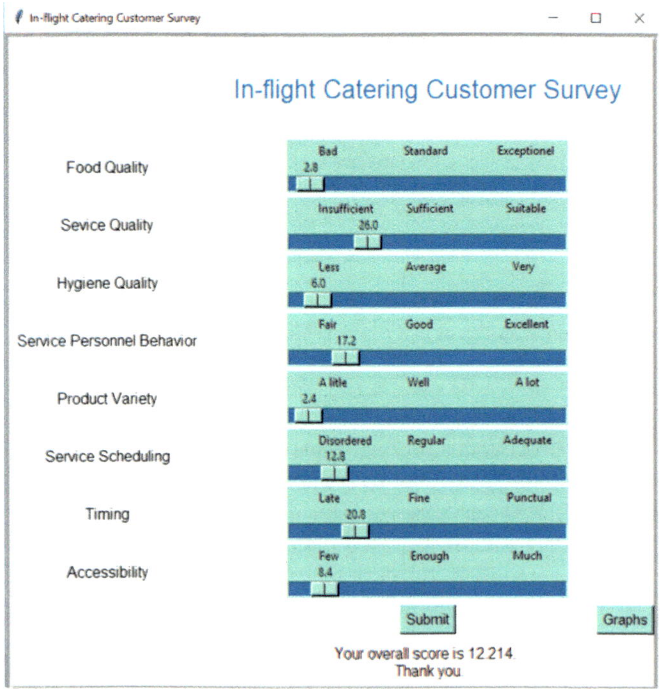

Fig. 24 In-flight customer survey of DM 5 with score 12.214

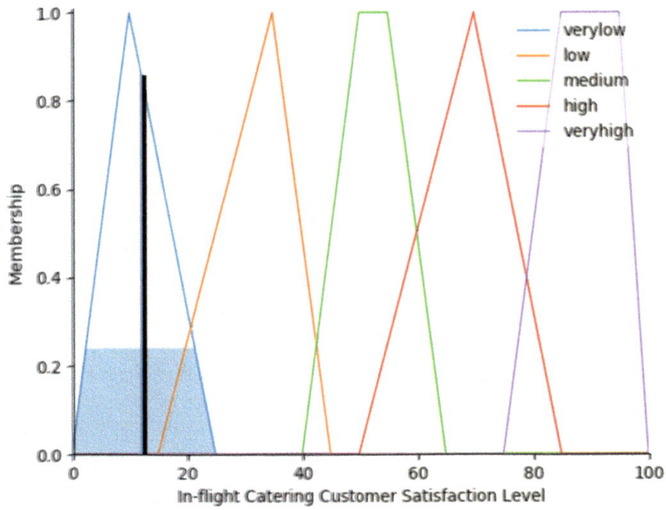

Fig. 25 Illustration of DM 5 customer satisfaction level fuzzy membership

Acknowledgements This work has been supported by the Scientific Research Projects Commission of Galatasaray University under grant number # FBA-2020-1036.

References

1. Can, A.V., Kıymaz, M.: Bilişim Teknolojilerinin Perakende Mağazacılık Sektörüne Yansımaları: Muhasebe Departmanlarında Endüstri 4.0 Etkisi, Süleyman Demirel Üniversitesi Sosyal Bilimler Enstitüsü Dergisi (CİEP Özel Sayısı), 107–117 (2016)
2. Atalık, Ö., Akan, Ş., Bakır, M.: Havacilik 4.0: Havayolu Ve Havaalani Endüstrisinde Güncel Endüstri 4.0 Uygulamalari, II. In: International Conference on Empirical Economics and Social Science (ICEESS' 19), pp. 879–891 (2019)
3. Özdoğan, O.: Endüstri 4.0: Dördüncü Sanayi Devrimi ve Endüstriyel Dönüşümün Anahtarları (2017)
4. Barbosa, G.F., Aroca, R.V.: Advances of industry 4.0 concepts on aircraft construction: an overview of trends. J Steel Struct. Constr. **3**, 125 (2017). https://doi.org/10.4172/2472-0437.1000125
5. AVRSPOT: How apps like AR Wayfinding Gatwick airport can help promote your brand. https://www.avrspot.com/apps-like-ar-wayfinding-gatwick-airport-can-help-promotebrand/ (2018)
6. Koh, D.: SATS introduces augmented reality technologies to Changi Airport operations. https://www.opengovasia.com/sats-introduces-augmented-reality-technologies-to-changiairport-operations/ (2019)
7. Robertson, A.: Air New Zealand imagines flight attendants using HoloLens to read your emotions. https://www.theverge.com/2017/5/24/15686740/microsoft-hololens-air-newzealand-augmented-reality-airplane-emotion-reading (2017)
8. Sillers, P.: Airport biometrics: how new customs technology is going to make security queues a thing of the past. https://www.independent.co.uk/travel/news-and-advice/airportbiometrics-trial-face-recognition-klm-brisbane-sita-a7813271.html

9. Pozzi, J.: Engine makers, MROs step up virtual reality platforms. https://aviationweek.com/awinmro/engine-makers-mros-step-virtual-reality-platforms (2017)
10. King, T.: In-flight catering. J. Tourism Hospitality Res. **3**(2), 181–184 (2001)
11. Oliver, R.L.: Measurement and evaluation of satisfaction processes in retail settings. J. Retail. **57**, 25–48 (1981)
12. Retnaningsih, W.: Effect on the quality of customer satisfaction of lion air airline in Surabaya City. Int. J. Bus. Manage. Invent. **2**(8), 16–25 (2013)
13. Chou, C., Liu, L., Hung, S., Yih, J., Han, T.: An evaluation of airline service quality using fuzzy weighted SERVQUAL method. J. Appl. Soft Comput. **11**, 2117–2128 (2011)
14. Hynes, J., Dredge, F.: Managing Customer Services. Gower Publishing Limited, Aldershot (1998)
15. Wang, R., Shu, L.H., Lin, Y.H., Tseng, M.L.: Evaluation of customer perception on airline service quality in uncertainty. Proc. Soc. Behav. Sci. **25**, 419–437 (2011)
16. Eric, T., Laws, Y.: Managing passenger satisfaction: some quality issues in airline meal service. J. Qual. Assur. Hospitality Tourism **6**(1/2), 89–113 (2005)
17. Bruce, S.: In-flight catering. Unpublished PhD dissertation, University of Helsinki, Finland (2001)
18. Kahraman, C., Cebeci, U., Ruan, D.: Multi-attribute comparison of catering service companies using fuzzy AHP: the case of Turkey. Int. J. Prod. Econ. **87**, 171–184 (2004)
19. Franklin, F.G.: History of in-flight catering: it all began in 1903 with Orville Wright. J. Airline Food Travel Service **16**(2), 21 (1999)
20. Dike, D.N.: Assessment of the quality of customer service in Nigeria's Airline industry. Int. J. Innov. Res. Stud. **2**(6), 519–538 (2013)
21. Mahato, A.K.: Assessing satisfaction: a study on international passengers using Nepal Airlines. AU J. Manage. **9**(1), 43–52 (2012)
22. Ammar, S., Moore, D., Wright, R.: Analysing customer satisfaction surveys using a fuzzy rule-based decision support system: enhancing customer relationship management. Database Market. Customer Strategy Manage. **15**(2), 91–105 (2008)
23. Ammar, S., Wright, R., Selden, S.: Ranking state financial management: a multilevel fuzzy rule based system. Decis. Sci. **31**, 449–481 (2000)
24. Ammar, S., Wright, R.: A hybrid system for modeling credit ratings for US airports. In: Proceedings of IEEE International Conference on Fuzzy Systems, Budapest (2000)
25. Narasimhan, R.: A fuzzy subset characterization of a site-selection problem. Decis. Sci. (2007)
26. Pedregosa, F., et al.: Scikit-learn: machine learning in Python. J. Mach. Learn. Res. **12**, 2825–2830 (2011)
27. Lundh, F.: An introduction to Tkinter (1999)
28. Zadeh, L.A.: Fuzzy sets. Inf. Control **8**, 338–353 (1965)
29. Warner, J., et al.: JDWarner/scikit-fuzzy: Scikit-fuzzy version 0.4.2 (Version v0.4.2). Zenodo. https://doi.org/10.5281/zenodo.3541386 (2019)
30. Matplotlib: Python plotting—Matplotlib 3.2.0 documentation. https://www.matplotlib.org. Retrieved 14 Mar 2020
31. NumPy—NumPy. https://www.numpy.org. NumPy developers

Analysis of Intelligent Software Implementations in Air Cargo Using Fermatean Fuzzy CODAS Method

Irem Ucal Sari, Dorota Kuchta, and Duygu Sergi

Abstract The chapter focuses on the problem of analyzing and selecting intelligent software in Air Cargo in the concept of Aviation 4.0. First, the notions, problems and challenges linked to air cargo are discussed. Recent developments, ongoing innovative projects and unfilled gaps in the area of intelligent air cargo software are presented. Next, the proposed method to analyze a select software to be used by air cargo companies is described. It is a modified version of one of the recent multi-criteria decision-making methods, called CODAS. Its original, crisp version and its existing fuzzy extensions are first presented. Next, an original extension of the method, using Fermatean fuzzy sets, is proposed. In the application section a logistics company is considered, which is facing the problem of selecting software supporting the air cargo process. The criteria are selected by experts holding various positions in the company, and three alternatives of air cargo software provider are determined. Then, the proposed method is applied to solve the intelligent software selection problem. Finally, conclusion and future research perspectives are given.

Keywords Fermatean fuzzy sets · Fermatean fuzzy CODAS method · Air cargo · Intelligent softwares

I. Ucal Sari (✉)
Industrial Engineering Department, Istanbul Technical University, Macka, Istanbul, Turkey
e-mail: ucal@itu.edu.tr

D. Kuchta
Wroclaw University of Science and Technology, Wyb. Wyspianskiego 27, 50-370 Wroclaw, Poland
e-mail: dorota.kuchta@pwr.edu.pl

D. Sergi
Department of Industrial Engineering, MEF University, Ayazağa Cad. No.4, 34396 Maslak, Sarıyer, Istanbul, Turkey
e-mail: sergid@mef.edu.tr

C. Kahraman and S. Aydın (eds.), *Intelligent and Fuzzy Techniques in Aviation 4.0*, Studies in Systems, Decision and Control 372,
https://doi.org/10.1007/978-3-030-75067-1_7

1 Introduction

Industrial revolutions lead to great changes and innovations in the aviation industry as in every sector. The first revolution in the aviation industry is referred as Aviation 1.0 that was the start of commercial aviation where flight developed under visual flight rules, following visuals pieces of information and signals and there was not really any instrumental guide for the pilots. In the subsequent stage, Aviation 2.0, the old system was replaced by electric devices. In the third stage, Aviation 3.0, commercial aviation included the huge incorporation of electronics in the cockpit, driven by the accessibility of reliable and usable digital data processing and data communication technology [1]. With the latest revolution, Industry 4.0, the aviation industry entered the era of Aviation 4.0. In the Aviation 4.0 concept, technologies such as artificial intelligence, internet of things, automation and digitization have started to be used in aviation processes such as ground services, cabin services, routing and cargo services, and significant improvements have been observed in processes and security.

Air cargo, which is one of the important industries in aviation, has been significantly improved by the implementations of Aviation 4.0. Internet of things (IoT) technology that is used to monitor changes in the temperature, humidity and shape of the cargo in air cargo tracking, cyber-physical systems (CPS) that are used for load optimization, aircraft fleet optimization services supported by IoT and CPS can be given as some examples of the Aviation 4.0 implementations in the air cargo industry.

There are several softwares developed to handle the entire system as a whole and provide solutions in various operations of an air cargo system. The features of the softwares have a high range but most of them focused on some of the operations in the air cargo industry. Operations that are desired to be digitized may also vary from business to business. Therefore, it is determined as the aim of this chapter to create a roadmap for the selection of smart software and systems in airline cargo companies. For this purpose, one of the widely used multi-criteria decision-making method that is CODAS is selected. To better model, the unknown characteristics of the ongoing technological developments and uncertainty in the linguistic evaluations of decision-makers, a new fuzzy extension of CODAS that is fuzzy Fermatean CODAS method is proposed for the first time and used in the analysis.

The rest of the chapter is organized as follows: In Sect. 2 literature review on Aviation 4.0 implementations on air cargo and intelligent logistics are given. Then, in Sect. 3 basics of the air cargo systems are detailed. In Sect. 4 basics of the CODAS method is given both for crisp and fuzzy versions. In Sect. 5 fermatean fuzzy CODAS method is introduced including the preliminaries. In Sect. 6 the proposed method is applied to determine the best intelligent system for an air cargo company. Finally, in Sect. 7, a chapter is concluded with the interpretations.

2 Literature Review

In the literature review, it is seen that air cargo is investigated with various authors mostly in terms of revenue management, route optimization and scheduling, but only a few of them are examined improvements of Aviation 4.0 implementations.

Durak and Tolga [2] proposed the weighted Euclidean distance-based approach method for selection of the most suitable robotic process automation in the air cargo sector. Tabares et al. [3] proposed a ground handling management structure that allows the automation of operations to face the growing demand for aircraft ground handling operations. Wong et al. [4] proposed a closed-loop dynamic air cargo loading digital twin system that utilizes simulation, multi-dimensional immersive virtual reality system, Internet of Things, and real-time sensors for connecting, monitoring and controlling the operations in physical operations and virtual space in air cargo loading in terminals. Tolga and Durak [5] analyzed six innovation projects that are digital dangerous goods declaration, new type pallet usage, software for stack planning and loading, wooden stack saving, unit load device online tracking, and providing compatibility of intelligent cargo with the dangerous goods declaration using fuzzy COPRAS method. Gumzej et al. [6] focused on the construction of an intelligent, safe and secure physical internet transport unit for air cargo that has advantages of improved safety and security and decreased personnel. Cimato et al. [7] proposed a cloud based-query system to balance cancellations for air cargo companies, which allows a more accurate estimate of the cancellation rates. Fang and Mao [8] investigated an energy-efficient elevating transfer vehicle routing problem, in which an elevating transfer vehicle serves a multi-level freight handling system to transport cargo containers between airside and landside in an air cargo terminal. Delgado et al. [9] investigated the effects of scheduling recovery to adapt last-minute demand changes using mixed-integer linear programming. Emde et al. [10] proposed a mixed-integer linear programming model to support airlines in realizing efficient unit load devices preparation processes. Giusti et al. [11] proposed a methodology that allows assessing the performances resulting from different RFID implementation set-ups within an air cargo handler's warehouse.

Besides that, there are lots of studies focused on intelligent logistics. Zhang and Chen [12] discussed intelligent logistics in terms of its driver points, the related technology and future developments in warehousing and distribution. Wang et al. [13] proposed an IoT-based intelligent logistics dispatching system, which enables dynamic coordination between customers, order-picking robots and cloud technology. Liu and Li [14] proposed logistics a service selection scheme that uses the traditional web service combination idea to solve the intelligent logistics service problem. Xue et al. [15] designed a hybrid many-objective competitive swarm optimization algorithm to solve many-objective optimization model for multi-robot task allocation problem in intelligent logistics. Tang et al. [16] designed a smart logistics system architecture based on edge computing that is used to preprocess the data, distribute the tasks, and ensure the security of sensitive data to realize the business security and data security of the intelligent logistics system. Vijay et al. [17] proposed an architecture that extends power line communications with Bluetooth Low Energy devices that are embedded on the cargo for continuous monitoring and tracking.

To the best of our knowledge, none of the papers studied on the selection criteria of intelligent air cargo systems. Therefore, in this chapter, first the decision criteria for intelligent air cargo systems are determined then the selection process is realized for an air cargo company using fuzzy CODAS method.

3 Air Cargo System

Let's start with basic notions and definitions [18] used with respect to air cargo to understand its dynamics better.

The transport by air is expensive, often 10–50 times the price of surface transport [18]. Therefore, it is usually only chosen for goods with at least one of the following four properties:

- *urgent*: The goods need to be delivered within a short time. They may be spare parts or living animals.
- *perishable*: The goods need special treatment (e.g. cooling). Examples are flowers or fresh food.
- *valuable*: The goods are of high value and have to be delivered without the risk of getting lost. Examples are banknotes or works of art.
- *dangerous*: The goods may harm their environment. Examples are chemicals or radioactive material.

A flight is a sequence of aircraft movements performed by the same aircraft. A leg is a nonstop aircraft movement between two airports. A segment is a subsequence of legs of a flight along which cargo can be transported.

The majority of the cargo transported by cargo airlines is at least the size of a wooden pallet, defined in ISO 2003 [19]. Most shipments are box-shaped but irregular shapes occur too. The cargo is often assembled onto special air cargo pallets or containers, generally named unit load devices (ULD). The ULDs are placed in the aircraft on positions assigned to them and fixed appropriately. There exist ULD types with different shapes. The ULDs are filled with goods to be transported, placed in the aircraft and downloaded after the landing. They may continue their journey in case of a segment or are emptied of the case of flights or legs.

The cargo transport comprises several processes and decision problems:

1. *Aircraft Configuration Problem (ACP)*: the decision which ULDs should be built for each flight, taking into account the customers' orders;
2. *Build-up Scheduling Problem (BSP)*: the decision at what time and in which location each ULD is built;
3. *Air Cargo Palletization Problem (APP)*: the decision about the assignment of items to and their placement inside the ULDs. The goal is to load as many of the given items as possible onto the given set of ULDs in a legal and safe way;

4. *Weight and Balance Problem (WBP)*: the decision about the assignment of ULDs to the loading positions inside the aircraft. In each aircraft there is distinguished a certain set of loading positions. The assignment is performed with the objective of maximization of space utilization in each loading position. At the same time, there is a maximum weight for each loading position and there also exist weight limits for various subsets of loading positions. On top of that, the cargo on the aircraft has to be balanced: the center of gravity (CG) of the aircraft has to belong to a predefined range.

There are certain processes which take place also before the building up of ULDs and after the aircraft lands. The customer drops his shipment at the export point, either as single items or pre-built ULDs. The cargo is checked and depending on its scheduled departure time either temporarily stored or directly sent to the build-up process. After unloading at the next airport, the ULDs are again stored or disassembled into their individual items. These items can then be picked up by the customer or continue their journey toward their final destination. In case of storage in the airport, the decision about storage place has to be taken (taking into account both the space usage of the storage rooms and their distance from the points where the customers drop or pick up their items, as well as the storage time).

Li [20] proposes a more detailed example description of the processes and activities that form the cargo transport:

1. market agency contracts goods
2. the shipper's letter of instruction is filled in
3. agents review the documents provided by the shipper
4. the documents are accepted, the airway bill is filled in
5. agent checks and acknowledges the receipt of goods from the consignor
6. labels or tags are produced (for the classification of goods)
7. a formal application for the airline cabin is sent
8. a formal application to the airline booking is sent
9. agents take care of cargo inspection, customs and other export procedures
10. the warehouse receipt is prepared
11. containers are provided and loaded
12. the airway bill is signed
13. the shipping is transferred to the airlines
14. the airlines care for the temporary storage of goods
15. the airlines assemble containers
16. the cargo is loaded and balanced
17. the cargo manifests, transfer documents, transfer port manifest are produced
18. goods leave the warehouse: the containers are loaded and shipped to the airfield
19. clearing cost takes place: settlement between airlines and agents.

It also has to be underlined that the above processes need a lot of paperwork. Many documents required by law [21, 22] has to be generated, checked for correctness, shared and approved by all parties involved, which consumes a considerable amount of time. Another important issue is the high number of parties involved in the process (exporters, forwarders, various agents and intermediaries, custom agencies, airlines, airports etc.), which requires an efficient communication and documents exchange among them. With the growing complexity and scale of cargo transport all over the world, the above-described processes cannot be handled manually anymore [23]. Manual handling of these processes results in numerous delays of individual items, losses, damages of perishable goods etc. Documentation generation and exchange process, as well as a communication process, is also time-consuming and prone to errors, which in their turn cause delays and damages. That is why the aircraft industry has been working intensively on innovations in the field of intelligent software support in the processes described above.

For example, following [24], let us consider the IATA's StB Cargo initiative. Its declared aim is "Making air cargo easier, smarter and faster". The program includes six projects: e-freight and e-AWB, ONE Record, Interactive Cargo, Smart Facility, Air Cargo Incident Database and Cargo Connect.

E-freight stands for the digitalization of paper-based documentation and e-AWB for the introduction of the electronic air waybill (e-AWB). However, these projects have been encountering obstacles. In 2005, IATA expected that within five years, by 2010, it would be able to offer paperless handling, but in 2019 it was still far from this goal [24].

The ONE Record project within the StB Cargo initiative is intended to help ensure that data is shared more intelligently and thus better between stakeholders. The project is still going on [24].

Within the framework of the "Interactive Cargo" project, IATA wants to improve the visibility of cargo. This is especially important for urgent, perishable and valuable goods. It should enable shippers and authorized recipients to query the status of a shipment at any time. The project is still going on [24].

With the establishment of the Air Cargo Incident Database (ACID), IATA is willing to gain the possibility to analyse incidents and irregularities in the transport of air cargo in a structured manner. Most airlines have been collecting data on incidents and irregularities in the past, but there is the need for a uniform procedure for collecting, evaluating and comparing data. The project is still going on [24].

The Cargo Connect initiative aims to simplify and improve electronic communication between airlines, ground handling agents (GHA), forwarders and customs agents. The project is still going on [24].

In the meantime, individual companies are trying to close the still existing gaps through individual initiatives. One example is the cooperation between the Swedish communications manufacturer Ericsson and Finnair and the freight forwarder Panalpina, who are working on a platform on which different parties can share shipment data [24]. Also, several documents have already been digitalized by air companies, e.g. [25]:

1. eAWB—the electronic Air Waybill, which constitutes the legal basis for the transport agreement between freight forwarder and airline;

2. eCSD—the IATA standard electronic consignment security declaration, the digital equivalent to the regular paper-security and screening information of shipments;
3. eHM—the electronic House Manifest, which has replaced paper House Manifest or Consol Manifest;
4. eDGD—the electronic Dangerous Goods Declaration: a new, digital document for the management of dangerous goods data.

A number of other initiatives are mentioned in the professional literature, some of which go beyond the processes described above, taking into account the whole environment of cargo loading and transport. Brett [26] mentions All Nippon Airways' (ANA) initiative with respect to the booking platform which will be able to provide real-time quotes for all available air cargo capacity. Online freight booking portal cargo.one and IBS Software are partnering to combine their platforms [27], which means that airlines that utilise IBS Software's iCargo platform will be able to offer their capacity to the freight forwarders that use the cargo.one booking portal without any major systems changes. Air cargo software firm Nallian has launched a road feeder management app that will allow trucking companies and ground handlers to co-ordinate airport-to-airport pick-ups and deliveries at Brussels airport [28]. Kale Logistics is working on a digital air waybill tool [29]. Hartsfield–Jackson Atlanta International Airport, the airfreight community in Atlanta, and IT business Kale Logistics Solutions, have collaborated to create North America's first air cargo community system (ACCS). The ACCS allows all the stakeholders (i.e. shippers, airlines, trucking companies, customs brokers, freight forwarders and cargo handlers) to communicate electronically with each other [30]. Teleport, AirAsia's logistics division, has launched Freightchain—a new digital air cargo platform operated using blockchain. Freightchain enables freight forwarders and shippers to instantly book and confirm cargo on any of AirAsia's 247 aircraft, without the need to go through traditional time-consuming sales or email channels [31]. Peli BioThermal applied the Internet of Things to control the temperature of perishable goods during the whole transport process [32]. There is also research conducted on real-time cargo tracking capability [33].

It has to be underlined that intelligent software in cargo delivery and loading would have to be able to solve several optimization problems [18], which range from selecting the containers to load over solving three-dimensional packing problems to balancing the aircraft for fuel efficiency. Some other examples are mentioned in [34]: the flight itinerary assignment to individual loads (where availability of flight itinerary options, cargo acceptance cut-off times, destination arrival times, flight itinerary costs, and tardiness penalties have to be taken into account) and the multi-vehicle routing to pick up customer loads and deliver them to the airports (here the depots, customers, airports, starting times of the selected flight itineraries, and the vehicle fleet size are important). It has to be stated that in the existing software for cargo transport and loading the optimization aspect is neglected. Only few examples are contradicting this statement, e.g. the advanced load planning tool, CHAMP's Weight and Balance [35]: an algorithm-based system which helps ensure that maximum space is used on an aircraft.

To sum up, it has to be underlined that advanced technology has been used for some time already to support air cargo loading and transport. However, the main aspects covered so far are the documentation and communication as well as on line booking and real time tracking. The decision making in the area of assignment of items to pallets, pallets to flights, legs or segments, pallets to location positions, to storage space etc., where there is a huge potential for optimization of time, cost, losses, space utilization etc., is still, to a large degree, left in the hands of humans. Here dwells a big potential for future developments in the context of Aviation 4.0.

4 CODAS Method

The CODAS (COmbinative Distance-Based ASsessment) method, proposed in [36] is a method which can be used to solve multicriteria decision making problems. We will start by presenting its basic, crisp version.

4.1 Crisp CODAS Method

We consider the decision-making matrix $\boldsymbol{X} = \left[x_{ij}\right]_{n \times m}$, whose rows $(i = 1, \ldots, n)$ stand for n alternatives and columns $(j = 1, \ldots, m)$ for m criteria. Each criterion has been assigned a weight w_j, $\sum_{j=1}^{m} w_{ij} = 1$. The matrix elements x_{ij} $(i = 1, \ldots, n, j = 1, \ldots, m)$ denote the performance of the respective alternative under the respective criteria.

The set of indices denoting the criteria, $\boldsymbol{J} = \{1, \ldots, j\}$ is composed of two disjunctive sets $\boldsymbol{J}^{\uparrow}$ and $\boldsymbol{J}^{\downarrow}$, $\boldsymbol{J} = \boldsymbol{J}^{\uparrow} \cup \boldsymbol{J}^{\downarrow}$, where $\boldsymbol{J}^{\uparrow}$ contains the indices corresponding to maximisation criteria and $\boldsymbol{J}^{\downarrow}$ to minimization criteria. Normalized matrix $\boldsymbol{N} = \left[n_{ij}\right]_{n \times m}$ is then defined, where:

$$n_{ij} = \begin{cases} \dfrac{x_{ij}}{\max\limits_{i=1,\ldots,n} x_{ij}} \text{if } j \in \boldsymbol{J}^{\uparrow} \\ \dfrac{\min\limits_{i=1,\ldots,n} x_{ij}}{x_{ij}} \text{if } j \in \boldsymbol{J}^{\downarrow} \end{cases} \tag{1}$$

In the next step the weighted normalized matrix $\boldsymbol{S} = \left[s_{ij}\right]_{n \times m}$ is determined, where

$$s_{ij} = w_j n_{ij} (i = 1, \ldots, n, j = 1, \ldots, m) \tag{2}$$

The negative-ideal solution is a vector NS composed of the worst weighted values of each criterion, we have thus

$$NS = \left[ns_j\right]_{1\times m}, ns_j = \min_{i=1,\ldots,n} s_{ij}, j = 1, \ldots, m \tag{3}$$

Two distance vectors are then constructed: vector $\boldsymbol{E} = [e_i]_{1\times n}$ and vector $\boldsymbol{T} = [t_i]_{1\times n}$, containing, respectively, the Euclidean and Taxicab distances of the respective alternative from the negative-ideal solution. We have thus:

$$e_i = \sqrt{\sum_{j=1}^{m} \left(s_{ij} - ns_j\right)^2} \tag{4}$$

$$t_i = \sum_{j=1}^{m} \left|s_{ij} - ns_j\right| \tag{5}$$

The ranking of the alternatives is determined on the basis of the relative assessment matrix $\mathrm{H} = \left[h_{ij}\right]_{n\times n}$, where

$$h_{ik} = (e_i - e_k) + \psi(e_i - e_k)(t_i - t_k) \tag{6}$$

Is a weighted sum of the differences between the two distances, in which the Taxicab distance is taken into account only if the Euclidean distance is considered not to distinguish both alternatives with indices i and k, $i, k = 1, \ldots, n$. The indistinguishability is decided upon by means of the function ψ, defined on the set of real numbers as;

$$\psi(x) = \begin{cases} 1 \text{ if } |x| \geq \tau \\ 0 \text{ otherwise} \end{cases} \tag{7}$$

where τ is a small number set by the decision maker.

The final outcome of the method is the ranking of the alternatives, based on the decreasing order of assessment scores, defined as

$$H_i = \sum_{k=1}^{n} h_{ik} \tag{8}$$

4.2 Fuzzy Extensions of the CODAS Method

In [37] a fuzzy extension of the CODAS method is proposed. The assumption is that matrix X has fuzzy elements, being trapezoidal fuzzy numbers. We have thus

$$\widetilde{\boldsymbol{X}} = \left[\tilde{x}_{ij}\right]_{n\times m} \tag{9}$$

where $\tilde{x}_{ij}$ are trapezoidal fuzzy numbers.

Now we will present which equivalents of crisp notions mentioned in section K1 have been assumed. Letters with tilde will stand for trapezoidal fuzzy numbers, e.g. $\widetilde{A} = (a_1, a_2, a_3, a_4)$ is a trapezoidal fuzzy number with support $[a_1, a_4]$ and core $[a_2, a_3]$.

- standard arithmetic operations on trapezoidal fuzzy numbers are used
- Equation (1) is replaced with

$$\tilde{n}_{ij} = \begin{cases} \dfrac{\tilde{x}_{ij}}{\max\limits_{i=1,\ldots,n} \Delta(\tilde{x}_{ij})} \text{ if } j \in \boldsymbol{J}^{\uparrow} \\ 1 - \dfrac{\tilde{x}_{ij}}{\max\limits_{i=1,\ldots,n} \Delta(\tilde{x}_{ij})} \text{ if } j \in \boldsymbol{J}^{\downarrow} \end{cases} \tag{10}$$

where

$$\Delta(\widetilde{A}) = \frac{a_1 + a_2 + a_3 + a_4 - \frac{a_3 a_4 - a_1 a_2}{(a_3 + a_4) - (a_1 + a_2)}}{3} \tag{11}$$

Any other defuzzification method could be used instead of Eq. (11). In Eq. (10) we get trapezoidal fuzzy numbers.

- In Eq. (2) we get trapezoidal fuzzy numbers

$$\tilde{s}_{ij} = w_j \tilde{n}_{ij} (i = 1, \ldots, n,\ j = 1, \ldots, m) \tag{12}$$

- Equation (3) is replaced with

$$\boldsymbol{NS} = \left[\widetilde{ns}_j\right]_{1 \times m}, \widetilde{ns}_j = \tilde{s}_{kj} \text{such that } \Delta(\tilde{s}_{kj}) = \min_{i=1,\ldots,n} \Delta(\tilde{s}_{ij}),\ j = 1, \ldots, m \tag{13}$$

- Equation (4) is replaced with

$$e_i = \sum_{j=1}^{m} \sqrt{\frac{\left((s_{ij})_1 - (ns_j)_1\right)^2 + 2\left((s_{ij})_2 - (ns_j)_2\right)^2 + 2\left((s_{ij})_3 - (ns_j)_3\right)^2 + \left((s_{ij})_4 - (ns_j)_4\right)^2}{6}} \tag{14}$$

- Equation (5) is replaced with

$$t_i = \sum_{j=1}^{m} \frac{\left|(s_{ij})_1 - (ns_j)_1\right| + 2\left|(s_{ij})_2 - (ns_j)_2\right| + 2\left|(s_{ij})_3 - (ns_j)_3\right| + \left|(s_{ij})_4 - (ns_j)_4\right|}{6} \tag{15}$$

Other existing fuzzy versions of the CODAS method use other types of fuzzy numbers instead of triangular fuzzy numbers. The authors of [38–44] chose various versions of intuitionistic fuzzy sets [45] for the representation of alternatives in Eq. (9). In [46–49], Pythagorean fuzzy sets are applied, defined in [50], and in [51, 52] spherical fuzzy sets [53] can be found in Eq. (9). In [54, 55] hesitant fuzzy sets are used [56], which are transformed into triangular fuzzy sets by means of the fuzzy envelope operation defined in [57]. In [58, 59] neutrosophic sets [60] were applied. The authors of [61], for their turn, used Z-fuzzy numbers in Eq. (9) [62], and the authors of [63] resorted to soft fuzzy sets [64]. Of course, the minimum, maximum and the distance notions are defined separately for each case.

5 Fermatean Fuzzy CODAS Method

Senapati and Yager [65] proposed Fermatean fuzzy sets as one of the most recent extensions of fuzzy set theory. Fermatean fuzzy sets are one of the orthopair fuzzy sets that are fuzzy sets in which each of their membership grade is described by a pair of values both from the unit interval, indicating the degree of membership and non-membership [66] Intuitionistic fuzzy sets are the first proposed orthopair fuzzy sets [45] in which sum of the membership degree and the non-membership degree should be at most 1. Pythagorean fuzzy sets are proposed by [50] based on the fact that intuitionistic fuzzy sets cannot be used when this sum is greater than 1. Pythagorean fuzzy sets in which squares of membership and non-membership degrees should be less than 1 could also be inapplicable when the sum of squares is greater than 1. Fermatean fuzzy sets, in other words 3-rung orthopair fuzzy sets, assign fuzzy parameters from a larger domain that enables analysts to handle higher levels of uncertainties. In Fermatean fuzzy sets, the sum of cubes of membership and non-membership functions should be less than 1. Therefore the main advantage of Fermatean fuzzy sets is that it can describe more uncertainties than intuitionistic fuzzy sets and pythagorean fuzzy sets, which can be applied in many decision-making problems [67].

A Fermatean fuzzy set (FFS) F can be represented on the universe of discourse X as in Eq. (16):

$$F = \{\langle x, \alpha_F(x), \beta_F(x)\rangle : x \in X\} \tag{16}$$

where the functions $\alpha_F(x)$ and $\beta_F(x)$ indicates the degree of membership and non-membership of x to in the set F respectively and each element of X is independently mapped to a value between 0 and 1 as $\alpha_F(x) : X \to [0, 1]$ and $\beta_F(x) : X \to [0, 1]$. The sum of membership and non-membership degrees' cubes can be at most equal to 1 as given in Eq. (17) [65]:

$$0 \leq (\alpha_F(\mathrm{x}))^3 + (\beta_F(\mathrm{x}))^3 \leq 1 \tag{17}$$

Furthermore, the degree of indeterminacy of x to F is stated in Eq. (18) for any FFS, F and $x \in X$.

$$\pi_F(x) = \sqrt[3]{1 - (\alpha_F(x))^3 - (\beta_F(x))^3} \tag{18}$$

When there is only one element in set X, this set $\langle \alpha_F(x), \beta_F(x) \rangle$ called as a Fermatean fuzzy number (FFN); denoted by $\mathrm{F} = (\alpha_F, \beta_F)$ [68].

5.1 Algebraic Operations of FFNs

Let $F = (\alpha_F, \beta_F)$, $F_1 = (\alpha_{F_1}, \beta_{F_1})$ and $F_2 = (\alpha_{F_2}, \beta_{F_2})$ be three FFNs, and λ is a scalar that is greater than 0. The algebraic operations are given in Eqs. (19–24) [66, 67]:

Summation:

$$\mathrm{F}_1 \boxplus \mathrm{F}_2 = \left(\sqrt[3]{\alpha_{F_1}{}^3 + \alpha_{F_2}{}^3 - \alpha_{F_1}{}^3 \alpha_{F_2}{}^3},\ \beta_{F_1}\beta_{F_2}\right) \tag{19}$$

Subtraction:

$$\mathrm{F}_1 \boxminus F_2 = \left(\sqrt[3]{\frac{\alpha_{F_1}{}^3 - \alpha_{F_2}{}^3}{1 - \alpha_{F_2}{}^3}},\ \frac{\beta_{F_1}}{\beta_{F_2}}\right) \text{ if } \alpha_{F_1} \geq \alpha_{F_2},\ \beta_{F_1} \leq \min\left\{\beta_{F_2}, \frac{\beta_{F_2}\pi_1}{\pi_2}\right\} \tag{20}$$

Multiplication:

$$\mathrm{F}_1 \boxtimes \mathrm{F}_2 = \left(\alpha_{F_1}\alpha_{F_2},\ \sqrt[3]{\beta_{F_1}{}^3 + \beta_{F_2}{}^3 - \beta_{F_1}{}^3 \beta_{F_2}{}^3}\right) \tag{21}$$

Scalar Product:

$$\lambda F = \left(\sqrt[3]{1 - (1 - \alpha_F^3)^\lambda},\ \beta_F^\lambda\right) \tag{22}$$

Exponentiation:

$$F^\lambda = \left(\alpha_F^\lambda,\ \sqrt[3]{1 - (1 - \beta_F^3)^\lambda}\right) \tag{23}$$

Division:

$$F_1 \boxslash F_2 = \left(\frac{\alpha_{F_1}}{\alpha_{F_2}} , \sqrt[3]{\frac{\beta_{F_1}^{\ 3} - \beta_{F_2}^{\ 3}}{1 - \beta_{F_2}^{\ 3}}} \right) \text{ if } \alpha_{F_1} \leq \min\left\{ \alpha_{F_2} , \frac{\alpha_{F_2}\pi_1}{\pi_2} \right\} , \ \beta_{F_1} \geq \beta_{F_2} \tag{24}$$

Equation (25) is proposed for the FFNs which do not satisfy the condition in Eq. (24):

$$F_1 \boxslash F_2 = \left(\frac{\alpha_{F_1}}{\alpha_{F_2}} , \sqrt[3]{\frac{\beta_{F_2}^{\ 3} - \beta_{F_1}^{\ 3}}{1 - \beta_{F_1}^{\ 3}}} \right) \text{ if } \alpha_{F_1} \leq \min\left\{ \alpha_{F_2} , \frac{\alpha_{F_2}\pi_1}{\pi_2} \right\} , \ \beta_{F_1} < \beta_{F_2} \tag{25}$$

5.2 Ranking of Fermatean Fuzzy Sets

The score function and the accuracy function are proposed in [65, 66] to rank FFSs. In Eq. (26) the score function of $F = (\alpha_F, \beta_F)$ is described:

$$score(F) = \alpha_F^3 - \beta_F^3 \quad \text{where } score(F) \in [-1, 1] \tag{26}$$

In Eq. (27) the accuracy function of $F = (\alpha_F, \beta_F)$ is represented:

$$acc(F) = \alpha_F^3 + \beta_F^3 \tag{27}$$

where $acc(F) \in [0, 1]$.

The ranking technique for any two FFNs can be defined as follows depending on the score and accuracy functions of the FFNs [68]:

Let $F_1 = (\alpha_{F_1}, \beta_{F_1})$ and $F_2 = (\alpha_{F_2}, \beta_{F_2})$ be two FFNs. $score(F_1)$, $score(F_2)$, $acc(F_1)$ and $acc(F_2)$ are the score values and accuracy values of F_1 and F_2, then

$$score(F_1) < score(F_2), \textit{then } F_1 < F_2; \tag{28}$$

$$score(F_1) > score(F_2), \textit{then } F_1 > F_2; \tag{29}$$

$$score(F_1) = score(F_2), \quad \text{then if} \quad \begin{array}{l} acc(F_1) < acc(F_2), \text{then } F_1 < F_2 \\ acc(F_1) > acc(F_2), \text{then } F_1 > F_2 \\ acc(F_1) = acc(F_2), \text{then } F_1 = F_2 \end{array} \tag{30}$$

Table 1 Fermatean fuzzy scale for rating the decision maker and criteria

Linguistic statement for the expertise level	Membership degree	Non-membership degree	Indeterminacy degree
Very low (VL)	0.05	0.95	0.52
Low (L)	0.25	0.80	0.78
Medium (M)	0.50	0.50	0.90
High (H)	0.80	0.15	0.78
Very high (VH)	0.95	0.01	0.52

5.3 *Proposed Methodology*

As it is noticed before, Fermatean fuzzy sets have a larger domain for membership and non-membership degrees which can handle higher levels of uncertainties. Therefore, Fermatean Fuzzy CODAS method is proposed in this section based on the steps defined in [70] that are detailed below:

Step 1. Determination of Decision Criteria, Alternatives and Fermatean Fuzzy Scales of Linguistic Statements.

Decision criteria (C_j where $j = 1, 2, \ldots, m$) could be defined based on the literature survey or by the experts using the consensus technique. A_i ($i = 1, 2, \ldots, n$) represents the alternatives that will be evaluated according to the decision criteria.

Fermatean Fuzzy Scale for membership and non-membership functions of linguistic statements is given in Table 1.

Fermatean Fuzzy CODAS Scale for the evaluation of the alternatives according to criteria is given in Table 2.

Step 2. Determination of the decision maker's weights.

In this step, weights are assigned to decision makers based on their level of expertise. The level of expertise of each decision maker is determined using linguistic statement given in Table 1 and normalized to find the weights for each decision maker. Equation (31) can be used for the normalization of the weights where λ_k represents the weight of decision maker k (k = 1, 2,..., K) with $\sum_{k=1}^{K} \lambda_k = 1$:

$$\lambda_k = \frac{(\alpha_k + \pi \alpha_k \beta_k)}{\sum\limits_{k=1}^{K} (\alpha_k + \pi \alpha_k \beta_k)} \tag{31}$$

Step 3. Determination of criteria weights.

Each decision maker determines the criteria weights which is denoted as w_{jk} with $k = 1, 2, \ldots, K$. Fermatean fuzzy weighted aggregation operator (FFWA) [65] could be used to calculate the aggregated criteria weights that is calculated by Eq. (32):

Table 2 Fermatean fuzzy scale for rating the alternatives

Linguistic statement	Membership degree	Non-membership degree	Indeterminacy degree
Absolutely weakly satisfactory (AWS)	0.05	0.95	0.5223
Very weakly satisfactory (VWS)	0.15	0.90	0.6444
Weakly satisfactory (WS)	0.25	0.80	0.7788
Fairly satisfactory (FS)	0.40	0.70	0.8401
Strongly satisfactory (SS)	0.50	0.60	0.8702
Very strongly satisfactory (VSS)	0.70	0.20	0.8657
Absolutely satisfactory (AS)	0.95	0.01	0.5224

$$w_j = FFWA\left(w_{j1}, w_{j2}, \ldots, w_{jK}\right) = \left(\sum_{k=1}^{K} \lambda_k \alpha_{w_{jk}}, \sum_{k=1}^{K} \lambda_k \beta_{w_{jk}}\right) \tag{32}$$

Step 4. Determination of individual decision matrix.

Individual decision matrix consisting of n alternatives and m criteria is constructed by each decision-maker using the scale of Fermatean fuzzy CODAS method that is given in Table 2. $\tilde{x}_{ij_k}$ denotes the performance value shown by the ith alternative on the jth criterion according to decision maker k.

$$\widetilde{X}_k = \left[\tilde{x}_{ijk}\right]_{n\times m} = \begin{bmatrix} \tilde{x}_{11_k} & \tilde{x}_{12_k} & \tilde{x}_{1m_k} \\ \tilde{x}_{21_k} & \tilde{x}_{22_k} & \tilde{x}_{2m_k} \\ \vdots & \vdots & \vdots \\ \tilde{x}_{n1_k} & \tilde{x}_{n2_k} & \tilde{x}_{nm_k} \end{bmatrix} \tag{33}$$

where $\tilde{x}_{ij_k} \geq 0$ and $0 \leq \left(\alpha_F\left(\tilde{x}_{ij_k}\right)\right)^3 + \left(\beta_F\left(\tilde{x}_{ij_k}\right)\right)^3 \leq 1$.

Step 5. Determination of aggregated decision matrix.

The individual alternative evaluations are combined into aggregated decision matrix by FFWA aggregation operator.

$$\widetilde{X} = \left[\tilde{x}_{ij} \right]_{n \times m} = \begin{bmatrix} \tilde{x}_{11} & \tilde{x}_{12} & \tilde{x}_{1m} \\ \tilde{x}_{21} & \tilde{x}_{22} & \tilde{x}_{2m} \\ \vdots & \vdots & \vdots \\ \tilde{x}_{n1} & \tilde{x}_{n2} & \tilde{x}_{nm} \end{bmatrix} \tag{34}$$

where

$$\tilde{x}_{ij} = \left(\sum_{k=1}^{K} \lambda_k \alpha_{x_{ijk}}, \sum_{k=1}^{K} \lambda_k \beta_{x_{ijk}} \right) \tag{35}$$

Step 6. Calculation of the Fermatean fuzzy normalized decision matrix.

In this step, linear normalization is performed using Eq. (36).

$$\tilde{n}_{ij} = \begin{cases} \frac{\tilde{x}_{ij}}{\max\limits_{i=1,\ldots,n} (\tilde{x}_{ij})} \text{if } j \in \boldsymbol{J}^{\uparrow} \\ \frac{\tilde{x}_{ij}}{\min\limits_{i=1,\ldots,n} (\tilde{x}_{ij})} \text{if } j \in \boldsymbol{J}^{\downarrow} \end{cases} \tag{36}$$

In order to find the maximum and minimum values with respect to a certain criterion, score and accuracy functions (Eqs. (26–27)) and their relations (Eqs. (28–30)) are used.

Step 7. Calculation of the Fermatean Fuzzy weighted normalized decision matrix.

After determining the weight coefficients ($\tilde{w}_j$; the weight of jth criterion), related to the evaluation criteria, the weighting of the matrix is performed by multiplying the column elements of the normalized decision matrix with the respective weight coefficients:

$$\tilde{s}_{ij} = \tilde{w}_j \tilde{n}_{ij} (i = 1, \ldots, n, j = 1, \ldots, m) \tag{37}$$

where $0 < \tilde{w}_j < 1$ and $\sum\limits_{J=1}^{m} \tilde{w}_j = 1$.

Step 8. Determination the Fermatean fuzzy negative-ideal solution.

In order to obtain the solution from the CODAS method, the negative-ideal solution point should be determined. Accordingly, the smallest of column values (minimums) in the weighted matrix are selected. Equation (38) is used to determine the negative-ideal solution point.

$$\boldsymbol{NS} = \left[\widetilde{ns}_j \right]_{1 \times m}, \widetilde{ns}_j = \min_i \tilde{s}_{ij} \tag{38}$$

In this step, the alternative that is at the farthest distance to the negative-ideal solution according to all criteria is the optimal alternative [38].

Step 9. Calculation of Fermatean fuzzy Euclidean and Taxicab distances.

In the CODAS method, Euclidean and Taxicab distance approaches are used to determine the distance of the criterion values related to alternatives from the negative-ideal solution point. The calculation of Euclidean distance (e_i) [69] and Taxicab distance (t_i) values of each alternative are shown in Eq. (39) and Eq. (40), respectively.

$$\mathrm{e_i} = \sqrt{\sum_{j=1}^{m}\left(\frac{1}{2\mathrm{m}}\left[\left(\alpha_{sij}^3 - \alpha_{nsj}^3\right)^2 + \left(\beta_{sij}^3 - \beta_{nsj}^3\right)^2 + \left(\pi_{sij}^3 - \pi_{nsj}^3\right)^2\right]\right)} \tag{39}$$

$$\begin{aligned} t_\mathrm{i} &= \frac{1}{m}\sum_{j=1}^{m}\left|\tilde{s}_\mathrm{ij} - \widetilde{ns}_\mathrm{j}\right| \\ &= \frac{1}{2m}\sum_{j=1}^{m}\left|\alpha_{sij} - \alpha_{nsj}\right| + \left|\beta_{sij} - \beta_{nsj}\right| + \left|\pi_{sij} - \pi_{nsj}\right| \end{aligned} \tag{40}$$

Step 10. Creating the comparative evaluation matrix.

In this step, each alternative is compared in terms of Euclidean and Taxicab distance to other alternatives. The creation of the comparative evaluation matrix is shown in Eqs. (41) and (42).

$$R_a = [h_{ik}]_{nxn} \tag{41}$$

$$h_{ik} = (E_i - E_k) + (\psi(E_i - E_k) \times (T_i - T_k)) \tag{42}$$

where the value ψ is a threshold value indicating the equality of the distances $\mathrm{E_i}$. Equation (43) is used in the calculation of this value.

$$\psi(\mathrm{x}) = \begin{cases} 1\ \mathrm{f}\ |\mathrm{x}| \geq \tau, \\ 0\ \mathrm{if}\ |\mathrm{x}| < \tau \end{cases} \tag{43}$$

The τ value in Eq. (43) is a parameter determined by the decision-maker. Accordingly, if the difference between the Euclidean distances of the two alternatives compared is less than τ, the corresponding alternatives are compared according to the Taxicab distances.

Step 11. Calculation of evaluation scores.

In the last step, the evaluation score of each alternative is calculated using Eq. (44).

$$H_{ik} = \sum_{k=1}^{n} h_{ik} \tag{44}$$

By ordering the H scores of the alternatives from largest to smallest, the alternatives are ranked from the best to the worst.

6 Application

For this study, a logistics company which is one of the most used air cargo companies in Turkey is chosen for the application. In addition to air cargo, the company also provides cargo delivery service from address to address. In the current situation, although the company uses digital tools, it has not yet switched to a combined smart system for air cargo operations. There are three intelligent air cargo system providers and each system has different features in different levels of detail. Alternative 1 uses a software which can share data with external systems such as intelligent production system of the customer or intelligent system used in road cargo delivery department. Besides that Alternative 1 provides more complex algorithms that result in closer solutions to optimal for Aircraft configuration problems. Higher initial cost and complexity of the usage are its disadvantages. Alternative 2 provides an interface for ease of use but the improvement on the processes are relatively low than other alternatives. External data share is not possible for this alternative. It has an anomaly detection feature for a few cargo properties and it is the minimum costly one among alternatives. Alternative 3 has a special feature for the anomaly detection which can control the temperature, shape deformities and positioning but it cannot be integrated with other intelligent systems. Ease of use is another advantage of this system.

To find out the best alternative for intelligent air cargo system implementation, the proposed fuzzy Fermatean CODAS method is used as follows.

To define the decision criteria, two managers from the air cargo department (Expert 1) and one manager from the road logistics department of the company (Expert 2), and one manager from one of the largest customers (Expert 3) are chosen. Since the managers of the air cargo department made the decisions together, they were handled as a single expert. The following seven criteria are determined by the consensus technique:

Low life cycle cost of the system (C_1): It is one of the important criteria that the managers take into account. The initial cost that has to be paid for the system and the operational costs of the software updates increase the total cost as well as the prices reflected to the customer.

Usability (C_2): Since the system will be used by both customers and employees, it has been decided that the duration and intensity of the training required in its use is an important criterion.

Improving the air cargo load factors (C_3): Cargo load factor is the ratio of the average load to total vehicle freight capacity. Therefore, improvement in the air cargo load factor will help to decrease costs and improve efficiency.

Implementation with intelligent cargo transport systems (C_4): This feature will be helpful for holistic supply chain management and will reduce the total cost of cargo delivery service from address to address.

Implementation with intelligent production systems (C_5)*:* This feature of the system allows customers to achieve predicted cargo loading schedules and produce according to these forecasts. This feature will increase the number of customers who wants to decrease warehousing costs.

Tracking and Anomaly Detection (C_6): This feature enables the carrier and the customer real-time tracking and it gives an alarm when the temperature or other conditions that are important for the cargo goes out of the specified limits. Although existing systems have a very different range of complexity and detail, they are often utilizing sensors, RFID and IoT technologies.

Improving cargo delivery lead time (C_7): With this feature, it is expected that the costs will decrease as the waiting times decrease. Also, shorter delivery times will increase the number of customers.

After determining the criteria, decision makers' weights are calculated according to their expertise levels that are both given in Table 3.

Table 4 shows the criteria weights of each decision maker and the aggregated weights which is calculated using Eq. (32).

After determining the criteria weights, experts evaluated the alternatives according to their meeting the criteria. The individual decision matrices obtained are given in the Table 5.

Table 3 Linguistic statements and weights of expert's level of expertise

	Expert 1 (E1)	Expert 2 (E2)	Expert 3 (E3)
Linguistic statement	VH	H	M
λ_k	0.37	0.35	0.28

Table 4 Criteria weights

	E1	E2	E3	$\tilde{w}_j$
C1	(0.5, 0.6)	(0.25, 0.8)	(0.15, 0.9)	(0.31, 0.75)
C2	(0.25, 0.95)	(0.4, 0.7)	(0.4, 0.7)	(0.34, 0.74)
C3	(0.95, 0.01)	(0.7, 0.2)	(0.05, 0.95)	(0.61, 0.34)
C4	(0.15, 0.9)	(0.15, 0.9)	(0.25, 0.8)	(0.18, 0.87)
C5	(0.05, 0.95)	(0.05, 0.95)	(0.7, 0.2)	(0.23, 0.74)
C6	(0.4, 0.7)	(0.5, 0.6)	(0.5, 0.6)	(0.46, 0.64)
C7	(0.7, 0.2)	(0.95, 0.01)	(0.95, 0.01)	(0.86, 0.08)

Table 5 Individual decision matrices

	C1	C2	C3	C4	C5	C6	C7
Expert 1							
A1	AWS	WS	VSS	AS	AS	FS	VSS
A2	AS	AS	VWS	AWS	AWS	SS	WS
A3	SS	AS	WS	AWS	AWS	AS	VWS
Expert 2							
A1	VWS	AWS	VSS	AS	AS	WS	AS
A2	AS	VSS	WS	AWS	AWS	FS	FS
A3	VSS	SS	WS	AWS	AWS	AS	WS
Expert 3							
A1	WS	VWS	SS	AS	AS	FS	AS
A2	VSS	SS	WS	AWS	AWS	SS	FS
A3	SS	VSS	FS	AWS	AWS	AS	FS

At this stage, individual decision matrices are aggregated, normalized, and weighted, respectively. Matrices belonging to these operations are shown in Table 6.

In order to obtain the solution, the negative-ideal solution point is determined using Eq. (38) as $= (0.16, 0.89)(0.21, 0.88)(0.33, 0.83)\ (0.05, 0.95)(0.05, 0.95)$ $(0.37, 0.73)(0.30, 0.81)$. After that, Euclidean and Taxicab distances are calculated using Eqs. 39–40 for each alternative and given in Table 7.

Using Eqs. (41–43), relative assessment matrix is constructed. The τ value is determined as 0.01 in the calculations (Table 8).

At the end final scores and ranks of the alternatives are calculated using Eq. (44) and represented in Table 9.

According to the results obtained, the alternative that meets the improvement at delivery time, improving air cargo load factor and having a tracking and anomaly detection criteria, which stands out in terms of criterion weights, relatively more than other alternatives, is preferred despite having the highest cost.

Of course, the decision to select Alternative 1 is closely related to the opinion, preferences and expertise of the selected experts. It might be considered to verify this decision, taking into account also the opinion of other experts. This verification might take place in the analogous form. However, the dominance of Alternative 1 is here clear (it is the only alternative with a positive score), thus, in this case, it can be safely assumed that Alternative 1 actually best fits the needs of the company in question. This conclusion is reinforced by the fact that the three experts selected for the decision-making process represent the main pillars of the company: the air cargo department, the road logistics department and the CRM department.

Table 6 Decision matrices that are used in the calculation

	C1	C2	C3	C4	C5	C6	C7
Aggregated decision matrix							
A1	(0.14, 0.89)	(0.15, 0.88)	(0.64, 0.31)	(0.95, 0.01)	(0.95, 0.01)	(0.35, 0.74)	(0.86, 0.08)
A2	(0.88, 0.06)	(0.74, 0.24)	(0.21, 0.84)	(0.05, 0.95)	(0.05, 0.95)	(0.46, 0.64)	(0.34, 0.74)
A3	(0.57, 0.46)	(0.72, 0.27)	(0.29, 0.77)	(0.05, 0.95)	(0.05, 0.95)	(0.95, 0.01)	(0.26, 0.81)
Normalized decision matrix							
A1	(0.16, 0.89)	(0.21, 0.88)	(1, 0)	(1, 0)	(1, 0)	(0.37, 0.73)	(1, 0)
A2	(1, 0)	(1, 0)	(0.33, 0.83)	(0.05, 0.95)	(0.05, 0.95)	(0.49, 0.63)	(0.40, 0.74)
A3	(0.65, 0.46)	(0.98, 0.18)	(0.45, 0.76)	(0.05, 0.95)	(0.05, 0.95)	(1, 0)	(0.29, 0.81)
Weighted normalized decision matrix							
A1	(0.05, 0.94)	(0.07, 0.93)	(0.61, 0.34)	(0.18, 0.87)	(0.23, 0.74)	(0.17, 0.82)	(0.86, 0.08)
A2	(0.31, 0.75)	(0.34, 0.74)	(0.20, 0.84)	(0.01, 0.98)	(0.01, 0.97)	(0.23, 0.77)	(0.34, 0.74)
A3	(0.20, 0.79)	(0.34, 0.74)	(0.28, 0.77)	(0.01, 0.98)	(0.01, 0.97)	(0.46, 0.64)	(0.26, 0.81)

Table 7 Euclidean and taxicab distances

	Euclidean distance	Taxicab distance
A1	0.34	0.30
A2	0.16	0.12
A3	0.15	0.11

Table 8 Relative assessment matrix

	A1	A2	A3
A1	0.00	0.09	0.19
A2	−0.09	0.00	0.02
A3	−0.19	−0.02	0.00

Table 9 Scores and ranks of the alternatives

	Scores	Ranks
A1	0.28	1
A2	−0.07	2
A3	−0.21	3

7 Conclusion

In this chapter, we consider the problem of analysing and selecting software supporting the air cargo process. This problem requires first of all the identification of alternatives accessible on the market and their characteristics. As we show in our analysis of the present situation in the field of intelligent software supporting the air cargo process, conducted against the background of the challenges related to air cargo today, it has to be stated that the accessible software is still far from satisfying the needs of all the air cargo stakeholders. Numerous innovative projects have been being implemented, but many of them have not led to a tangible result yet. On top of that, there is still a great potential for new innovations which would lead to speeding up the air cargo processes and reducing their cost, while observing the variety of constraints linked to air cargo and fulfilling the quality requirements. On the other hand, a number of intelligent software solution is already used all over the world, although there is still huge potential for improvement, for application of existing theoretical research results and recent technologies.

In the application part of the chapter, we consider a Turkish large logistics company which is facing the problem of choosing a software for the support of their air cargo related activities. The alternatives analysed correspond to three air cargo software providers present on the market.

The next step in the decision-making process is the selection of criteria. In the case study presented in the chapter, the experts chose criteria linked to the cost of the software, to its capacity of improving the parameters of the air cargo process, to its capacity of detecting problems occurring in this process, to its compatibility with other software used in the company and to the difficulty of using the software. Of course, the selection of these criteria is highly subjective. In the future, when the alternatives accessible on the market of Aviation 4.0 related software have more functionalities and are more sophisticated, the selection criteria are bound to change.

As it can be concluded from the analysis of the air cargo process and the developments of related software, in the future more and more stakeholders will have to be involved in the process of selecting air cargo supporting software. In the present market situation, it was justified to select only three experts from the considered logistics company, but the air cargo supporting intelligent software in the future will serve many stakeholders: customers, agents, intermediaries, airports etc. All of them will have to be taken into account—both while choosing the criteria for the software selection and while evaluating them. An important criterion will have to be the expansion possibilities of the software, the capacity to integrate new optimization and automation related intelligent solutions.

Once the alternatives and the evaluation criteria are set, the method of determining the optimal alternative has to be selected. There exists a variety of multicriteria decision-making methods. We have decided to use a recently proposed method, called CODAS. The reason for this decision was the simplicity of the method and its intuitiveness. At the same time, we decided to use a fuzzy approach. This choice was due to the fact that the above-listed software evaluation criteria, chosen by the experts, are basically (apart from software cost) of qualitative nature, thus it is natural to assess the alternatives with respect to the different criteria by means of human language-related terms, which can be conveniently modelled by means of fuzzy sets. We conducted a literature research on the existing fuzzy extensions of the CODAS method and we found out that a recently proposed type of fuzzy sets, which offers broader modelling possibilities than the existing types, the Fermatean fuzzy sets, has not yet been used in the framework of the CODAS approach yet. Thus, from the point of view of multicriteria decision-making theory, the main novelty of the chapter is the proposal of the Fermatean fuzzy sets-based CODAS method.

The method was applied to the case study discussed in the chapter. The experts were presented with a set of linguistic terms to use in their assessment—which were in fact Fermatean fuzzy sets. The experts were not required to understand the theory behind these notions. An important limitation of this approach was the fact that the experts were not able to create their own "language", their own linguistic terms, but were forced to use the imposed "vocabulary". Future research should incorporate the generation of linguistic terms to be used in an interactive process with the decision makers, so that the fuzzy set modelled linguistic terms used in the decision-making process correspond exactly to the opinions and preferences of the participants of the decision-making process. In the ideal case, the experts should know basic elements of fuzzy set theory and be able to choose between different types of fuzzy sets and different membership functions for the modelling of the linguistic terms to be used by them.

To sum up, the future research step will include:

- an on-going analysis of innovations and new products in the area of intelligent software related to air cargo;
- identification of stakeholders of such software;
- an on-going identification of criteria of assessment of such software, corresponding to the expectations of the stakeholders of the air cargo process;
- elaboration of interacting procedures of identification of linguistic terms to be used in the process of assessment of such software, making use of the modelling by means of different types of fuzzy numbers;
- elaboration of case studies related to the analysis and selection of intelligent software related to the air cargo process.

The outcomes of these research steps will largely depend on the development in the field and on the market of intelligent software related to air cargo. It can be assumed that this development will be inevitably speeding up and taking larger and larger steps in the context of Aviation 4.0. Thus, a lot of research will be needed in order to elaborate suitable tools which would effectively guide companies in the selection of optimal software supporting air cargo processes.

References

1. Valdes, R.A., Comendador, V.F.G., Sanz, A.R., Castan, J.P.: Aviation 4.0 more safety through automation and digitization. Aircraft Technol. **2**(4), 25–41 (2018). https://doi.org/10.5772/intechopen.73688
2. Durak, G., Tolga, A.C.: Process robot automation selection with MADM in airline cargo sector. In: International Conference on Intelligent and Fuzzy Systems, pp. 525–533. Springer, Cham (2020)
3. Tabares, D.A., Mora-Camino, F., Drouin, A.: A multi-time scale management structure for airport ground handling automation. J. Air Transp. Manage. **90**, 101959 (2021)
4. Wong, E.Y., Mo, D.Y., So, S.: Closed-loop digital twin system for air cargo load planning operations. Int. J. Comput. Integr. Manuf. 1–13 (2020). https://doi.org/10.1080/0951192X.2020.1775299
5. Tolga, A.C., Durak, G.: Evaluating innovation projects in air cargo sector with fuzzy COPRAS. In: International Conference on Intelligent and Fuzzy Systems, pp. 702–710. Springer, Cham (2019)
6. Gumzej, R., Komkhao, M., Sodsee, S.: Design of an intelligent, safe and secure transport unit for the physical internet. In: International Conference on Computing and Information Technology, pp. 60–69. Springer, Cham (2020)
7. Cimato, S., Gianini, G., Sepehri, M., Asal, R., Damiani, E.: A cryptographic cloud-based approach for the mitigation of the airline cargo cancellation problem. J. Inform. Secur. Appl. **51**, 102462 (2020)
8. Fang, Z., Mao, J.: Energy-efficient elevating transfer vehicle routing for automated multi-level material handling systems. IEEE Trans. Autom. Sci. Eng. **17**(3), 1107–1123 (2019)
9. Delgado, F., Sirhan, C., Katscher, M., Larrain, H.: Recovering from demand disruptions on an air cargo network. J. Air Transp. Manage. **85**, 101799 (2020)
10. Emde, S., Abedinnia, H., Lange, A., Glock, C.H.: Scheduling personnel for the build-up of unit load devices at an air cargo terminal with limited space. OR Spectrum **42**(2), 397–426 (2020)
11. Giusti, I., Cepolina, E.M., Cangialosi, E., Aquaro, D., Caroti, G., Piemonte, A.: Mitigation of human error consequences in general cargo handler logistics: impact of RFID implementation. Comput. Ind. Eng. **137**, 106038 (2019)
12. Zhang, W., Chen, Y.: Intelligent technology related to warehousing and distribution in intelligent logistics. In: 2020 International Conference on Wireless Communications and Smart Grid (ICWCSG), pp. 175–177. IEEE (2020)
13. Wang, J., Lim, M.K., Zhan, Y., Wang, X.: An intelligent logistics service system for enhancing dispatching operations in an IoT environment. Transp. Res. Part E Logist. Transp. Rev. **135**, 101886 (2020)
14. Liu, R., Li, H.: Intelligent logistics service combination algorithm based on internet of things. J Intel Fuzzy Syst (Preprint), 1–8 (2020). https://doi.org/10.3233/JIFS-179854
15. Xue, F., Dong, T., You, S., Liu, Y., Tang, H., Chen, L., Li, J. et al.: A hybrid many-objective competitive swarm optimization algorithm for large-scale multirobot task allocation problem. Int. J. Mach. Learn. Cybern. 1–15 (2020). https://doi.org/10.1007/s13042-020-01213-4
16. Tang, Y., Zhang, J., Yuan, X., Hao, H., Wang, J., Zuo, Y., Wan, Y. et al.: Intelligent logistics system architecture design based on edge computing. In: 2019 Chinese Automation Congress (CAC), pp. 1682–1685. IEEE (2019)
17. Vijay, R., Prabhakar, T.V., Hegde, V., Rao, V.S., Prasad, R.V.: A heterogeneous PLC with BLE Mesh network for reliable and real-time smart cargo monitoring. In: 2019 IEEE International Symposium on Power Line Communications and its Applications (ISPLC), pp. 1–6. IEEE (2019)
18. Brandt, F., Nickel, S.: The air cargo load planning problem-a consolidated problem definition and literature review on related problems. Eur. J. Oper. Res. **275**(2), 399–410 (2019)
19. URL-1, https://www.iso.org/ics/55.180.20/x/

20. Li, Z.Z.: Based on value analysis process optimization of air cargo transport. in advanced materials research, vol. 468, pp. 689–693. Trans Tech Publications Ltd. (2012). https://doi.org/10.4028/www.scientific.net/AMR.468-471.689
21. URL-2, https://www.twill.net/blog/freight-documents/
22. URL-3, https://www.shipafreight.com/learn-more/documents-list/
23. Tian, C., Zhang, H., Li, F., Liu, T.: Air cargo load planning system: a rule-based optimization approach. In: 2009 IEEE/INFORMS International Conference on Service Operations, Logistics and Informatics, pp. 454–459. IEEE (2009)
24. Schäfer, J.D.: Luftfracht: Akteure–Prozesse–Märkte–Entwicklungen. Springer Fachmedien Wiesbaden GmbH (2019)
25. URL-4, https://lufthansa-cargo.com/eservices/efreight
26. Brett, D.: ANA Cargo expands digital booking capabilities. Aircargo news (2020). https://www.aircargonews.net/airlines/ana-cargo-expands-digital-booking-capabilities/?fbclid=IwAR2x7gmQhO6ViNSRJrXOT4VATkYRhCd38rcYQJ_QfX3OrUCc_fjjuXhSPBM
27. Brett, D.: Cargo.one the latest to integrate with IBS. Aircargo news (2020). https://www.aircargonews.net/technology/cargo-one-the-latest-to-integrate-with-ibs/
28. Brett, D.: Nallian to offer RFS slot booking app at Brussels Airport. Aircargo news (2020). https://www.aircargonews.net/cargo-airport/nallian-to-offer-rfs-slot-booking-app-at-brussels-airport/
29. Harry, R.: Kale logistics offers free trial of air waybill processing tool. Aircargo news (2020). https://www.aircargonews.net/technology/e-air-waybill/kale-logistics-offers-free-trial-of-air-waybill-processing-tool/
30. Harry, R.: Air cargo community system launched at Atlanta Airport. Aircargo news (2020). https://www.aircargonews.net/technology/logistics-automation/air-cargo-community-system-launched-at-atlanta-airport/#:~:text=17%20%2F%2012%20%2F%202019&text=Hartsfield%E2%80%93Jackson%20Atlanta%20International%20Airport,cargo%20community%20system%20(ACCS)
31. Harry, R.: AirAsia launches blockchain-based cargo booking platform. Aircargo news (2020). https://www.aircargonews.net/technology/logistics-automation/airasia-launches-blockchain-based-cargo-booking-platform/
32. Harry, R.: Peli biothermal software update enables temperature tracking of mass shipments. Aircargo news (2020). https://www.aircargonews.net/technology/peli-biothermal-software-update-enables-temperature-tracking-of-mass-shipments/
33. Chen, S.L.: Aerial logistics management for carrier onboard delivery. Naval Postgraduate School Monterey United States (2016)
34. Azadian, F., Murat, A., Chinnam, R.B.: An unpaired pickup and delivery problem with time dependent assignment costs: application in air cargo transportation. Eur. J. Oper. Res. **263**(1), 188–202 (2017)
35. Harry, R.: CargoLogicAir operations boosted with CHAMP's load planning tool. Aircargo news (2020). https://www.aircargonews.net/airlines/cargologicair-operations-boosted-with-champs-load-planning-tool/
36. Keshavarz Ghorabaee, M., Zavadskas, E.K., Turskis, Z., Antucheviciene, J.: A new combinative distance-based assessment (CODAS) method for multi-criteria decision-making. Econ. Comput. Econ. Cybern. Stud. Res. **50**(3) (2016)
37. Ghorabaee, M.K., Amiri, M., Zavadskas, E.K., Hooshmand, R., Antuchevičienė, J.: Fuzzy extension of the CODAS method for multi-criteria market segment evaluation. J. Bus. Econ. Manage. **18**(1), 1–19 (2017)
38. Dahooei, J.H., Zavadskas, E.K., Vanaki, A.S., Firoozfar, H.R., Keshavarz-Ghorabaee, M.: An evaluation model of business intelligence for enterprise systems with new extension of codas (CODAS-IVIF) (2018)
39. Ren, J.: Sustainability prioritization of energy storage technologies for promoting the development of renewable energy: a novel intuitionistic fuzzy combinative distance-based assessment approach. Renew. Energy **121**, 666–676 (2018)

40. Bolturk, E., Kahraman, C.: Interval-valued intuitionistic fuzzy CODAS method and its application to wave energy facility location selection problem. J. Intel. Fuzzy Syst. **35**(4), 4865–4877 (2018)
41. Yeni, F.B., Özçelik, G.: Interval-valued Atanassov intuitionistic Fuzzy CODAS method for multi criteria group decision making problems. Group Decis. Negot. **28**(2), 433–452 (2019)
42. Roy, J., Das, S., Kar, S., Pamučar, D.: An extension of the CODAS approach using interval-valued intuitionistic fuzzy set for sustainable material selection in construction projects with incomplete weight information. Symmetry **11**(3), 393 (2019)
43. Karagoz, S., Deveci, M., Simic, V., Aydin, N., Bolukbas, U.: A novel intuitionistic fuzzy MCDM-based CODAS approach for locating an authorized dismantling center: a case study of Istanbul. Waste Manage. Res. **38**(6), 660–672 (2020)
44. Seker, S.: A novel interval-valued intuitionistic trapezoidal fuzzy combinative distance-based assessment (CODAS) method. Soft. Comput. **24**(3), 2287–2300 (2020)
45. Atanassov, K.: Intuitionistic fuzzy sets. Int. J. Bioauto. **20**(1) (2016)
46. Bolturk, E.: Pythagorean fuzzy CODAS and its application to supplier selection in a manufacturing firm. J. Enterp. Inf. Manag. **31**(4), 550–564 (2018)
47. Bolturk, E., Kahraman, C.: A modified interval-valued pythagorean fuzzy CODAS method and evaluation of AS/RS technologies. J. Multiple-Valued Logic Soft Comput. **33**, 415–429 (2019)
48. Peng, X., Ma, X.: Pythagorean fuzzy multi-criteria decision making method based on CODAS with new score function. J. Intell. Fuzzy Syst. **38**(3), 3307–3318 (2020)
49. He, T., Zhang, S., Wei, G., Wang, R., Wu, J., Wei, C.: CODAS method for 2-tuple linguistic Pythagorean fuzzy multiple attribute group decision making and its application to financial management performance assessment. Technol. Econ. Develop. Econ. **26**(4), 920–932 (2020)
50. Yager, R.R.: Pythagorean membership grades in multicriteria decision making. IEEE Trans. Fuzzy Syst. **22**(4), 958–965 (2013)
51. G¨ʊNDOĞDU, F.K., Kahraman, C.: Extension of CODAS with spherical fuzzy sets. J. Multiple-Valued Logic Soft Comput. **33**, 481–505 (2019)
52. Karaşan, A., Boltürk, E., Gündoğdu, F.K.: Assessment of livability indices of suburban places of istanbul by using spherical fuzzy CODAS method. In: Decision Making with Spherical Fuzzy Sets, pp. 277–293. Springer, Cham (2020)
53. Kutlu Gündoğdu, F., Kahraman, C.: Spherical fuzzy sets and spherical fuzzy TOPSIS method. J. Intel. Fuzzy Syst. **36**(1), 337–352 (2019)
54. Yalçın, N., Yapıcı Pehlivan, N.: Application of the fuzzy CODAS method based on fuzzy envelopes for hesitant fuzzy linguistic term sets: a case study on a personnel selection problem. Symmetry **11**(4), 493 (2019)
55. Karasan, A., Zavadskas, E.K., Kahraman, C., Keshavarz-Ghorabaee, M.: Residential construction site selection through interval-valued hesitant fuzzy CODAS method. Informatica **30**(4), 689–710 (2019)
56. Torra, V.: Hesitant fuzzy sets. Int. J. Intell. Syst. **25**(6), 529–539 (2010)
57. Liu, H., Rodríguez, R.M.: A fuzzy envelope for hesitant fuzzy linguistic term set and its application to multicriteria decision making. Inf. Sci. **258**, 220–238 (2014)
58. Boltürk, E., Karaşan, A.: Interval valued neutrosophic CODAS method for renewable energy selection. In: Liu, J., Lu, J., Xu, Y., Martinez, L., Kerre, E. (eds.) Data Science and Knowledge Engineering for Sensing Decision Support, pp. 1026–1033 (2018)
59. Karasan, A., Bolturk, E., Kahraman, C.: An integrated methodology using neutrosophic CODAS & fuzzy inference system: assessment of livability index of urban districts. J. Intel. Fuzzy Syst. **36**(6), 5443–5455 (2019)
60. Rivieccio, U.: Neutrosophic logics: prospects and problems. Fuzzy Sets Syst. **159**(14), 1860–1868 (2008)
61. Tüysüz, N., Kahraman, C.: CODAS method using Z-fuzzy numbers. J. Intel. Fuzzy Syst. **38**(2), 1649–1662 (2020)
62. Zadeh, L.A.: A note on Z-numbers. Inf. Sci. **181**(14), 2923–2932 (2011)
63. Peng, X., Garg, H.: Algorithms for interval-valued fuzzy soft sets in emergency decision making based on WDBA and CODAS with new information measure. Comput. Ind. Eng. **119**, 439–452 (2018)

64. Ahmad, B., Kharal, A.: On fuzzy soft sets. Adv. Fuzzy Syst. 586507 (2009). https://doi.org/10.1155/2009/586507
65. Senapati, T., Yager, R.R.: Fermatean fuzzy weighted averaging/geometric operators and its application in multi-criteria decision-making methods. Eng. Appl. Artif. Intell. **85**, 112–121 (2019)
66. Du, W.S.: Weighted power means of q-rung orthopair fuzzy information and their applications in multiattribute decision making. Int. J. Intell. Syst. **34**(11), 2835–2862 (2019)
67. Liu, D., Liu, Y., Chen, X.: Fermatean fuzzy linguistic set and its application in multicriteria decision making. Int. J. Intell. Syst. **34**(5), 878–894 (2019)
68. Senapati, T., Yager, R.R.: Some new operations over Fermatean fuzzy numbers and application of Fermatean fuzzy WPM in multiple criteria decision making. Informatica **30**(2), 391–412 (2019)
69. Senapati, T., Yager, R.R.: Fermatean fuzzy sets. J. Ambient Intel. Human. Comput. **11**(2), 663–674 (2020)
70. Buyukozkan, G., Göçer, F.: Prioritizing the strategies to enhance smart city logistics by intuitionistic fuzzy CODAS. In: 2019 Conference of the International Fuzzy Systems Association and the European Society for Fuzzy Logic and Technology (EUSFLAT 2019). Atlantis Press (2019)

Intelligent and Fuzzy Applications in Aircraft Handling Services with Aviation 4.0

Diego Alonso Tabares and Felix Mora-Camino

Abstract This chapter introduces and analyses the state of the art of aircraft ground handling automation at airports. With the perspective of full automation of the air transport industry (Aviation 4.0), a management structure compatible with on-line automation of ground handling vehicles is proposed. This leads to identifying different decision problems to be tackled to generate safe and efficient automated ground handling operations. Solution algorithms are proposed where inherent uncertainty is modelled using fuzzy dual numbers demonstrating the feasibility of higher automation of ground handling through Aviation 4.0.

Keywords Aviation 4.0 · Airport ground operations · Aircraft ground handling · Fleet management · Automation · Ground handling innovation

1 Introduction

Industry 4.0 is associated with the fourth industrial revolution. After the industrial revolutions based successively on steam power, on electricity and on electronics, new technologies such as artificial intelligence, data science, internet of things and others contribute to the generation of smarter production processes characterizing Industry 4.0. In the field of Aviation, the aeronautical industry has become a rich field of application of these new concepts and approaches, giving way to Aviation 4.0 [1]. However, it appears that to fully meet its objectives of efficiency and safety, Aviation 4.0 has to cover the whole life cycle of these new products and processes, including their operation.

Ground handling operations represent the airside activities at airports in charge of processing passengers, cargo, facilities and supplies at and around parked aircraft.

D. Alonso Tabares
Chair of ISO TC 20/SC 9 – Air Cargo and Ground Equipment, Geneva, Switzerland
e-mail: diego.alonso-tabares@airbus.com

F. Mora-Camino (✉)
DUT, South Africa, Neometsys and ENAC, Toulouse, France, Fluminense Federal University, Niterói, Brazil

C. Kahraman and S. Aydın (eds.), *Intelligent and Fuzzy Techniques in Aviation 4.0*,
Studies in Systems, Decision and Control 372,
https://doi.org/10.1007/978-3-030-75067-1_8

Most of these operations are performed by different service providers, using vehicles called Ground Support Equipment (GSE) which are specific to each type of operation. Ground handling has not been considered a prominent activity within the Air Transportation System (ATS); however, this activity is an important enabler for efficient airport operation and its management is an important issue. Over the last decades, the complexity of ATS has increased to cope with the worldwide growth of air traffic. Today the operation of this system involves global actors (airports, airlines, Air Traffic Control (ATC), Air Traffic Management (ATM)) as well as local actors (ground handlers, local suppliers …) whose coordination, while pursuing different and sometimes contradictory objectives, is difficult to achieve. Ground handling operations are at the interface of air and ground operations of air transport and play a critical role for its efficiency and safety, then in the perspective of a global implementation of Aviation 4.0, its automation results essential.

The current chapter is composed of the following sections:

Section 2 introduces the current practice of ground handling operations at airports; then in Sect. 3, after considering some of the few studies existing about automation of ground handling, a general management structure is proposed with its different time scales and elementary decision problems covering the whole set of ground handling activities. Then, in Sect. 4, the corresponding decision problems are formulated as optimization problems over a sliding time window and solution algorithms are proposed where the uncertainty about the duration of each elementary process is coped by using fuzzy dual numbers. Finally, the last section brings conclusions and perspectives with respect to Ground Handling automation in the context of Aviation 4.0.

2 Aircraft Ground Handling Operations

The ground handling tasks are usually defined and contracted between the airline and the provider using a service level agreement that defines the scope, price and quality level desired and key performance indicators as well. It has been described an analyzed in the perspective of efficiency [2, 3] and [4] and more specifically in the perspective of increased automation [5]. The ground handling activities are usually divided in:

"Over the wing" activities, meaning aircraft passenger cabin related activities: passenger boarding and deplaning, catering galleys, cabin cleaning and preparation, safety and security checks as needed.

"Under the wing" activities, which are cargo (both containerized and bulk) unloading and loading and other ramp activities: provide electricity (Ground Power Unit (GPU)), condition aircraft cabin temperature (Pre-Conditioned Air unit (PCA)), chocking of landing gears, use of cones, marshalling, refueling, potable water and toilet servicing, towing/pushback and provide access means to the passengers (via stairs, ramps or Passenger Boarding Bridge (PBB)).

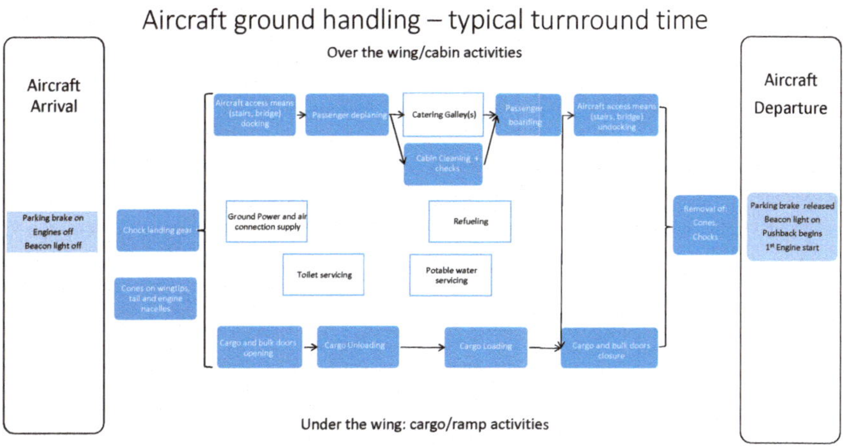

Fig. 1 Typical turnaround time. Tasks sequence

The generic sequence of these activities, shown in Fig. 1, is established depending on:

- Those tasks related to aircraft arrival: passenger deplaning, cargo unloading, toilet servicing, removal of catering, cabin cleaning.
- Tasks related to aircraft departure: passenger boarding, cargo loading, refueling, potable water, provide catering, cabin preparation.
- Safety and security requirements for the overall operation.
- Aircraft accessibility to doors and servicing panels.
- Links and constraints (represented with arrows in Fig. 1). These constraints can be hard or soft. Examples of hard constraints are: need to unload the cargo before loading or to deplane passengers before you enplane them. Examples of soft constraints, which will be highly depending on time pressure (degraded standards as risk mitigation) and operator policies, are: Catering start in some galleys while passengers deplaning has not yet finished. Some ground operators start approaching the PBB even with the aircraft not fully on-chocks.

As a result, the aircraft ramp handling sequence is composed of mostly independent chains of tasks for current aircraft from the moment the aircraft is "on-chocks", namely:

- Passenger cabin related: passenger deplaning, catering and cleaning, cabin preparation and checks and passenger boarding
- Cargo and bulk unloading/loading
- Refueling (it is widely accepted to have refueling with passenger onboard the aircraft)
- Potable water servicing
- Toilet servicing

- Ground power and air-conditioned supply (depending on pilot request, availability and regulations of use for aircraft Auxiliary Power Unit).

Not all of these activities are carried out at each aircraft stop. First and last flight of the day do only the tasks related to departure and arrival respectively. For the flights in between, some tasks may or may not be done depending on aircraft status, next assigned mission and infrastructure available at the airport (in Fig. 1 these optional tasks are shown with light background). Figure 2 represent a typical layout of the aircraft ramp.

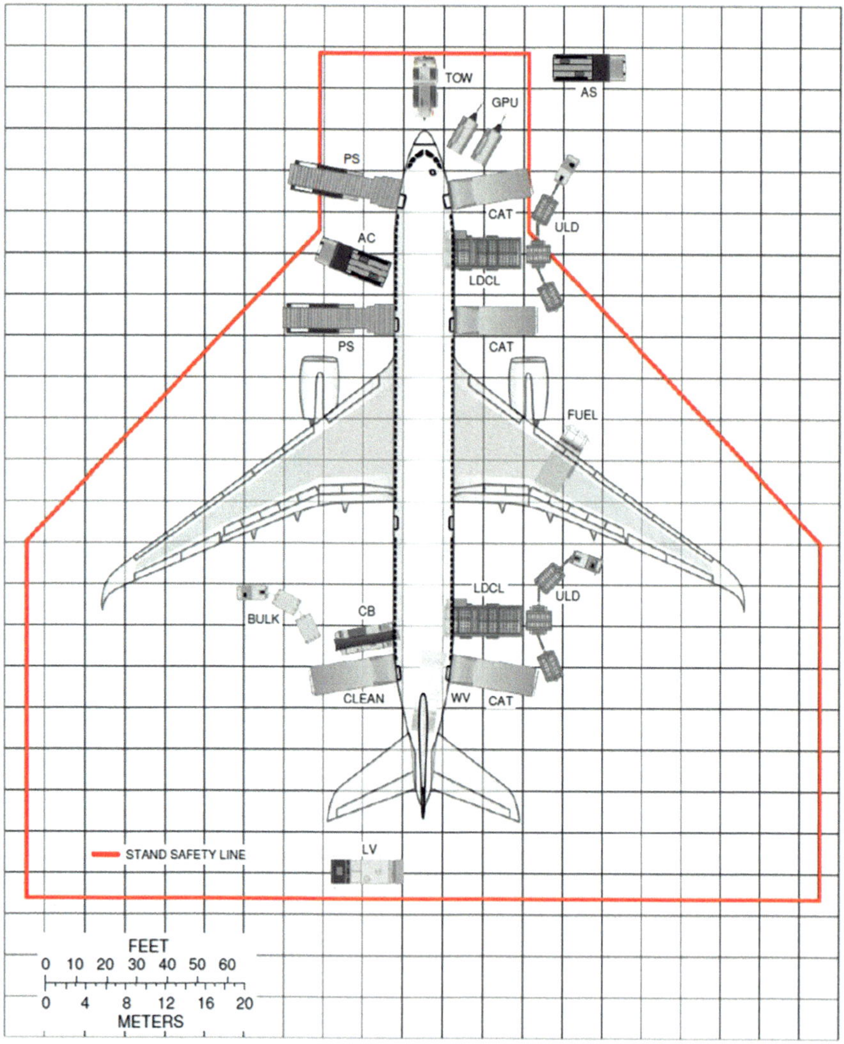

Fig. 2 Aircraft typical ramp layout

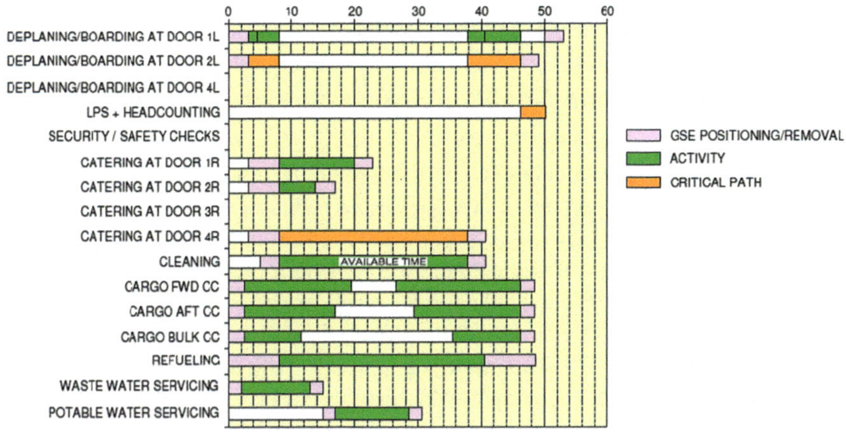

Fig. 3 Example of aircraft turnaround time

The turnaround time (TRT) can be defined as the time between the aircraft is on-chocks until it is off-chocks. It can be estimated following a scenario and some assumptions and a set of parameters and rates. An example of TRT result for a given scenario is presented in Fig. 3. The vast majority of the cases, the critical path, the sequence of activities that will determine the duration of TRT, is given by the cabin related activities chain (passenger deplaning, catering, cleaning, cabin preparation and checks and passenger boarding). Each activity, unless specified differently, comprises the phases of GSE docking or supply connection to the aircraft, the task itself and the GSE removal or disconnection of the utility. Each activity requires a specific GSE, with its own design, shape and constraints (See Fig. 2). It is important to highlight that even for the same aircraft type, the turnaround time is not a fix value and the schedule and the duration of each task and in consequence of the global TRT will vary, because it also depends on aircraft cabin layout, airline operation, infrastructure…

In addition, there are several providers in charge of handling activities for the same aircraft turnaround. Depending on airport size and market regulations, the ground handling companies that can typically be found for the same aircraft turnaround are specialized in refueling and ramp handling activities. To complete the list of actors for ground handling, it should be added the airline, the ATC, the airport, the cockpit crew, the cabin crew and the ground handling crew (from five to dozens). The high number of actors involved adds to the complexity and management of the process.

3 Ground Handling Fleets Automation

3.1 Previous Ground Handling Automation Studies

The ground handling process has been considered to be a minor activity at airports separated from other airport decision processes and which should deliver on time all services required for an arriving or departing aircraft in an assigned time window. As a result, the Operations Research literature presents a small number of published works in this field of application although the management of ground handling fleets can be related with more classical routing and scheduling problems for fleets of vehicles for which an extended literature is available [6–10]. Surprisingly, there exists also only a rather limited bibliography, offering very partial solutions, with regards to information processing and decision making for real time automated ground handling operations. Some published papers related to Automated Ground Vehicles (AGV) consider scheduling problems in manufacturing environments [11–13] which present some common characteristics with GSE fleets management problems.

The first published studies were focusing on one type of ground handling fleet or airport vehicles (passenger buses, catering vehicles, fuel trucks, etc.) in a static and deterministic Vehicle Routing Problem (VRP) framework [14]. Works have been published with respect to each type of ground handling fleets or airport vehicle fleet: catering vehicles staff allocation, fuel truck fleets, connecting baggage fleets, passenger bus fleets and de-icing fleets. In these studies, it has been assumed in general that aircraft and airlines respect perfectly scheduled arrival and departure times, then each GSE should be assigned to successive tasks on different aircraft along the operations period. In general, this literature did not elaborate about how to cope with perturbations and disruptions.

Mao [15] proposed to solve the GH scheduling problem under uncertainty by considering it as a multi-project scheduling approach. The aircraft is taken as a project agent which is composed by a set of activities and the ground handling providers as resource agents. Two approaches, non-cooperative and cooperative, compared with three heuristic methods: first come first served, maximum total travel work content first and shortest activity from the shortest project. The results showed that for the five approaches the total project delay decreased with the increase of the delay cost per time unit. More recently some works [16, 17] have introduced global tactical approaches considering static and deterministic VRP frameworks. These approaches lead to the formulation of static large mixed variables optimization problems presenting many constraints where global objective functions are considered to assess the performance of the solutions. However, in both, the corrective process, i.e. the re-planning actions to face short term changes in the aircraft schedules, is not developed in the perspective of automation of GH. Other more recent development was undertaken by Andreatta [18] where the fleet assignment and staff management problem are treated using heuristic based on task clustering and earliest due time priority and applying them together with an existing commercial dispatch software to Berlin Tegel airport.

Ground handling activities are subject to the effect of many upstream and endogenous perturbations and their management has been in general considered as a reaction to the resulting situations. Classical approaches through stochastic processes theory lead to very complex problems which also require large volumes of statistical data to be applied on current conditions following theoretical assumptions which rarely are met. So, this area looks to be a potential field of application of the fuzzy modelling approach and decision making. However, very few works have been published where fuzzy modelling is adopted to tackle the management of GSE fleets. One of the few is relative to the central management of ground handling after a disruption situation [19] where fuzzy dual numbers are adopted to represent timing uncertainty.

3.2 ACDM and Ground Handling Decision Making

The recent development and implementation of ACDM (Airport Collaborative Decision Making) at airports favor increased GH automation by turning available up-to-date information about GH demands at airports. ACDM is a relatively recent concept [20] aiming at improving the overall efficiency of airport operations by allowing the optimization of resources and improving the predictability of events through the organization of an efficient flow of information between all airport and traffic operators. It focuses especially on aircraft turn-round and pre-departure sequencing processes. Increased predictability can be of significant benefit for all major airport and network operations impacting in both productivity and cost-efficiency. The ACDM has introduced a milestones approach (for the whole turn-round process of each aircraft at airports) which aims to achieve common situational awareness by tracking the progress of a flight from the initial planning to the take off. Among these milestones the updated information of Actual In Block Time (AIBT), Actual Commence of Ground Handling Time (ACGT), Estimated Off-Block Time (EOBT) and Target Off-Block Time (TOBT) are of particular interest for this study.

In the ACDM framework, on-line decision processes responsible for assigning tasks to the vehicles of the different GSE fleets may be designed to allow facing any perturbation (late arrival of aircraft, GSE breakdown…). Within the ACDM framework it is possible to generate for the parking positions associated with each airport terminal and for a given time window, an updated demand plan for GH tasks which will be realized using the different GSE fleets. These set of tasks A are a compilation from all the TRT that are ongoing or to come in the moving horizon time window (H) which covers the remaining operations period of the day. In that case, assumptions must be adopted concerning the GSE used for each chain of activities. Follows here a common operations scenario:

- Passenger cabin related activities: both boarding bridges and passenger stairs, as needed, have to be in the stand for the full TRT duration and the same means used for deplaning are used for boarding the passengers as well. - Cabin activities (catering and cleaning): the trucks may arrive at any time after aircraft arrival.

- Cargo and bulk unloading/loading: both cargo loaders and belt loaders arrive as soon as possible when the aircraft arrives. Unless the TRT duration is very long and would have the time to make other full aircraft TRT after unloading (TRT > 2 h) these GSE stay in place.
- Baggage tractors and dollies: these are the only GSE where precedence and consequence link could make sense.
- Refueling, potable water and toilet servicing: the corresponding trucks arrive at any given time after aircraft arrival.
- Ground power and air supply: both independent between them from GSE point of view, can arrive at any time, even being on hold waiting for aircraft arrival.
- Pushback towing: associated with aircraft departure, it arrives before scheduled aircraft departure and stays busy until the end of the pushback.

These tasks can be set-up with templates coming from operational or historical experience. For each of these tasks type, their number for a given aircraft arrival and/or departure operation can be determined comparing the size of the service demanded: configuration of the plane with respect to ground handling activities, volume of baggage to be loaded/unloaded, cabin area complexity to be cleaned and other elements…with the capacity of the corresponding GSE types and eventually specialized staff (cleaning, cargo crew, etc. …). This allows to define the whole set of elementary GH tasks attached to each aircraft arrival and/or departure operation. Then, for the next time period it will be possible to characterize with accuracy the current timetable for elementary GH tasks (A_k) during period $[t_k, t_{k+H}]$.

During GH operations, a GH manager will on one side forward the current time table to the GSE fleet assignment processes and on the other side feedback information to the ACDM, in general updating the AIBT, ACGT, EOBT and the TOBT for each parked aircraft under its control.

3.3 Ground Handling Management Structures

Management scheme of GH activities is in general organized so that no additional delays are generated while operations costs (possible global indexes for a GSE fleet are the total travelled distance and the total operations time) are under control. Planning and tactical decisions which contribute to these objectives are dependent on the adopted management structure while potential automation is relative to the operations level.

The management of GH operations can be either organized by aircraft parking stand, where a local manager is responsible for all GH activities around any aircraft at that stand including the assigned GSE to that stand, or globally at the level of an apron with different stands. In that second case, a GH manager is responsible for all the stands of this apron while GSE fleet managers are responsible of the different GSE fleets. When thinking about automation, the first management scheme leads to consider GH operations at each stand as a workshop with its own resources while in the second scheme resources are distributed and assigned to stands according to their needs. This second scheme which makes a better use of available resources is adopted in many large aprons and even complete airports, and is retained here. It leads to the global management of each GSE fleet involved in GH activities at a given apron area.

A necessary condition to avoid additional delays at GH is to size adequately the different resources and especially the different GSE fleets. Fluid operations are achieved when ground handling tasks do not have to wait for a vehicle to be freed from a previous ground handling assignment operation before starting its new assignment. In general, the unit price of these vehicles is small compared with the global economic airport's operations at stake, mainly the cost of delays. Therefore, it is assumed the fleet sizing issue is addressed by the service provider by estimating the necessary GSE fleets on peak day/hours of the considered season without bargaining any trade-offs with performances indexes such as total delays or total number of delayed departure flights. This can be done periodically using predictions of demand for an entire season through the resolution of large optimization problems or through the use of simulation models. Provisions are also taken to size technical reserves to face GSE programmed maintenance as well as possible vehicle failure events. The sizing of each GSE fleet type size can be performed separately once traffic congestion effects are avoided by the configuration of the different dedicated routes and traffic rules.

Total travel distance has a direct incidence on the fleet operations cost and on the environment impact (pollution) and can be adopted as an optimization criterion. A way to reduce these costs and impacts is to position the components of the different GSE fleets along the parking area of the considered apron in accordance with short term predictions of demand, for example on a daily basis using nominal timetables of GH operations for the next day. There it is supposed that a reduced network of feasible direct paths has been establish to connect GSE and aircraft parking positions so that there is no routing issue to be solved when a GSE is assigned to a GH task. At this stage specialized workforce can be assigned to the different non automated GSE which are planned to be used during the next day, they are to remain attached to their assigned GSE during real operations where perturbations are frequent.

At the operations level, increased automation of GH activities supposes the on-line availability of solutions for the scheduling and routing of the different GSE fleets. Once the GSE fleets are positioned and their operators are assigned, to take into account current conditions of operations, using information generated by the ACDM, on-line assignment of GSE vehicles (and attached operators if any) to GH tasks is performed. Figure 4 displays the whole general process for GH and GSE fleet management:

Four types of fleet assignment problems have to be solved according to the characteristics of the tasks performed by each type of GSE:

- *Type I*: A closed loop from the GSE parking, until the Aircraft stand coming back to the same GSE parking it departed. This case is applicable to cargo loaders, belt loaders, passenger stairs, tow tractors, fuel hydrant dispenser vehicles (that

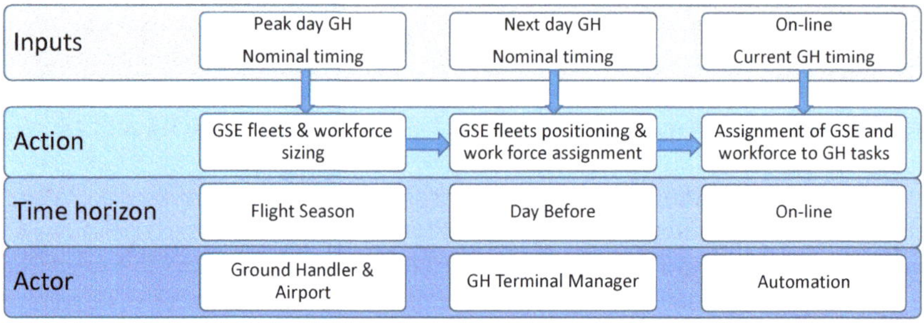

Fig. 4 GH and GSE fleet management

take the fuel from underground piping facility); mobile self-propelled units for electricity, air and engine start.

- *Type II*: A closed loop from a fix facility, then going to aircraft stand i, aircraft stand j … aircraft stand n and finally coming back to the same fix facility the GSE departed. This case is applicable to catering trucks, cleaning trucks, refuel trucks, potable water truck, toilet servicing trucks and cargo tractors.
- *Type III*: A double closed loop with idle parking. This case is applicable to baggage and container dollies, which are pulled in a train by tractors. (The tractors themselves may follow circulation model type IV, when they tow other GSE like stairs, or air/electricity units to the stand). This circuit may be also used for container and baggage dollies being AGV themselves.
- *Type IV*: An open circuit. Several origins, stands and potential destinations. It is the case of passenger buses, crew buses, passenger with reduced mobility vehicles and baggage tractors. For the specific case of the baggage tractors, it applies when they tow a towable or non-self-propelled unit like electricity generator, stairs, air unit, air start unit, towbars … on demand from a GSE parking area to the aircraft parking stand.

From the point of view of mathematical programming, when considering that all time durations are deterministic, this will lead to the formulation of large combinatorial optimization problems with many constraints, which are expected to be NP-Hard by analogy with other assignment problems [20]. These problems can be solved exactly using optimization solvers, however in the case of on-line assignment of GSE vehicles, the exact solution from the mathematical point of view is not essential while availability of a feasible solution despite perturbations is a critical issue. In the next section, the construction of an updated timing for GH tasks which takes into account uncertainty in task duration is discussed while the corresponding mathematical formulations and solutions algorithms are produced.

4 Mathematical Formulation of GSE Decision Problems

The description of the necessary data to formulate the fleet location and assignment problems is given here.

4.1 Resources and Demand for Ground Handling Services

The following parameters characterize the GH offer at an airport apron:

Sizing: S, number of aircraft parking stands impacted, indexed by s, $s = 1$ to S. P: number of GSE parking bases, indexed by p, $p = 1$ to P. N: total number of GSE vehicles of a given type, indexed by n, $n = 1$ to N. N_{maxp}: capacity of the pth GSE parking base. p_n: parking base of the nth GSE vehicle.

Locations: Coordinates of each aircraft stand reference point and GSE parking base (x, y) given in a reference system.

Operations parameters: Speeds for GSE vehicles. Nominal values for servicing roads and inside aircraft parking stands. D_{ij}^{k} distance between element i and element j travelled at speed k according to their coordinates and available service routes. Then, T_{ij} the nominal time it takes to go from element i to element j, can be easily computed. Here it is considered that there an uncertainty in the real value of this duration and a dual fuzzy value [22] is attached to this time:

$$\tilde{T}_{ij} = T_{ij} + d_{ij} + \varepsilon \cdot d_{ij} \tag{1}$$

where d_{ij} is the mean overtime and where the operators *nom* and *maj* are defined by:

$$nom(\tilde{T}_{ij}) = T_{ij} \tag{2}$$

and

$$maj(\tilde{T}_{ij}) = T_{ij} + 2 \cdot d_{ij} \tag{3}$$

The dual fuzzy number $\tilde{T}_{ij}$ is represented in Fig. 5.

Let *Optime* be the total optimization period (no longer than one day) with a moving overlook time window H. For example, H could be set to two hours to cover typical TRTs for most of short-range aircraft. kth is the time period considered: $[t_k, t_{k+H}]$. A is the set of all ground handling tasks planned for the whole period (typically, a day). A_k is the set of ground handling tasks planned for period $[t_k, t_{k+H}]$, A_{ks} is the set of ground handling tasks planned at aircraft stand s for period $[t_k, t_{k+H}]$ where $A_k = {}_{s\in\{1,\ldots,S\}}{}^{\cup} A_{ks}$. A_k can also be expressed as the union of the ground handling tasks associated to aircraft arrival (forward scheduled) AA_k, and those linked to aircraft departure (backward scheduled), AD_k:$A_k = AA_k \cup AD_k$. Task $a \in A_k$ is

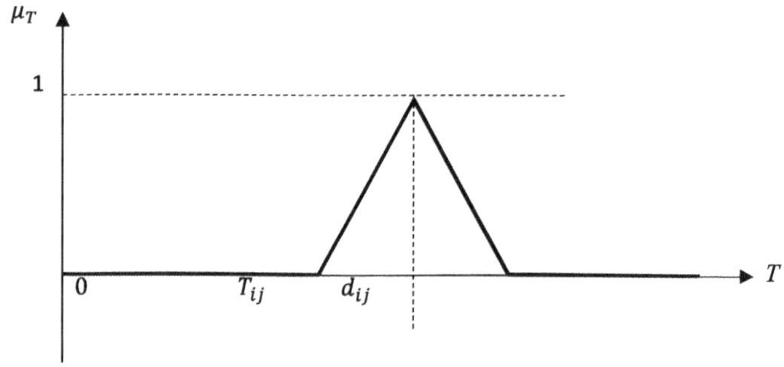

Fig. 5 Fuzzy dual representation of an uncertain duration

characterized by the corresponding aircraft parking stand s_a for a given aircraft α_a assigned to this parking stand at arrival or departure for the considered GSE type. s_a is the aircraft stand where the activity a is planned to be performed. Δa is the nominal duration of task $a \in A$ and $\widetilde{\Delta}_a$ its fuzzy dual duration. $\tilde{T}_a^{init}$ are the scheduled time for beginning of activity a. $\tilde{T}_a^n$ is the scheduled total service time for activity a, including the times of moving a given GSE n, to go from its parking base to an aircraft stand and back (assumed to be the same base) and doing the a task in stand s:

$$\tilde{T}_a^n = 2\tilde{T}_{p_n s_a} + \widetilde{\Delta}_a a A k, n\{1, \ldots, N\} \tag{4}$$

F_{na} is the set of activities that cannot be done by GSE n because it has already been assigned task a. $|F_{na}|$ = cardinal (F_{na}) $n \in \{1, \ldots, N\}$, $a \in A_k$. A GSE can perform task b when it has already been assigned to task a if:

$$(\tilde{T}_b^{init} - \tilde{T}_{p_n s_b} < \tilde{T}_a^{init} - \tilde{T}_{p_n s_a} \text{andor} (\tilde{T}_{an}^{end} < \tilde{T}_{bn}^{init}) \tag{5}$$

with

$$\tilde{T}_{an}^{end} = \tilde{T}_a^{init} + \tilde{T}_{p_n s_a} + \widetilde{\Delta}_a \tag{6}$$

and

$$\tilde{T}_{bn}^{end} = \tilde{T}_b^{init} + \tilde{T}_{p_n s_b} + \widetilde{\Delta}_b \tag{7}$$

Otherwise, that GSE cannot perform task b. Observe that the solution of the assignment problem of GSE to tasks will provide a resulting scheduling (timing) of tasks for each GSE. If GSE n is assigned to task a, it should be available for this task at time $nom(\tilde{T}_a^{init} - maj(\tilde{T}_{p_n s_a}))$.

4.2 Formulation of the GSE Assignment Problem for Type I Operations

In this case, GSEs perform a closed loop from the GSE parking until the aircraft parking stand and coming back to the same GSE parking they departed. The GSE parking is subject to capacity constraints.

At the tactical level, a first optimization problem associated with the GSE fleet location in the P bases with an embedded assignment problem to a set of the GH tasks A of a given day can be formulated considering the decision variables:

$$x_{na} = \begin{cases} 1 \text{ if GSE } n \text{ is assigned to task a} \\ 0 \text{ otherwise} \end{cases} \quad n \in \{1, \ldots, N\}, a \in A \tag{8}$$

$$y_{np} = \begin{cases} 1 \text{ if GSE } n \text{ is based in parking } p \\ 0 \text{ otherwise} \end{cases} \quad n \in \{1, \ldots, N\}, \ p \in \{1, \ldots, P\} \tag{9}$$

Here the optimization objective is to minimize the total distance travelled by the GSE to perform the ground handling tasks during a whole day. Distances for both travels (from GSE parking to aircraft stand and back) are assumed to be the same. In the proposed solution framework, A is the set of the GH tasks planned for the next day and N is the size of the available considered fleet. This leads to the real valued GSE location and assignment problem formulation:

$$\min_{x_{na}, y_{np}, n \in \{1,\ldots,N\}, a \in A, p \in \{1,\ldots,P\}} \sum_{n=1}^{N} \sum_{a A} 2 D^1_{p_n s_a} x_{na} y_{np} \tag{10}$$

Subject to constraints:

$$\sum_{n=1}^{N} x_{na} = 1, \forall a \in \{1, \ldots, A\} \tag{11}$$

$$\sum_{p=1}^{P} y_{np} \left(\sum_{\overline{a} \in F_{na}} x_{n\overline{a}} \right) + x_{na} \leq 1, \forall a \in A, \forall n \in \{1, \ldots, N\} \tag{12}$$

$$\sum_{p=1}^{P} \sum_{a=1}^{A} maj(\tilde{T}^n_a) \cdot x_{na} \cdot y_{np} \leq Optime, \forall n \in \{1, \ldots, N\} \tag{13}$$

$$\sum_{n=1}^{N} y_{np} \leq Nmax_p, \forall p \in \{1, \ldots, P\} \tag{14}$$

$$\sum_{n=1}^{N} y_{np} \leq 1, \forall n \in \{1, \ldots, N\} \tag{15}$$

$$(x_{na}, y_{np}) \in \{0, 1\}^2, \forall n \in \{1, \ldots, N\}, \forall a \in A, \forall p \in \{1, \ldots, P\} \tag{16}$$

Constraint in Eq. 11 expresses that each ground handling activity is performed by a single GSE. The constraint in Eq. 12 expresses that a GSE cannot be assigned to conflicting tasks whose realization by the same vehicle is unfeasible (time overlapping tasks, distance between activity locations, round trip duration, etc.). Constraint in Eq. 13 requests that the total service time (travel times + ground handling at stands) assigned to a single GSE must be less than the duration of the moving look ahead time window. Constraint in Eq. 14 expresses the capacity of parking base p; constraint in Eq. 15 considers that a GSE vehicle is at most located at one parking base and constraint in Eq. 16 expresses the binarity of the decision variables x_{na} and y_{np}.

Other optimization criteria than the sum of travelled distances by GSE could be considered. However, since it is expected that ground handling activities will suffer in permanence perturbations, the optimization criteria is used here as a guideline towards global efficiency rather than as a mathematical measure of the optimality of the solution of a problem which is in permanence resettled in the considered on-line environment. The above formulation can be used to solve the fleet sizing problems by considering that A represents for a season the set of GH activities to be performed on a peak demand day while N is a parameter which can be increased from low values to obtain first feasibility of constraints Eqs. 11–16 and then to decrease the minimum value of criterion in Eq. 10. A sizing solution is obtained when the value of an additional vehicle is higher than the resulting diminution of Eq. 10 when solving the defined problem statement. Here it is considered that the number of GSE located at a parking position is fixed for a whole day, while the assignment of a given vehicle to a ground handling task, meanwhile not effective, can be changed according to task schedule perturbations.

Therefore, the GSE fleet location and assignment problem can be split in two stages: one which considers the assignment of GSE to parking positions for the whole day and a second one, working on a moving time window to assign vehicles from their parking positions to ground handling tasks at the different aircraft stands. Therefore, first the decision variables y_{np} can be chosen on a static framework and in a second stage the decision variables x_{na} can be established on a dynamic framework. To face perturbations, the fleet assignment will be restarted every time t_k a relevant scheduled time for a task for which no GSE is already under way, is modified in the ACDM information system. In this case the new time window is $[t_k, t_{k+H}]$. Observe that once fixed the y_{np} variables, problem in Eqs. 10–16 becomes a Linear Integer Programming (LIP) problem.

4.2.1 Location of GSE at Parking Bases

For the first problem the y_{np}, must be chosen, assigning vehicles to GSE parkings. Here the horizon of optimization is the whole day period of operation of the considered aircraft parking apron (*Optime*). The following nomenclature for Problem 1 is adopted: g_p = number of GSE assigned to parking base *p*; A_s is the number of GH activities planned at stand *s* during the considered day. V_{ps} represents the number of times a GSE from parking *p* goes to aircraft stand *s*. The objective function used in the location problem considers, on a daily basis, the minimization of the total travelled distance by the considered GSE fleet. This leads to the formulation of Problem 1:

$$\min_{g_p, V_{ps}, p\in\{1,\ldots,P\}, s\in\{1,\ldots,S\}} \sum_{p=1}^{P}\sum_{s=1}^{S} V_{ps} * 2D_{ps} \tag{17}$$

under constraints:

$$\sum_{p=1}^{P} V_{ps} = A_s, \forall s \in \{1, \ldots, S\} \tag{18}$$

$$\sum_{a=1}^{A} V_{ps_a} \cdot maj(\tilde{T}_{pa}) \leq g_p \cdot Optime, \forall p \in \{1, \ldots, P\} \tag{19}$$

$$g_p \in \{0, \ldots, N_{max_p}\} \tag{20}$$

$$V_{ps} \geq 0, V_{ps} \in \mathbb{N}, \forall p \in \{1, \ldots, P\}, \forall s \in \{1, \ldots, S\} \tag{21}$$

The constraints are explained as follows: Eq. 18 considers that the aircraft stand *s* is visited by the GSE vehicle as many times as activities are scheduled at this stand, A_s; Eq. 19 ensures that the mean occupation of a GSE from any stand is less or equal to the whole operations period; constraint in Eq. 20 expresses that the number of vehicles g_p (non-negative integer) on every GSE parking, p, is less or equal than the maximum capacity of that parking N_{maxp}, and Eq. 21 considers that the number of times a GSE from parking *p* goes to stand *s* is a non-negative integer. Once this LIP problem is solved V_{ps} and g_p will be available. Then to each GSE parking p is assigned the number of GSE units, g_p that will be based there.

4.2.2 Formulation of the GSE Assignment Problem

As a consequence of solving the previous problem, for every GSE *n*, and for every parking *p*, y_{np} can be defined and therefore it be can proceeded to the assignment of each GSE to the requested tasks of A_k where A_k is the set of tasks to be executed during the current time window $[t_k, t_{k+H}]$, by solving the reduced problem, Problem 2:

$$\min_{x_{na}, n\in\{1,\ldots,N\}, a\in A_k} \sum_{n=1}^{N}\sum_{a=1}^{A_k} 2D^1_{p_n s_a} x_{na} \tag{22}$$

Subject to constraints:

$$\sum_{n=1}^{N} x_{na} = 1, \forall a \in \{1, A_k\} \tag{23}$$

$$\sum_{\overline{a}\in F_{na}} x_{n\overline{a}} + x_{na} \leq 1, \forall a \in A_k, \forall n \in \{1, \ldots, N\} \tag{24}$$

$$\sum_{a=1}^{A_k} maj(\tilde{T}^n_a) \cdot x_{na} \leq H, \forall n \in \{1, \ldots, N\} \tag{25}$$

$$x_{na} \in \{0, 1\}, \forall n \in \{1, \ldots, N\}, \forall a \in A_k \tag{26}$$

From the point of view of mathematical programming, this leads to consider at time t_k, over the new time window $[t_k, t_{k+H}]$, a reduced but still large combinatorial optimization problem with many constraints, which also is expected to be NP-Hard. It is intended here to get a very quick way of computing a feasible solution with acceptable performance since the operational environment of the problem is subject to frequent perturbations.

4.2.3 Heuristics for the GSE Assignment Problem Type I

Two solutions algorithms, based on greedy approaches, are proposed in this section to solve on-line the assignment decisions for each type of GSE that fits in Type I problems. The first assignment heuristic, Earliest Available GSE (EAG) is a straightforward greedy heuristic based on the ranking of available GSE with respect to their estimated arrival time at ground handling position. Since this approach does not consider the distribution of demand along the moving time horizon, there is a risk to make use of more GSE than really necessary. The second heuristic approach, Less Constrained GSE (LCG) assigns the GSE by considering the perspective of their reutilization during the considered moving time horizon.

Heuristic EAG: Earliest available GSE

Initialization:

k = 0, do for $a \in A, n = 1toN : x_{na} = 0, T_n^{free} = t_0$

Time loop: For t_k Do:

Let A_k^{free} be the set of tasks starting during time window $[t_k, t_{k+H}]$ with no GSE already assigned

$$\forall n \in \{1, \ldots, N\} : x_{na} = 0$$

Rank by increasing $nom(\tilde{T}_a^{init})$ the different GH tasks in A_k^{free}.

For each of these ranked tasks $a A_k^{free}$ do:

For n = 1 to N, let T_n^{free} be the time at which GSE n is free from any task at its parking base pn and compute:

$$\tilde{t}_n^a = \tilde{T}_n^{free} + \tilde{T}_{(p_n s_a)} \text{ with } \tilde{T}_n^{free} = \tilde{T}_n^{free} + 0 \cdot \varepsilon$$

Choose $n* = \min_{n \in \{1,\ldots,N\}, T_n^{free} < t_k + H} maj(\tilde{t}_n^a)$

$$\text{Make } x_{n*a} = 1 \text{ and } \tilde{T}_{n*}^{free} = \tilde{T}_{n*}^{free} + \tilde{T}_a^{n*}$$

Return to Time Loop

This heuristic provides with a reduced amount of calculations a feasible solution, however travelled distances are only indirectly taken into account when computing the $\tilde{t}_n^a$.

Heuristic LCG: Less Constrained GSE

Consider F_{na}be the set of unassigned tasks that cannot be done by GSE n due to its previous assignment to task a. The idea here is to assign task a to the GSE n which minimizes $|F_{na}|$, so that opportunities to reassign this GSE are preserved. The corresponding algorithm is given by:

Initialization:
k=0, do for $a \in A, n = 1 \text{ to } N: x_{na} = 0$
Time loop: For t_k Do:
Let A_k^{free}be the set of tasks starting during time window $[t_k, t_{k+H}]$ and for which no GSE has been already assigned
$\forall n \in \{1, \ldots, N\}, \forall a \in A_k^{free}: x_{na} = 0$

Compute F_{na} and $|F_{na}|$, for $a \in A_k^{free}$and $n \in \{1, \ldots, N\}$.
For each of these ranked tasks $a \in A_k^{free}$do:
For each of the n vehicles do:
If vehicle n can do task a, then
Make $x_{n*a} = 1$
Update assigned tasks with $|F_{na}|$ by vehicle n
Wait for t_{k+1} to Return to Time Loop

An up to date set of unfeasible tasks is kept: when the nth vehicle is assigned a task a, the unfeasible tasks file is updated by the addition of $|F_{na}|$ tasks.

Comparing the algorithms of Heuristic EAG and Heuristic LCG, it appears that the second heuristic generates a larger volume of computation to provide a solution (repetitive computation of F_{na}) but its complexity remains polynomial as for heuristic A. Observe that by construction, both heuristics comply with restrictions in Eqs. 23–26. Also, it can be considered that if $|F_{na}|$ is small, that may be either because GSE n is not far from aircraft stand sa when service a is requested and has not a large time span to interfere with other demands, or because ground handling activity at that time is reduced. Therefore, it could be of interest to use Heuristic EAG for periods of low level of activity and Heuristic LCG for periods when demand for ground handling services is high.

A numerical application and comparison of heuristics on an actual airport can be found in a recent publication [23].

4.3 Formulation of the GSE Assignment Problem Type II

In this case, the GSE perform loops such as starting from a GSE Fix facility, then going to a first aircraft parking stand, then to another aircraft stand and so on until it finishes their circuit coming back to the GSE Fix facility (with no explicitly considered capacity constraint). As a reminder this circuit is applicable to the following GSE: catering trucks, cleaning trucks, refuel trucks, potable water truck, and toilet servicing truck and cargo tractors.

This case will be illustrated with the example of a fleet of refueling trucks. It will be supposed to have a limited capacity for each vehicle n, C_n expressed in tons. These trucks refill fuel at a unique fixed facility (in this example a fuel farm, indexed by 0 in the following) and go from one aircraft to the next at parking stands until they have to return to the fix facility if and only if they do not have any pending task in the considered horizon or if their fuel quantity still onboard is below a given threshold (C_{min} the common minimum uplift, e.g. 1 ton for fuel).

The demand D_a is supposed to be known at the start of the current horizon $[t_k, t_{k+H}]$ for each refueling operation associated with aircraft α_a. As a minimum, this demand will be bounded: lower bounded by the minimum delivery (usually by a pricing policy, e.g. no less than 0.5 tons), upper bounded by the aircraft fuel tank capacity. It can be also be estimated from operations and historical data (e.g. aircraft type A, from carrier C, doing a route X–Y gets two tons uplift).

Contrarily to the Type I case, each truck is again available for assignment once a fueling operation is completed. Refueling rate of trucks is supposed to take an approximatively constant time R_n. The energy to operate these trucks is supposed to be loaded (fuel or eventually battery or fuel cell in the future) during the payload refueling, the corresponding time is considered to be negligible or masked with respect to the payload refueling time. Refuel trucks are filled to their maximum capacity C_n when they are in the fix facility (fuel farm, in their case), and they are only dispatched under maximum payload C_n.

Let E_n be the current fuel payload of truck n whose last position is L_n.

The previous heuristics can be adapted directly to this case by taking care of truck capacity and satisfaction of demands. For instance, Heuristic EAG can be modified to Heuristic C such as follows.

Heuristic C: Earliest available GSE for Refueling Tours with Capacity

Initialization:

k=0,

Do for $\in A, n = 1 \; to \; N: x_{na} = 0,\; E_n = C_n,\; T_n^{free} = t_0$, L_n=0

Time loop:

For t_k Do:

Let B_k^{free} be the set of refueling tasks starting during time window $[t_k, t_{k+H}]$ and for which no truck has been already assigned

$$\forall n \in \{1, \ldots, N\}, \forall a \in B_k^{free}: x_{na} = 0$$

Rank by increasing T_a^{init} the different GH tasks in B_k^{free}

For each of these ranked tasks $a \in B_k^{free}$ do:

For n=1 to N, if $E_n > D_a$, let T_n^{free} be the time at which truck n is free from any task while at position L_n and

Compute $t_n^a = T_n^{free} + T_{L_n s_a}$

Choose n*= $\min_{n \in \{1,\ldots,N\},\, E_n > D_a,\; T_n^{free} < t_k + H} \{ t_n^a \}$

Make $x_{n*a} = 1$

$T_{n*}^{free} = T_{n*}^{free} + T_{L_{n*} s_a} + \Delta_a$ and

En=En-Da

If En< C_{min}

Then

$T_{n*}^{free} = T_{n*}^{free} + \Delta_a + T_{s_a 0} + R$,

$E_n = C_n$ and Ln*=0

Else $T_{n*}^{free} = T_{n*}^{free} + \Delta_a$ and Ln* = sa

Return to Time Loop

This heuristic presents a rather limited increase in the amount of necessary computation to get a solution over a time window with respect to Heuristic EAG. A similar approach can be adopted for delivery of potable water, as well as for toilet servicing (with dumping instead of filling operations).

4.4 Formulation of the GSE Assignment Problem Type III

As a reminder this circuit type is applicable to baggage and container dollies, which are pulled in a train by tractors (The tractors themselves may follow circulation model type IV, when they tow other GSE like stairs, PCA, GPU to the stand).

This circuit may be also used for container and baggage dollies being AGV themselves.

These baggage and container dollies follow the circuit:

- First Loop: from idle or overflow parking to the departure make-up baggage area I. Then to aircraft parking departing stand I, then to idle/overflow parking.
- Second Loop: from idle or overflow parking to the arrival aircraft parking stand; then to the arrival/transfer baggage area I, coming back to the idle/overflow parking.

Both loops are independent, i.e. one dolly may always stay in the departures loop.

Figure 6 provides an example of management of dollies stocks between aircraft stands and passengers terminals, showing the two closed circuits having the idle/overflow parking in common.

Baggage and container dollies are passive vehicles towed by tractors. In the following formulation, it is supposed that there are two types of dollies (for bulk baggage and for ULD's) parking and that trains of dollies pulled by a tractor (or being individual AGV) are operated in closed loop from their parking:

Fig. 6 Circuit for baggage and container dollies

Those associated with an aircraft arrival. Idle dollies from their parking will be dispatched to the aircraft arrival parking stand. They will wait to be loaded with baggage from this arriving aircraft. Once they are loaded with arrival baggage, they leave to the baggage reclaim areas, where its content is dumped into the appropriate carrousel. Then, being empty they will be taken to the dolly parking area and eventually charge their batteries if needed. This configures a closed tour for the aircraft arrival activity.

Those associated with an aircraft departure. The dollies will be taken from the parking to the baggage chute make-up area (typically 2–3 h before flight departure depending on the check-in and flight closure policy of the airline); there, they are filled with the hold baggage for the departing aircraft. Then, they may start to be dispatched to the departure aircraft parking stand when they are loaded. Once the flight has been closed and the last baggage has been treated, the last dolly can proceed to the aircraft stand. These dollies are unloaded in the stand to the aircraft cargo hold; then, the empty dollies can be taken to the dollies parking. This configures a closed tour for the aircraft loading activity.

Solution algorithm for the GSE assignment problem Type III

A solution algorithm based on greedy heuristics is proposed in this section to solve in real time the assignment decisions for each type of GSE that fits in Type III circuits.

Let be A, the set of all ground handling tasks planned for the whole period (a day in general) is split in two:

$$A = AA \cup AD \tag{21}$$

AA: the ground handling tasks associated with aircraft arrival.
AD: the ground handling tasks associated with aircraft departure.

According to the baggage volume to be unloaded/loaded from an arrival/departing aircraft, the number and the size of the dolly trains necessary to perform the baggage processing can be known beforehand. This volume is transmitted from an IATA movement message (arrival aircraft) or from the airline departure control system (departure aircraft).

The different dolly trains are treated without distinction, so no nominal assignment of them is considered here. It is only supposed that the loading capacity of the stocks of dollies available at their parking is sufficient to process the baggage demand.

The number of unloading/loading tasks associated with a flight can be defined as well as their parameters. Any task a in *AA* ($a \in AAA$) is characterized by the parameters:

- ss_a: the arrival aircraft parking position of task a
- sf_a: the baggage carrousel position of task a
- T_a^{init} :the planned start time of task a
- ΔS_a: the duration of loading task a at aircraft stand
- ΔF_a: the duration of unloading task a at baggage reclaim/ transfer area

- Ta: the cycle duration for the time of transfer from ss_a to sf_a and from sf_a to ss_a including the transfer time from dollies parking to aircraft arrival stand, and the time to go through dollies parking after baggage carrousel drop-off.

Any task a in AD ($a \in ADA$) is characterized by the parameters:

- ssa: the starting baggage make up area of task a
- sfa: the final aircraft departure stand of task a
- T_a^{init}: the planned start time of task a
- ΔS_a: the duration of loading task a at make-up area position
- ΔF_a: the duration of unloading task a at aircraft stand
- T_a: the cycle duration, time of transfer from ss_a to sf_a and from sf_a to ss_a passing by the dollies parking.

For all AA and AD tasks, tractors and dollies are supposed to start and finish the tasks at the dollies parking. During their operations tractors will be assigned indifferently to aircraft unloading or aircraft loading activities according to the GSE to tasks assignment logic adopted.

When considering the time window $[t_k, t_{k+H}]$ let:

AA_k is the set of ground handling tasks of type AA planned for period $[t_k, t_{k+H}]$,
AD_k is the set of ground handling tasks of type AD planned for period $[t_k, t_{k+H}]$.

Then:

$$A_k = AA_k \cup AD_k \tag{22}$$

Let N be, like for Type I and Type II problems, the size of the considered GSE fleet type (here tractors for baggage dollies, or AGV equivalent for dollies).

Let L_n^0 be the position of the nth GSE (tractor or AGV) vehicle at start of day operation.

Heuristic D: Earliest available GSE for successive closed tours tasks

Initialization:

k=0,

Do for $\in A, n = 1\ to\ N: x_{na} = 0,\ T_n^{free} = t_0$, Ln=L_n^0

Time loop:

For t_k Do:

Let C_k^{free} be the set of tasks starting during time window $[t_k, t_{k+H}]$ and for which no GSE has been already assigned:

$$\forall n \in \{1, \dots, N\}, \forall a \in C_k^{free}: x_{na} = 0$$

Rank by increasing T_a^{init} the different ground handling tasks in C_k^{free}

For each of these ranked tasks $a \in C_k^{free}$ do:

For n=1 to N, let T_n^{free} be the time at which GSE n is free from any task while at position L_n and compute

$t_n^a = T_n^{free} + T_{L_n ss_a}$

Choose n*= $\min_{n \in \{1,\dots,N\},\ T_n^{free} < t_k + H} \{ t_n^a \}$

Make $x_{n*a} = 1$

$T_{n*}^{free} = T_{n*}^{free} + \Delta S_a + 2T_a + \Delta F_a$

and $L_{n*} = ss_a$

Return to Time Loop

This heuristic is similar to Heuristic EAG in the amount of necessary computation to get a solution over a time window.

4.5 *Formulation of the GSE Assignment Problem Type IV*

This model is applicable to: passenger buses, crew buses, PRM vehicles and baggage tractors. For the specific case of the baggage tractors applies when they tow a towable or non-self-propelled unit like GPU, stairs, PCA, ASU, towbars … on demand from a GSE parking area to the aircraft parking stand.

There is typically only one payload that would be taken in each trip. Both origin and destinations can change at every trip, becoming an open circuit. It is a VRP with pick-up and delivery problem with time windows and vehicles capacities (in this case: car, bus or truck), knowing that again the route to take is known between the several origins, stands and potential destinations. Heuristic C, with no capacity constraint can be easily adapted, to propose a solution in this case.

5 Conclusions

The objective of this chapter has been to contribute to the automation of ground handling at airports. For this, a multi-time scale airport ground handling management structure compatible with increased ground handling automation at airports in the context of ACDM operations, has been proposed. Mathematical formulations of the corresponding GSE location and assignment problems have been introduced and a strategy to provide solutions for these recurrent problems, where time uncertainties are represented by fuzzy dual numbers and where tasks to be performed are produced on-line, has been developed.

To put effectively into operation the proposed decision structure, the displayed heuristics could be improved using a layer of artificial intelligence techniques to cope with severe perturbations (break down of already assigned GSE, abnormal delay in performing a task), internet of things could generate a complete view of the current GH processes and data science techniques could update the values of the GH related operations parameters. All of this would configurate a strategic application of Aviation 4.0 techniques in the operations of airports.

With the COVID-19 crisis which has been affecting deeply the air transport industry, it appears essential that GH activities must be made pathogen-proof with respect to passengers, giving through full automation an important contribution to a pandemic free air transportation [24, 25]. Aviation 4.0 should turn feasible from the technical as well as the economic points of view solutions to this new challenge.

References

1. Arnaldo Valdés, R., et al.: Industrial IoT solutions—remote monitoring & control of mission critical environments, Chapter Aviation 4.0: More Safety through Automation and Digitization (2018). https://doi.org/10.5772/intechopen.73688
2. Beelaerts, W., et al.: Future airport turnaround ground handling processes. How to reduce the turnaround time of aircraft at the airport. In: 10th TRAIL Congress, Delft (2008). Available at: https://resolver.tudelft.nl/uuid:da6f9bb6-5306-4b73-af68-efe4e00fe01b
3. Wu, C., Caves, R.: Aircraft operational costs and turnaround efficiency at airports. J. Air Transp. Manage. **6**, 201–208 (2000). https://doi.org/10.1016/S0969-6997(00)00014-4
4. Wu, C., Caves, R.: Modelling and simulation of aircraft turnaround operations at airports. Transp. Plan. Technol. **27**(1), 25–46 (2004). https://doi.org/10.1080/0308106042000184445
5. Alonso Tabares, D., Mora-Camino, F.: Aircraft ground operations: steps towards automation. CEAS Aeronaut. J. **10**(3), 965–974 (2019). https://doi.org/10.1007/s13272-019-00390-5
6. Desrosiers, J., et al.: Routing with time windows by column generation. Networks **14**(4), 545–565 (1984). https://doi.org/10.1002/net.3230140406
7. Kolen, A.W.J., et al.: Vehicle routing problem with time windows. Op. Res. **35**, 266–273 (1987). https://doi.org/10.1287/opre.35.2.266
8. Kohl, N., Madsen, O.: An optimization algorithm for the vehicle routing problem with time windows based on Lagrangean relaxation. Op. Res. **45**(3), 395–406 (1997). https://doi.org/10.1287/opre.45.3.395
9. Potvin, J.Y.: An exchange heuristic for routing problems with time windows. J. Op. Res. Soc. **46**, 1422–1446 (1995). https://doi.org/10.2307/2584063

10. Prins, C.: Efficient heuristics for the heterogeneous fleet multi-trip vrp with application to a large-scale real case. J. Math. Model. Alg. **1**, 135–150 (2002). https://doi.org/10.1023/A:1016516326823
11. Valmiki, P., et al.: A study on simulation methods for AGV fleet size estimation in a FMS. Mater. Today Proc. **5**, 3994–3999 (2018). Available at: https://isiarticles.com/bundles/Article/pre/pdf/86906.pdf
12. Nageswararao, M., et al.: Simultaneous scheduling of machines and AGV in FMS with minimization of tardiness criterion. Proc. Mater. Sci. **5**, 1492–1501 (2014). https://doi.org/10.1016/j.mspro.2014.07.336
13. Nouri, H.E., et al.: Simultaneous scheduling of machines and transport robots in flexible job shop environment using hybrid metaheuristics based on clustered holonic multiagent model. Comput. Ind. Eng. **102**, 488–501 (2016). https://doi.org/10.1016/j.cie.2016.02.024
14. Kuhn, K., Loth, S.: Airport Service Vehicle Scheduling, Eighth USA/ Europe Air Traffic Management Research and Development Seminar (ATM) (2009). Available at: https://www.researchgate.net/publication/225002946_Airport_Service_Vehicle_Scheduling
15. Mao, X., et al.: Stable multi-project scheduling of airport ground handling services by heterogeneous agents (2007). Available at: https://dke.maastrichtuniversity.nl/nico.roos/wp-content/uploads/2016/01/MRS09AAMAS.pdf
16. Fitouri-Trabelsi, S.: Contribution to the organization of ground handling management at airports. Ph.D. Thesis (2013). Available at: https://www.theses.fr/2013INPT0118
17. Padrón Astorga, S.: A multi-objective optimization approach to the ground handling scheduling problem, Ph.D. Thesis (2014). Available at: https://ddd.uab.cat/record/128757
18. Andreatta, G., de Giovanni, L.: A fast heuristic for airport ground-service equipment-and-staff allocation, Proc. Soc. Behav. Sci. **108** (2014). https://doi.org/10.1016/j.sbspro.2013.12.817
19. Fitouri-Trabelsi, S., et al.: An operational approach for ground handling management at airports with imperfect information. In: ICIEOM 2013, 19th International Conference on Industrial Engineering and Operations Management (2013). Available at: https://hal-enac.archives-ouvertes.fr/hal-00925045/document
20. Eurocontrol: Airport CDM Implementation Manual v5.0 (2017). Available at: https://www.eurocontrol.int/sites/default/files/publication/files/airport-cdm-manual-2017.PDF
21. Cordreau, J.F.: A branch-and-cut algorithm for the dial a ride problem. Op. Res. **54**, 573–586 (2006). https://doi.org/10.1287/opre.1060.0283
22. Mora-Camino, F., Nunes, C.A.: Fuzzy dual numbers: theory and applications. Stud. Fuzziness Soft Comput. (2018). https://doi.org/10.1007/978-3-319-65418-8
23. Alonso Tabares, D., et al.: A multi-time scale management structure for airport ground handling automation. J. Air Transp. Manage. **90**(C) (2020)
24. https://doi.org/10.1016/j.jairtraman.2020.101959
25. Drljaca, M., et al.: The role and influence of industry 4.0. in airport operations in the context of COVID-19. Sustainability (2020). https://doi.org/10.3390/su122410614
26. Alonso Tabares, D.: An airport operations proposal for a pandemic-free air travel. J. Air Transp. Manage. **90**(C) (2020). https://doi.org/10.1016/j.jairtraman.2020.101943

Novel Spherical Fuzzy Eco-holonic Concept in Sustainable Supply Chain of Aviation Fuel

Fariba Farid and Yaser Donyatalab

Abstract The objective of this chapter is to extend and combine two concepts of Eco-holonic structures and spherical fuzzy sets to introduce a novel concept of spherical fuzzy Eco-holonic structures. Both concepts consider different aspects of sustainability around the systems of real-world cases, so that caused an increase in a comprehensive analysis of the system. Eco-holonic structures help to consider many details of the system together with fuzzy logic, which assists to consider and evaluate the uncertainty around the system. Then, a multi attribute group decision making (MAGDM) method based on the spherical fuzzy Eco-holonic structure is proposed. Supply chain selection of aviation fuels is a hot topic these days and has very complexities. Sustainable supply chain (SSC) selection of aviation fuels is the most significant field of study among all aviation fuel supply chain problems because of many intuitive criteria that have to be considered through the decision-making procedure. Constructing and defining such an intuitive and comprehensive model for the aviation fuel supply chain is still out of sight, even though the significant research that have been done in this field. Another hardness of SSC in the aviation fuel is the satisfaction of all criteria based on such a complicated model. In this chapter, a spherical fuzzy Eco-holonic structure is defined for the sustainable supply chain of aviation fuels problem. Then, the proposed MAGDM method in SF Eco-holonic structure is applied to solve the SSC of aviation fuel problem. To show the feasibility and applicability of the proposed SF eco-holonic structure, it is applied in the aviation fuel SSC problem.

Keywords Spherical fuzzy eco-holonic concept · Spherical fuzzy eco-holonic structure · Spherical fuzzy eco-holarchy · Sustainable supply chain of aviation fuel · MAGDM model in SF eco-holonic structure

F. Farid (✉) · Y. Donyatalab
Industrial Engineering Department, Istanbul Technical University, Macka, Istanbul, Turkey
e-mail: farid20@itu.edu.tr

Y. Donyatalab
e-mail: donyatalab18@itu.edu.tr

C. Kahraman and S. Aydın (eds.), *Intelligent and Fuzzy Techniques in Aviation 4.0*, Studies in Systems, Decision and Control 372,
https://doi.org/10.1007/978-3-030-75067-1_9

1 Introduction

In this section, we will provide a clear introduction to the study, the approach, and the methodology. Also, in this chapter, it is tried to prepare a strong relation among the derived points from literature with what is proposed in this study. One of the basics of globalization is transportation; specifically, international transportation has much more significant. Based on aviation industries' future growth of at least 5% per year [1], this fact is entirely discernible by considering the International Energy Agency (IEA) yearly reports about aviation fuels. Assume aviation industries as looking like a factory, by producing some goods that are needed some resources (all kinds like materials or human resources), increasing the amount of production defiantly; it will be required more resources. This fact is evident in IEA's yearly published report, "World Energy Outlook (WEO)" [2, 3]. Globalization is needed for more transportation, mostly international transportation, and this is directly related to the global economy growing, so it could be derived that aviation fuel increased significantly.

Zadeh [4] introduced fuzzy logic and fuzzy set theory, which was established a new foundation in stating uncertainty of the real world scenarios and attract many researchers to the concept. So, fuzzy logic has appeared as one of the significant techniques to reveal and, at the same time, manipulate the vagueness and imprecision of the real scenarios. Defining the linguistic variables by Zadeh in [5] extended the pertinence of fuzzy in different problems because fuzzy linguistic variables directly could acquire the human brain's ability to consider the uncertainty of the real world scenarios. This way could directly state the human brain's analysis of the uncertainty of cases into the mathematical nature of those cases [6]. Additionally, the revolution of fuzzy set theory and introducing many extensions of fuzzy sets is another considerable point, which increased the applicability of fuzzy logic theory [7].

The content of the fuzzy extensions will be discussed in detail in the literature review part. Here it is just referred to some of those essential extensions like ordinary fuzzy sets, type-2 fuzzy sets, interval-valued fuzzy sets, intuitionistic fuzzy sets, hesitant fuzzy sets, Pythagorean fuzzy sets, picture fuzzy sets, q-rung ortho pair fuzzy sets, and spherical fuzzy sets [8].

There are different alternatives for aviation fuels to be selected among as power of aircraft. In [9], it is demonstrated that refinement of some new feedstocks for some alternatives will generate similar or slightly higher lifecycle emissions than the refinement of traditional fuels based on crude oil resources.

Therefore, the traditional problem is a complicated decision making among different alternatives for aviation fuels to select the optimal fuel based on many attributes. Technological performance, availability, costs, safety, and sustainability are some of the most famous attributes regarded in many manuscripts and also are signs that we will consider in this manuscript to make a clear choice in pace forward constructing a sustainable supply chain for aviation fuel. Motivated by the above discussion, first, we defined a novel spherical fuzzy Eco-holonic structure and then proposed a MAGDM method in the proposed spherical fuzzy Eco-holonic environment. The proposed MAGDM method in the spherical fuzzy Eco-holonic environment is applied to the sustainable supply chain of aviation fuels. The sustainable supply selection problem in aviation fuel is defined based on Spherical Fuzzy Eco-holonic structure and solved by the suggested MAGDM algorithm. The rest of the chapter is designed as follows: Sect. 2 prepares a literature review about the concept of Holarchy and Holonic structures, then it is reviewed the fuzzy logic and fuzzy sets theory, and afterward, it is presented an overview of decision-making problems. In the following section aviation industries, and aviation fuels are studied, and the sustainable supply chain (SSC) of aviation fuels is reviewed in detail. In Sect. 3, the concept of Eco-holarchy and Eco-holonic structure is presented mathematically, and then the spherical fuzzy concept is also discussed. In Sect. 4, the novel concept of spherical fuzzy Eco-holonic structure is defined and discussed in detail together with properties of the spherical fuzzy Eco-Holons and spherical fuzzy Eco-holarchy. Next, in Sect. 4, a MAGDM method is proposed based on novel SF Eco-holonic structures and tried to show the advantages of applying the SF Eco-holonic structures in the decision-making area. Section 5 discussed the sustainable supply chain selection problem with targeted the aviation industries and aviation fuels and suppliers. In this section, the issue of sustainable aviation fuel supplier selection is defined based on SF Eco-holonic structure and used the recommended MAGDM algorithm in SF Eco-holonic structure to solve it. In Sect. 6, the results are summarized, and the chapter is concluded.

2 Literature Review

In this section, it is tried to review some points which will be used in the methodology section from different perspectives based on many researchers that work in areas related to the subject of the manuscript. So, the literature review section is divided into five main sub-sections and, each sub-section is assigned to overview significant subjects that are applied in the methodology and application sections. In the following of this section, first, it is discussed Holarchy and Holonic structure then the fuzzy logic concept is reviewed, and then the decision-making subject is reviewed. Afterward, in Sect. 4 of this section, Aviation industries, and aviation fuels are discussed in detail After that last sub-section of Sect. 2, the sustainability of the supply chain (SSC) in the area of aviation fuel is reviewed.

2.1 Holarchy and Holonic Structures

The word "Holon," which is coined by Arthur Koestler in his fundamental works The Ghost in the Machine (1967) and Janus; a summing up (1978), comes from a combination of two words: holos, whole, and the suffix on. To be more precise,

"Holon" denotes the neuter form, and means particle or part (as in proton, neutron and electron) [10].

Koestler described the holon as an entity distinguished by the set rules of behavior and is, at the same time, autonomous, self-reliant, and dependent; interactive vertically. Therefore, on account of these characteristics, Koestler indicated that the holon is a whole, which characterized by a self-assertive tendency (stems from its autonomy), the integrative tendency (derives from its ability to interact and coordinate with other superordinate and subordinate holons; and being at the same time both included and inclusive), and self-preservation tendency dynamically (comes from a context of vertical interrelationships) [10]. However, this term indicates a systemic entity—conceptual or observational—and is spreading to all types of disciplines. Arthur Koestler noted that a systemic entity of holon rooted in its features of being both autonomous and dependent since it is both a whole (autonomous and containing) and a part (dependent and contained) of a modified hierarchy called a holarchy (for example, quark, protons, atoms, molecules, cells, tissues, organs, organisms, social groups, populations, federations, planets, solar systems, galaxies, galaxy clusters, etc.) [10].

In recent years, the concept of holon and holarchy have been studied by many researchers in different types of contexts; such as physics [11], engineering [12, 13], robotics, biology [14], organizational studies, management science [15, 16], business administration and entrepreneurship [17], production and supply chain systems [18–20].

The holon is part of a structure centered on relationships with the other component, subordinate and composed, and superordinate structures [10]. According to the various interpretations of holons, the following three types of holarchies can be identified:

Structural holarchies (modular holarchies and fractal holarchies):

These types of holarchies delineate the preliminary idea correspond to which holons are viewed uniquely on account of their qualitative and structural features and their similarities of genus and species [21]. Structural holarchies stem from the holons increasing size arrangement into subgroups, which are considered similar "modules" that form a more or less vertically extended structure [21].

"The modular holarchy," the simplest form of structural holarchy, is an ordered arrangement of holons conceived of as component module-entities, composed of smaller modules that form larger modules. And, "Fractal holarchies" are special forms of multi-strata modular holarchies. They are composed of similar elements arranged in levels that, in the end, form a fractal holarchy.

Cognitive holarchies (Koestler's Open Hierarchic Systems (OHS) or Self organizing Open Hierarchical Order, (SOHO), Wilber's Kosmos, and Beer's Viable System Model):

These cases come from the belief that the incremental size of holons in groups considered as autonomous "cognitive entities," which are interconnected through relations of programming, coordination, and control (typical of sentient, individual and social holons) to form larger entities [22]. Koestler represents the concept of the Self-Organizing Open Hierarchical Order (SOHO). He used this term to demonstrate that a holarchy comprehends as both an ordered arrangement of modules and a vertical system of ever-larger cognitive units possessing consciousness (Holons as an aggregate of consciousness working in some fashion as one unit).

With the holonic view, Ken Wilber (1995) demonstrates the evolutionary path of "Nature" toward the consciousness that characterizes man and his social groupings. To be more specific, for this reason, he has adopted a particular notion of holon—as an entity that in varying degrees possesses interiority and consciousness—as an individual (individual holon), or social grouping of individual holons (social holons), which form a qualitatively defined holarchy [23] named as Kosmos. Stafford Beer's model, which is universally known as the Viable System Model (VSM) [24, 25], interprets organizations as viable systems. This model indicates that organizations are open, recursive, and adaptable and, because of their cognitive and control structure, they have capable of communicating with the environment (economic and non-economic), and tend to have a long-term existence thanks to continual adaptation, even in the presence of disturbances that were not foreseen when the system was designed and carried out.

Operational Holarchies (Shimizu's Autonomic Cognitive Computer, the Holonic Manufacturing System, and the Holonic Control System (HCS)):

The third sort of holarchies derives from the arrangement of holons sentient or artificial in subgroups of an increasing size considered as "processors" or "processes," interconnected in ever-larger operational structures through their inputs and outputs [26]. Shimizu described an attractive multi-level operational holarchy where a complex task is decomposed into partial tasks, carried out by operative elements that make up a complete machine. He introduced bio-holon and bio-holonic as a discipline that studies holonic applications in biology [27]. The Holonic Manufacturing Systems (HMS) are operational reticular holarchies [28] that typically operate in the manufacturing or transport sectors [29], where the holons are machines that makeup increasingly larger structures (parts of successive structures) that carry out elementary processes which are often arranged in modules of identical machines. The method of control in Holonic Control Systems (HCS) can be understood as the gradual approximation of the movement of a variable (under control) toward desired values (objectives), made possible by the action of one or more other variables (the control levers) whose values are recalculated, at every instant, to gradually counter any shifting away for the objective [30].

2.2 Decision Making

The procedure of decision making is alongside selecting the best alternative from all possible alternatives. In such problems, named decision-making problems, various approaches were introduced to choose the best alternative. This procedure gets more complicated when the issue is considered in an uncertain condition. Consequently, from different points of view could consider and evaluate a decision-making process. Researchers with many perspectives try to come up with some contributions in many articles and books published about decision making. The significant point is that each of the alternatives has different features based on other criteria and sub-criteria to decide. It means that an alternative related to the criteria that will be considered could have various feedbacks, positive or negative, with different magnitudes. Indeed, this is the point that through whole decision-making procedures are considered and evaluated different alternatives. So, it could be concluded that the most crucial part of each decision-making process is the criteria considered in the problem. Also, criteria are not equal in the decision-making procedure, related to the nature of each criterion. Some have a more important level than others shown by different weights for each criterion [31]. In the following, it is introduced the weighting methods, particularly objective weighting methods.

It is straightforward that different criteria have different impacts. Some of them are much more important through the process, and while some others have less significance in detecting and evaluating alternatives [31]. So, the determination of weights for criteria in a MAGDM problem is one of the momentous steps. Ma et al. categorize the weighting method of criteria as: determine weights of criteria by subjective methods or objectively by considering their rate of importance and transition of information through the process of decision making [32]. Subjective methods could refer to DMs express the importance of criteria subjectively, Roostaee et al. in [33] and Tavana et al. in [34], which some subjective methods as AHP [35] and Delphi method [36] determine the weights of criteria. Chu et al. introduced the least square method [37]. For objective methods also could refer to weighting methods based on mathematical programming methods [38, 39], an integrated approach of subjective information in an objective method is also introduced by Ma et al. in [32]. Also, for objectively determining weights of criteria could refer to weighting based on entropy methods like [40, 41]. High entropy value related to a criterion represents the evaluation of alternatives based on that criterion is highly similar and has much significance, which means the criterion could have more information in it to transmit and evaluate alternatives [42]. Some other researchers used entropy measures to determine the weights of criteria [43–45].

2.3 Aviation Industry and Aviation Fuel Supply Chain

Nowadays, one of the most important role players in the world's economy is the Aviation industries [46], because are supporting 87.7 million jobs all over the world, and 11.3 million people directly work in the aviation industry. Additionally, 4.5 billion passengers carried in 2019 by aviation industries and different airlines all over the world. The global aviation industry produced just 2% of the world's emitted carbon dioxide (CO_2) across the globe.

Among all transportation industries, the aviation industries are just emitting 12% of the CO_2, which is so considerable in front of the 74% emitted CO_2 by road transporters. The overall (average) occupancy of aircraft is 82% greater than the other forms of transports. A very considerable point is that if aviation industries assume as a country, it will rank 17th world wide based on Gross Domestic Product (GDP) by generating $691.3 billion of GDP per year, which shows that it is even more remarkable that some members of G20 countries.

Human needs more and more to aviation, on account of the increase in the number of passengers and goods expected to be carried by aviation. Meanwhile, the aviation and aviation industries, like all terrestrial systems, have needs necessary for these industries to continue developing aviation's goals. The main areas of need could be referred to as Aircraft and aircraft components, Aviation infrastructures (Airports, Air navigation service providers (ANSPs), Aviation communications and providers of these communications), Carters, Insurance companies, Service companies [47].

But a big and significant piece of the above-presented puzzle is fuel for aviation industries, which will bring out the suppliers of fuels because our needing for aviation is enormously increasing, so aviation's need to fuel will also increase on the same scale [1].

Aviation fuels categorized into two main types based on [48]:

Jet fuel, the leading aviation fuel type, is mainly used in turbine engines or, in other words, a fuel for large aircraft. This type of fuel, which is producing by crude oil, which have restricted natural sources and also like other fossil fuels causes carbon dioxide (CO_2) emissions. Jet fuel, the leading aviation fuel type, is mainly used in turbine engines or, in other words, a fuel for large aircraft. This type of fuel, produced by crude oil, has restricted natural sources and, like other fossil fuels, causes carbon dioxide (CO_2) emissions. Furthermore, the use of jet fuels could cause some other type of environmental issues like land depletion, natural resource depletion, and water pollution together with dependency on a single source of fuel for the aviation industry is constructing a grey and full of uncertainty about the aviation fuel supply chain [49].

The invention of gas turbine engines (also called turbojets) developed aviation fuel technology. Due to the reason that turbojets have better usage properties than old generation engines, the aviation industry has been improved in the fuel system of aviation and the operational requirements [70]. And this type of fuel offers some other alternative sources for fuel production as a conversion of coal, gas, biomass to aviation fuels rather than traditional distilled crude sources.

The other aspect of having such diverse alternatives for aviation fuels could help the supply chain management in aviation industries cross from aviation fuel supply security and efficient air traffic management to achieve a sustainable supply chain of aviation fuel [69]. The literature has been completed by referring to some manuscripts based on multi-criteria decision making (MCDM) methods to select the optimal fuel. An integrated MCDM model is proposed by Gudiel Pineda et al. in [50] to identify critical factors for airlines improvement by combining dominance based rough set approach, DANP and VIKOR to select the most appropriate improvement alternative. In [51] Lee et al. proposed a MCDM method for selecting green aviation fleet programs by integrating DEMATEK, ANP, ZOGP methods while considering many attributes contains biofuels alternative criteria. Chen and Ren in [48] applied fuzzy MCDM methods (FANP and fuzzy GRA) to assessing the alternative aviation fuels. Also, many researches consider different MCDM methods to aviation industry problems in [52–57].

2.4 Sustainable Supply Chain (SSC) for Fuel of Airlines

Various suppliers for aviation fuels makes the problem more complicated. The reason is that there will be internationally diversified suppliers for airline fuel together with different fuel types, many features, and constraints that are related to each supplier and also fuels kinds. Besides the complexities mentioned above around the aviation fuel's supply chain problem, some economic, environmental, and social impacts of those fuels and suppliers have to be considered [58]. The mentioned impacts are also called triple bottom line (TBL) approach to achieve a sustainable supply chain proposed by Elkington in [59]. In [78], an integrated model based on TBL is suggested to show sustainability in reverse logistic systems. In [60], is suggested an integrated model based on TBL to show the sustainability in a reverse logistic systems. Slaper and Hall in [61] introduced a framework by considering social, environmental, and financial aspects to demonstrate sustainability. Furthermore, in [62], the supply chain's sustainability investigated to identify an effective model based on TBL approach by applying a fuzzy MCDM method. Based on different studies, TBL approach is illustrated in providing a sustainable supply chain of aviation fuels for airlines. Because of that, we consider three impacts of economic, environmental, and social in this study to investigate the sustainable supply chain of aviation fuel.

3 Preliminaries

In this section, first, the concept of Holonic structure and Holarchy will be defined, and then spherical fuzzy theory is studied. Both preliminary concepts are discussed and presented with mathematical definitions.

3.1 Eco-holarchy and Eco-holonic Structure

The Holonic structure constitutes a type of knowledge framework for modeling and formulating more complex structures of real-world cases. Holonic structures or holarchies were intentionally designed to create more sustainability around activities [63]. In other words, holonic structures integrate the concepts of sustainability and holonic systems. Thus, with a holonic structure of a system, it would be possible to hit the goal of having more sustainable systems [82]. This type of holonic structure of systems, which brings out more sustainability, is also called Eco-holonic, and those systems will be Eco-holonic systems [64]. In the following, we will give the Eco-holonic structures and Eco-Holarchies.

Some basic concepts and ingredients of Eco-holonic structures defined as below [64].

Eco-holonic units (Holons):

Definition 1 Eco-holonic units are referred to as all needed units in modeling a system, and every unit is made up of a real part and virtual part. The model will be extended by using the relationship between different units. Assume jth Eco-Holon at level i in previous defined Eco-holarchy which contains n_i real Eco-Holons and m_i virtual Eco-holons at level i:

$$EH_j^i = \left\{EH_1^i, EH_2^i, \ldots, EH_{n_i}^i\right\}_{Real} \cup \left\{EH_1^i, EH_2^i, \ldots, EH_{m_i}^i\right\}_{Virtual}, \quad (1)$$

Eco-holarchy:

Definition 2 Eco-Holarchies are defined and containing different Eco-Holons and interactions among them. Also, based on what explained for Eco-Holarchies, we could give the mathematical presentation of the Eco-Holarchies denoted by EH^i with h level as follows:

$$\begin{aligned} EH^i &= \left\{EH_{Real}^i\right\} \cup \left\{EH_{Virtual}^i\right\} \\ &= \left\{EH^0, EH^1, \ldots, EH^{h-1}\right\}_{Real} \cup \left\{EH^0, EH^1, \ldots, EH^{h-1}\right\}_{Virtual}, \end{aligned} \quad (2)$$

In Eq. 1, the EH^i is the set of Eco-Holons in both realistic and virtuality, and also could see that the real and virtual part contain possible holons. Also, we could show a specific Eco-Holon of this Eco-holarchy.

Remark 1 According to definition, Eco-Holons at level 0, EH^0 include all fundamental Eco-Holons which also called atomic Eco-Holons:

$$EH^0 = \left\{a_1^0, a_2^0, \ldots, a_n^0\right\}_{Real} \cup \left\{a_1^0, a_2^0, \ldots, a_m^0\right\}_{Virtual}, \quad (3)$$

Remark 2 In the given definition of Eco-Holons, we assumed that each Eco-Holon at each level of the Eco-holarchy might have overlapped. In other words, maybe there exist some common units between two different Eco-holons at a same level of Eco-holarchy.

3.2 Spherical Fuzzy Sets

Spherical Fuzzy Set (SFS) theory is introduced by Gundogdu and Kahraman in [65]; SFSs are constructed based on IFS and IFS-2; thus, it is capable of considering the membership, non-membership degrees, and it is considering hesitancy degree independently as well. The SFS is designed to develop a 3D construction for fuzzy sets by using a sphere surface (see Fig. 1), and help experts to be more flexible in giving a wide range of values for uncertainty degrees (membership degree, non-membership degree, and hesitancy degree) independently. The truth premise that real life cannot be explained by one or two uncertainty degrees makes SFSs more applicable for business [66]. Also, SFS domain is generalized by considering the sum square of the root of membership, non-membership and hesitancy degrees have to lie in interval [0, 1]. They introduced interval-valued SFS and some aggregation operators for SFSs and IVSFSs [67, 68], in the same year, 2019. Also, Ashraf and Saleem [69] proposed some new aggregation operators for SFS. Different types of decision-making techniques like Analytic Hierarchy Process (AHP), TOPSIS, WASPAS, and Gray Relational Analysis (GRA) methods in the SFS environment are developed by the same authors [65, 68, 70–72].

Definition 3 [68] Let X be the universal set and $x_i \in X$; $\forall i = 1, 2, \ldots n$, the SFS $\tilde{A}_s$ of the universe discourse X is defined as x_i as an element of $\tilde{A}_s$ with membership, non-membership and hesitancy degree values, then $\tilde{A}_s$ will be expressed mathematically in the form of:

$$\tilde{A}_s = \left\{ \langle x_i; \mu_{\tilde{A}_s}(x_i), \vartheta_{\tilde{A}_s}(x_i), h_{\tilde{A}_s}(x_i) \rangle | x_i \in X \right\}, \tag{4}$$

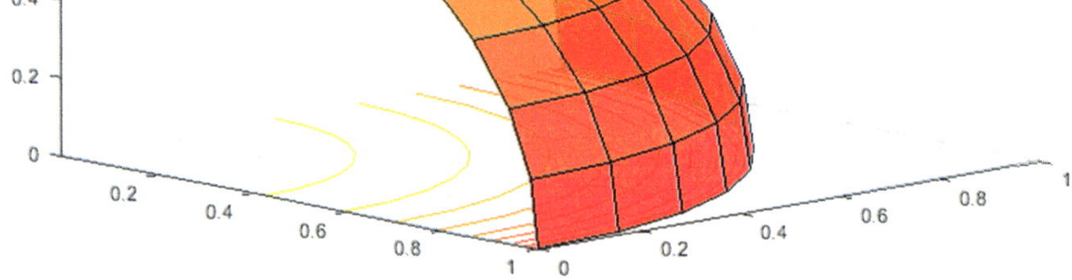

Fig. 1 Operating domain of SFSs

where $\mu_{\tilde{A}_s}(x_i), \vartheta_{\tilde{A}_s}(x_i), h_{\tilde{A}_s}(x_i)$ stands for membership degree, non-membership degree and hesitancy degree respectively and:

$$\begin{cases} \mu_{\tilde{A}_s}: X \to [0,1],\ x_i \in X \to \mu_{\tilde{A}_s}(x_i) \in [0,1] \\ \vartheta_{\tilde{A}_s}: X \to [0,1],\ x_i \in X \to \vartheta_{\tilde{A}_s}(x_i) \in [0,1], \\ h_{\tilde{A}_s}: X \to [0,1],\ x_i \in X \to h_{\tilde{A}_s}(x_i) \in [0,1], \end{cases} \tag{5}$$

which satisfies the condition that the sum square of these values to be between 0 and 1:

$$S_{\tilde{A}_s}(x_i) = \mu^2_{\tilde{A}_s}(x_i) + \vartheta^2_{\tilde{A}_s}(x_i) + h^2_{\tilde{A}_s}(x_i) \to 0 \le S_{\tilde{A}_s}(x_i) \le 1, \tag{6}$$

then refusal degree $R_{\tilde{A}_s}$ of u in the spherical fuzzy set $\tilde{A}_s$ will be as follows:

$$R_{\tilde{A}_s}(x_i) = \sqrt{1 - S_{\tilde{A}_s}(x_i)}, \tag{7}$$

Definition 4 [68] Let X be the universal set and $x_i \in X; \forall i = 1, 2, \ldots n$ then $\tilde{A} = \{\langle x_i; \mu_{\tilde{A}}(x_i), \vartheta_{\tilde{A}}(x_i), h_{\tilde{A}}(x_i)\rangle | x_i \in X\}$ and $\tilde{B} = \{\langle x_i; \mu_{\tilde{B}}(x_i), \vartheta_{\tilde{B}}(x_i), h_{\tilde{B}}(x_i)\rangle | x_i \in X\}$ be two spherical fuzzy sets (SFSs), then will have:

$$\widetilde{A}^c = \{\langle x_i; \vartheta_{\widetilde{A}}(x_i), \mu_{\widetilde{A}}(x_i), h_{\widetilde{A}}(x_i)\rangle | x_i \in X\}, \tag{8}$$

$$\widetilde{A} \subseteq \widetilde{B}: \begin{cases} \mu^2_{\widetilde{A}}(x_i) \le \mu^2_{\widetilde{B}}(x_i) \\ \vartheta^2_{\widetilde{A}}(x_i) \ge \vartheta^2_{\widetilde{B}}(x_i), \\ h^2_{\widetilde{A}}(x_i) \le h^2_{\widetilde{B}}(x_i) \end{cases} \tag{9}$$

$$\widetilde{A} = \widetilde{B} \text{ iff}: \widetilde{A} \subseteq \widetilde{B} \wedge \widetilde{B} \subseteq \widetilde{A}, \tag{10}$$

Definition 5 [68] Let X be the universal set and $x_i \in X; \forall i = 1, 2, \ldots n$ then $\tilde{A} = \{\langle x_i; \mu_{\tilde{A}}(x_i), \vartheta_{\tilde{A}}(x_i), h_{\tilde{A}}(x_i)\rangle | x_i \in X\}$ and $\tilde{B} = \{\langle x_i; \mu_{\tilde{B}}(x_i), \vartheta_{\tilde{B}}(x_i), h_{\tilde{B}}(x_i)\rangle | x_i \in X\}$ be two spherical fuzzy sets (SFSs) and $K > 0$. Therefore, some basic operators are defined as below:

$$\begin{aligned} \widetilde{A} \cup \widetilde{B} = \{\langle x_i; & max\{\mu_{\widetilde{A}}(x_i), \mu_{\widetilde{B}}(x_i)\}, min\{\vartheta_{\widetilde{A}}(x_i), \vartheta_{\widetilde{B}}(x_i)\}, \\ & min\{(1 - (max\{\mu_{\widetilde{A}}(x_i), \mu_{\widetilde{B}}(x_i)\})^2 - (min\{\vartheta_{\widetilde{A}}(x_i), \vartheta_{\widetilde{B}}(x_i)\})^2)^{0.5}, \\ & max\{h_{\widetilde{A}}(x_i), h_{\widetilde{B}}(x_i)\}\}\rangle | x_i \in X\}, \end{aligned} \tag{11}$$

$$\begin{aligned} \widetilde{A} \cap \widetilde{B}_s = \{\langle x_i; & min\{\mu_{\widetilde{A}}(x_i), \mu_{\widetilde{B}}(x_i)\}, max\{\vartheta_{\widetilde{A}}(x_i), \vartheta_{\widetilde{B}}(x_i)\}, \\ & max\{(1 - (max\{\mu_{\widetilde{A}}(x_i), \mu_{\widetilde{B}}(x_i)\})^2 - (min\{\vartheta_{\widetilde{A}}(x_i), \vartheta_{\widetilde{B}}(x_i)\})^2)^{0.5}, \end{aligned}$$

$$min\{h_{\widetilde{A}}(x_i), h_{\widetilde{B}}(x_i)\}\}\rangle|x_i \in X\}, \tag{12}$$

$$\widetilde{A} \oplus \widetilde{B} = \{\langle x_i ; \left(\mu_{\widetilde{A}}^2(x_i) + \mu_{\widetilde{B}}^2(x_i) - \mu_{\widetilde{A}}^2(x_i).\mu_{\widetilde{B}}^2(x_i)\right)^{0.5}, \vartheta_{\widetilde{A}}(x_i).\vartheta_{\widetilde{B}}(x_i), \\ \left(\left(1 - \mu_{\widetilde{B}}^2(x_i)\right)h_{\widetilde{A}}^2(x_i) + \left(1 - \mu_{\widetilde{A}}^2(x_i)\right)h_{\widetilde{B}}^2(x_i) - h_{\widetilde{A}}^2(x_i).h_{\widetilde{B}}^2(x_i)\right)^{0.5}\rangle|x_i \in X\}, \tag{13}$$

$$\widetilde{A} \otimes \widetilde{B}_s = \{\langle x_i ; \mu_{\widetilde{A}}(x_i).\mu_{\tilde{B}}(x_i), \left(\vartheta_{\widetilde{A}}^2(x_i) + \vartheta_{\mathrm{B}}^2(x_i) - \vartheta_{\widetilde{A}}^2(x_i).\vartheta_{\widetilde{B}}^2(x_i)\right)^{0.5}, \\ \left(\left(1 - \vartheta_{\widetilde{B}}^2(x_i)\right)h_{\widetilde{A}}^2(x_i) + \left(1 - \vartheta_{\widetilde{A}}^2(x_i)\right)h_{\widetilde{B}}^2(x_i) - h_{\widetilde{A}}^2(x_i).h_{\widetilde{B}}^2(x_i)\right)^{0.5}\rangle|x_i \in X\}, \tag{14}$$

$$k\widetilde{A} = \{\langle x_i ; \left(1 - \left(1 - \mu_{\widetilde{A}}^2(x_i)\right)^k\right)^{0.5}, \left(\vartheta_{\widetilde{A}}(x_i)\right)^k, \\ \left(\left(1 - \mu_{\widetilde{A}}^2(x_i)\right)^k - \left(1 - \mu_{\widetilde{A}}^2(x_i) - h_{\widetilde{A}}^2(x_i)\right)^k\right)^{0.5}\rangle|x_i \in X\}, \tag{15}$$

$$\widetilde{A}^k = \{\langle x_i ; \left(\mu_{\widetilde{A}}(x_i)\right)^k, \left(1 - \left(1 - \vartheta_{\widetilde{A}}^2(x_i)\right)^k\right)^{0.5}, \\ \left(\left(1 - \vartheta_{\widetilde{A}}^2(x_i)\right)^k - \left(1 - \vartheta_{\widetilde{A}}^2(x_i) - h_{\widetilde{A}}^2(x_i)\right)^k\right)^{0.5}\rangle|x_i \in X\}, \tag{16}$$

Remark 3 Assume $\tilde{A}_j = \langle \mu_{\tilde{A}_j}, \vartheta_{\tilde{A}_j}, h_{\tilde{A}_j} \rangle; \forall j = 1, 2, \ldots n$ be a collection of spherical fuzzy numbers (SFNs), the generalized addition will be as follows:

$$\underset{j=1}{\overset{n}{\oplus}} \widetilde{A}_j = \left(\widetilde{A}_1 \oplus \widetilde{A}_2 \oplus \cdots \oplus \widetilde{A}_n\right) \\ = \left\langle \begin{array}{l} \left(1 - \prod_{j=1}^n \left(1 - \mu_{\tilde{A}_j}^2\right)\right)^{0.5}, \prod_{j=1}^n \vartheta_{\tilde{A}_j}, \\ \left(\prod_{j=1}^n \left(1 - \mu_{\tilde{A}_j}^2\right) - \prod_{j=1}^n \left(1 - \mu_{\tilde{A}_j}^2 - h_{\tilde{A}_j}^2\right)\right)^{0.5} \end{array} \right\rangle, \tag{17}$$

Remark 4 Assume $\tilde{A}_j = \langle \mu_{\tilde{A}_j}, \vartheta_{\tilde{A}_j}, h_{\tilde{A}_j} \rangle; \forall j = 1, 2, \ldots n$ be a collection of spherical fuzzy numbers (SFNs), the generalized multiplication will be as follows:

$$\underset{j=1}{\overset{n}{\otimes}} \tilde{A}_j = \left(\tilde{A}_1 \otimes \tilde{A}_2 \otimes \cdots \otimes \tilde{A}_n\right) \\ = \left\langle \begin{array}{c} \prod_{j=1}^n \mu_{\tilde{A}_j}, \left(1 - \prod_{j=1}^n \left(1 - \vartheta_{\tilde{A}_j}^2\right)\right)^{0.5}, \\ \left(\prod_{j=1}^n \left(1 - \vartheta_{\tilde{A}_j}^2\right) - \prod_{j=1}^n \left(1 - \vartheta_{\tilde{A}_j}^2 - h_{\tilde{A}_j}^2\right)\right)^{0.5} \end{array} \right\rangle, \tag{18}$$

Definition 6 [68] Score (Sc) and accuracy (Ac) functions of sorting a SFN are defined respectively as follows:

$$Sc\left(\widetilde{A}\right)=\left(\mu_{\widetilde{A}}-\frac{h_{\widetilde{A}}}{2}\right)^2-\left(\vartheta_{\widetilde{A}}-\frac{h_{\widetilde{A}}}{2}\right)^2, \tag{19}$$

$$Ac\left(\widetilde{A}\right)=\mu_{\widetilde{A}}^2+\vartheta_{\widetilde{A}}^2+h_{\widetilde{A}}^2, \tag{20}$$

Remark 5 Note if $\tilde{A}$ and $\tilde{B}$ are SFNs then $\tilde{A}<\tilde{B}$ if and only if:

i. $Sc\left(\tilde{A}\right)<Sc\left(\tilde{B}\right)$,

or

ii. $Sc\left(\tilde{A}\right)=Sc\left(\tilde{B}\right)$ and $Ac\left(\tilde{A}\right)<Ac\left(\tilde{B}\right)$.

We will introduce two aggregation operators; the spherical fuzzy weighted arithmetic mean operator and spherical fuzzy weighted geometric mean operator, which are proposed in [65].

Definition 7 [65] Let X be the universal set and $x_i\in X;\forall i=1,2,\ldots n$ then $\tilde{A}=\left\{\langle x_i;\mu_{\tilde{A}}(x_i),\vartheta_{\tilde{A}}(x_i),h_{\tilde{A}}(x_i)\rangle|x_i\in X\right\}$ be a spherical fuzzy set (SFS) with corresponding weight vector $w=\{w_1,w_2,\ldots,w_n\};\ w_i\in[0,1];\ \sum_{i=1}^n w_i=1$. Spherical Fuzzy Weighted Arithmetic Mean ($SFWAM$) is defined as:

$$\begin{aligned}SFWAM_w\left(\widetilde{A}\right)&=\underset{i=1}{\overset{n}{\oplus}}w_i\widetilde{A}(x_i)=w_1\widetilde{A}(x_1)+w_2\widetilde{A}(x_2)+\cdots+w_n\widetilde{A}(x_n)\\&=\left\{\left(1-\prod\nolimits_{i=1}^n\left(1-\mu_{\widetilde{A}}^2(x_i)\right)^{w_i}\right)^{0.5},\prod\nolimits_{i=1}^n\left(\vartheta_{\widetilde{A}}(x_i)\right)^{w_i},\right.\\&\left.\left(\prod\nolimits_{i=1}^n\left(1-\mu_{\widetilde{A}}^2(x_i)\right)^{w_i}-\prod\nolimits_{i=1}^n\left(1-\mu_{\widetilde{A}}^2(x_i)-h_{\widetilde{A}}^2(x_i)\right)^{w_i}\right)^{0.5}\right\},\end{aligned} \tag{21}$$

Definition 8 [65] Let X be the universal set and $x_i\in X;\forall i=1,2,\ldots n$ then $\tilde{A}=\left\{\langle x_i;\mu_{\tilde{A}}(x_i),\vartheta_{\tilde{A}}(x_i),h_{\tilde{A}}(x_i)\rangle|x_i\in X\right\}$ be a spherical fuzzy set (SFS) with corresponding weight vector $w=\{w_1,w_2,\ldots,w_n\};\ w_i\in[0,1];\ \sum_{i=1}^n w_i=1$. Spherical Fuzzy Weighted Geometric Mean ($SFWGM$) is defined as:

$$\begin{aligned}SFWGM_w\left(\widetilde{A}\right)&=\underset{i=1}{\overset{n}{\otimes}}\left(\widetilde{A}(x_i)\right)^{w_i}=\left(\widetilde{A}(x_1)\right)^{w_1}\times\left(\widetilde{A}(x_2)\right)^{w_2}\times\cdots\times\left(\widetilde{A}(x_n)\right)^{w_n}\\&=\left\{\prod\nolimits_{i=1}^n\left(\mu_{\widetilde{A}}(x_i)\right)^{w_i},\left(1-\prod\nolimits_{i=1}^n\left(1-\vartheta_{\widetilde{A}}^2(x_i)\right)^{w_i}\right)^{0.5},\right.\\&\left.\left(\prod\nolimits_{i=1}^n\left(1-\vartheta_{\widetilde{A}}^2(x_i)\right)^{w_i}-\prod\nolimits_{i=1}^n\left(1-\vartheta_{\widetilde{A}}^2(x_i)-h_{\widetilde{A}}^2(x_i)\right)^{w_i}\right)^{0.5}\right\},\end{aligned} \tag{22}$$

In the following, we will introduce an entropy measure for spherical fuzzy sets (SFSs), which are proposed by Aydogdu and Gul in [73].

Definition 9 [73] Let X be the universal set and $x_i \in X; \forall i = 1, 2, \ldots n$ then $\tilde{A} = \left\{ \langle x_i; \mu_{\tilde{A}}(x_i), \vartheta_{\tilde{A}}(x_i), h_{\tilde{A}}(x_i) \rangle | x_i \in X \right\}$ be a spherical fuzzy set (SFS). Spherical Fuzzy entropy measure $E_{SFS}\left(\tilde{A}\right)$ is defined as:

$$E_{SFS}\left(\widetilde{A}\right) = \frac{1}{n} \sum\nolimits_{i=1}^{n} \left[1 - \frac{4}{5}\left(\left| \mu_{\tilde{A}}^2(x_i) - \vartheta_{\tilde{A}}^2(x_i) \right| + \left| h_{\tilde{A}}^2(x_i) - 0.25 \right| \right) \right], \quad (23)$$

4 Methodology

In this section, we first will introduce the novel spherical fuzzy Eco-holonic structure then it will offer a multi-attribute group decision making (MAGDM) model by using the proposed spherical fuzzy Eco-holarchy. In this way we will be able to show the applicability of the proposed SF Eco-holarchy concept.

4.1 Spherical Fuzzy Eco-holarchy (SF Eco-holarchy)

Spherical Fuzzy Eco-holonic units are all needed units in modeling a system in a spherical fuzzy environment. Units real and virtual parts are stated based on spherical fuzzy uncertainty, which shows the concept of possibility to have such a unit in real or not. Also, the virtual part of each unit is a kind of ambiguity around the nature of that unit. Then, we could conclude that the virtual part of units will be available to define based on spherical fuzzy sets. Spherical Fuzzy Eco-Holons are defined based on Spherical Fuzzy Eco-holonic units, which interact with each other. So, Eco-holonic units in a spherical fuzzy environment will contribute to the novel concept of Spherical Fuzzy Eco-holonic units as follows.

Definition 10 Let X be the universal set of units, then Eco-holonic units of the universal set X in spherical fuzzy environment are defined as Spherical Fuzzy Eco-holonic units denoted by $\widetilde{EH}_{sj}^{i}$, which determine jth spherical fuzzy Eco-holonic unit at ith level.

Definition 11 Let X be the universal set of units, then spherical fuzzy Eco-holonic units consists the spherical fuzzy Eco-Holons $\widetilde{EH}_{s}^{i}$. So, we will have spherical fuzzy Eco-holonic unit $\widetilde{EH}_{sj}^{i}$ be an element of spherical fuzzy Eco-Holon $\widetilde{EH}_{s}^{i}$ with membership, non-membership and hesitancy degree values. Then spherical fuzzy Eco-Holon $\widetilde{EH}_{s}^{i}$ will be expressed mathematically in the form of Spherical Fuzzy Eco-holonic units as follows:

$$\widetilde{EH}_{s}^{i} = \left\{ \left\langle \widetilde{EH}_{sj}^{i}; \mu_{\widetilde{EH}_{s}^{i}}\left(\widetilde{EH}_{sj}^{i}\right), \vartheta_{\widetilde{EH}_{s}^{i}}\left(\widetilde{EH}_{sj}^{i}\right), h_{\widetilde{EH}_{s}^{i}}\left(\widetilde{EH}_{sj}^{i}\right) \right\rangle | \, \widetilde{EH}_{sj}^{i} \in X \textit{ and}$$

$$i = 1, 2, \ldots, h \; and \; j = 1, 2, \ldots, n_i\}, \tag{24}$$

where, $\widetilde{EH}_{sj}^{i}$ is consists of n_i lower-level units or sub units so: $j = 1, 2, \ldots, n_i$ and there is h different levels of SF Eco holons so, $i = 1, 2, \ldots, h$.

Remark 6 It is assumed that jth SF Eco-holonic unit at level i contains n_i spherical fuzzy Eco-holonic units from lower-level SF Eco-Holons. It means that each SF Eco-holonic unit contains some SF Eco Holonic units from lower level (previous level) of SF Eco Holons.

Remark 7 According to definition, SF Eco-holonic units which consists the SF Eco-Holon at level $0(h = 0)$, $\widetilde{EH}_{s}^{0}$ are assumed as Atomic SF Eco-holonic units $\{u_k | u_k \in X \; and \; k = 1, 2, \ldots, a\}$ and fundamental SF Eco-holonic units $\widetilde{EH}_{sj}^{0}$, respectively. So, we will have as follows:

Atomic SF Eco-holonic units:

$$\widetilde{A}_s = \{u_k | u_k \in X \; and \; k = 1, 2, \ldots, a\}, \tag{25}$$

jth Fundamental SF Eco-holonicunit:

$$\widetilde{EH}_{sj}^{0} = \left\{ \left\langle u_k ; \mu_{\widetilde{EH}_{sj}^{0}}(u_k), \vartheta_{\widetilde{EH}_{sj}^{0}}(u_k), h_{\widetilde{EH}_{sj}^{0}}(u_k) \right\rangle | u_k \in X \; and \; k = 1, 2, \ldots, a \right\}, \tag{26}$$

Fundamental SF Eco Holon at level 0 (h = 0):

$$\widetilde{EH}_{s}^{0} = \left\{ \begin{array}{l} \widetilde{EH}_{sj}^{0} ; \mu_{\widetilde{EH}_{s}^{0}}\left(\widetilde{EH}_{sj}^{0}\right), \vartheta_{\widetilde{EH}_{s}^{0}}\left(\widetilde{EH}_{sj}^{0}\right), h_{\widetilde{EH}_{s}^{0}}\left(\widetilde{EH}_{sj}^{0}\right) | \widetilde{EH}_{sj}^{0} \in X \\ and \; i = 0 \; and \; j = 1, 2, \ldots, n_0 \end{array} \right\}. \tag{27}$$

In above definition as it is mentioned, $\widetilde{EH}_{sj}^{0}$ are fundamental SF-Eco-Holons and u_k are atomic SF Eco-holonic Units.

Definition 12 Let X be the universal set of units, then spherical fuzzy Eco-holonic units $\widetilde{EH}_{sj}^{i}$ and SF Eco-Holons $\widetilde{EH}_{s}^{i}$ and interactions of them consists the complexity that is needed to model and represent a system which is called spherical fuzzy Eco-holarchy $\widetilde{EHy}_s$. So, we will have:

$$\widetilde{EHy}_s = \left\{ \widetilde{EH}_{s}^{1}, \widetilde{EH}_{s}^{2}, \ldots, \widetilde{EH}_{s}^{h} \right\} = \left\{ \widetilde{EH}_{s}^{i} | \forall i = 1, 2, \ldots, h \right\}, \tag{28}$$

$\widetilde{EHy}_s$ is called h level SF Eco-holarchy where, $i = 1, 2, \ldots, h$.

Spherical fuzzy Eco-holarchy is composed of spherical fuzzy Eco-Holons, and each SF Eco-Holons contain some sub level spherical fuzzy Eco-Holons. So, in this

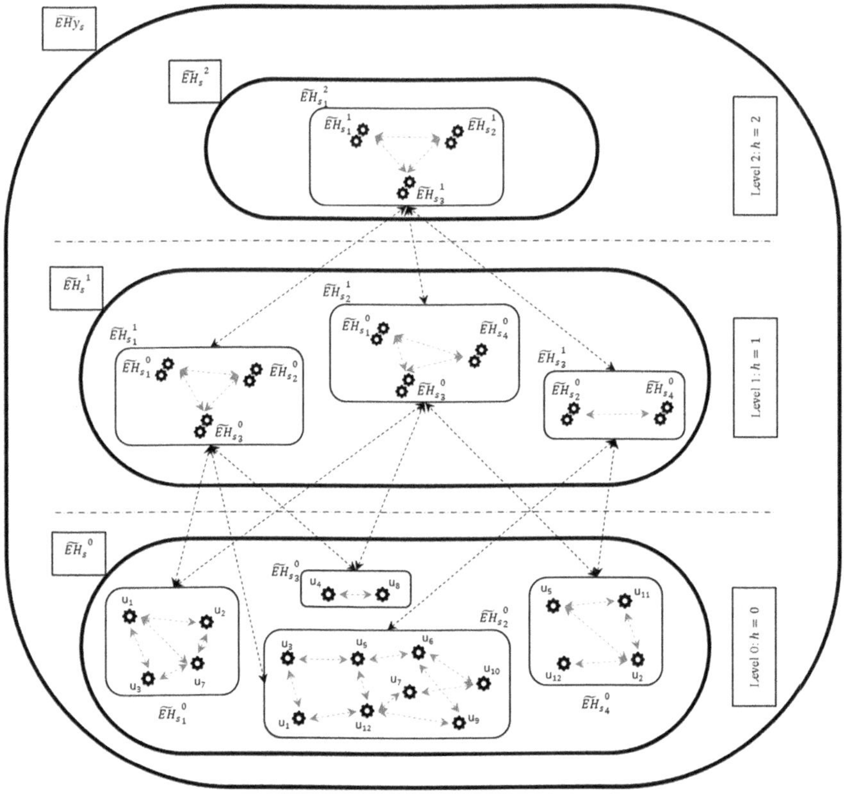

Fig. 2 The spherical fuzzy eco-holonic structure (containing just one SF eco-holarchy)

way, a SF Eco-holarchy is capable of defining a real-world system with all uncertainties around its components. A spherical fuzzy Eco-holonic structure is an architecture composed of one or more parallel spherical fuzzy Eco-Holarchies together, which comprehensively describe the integrity of the system with all possible states.

In the following, it is tried to figure out a Spherical Fuzzy Eco-holonic structure with a Spherical Fuzzy Eco-holarchy with in detail through the above definitions in Fig. 2. The details about the illustrated Fig. 2, a spherical fuzzy Eco-holarchy $\tilde{EHy}_s$ with three levels ($h = 0, 1, 2$) and different atomic SF Eco-holonic units, Fundamental SF Eco-holonic units, and SF Eco-Holons are shown in the Table 1.

4.2 SF Eco-holarchy Application in a MAGDM Problem

Complexity is an inherent part of real-world problems; hence, the challenge of implementing the best strategies to make a decision is intensified. Decision making is a process that aims to find an appropriate way to control a process via another process

Table 1 Spherical fuzzy eco-holonic structure $\left(\tilde{EHS}_s\right)$ and contained spherical fuzzy eco-holarchy $\left(\tilde{EHy}_s\right)$

Level (h)	Component title	Mathematical expression
$h = 0$	Atomic SF eco-holonic units $\{u_k\}$	$\tilde{A}_s = \{u_k; k = 1, \ldots, 12\}$
	Fundamental SF eco-holonic units $\left\{\widetilde{EH}_{s1}^{0}, \widetilde{EH}_{s2}^{0}, \widetilde{EH}_{s3}^{0}, \widetilde{EH}_{s4}^{0}\right\}$	$\widetilde{EH}_{s1}^{0} = \left\{\left\langle u_k; \mu_{\widetilde{EH}_{s1}^{0}}(u_k), \vartheta_{\widetilde{EH}_{s1}^{0}}(u_k), h_{\widetilde{EH}_{s1}^{0}}(u_k)\right\rangle, \lvert u_k \in X \text{ and } k = 1, 2, 3, 7\right\}$ $\widetilde{EH}_{s2}^{0} = \left\{\left\langle u_k; \mu_{\widetilde{EH}_{s2}^{0}}(u_k), \vartheta_{\widetilde{EH}_{s2}^{0}}(u_k), h_{\widetilde{EH}_{s2}^{0}}(u_k)\right\rangle, \lvert u_k \in X \text{ and } k = 1, 3, 5, 6, 7, 9, 10, 12\right\}$ $\widetilde{EH}_{s3}^{0} = \left\{\left\langle u_k; \mu_{\widetilde{EH}_{s3}^{0}}(u_k), \vartheta_{\widetilde{EH}_{s3}^{0}}(u_k), h_{\widetilde{EH}_{s3}^{0}}(u_k)\right\rangle, \lvert u_k \in X \text{ and } k = 4, 8\right\}$ $\widetilde{EH}_{s4}^{0} = \left\{\left\langle u_k; \mu_{\widetilde{EH}_{s4}^{0}}(u_k), \vartheta_{\widetilde{EH}_{s4}^{0}}(u_k), h_{\widetilde{EH}_{s4}^{0}}(u_k)\right\rangle, \lvert u_k \in X \text{ and } k = 2, 5, 11, 12\right\}$
	Fundamental SF eco-holon $\widetilde{EH}_s^{0}$	$\widetilde{EH}_s^{0} = \left\{\left\langle \widetilde{EH}_{sj}^{0}; \mu_{\widetilde{EH}_s^{0}}\left(\widetilde{EH}_{sj}^{0}\right), \vartheta_{\widetilde{EH}_s^{0}}\left(\widetilde{EH}_{sj}^{0}\right), h_{\widetilde{EH}_s^{0}}\left(\widetilde{EH}_{sj}^{0}\right)\right\rangle \vert i = 0 \text{ and } j = 1, 2, 3 \text{ and } 4\right\}$
$h = 1$	SF eco-holonic units $\left\{\widetilde{EH}_{s1}^{1}, \widetilde{EH}_{s2}^{1}, \widetilde{EH}_{s3}^{1}\right\} \widetilde{EH}_{s3}^{1}$	$\widetilde{EH}_{s2}^{1}$ is clarified as an example in diagram: $\widetilde{EH}_{s2}^{1} = \left\{\left\langle \widetilde{EH}_{sj}^{0}; \mu_{\widetilde{EH}_{s2}^{1}}\left(\widetilde{EH}_{sj}^{0}\right), \vartheta_{\widetilde{EH}_{s2}^{1}}\left(\widetilde{EH}_{sj}^{0}\right), h_{\widetilde{EH}_{s2}^{1}}\left(\widetilde{EH}_{sj}^{0}\right)\right\rangle \vert j = 1, 3, 4\right\}$ $= \left\{\begin{array}{l}\left\langle \widetilde{EH}_{s1}^{0}; \mu_{\widetilde{EH}_{s2}^{1}}\left(\widetilde{EH}_{s1}^{0}\right), \vartheta_{\widetilde{EH}_{s2}^{1}}\left(\widetilde{EH}_{s1}^{0}\right), h_{\widetilde{EH}_{s2}^{1}}\left(\widetilde{EH}_{s1}^{0}\right)\right\rangle, \\ \left\langle \widetilde{EH}_{s3}^{0}; \mu_{\widetilde{EH}_{s2}^{1}}\left(\widetilde{EH}_{s3}^{0}\right), \vartheta_{\widetilde{EH}_{s2}^{1}}\left(\widetilde{EH}_{s3}^{0}\right), h_{\widetilde{EH}_{s2}^{1}}\left(\widetilde{EH}_{s3}^{0}\right)\right\rangle, \\ \left\langle \widetilde{EH}_{s4}^{0}; \mu_{\widetilde{EH}_{s2}^{1}}\left(\widetilde{EH}_{s4}^{0}\right), \vartheta_{\widetilde{EH}_{s2}^{1}}\left(\widetilde{EH}_{s4}^{0}\right), h_{\widetilde{EH}_{s2}^{1}}\left(\widetilde{EH}_{s4}^{0}\right)\right\rangle\end{array}\right\}$
	SF eco holon $\tilde{EH}_s^{1}$	$\widetilde{EH}_s^{1} = \left\{\widetilde{EH}_{sj}^{1}; \mu_{\widetilde{EH}_s^{1}}\left(\widetilde{EH}_{sj}^{1}\right), \vartheta_{\widetilde{EH}_s^{1}}\left(\widetilde{EH}_{sj}^{1}\right), h_{\widetilde{EH}_s^{1}}\left(\widetilde{EH}_{sj}^{1}\right) \vert i = 1 \text{ and } j = 1, 2, 3\right\}$
$h = 2$	SF eco-holonic units $\left\{\widetilde{EH}_{s1}^{2}\right\}$	$\widetilde{EH}_{s1}^{2} = \left\{\widetilde{EH}_{sj}^{1}; \mu_{\widetilde{EH}_{s1}^{2}}\left(\widetilde{EH}_{sj}^{1}\right), \vartheta_{\widetilde{EH}_{s1}^{2}}\left(\widetilde{EH}_{sj}^{1}\right), h_{\widetilde{EH}_{s1}^{2}}\left(\widetilde{EH}_{sj}^{1}\right) \vert j = 1, 2 \text{ and } 3\right\}$ $= \left\{\begin{array}{l}\left\langle \widetilde{EH}_{s1}^{1}; \mu_{\widetilde{EH}_{s1}^{2}}\left(\widetilde{EH}_{s1}^{1}\right), \vartheta_{\widetilde{EH}_{s1}^{2}}\left(\widetilde{EH}_{s1}^{1}\right), h_{\widetilde{EH}_{s1}^{2}}\left(\widetilde{EH}_{s1}^{1}\right)\right\rangle, \\ \left\langle \widetilde{EH}_{s2}^{1}; \mu_{\widetilde{EH}_{s1}^{2}}\left(\widetilde{EH}_{s2}^{1}\right), \vartheta_{\widetilde{EH}_{s1}^{2}}\left(\widetilde{EH}_{s2}^{1}\right), h_{\widetilde{EH}_{s1}^{2}}\left(\widetilde{EH}_{s2}^{1}\right)\right\rangle, \\ \left\langle \widetilde{EH}_{s3}^{1}; \mu_{\widetilde{EH}_{s1}^{2}}\left(\widetilde{EH}_{s3}^{1}\right), \vartheta_{\widetilde{EH}_{s1}^{2}}\left(\widetilde{EH}_{s3}^{1}\right), h_{\widetilde{EH}_{s1}^{2}}\left(\widetilde{EH}_{s3}^{1}\right)\right\rangle\end{array}\right\}$
	SF eco-holon $\widetilde{EH}_s^{2}$	$\widetilde{EH}_s^{2} = \left\{\widetilde{EH}_{sj}^{2}; \mu_{\widetilde{EH}_s^{2}}\left(\widetilde{EH}_{sj}^{2}\right), \vartheta_{\widetilde{EH}_s^{2}}\left(\widetilde{EH}_{sj}^{2}\right), h_{\widetilde{EH}_s^{2}}\left(\widetilde{EH}_{sj}^{2}\right) \vert i = 2 \text{ and } j = 1\right\}$ $= \left\{\widetilde{EH}_{s1}^{2}; \mu_{\widetilde{EH}_s^{2}}\left(\widetilde{EH}_{s1}^{2}\right), \vartheta_{\widetilde{EH}_s^{2}}\left(\widetilde{EH}_{s1}^{2}\right), h_{\widetilde{EH}_s^{2}}\left(\widetilde{EH}_{s1}^{2}\right)\right\}$
Whole system	SF eco-holarchy	$\widetilde{EHy}_s = \left\{\widetilde{EH}_s^{0}, \widetilde{EH}_s^{1}, \widetilde{EH}_s^{2}\right\}$
	SF eco-holonic structure	$\tilde{EHS}_s = \left\{\tilde{EHy}_s\right\}$

[74]. Each MAGDM method consists of two main parts: aggregation and exploitation [75], which can be done through the following two steps. In the first step, data of all criteria and alternatives are combined by using the weight vectors. In the second step, the exploitation and qualification actions like ranking will be carried out.

In this section, we will introduce our offered MAGDM method and then show the steps of MAGDM algorithm in spherical fuzzy Eco-holarchy.

Proposed MAGDM Algorithm in SF Eco-holarchy.

Selecting the best alternative among the possible choices is a complex task since each alternative might be more desired according to a specific feature of criteria or sub-criteria. The process of choosing the best alternative is known as decision-making problems. The diversity of the features and the relations between them in each alternative, along with different criteria and sub-criteria, is combined with the number of DMs as well as their opinion and increases the complexity of the decision-making process. Further complexity is added in by considering the uncertain environment. Moreover, in the literature part is noted that the Eco-holonic structures are a knowledge framework for making sensible, modeling, and formulating and extra complex real cases of the world [63]. So, Holonic structures provide much more detailed models, which also brings us a sustainable model of the real case scenario [81]. A combination of the Eco-holonic concept with fuzzy logic will bring together a much more reliable comprehensive structure and increase the defined model's applicability as a real face of scenarios in the world. As explained above, a spherical fuzzy Eco-holonic structure consists of SF Eco-holarchies describes the integrity of a system on account of its many spherical fuzzy Eco-Holons. SF-Eco-Holons are interactive decisional units, which acts like a small-sized system by itself (means every SF Eco-Holons have defined inputs, outputs, and its own rules). Therefore, it would be much trustable to decide in such an environment. So, we proposed a MAGDM method to make a better decision in a spherical fuzzy Eco-holonic structure. And then, in the following, we define a MAGDM problem and show the steps of the MAGDM algorithm under SF Eco-holonic structure.

Suppose that n decision-makers $D = \{D_1, D_2, \ldots, D_n\}$ with related weight vector $w = \{w_1, w_2, \ldots, w_n\}, \sum_{i=1}^{n} w_i = 1, w_i \geq 0$ comment on m alternatives $A = \{A_1, A_{s2}, \ldots, A_m\}$ based on k criteria $C = \{C_1, C_2, \ldots, C_k\}$ with corresponding weight vector $\omega_j = (\omega_1, \omega_2, \ldots, \omega_k), \sum_{j=1}^{k} \omega_j = 1, \omega_j \geq 0$. Decision makers consider each alternative under a set of criteria and evaluate alternatives to strengthen under different criteria represented in linguistic terms. Accordingly, decision matrix of alternatives based on criteria for each decision-maker will be constructed, and then to quantifying these scales, they defined in a spherical fuzzy environment to consider the uncertainty of the real-world cases. Then, all criteria are considered based on an Eco-holarchy structure. The reason is that defining some criteria to evaluate alternatives of a real case is not enough; even it does not make good sense by considering a group of sub-criteria for those criteria.

So, in this way it is proposed MAGDM method in spherical fuzzy Eco-holonic structure and steps will be as follows.

Step 1. Problem statement and determination of goals, alternatives, and criteria:

In this step, like in every scientific field, we determine the issue's goals and significance. This step is known as the problem statement stage because here we have a decision-making problem; besides goals, the alternatives and criteria will be determined.

Step 2. Construct the Eco-Holons, Eco-holarchy, and Eco-holonic structure for the problem:

In this step, the criteria for the model will be defined based on SF Eco-Holons, SF Eco-Holarchies, and then SF Eco-holonic structure will be presented. Based on Definition 10, Definition 11, and Definition 12 we will construct a SF Eco-holonic structure for our model's criteria.

Step 3. Construct the individual decision matrices:

Every decision-maker construct an alternative-criteria decision matrix, based on Eco-Holons, which is defined in Step 2.

Step 3.2. Construct the Spherical Fuzzy Eco-holonic decision matrices:

Use spherical fuzzy linguistic scales defined in Table 2 and turn linguistic Eco-holonic decision matrices of experts from step (3.1) to spherical fuzzy Eco-holonic decision matrices.

Step 3.3. Normalize the constructed spherical fuzzy Eco-holonic decision matrices:

Normalize the constructed decision matrices, if it is not uniform (if different types of criteria in different scales exist, decision matrices should be normalized with considering cost or benefit attribute of the features).

Step 4. Determine the weights of DMs in SF Eco-holarchy:

Table 2 Spherical fuzzy linguistic scales

Spherical fuzzy linguistic scales $\left(\tilde{LS}_s\right)$	SFNs (μ, ϑ, h)
Absolutely high possible (AHP)	(0.9, 0.1, 0.1)
Very high possible (VHP)	(0.8, 0.2, 0.2)
High possible (HP)	(0.7, 0.3, 0.3)
Slightly high possible (SHP)	(0.6, 0.4, 0.4)
Equally possible (EP)	(0.5, 0.5, 0.5)
Slightly low possible (SLP)	(0.4, 0.6, 0.4)
Low possible (LP)	(0.3, 0.7, 0.3)
Very low possible (VLP)	(0.2, 0.8, 0.2)
Absolutely low possible (ALP)	(0.1, 0.9, 0.1)

It is supposed n decision-makers $D = \{D_1, D_2, \ldots, D_n\}$, which means that there will exist different experts with different specialties and experiences. So, it is necessary to evaluate the relative importance of each expert. This study tried to evaluate experts' weights based on mathematical background and freely without considering the subjective impacts like psychological effects, position, or post of an expert. So, it is proposed an objective method to determine experts' weights based on their experience in those fields. The weighting method is designed based on how much the uncertainty of an expert's decision matrix is lower, his/her individual experience in that field is much, which means that experts transmitted more information in the process of decision making. It is proposed the entropy measure to determine the decision makers' weights based on individual experience and transmitted information rate of experts. It is applied the entropy measure for spherical fuzzy sets to determine the weights of DMs in Eco-Holarchies because, in different Eco-Holons, the experts could have other performances based on their experiences in those situations and fields. The proposed weighting method will support and guarantee that it determines the weights of DMs objectively in SF Eco-holarchy.

Step 4.1. Determine weights of DMs in SF Eco-holarchy objectively:

We used the spherical fuzzy entropy measure to determine the weights, which is defined in Definition 9. After determining the entropy values of decision makers, then the divergence will be calculated for each expert by using Eq. (29) and finally get the weights of DMs by using Eq. (30).

$$D_{SFS}(D_n) = 1 - E_{SFS}(D_n), \tag{29}$$

where, $E_{SFS}(D_n)$ is showing the spherical fuzzy entropy of D_n (nth decision maker) and $D_{SFS}(D_n)$ is divergency of D_n.

$$W_{SFS}(D_n) = \frac{D_{SFS}(D_n)}{\sum_{i=1}^{n} D_{SFS}(D_n)}, \tag{30}$$

Step 5. Construct GDM matrix for SF Eco-holarchy:

Aggregate the decision matrices for each decision-maker in SF Eco-holarchy by using weighted aggregation operators, which are shown in Eqs. (21) and (22), by considering the corresponding weight vector of decision-makers. So, it is combining different ideas from different experts while considering their objective weights through whole SF Eco-holarchy to get the GDM matrix for SF Eco-holarchy.

Step 6. Determine the weights of criteria in SF Eco-holarchy:

As mentioned in the literature and the same in step (4.1), two main categories determine the weights of criteria as proposed by different research studies, subjective and objective weighting methods. Here it is applied an objective method to determine the weights of a criterion based on the practical information transmitted by that criterion in the process of decision making. So, it is applied spherical fuzzy entropy method to determine the weights of criteria in SF Eco-holarchy. Spherical fuzzy entropy measure is calculated for each criterion in SF Eco-Holons separately by using Eq. (28). The divergence will be calculated by Eq. (29) for each criterion and finally get the weights of criteria by using Eq. (30).

Step 6.1. Determine the weights of criteria in SF Eco-holarchy objectively:

Then weights of criteria will be determined in this step for SF Eco-holarchy because, in each SF Eco-holarchy, a specific criterion could carry a different amount of significance by considering its role in the area that Eco-holarchy covered and illustrated. In this way, it is regarded as the weights of criteria in each SF Eco-holarchy through the whole SF Eco-holonic structure and guarantees that it determines weights of criteria objectively in SF Eco-holarchy.

Step 7. Construct the weighted aggregated GDM matrix for SF Eco-holarchy:

Aggregate the decision matrices for each decision-maker in SF Eco-holarchy using weighted aggregation operators shown in Eqs. (21) and (22) by considering the corresponding weight vector of criteria.

Step 8. Evaluate score (Sc) and accuracy (Ac) functions of the weighted aggregated GDM matrix for SF Eco-Holarchies.

5 Application

Combustion of fossil fuels beside human activities yields a massive amount of methane, carbon dioxide, and nitrous oxide, which all are greenhouse gasses (GHGs) and polluting the environment [76, 77]. The requirement of oils for the transportation sector is severely increasing, and this sector's carbon emission will reach more than 80% [78]. Since air transportation's significant role-playing in modern life, it is also true for air transportation and aviation industries will also have a substantial impact on emission and environmental effects. Based on the U.S. Energy Information Administration and report of Annual Energy Outlook of 2020 [79, 80] the energy production (based on different sources) and energy consumption (based on other consuming sectors) are shown in Fig. 3.

It goes worse that, as mentioned before, the continuously increasing importance of aviation (see Fig. 4) and its rigor need to aviation fuel brings up the considerable need to fuel; besides that, it is considered that there is a problematic fuel supply chain issue for aviation industry. Figure 5 shows the energy consumption of the

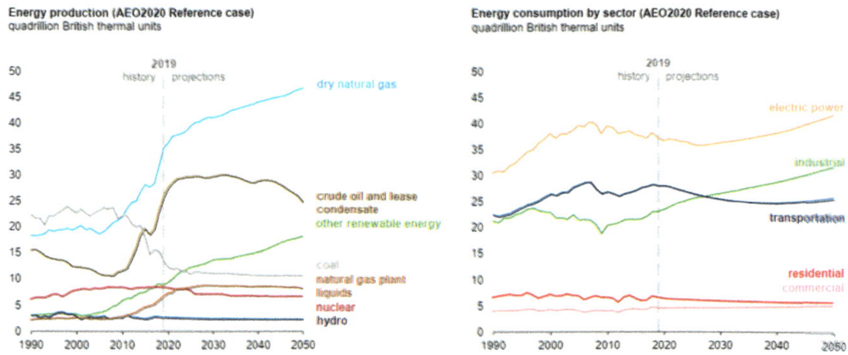

(a) (b)

Fig. 3 The energy production (based on different sources) and energy consumption (based on different sectors) [80] **a** energy production (1990–2050), **b** energy consumption (1990–2050)

Fig. 4 The passenger travel in transportation sector from 2010 to 2050 (based on type) [80]

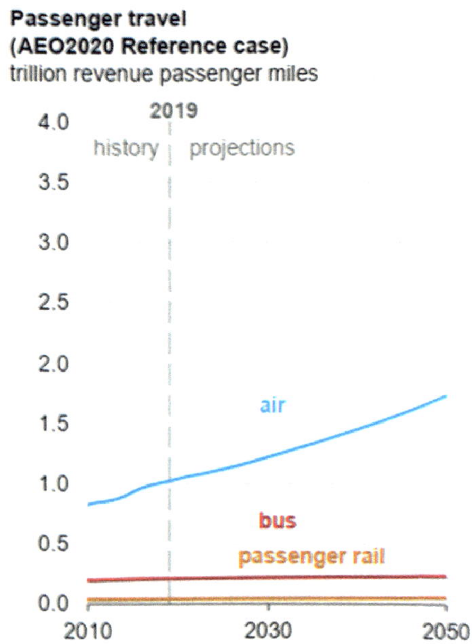

transportation sector based on different types of transportation and various fuels. The place of aviation and air transportation is significant and noticeable to see this sector is continuously increasing among different types of transportations. Also, could compare the amount of different fuels to be consumed in the transportation sector, especially in air transportation type, the fuels that able to be used as aviation fuels (jet fuel, motor gasoline, distilled fuel oil) with considering the ratio of each type. So, it is straightforward that selecting a suitable fuel supplier for this valuable

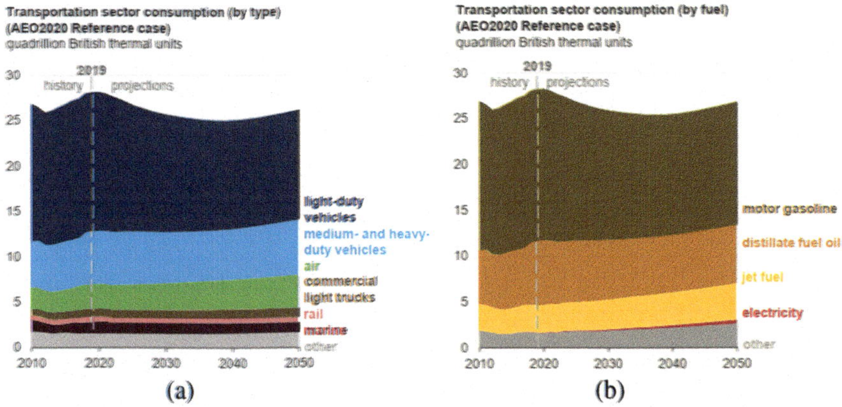

Fig. 5 The energy consumption in transportation sector (based on type and fuel) [80] **a** transportation energy consumption based on type (2010–2050), **b** transportation energy consumption based on fuel (2010–2050)

industry is one of the significant problems in both the supply selection area and the aviation industries.

While it is needed to detect the suitable fuel and thereby the best fuel supplier for aviation industries, some additional factors like considering the economic benefits, social acceptability, and environmental impacts of those fuels and suppliers of the fuels have to be considered. The three factors referred to above are providing sustainability for the system; when these factors are considered in a supply chain problem, it will turn into a sustainable supply chain [81]; the three factors are also called "triple bottom lines (TBL)" [58].

To sum it up, it is assumed an aviation fuel sustainable supply chain problem in an uncertain environment considering all aspects and facets that could have for the criteria of the problem. So, our alternatives are aviation fuels and their suppliers. In this problem, based on some researches discussed in the literature review part, it is considered four aviation fuels based on their structure (Algal based, Soybean based, refined petroleum, natural gas-based) as follows:

A_1: Algal based fuels which are using crop-algal to produce crude oil and then processed it to get aviation fuel. Algal based fuels are produced through hydrogenation and hydro-processing of renewable jet fuels, generally like paraffinic liquids in chemical formula of C_nH_{2n+2} [82].
A_2: Soybean based aviation fuels are a type of biofuels which are produced through the process of transesterification to yield an alkyl esters of fatty acid [83] then by hydrogenation will be proper to use as a aviation fuel [82].
A_3: Refined petroleum-based aviation fuels (like Aviation Gasoline (AVGAS)), which is a type of aviation fuel that use the spark igniting in internal combustion engines to provide the needed propel force for aircraft. Alkylate is the base component of this type of an aviation fuel [84].

Table 3 Different suppliers of different aviation fuels

		Fuels supplied by each supplier			
		A_1	A_2	A_3	A_4
Suppliers	S_1	A_1	–	A_3	–
	S_2	–	–	A_3	A_4
	S_3	A_1	A_2	–	–
	S_4	–	–	A_3	A_4
	S_5	–	A_2	–	–

A_4: Natural gas-based aviation fuels that are produced by using synthesis gas (also called syngas, which is a fuel gas mixture of carbon monoxide (CO) and hydrogen (H2)). So, aviation fuels will be yield by catalytic conversion of syngas, which is synthesized from natural gas [82, 83].

The are many suppliers all over the world, which provide the above fuels for aviation industries. Many companies in different countries through the whole five continents are supplying aviation fuels, to refer to a long list of them you could see [85]. Among those companies all over the world, could refer to top suppliers like British Petroleum (BP), Shell, Exxon Mobil, Chevron, Total, and Gazprom [86]. In this study, it is considered top supplier for each fuel, so a set of suppliers constitute of top aviation fuels suppliers, to determine sustainable supplier for appropriate fuel among all suppliers, as $S = \{S_1, S_2, \ldots, S_5\}$.

Accordingly, Table 3 could find out which company produce and supplies which type of aviation fuels. As an example, provider S_1 supplies algal based fuels (A_1) and AVAGAS (A_3) aviation fuel types, which could show as:$S_1 = \{A_1, A_3\}$.

As mentioned before, many essential factors indicate the appropriate fuel and then relevant supplier related to that fuel. The factors to determining the proper alternative are called indicators in this study; some of the indicators for preparing the most appropriate decision is related to the fuels nature, some about the reliability on providing and accessibility of the fuel, financial factors, and social point of view to fuel and rate of acceptance of the fuel among societies. A new sustainability assessment model is proposed for evaluating the sustainable supply chain of aviation fuels in this article. The proposed multi-indicator sustainability assessment model is considering four main indicators as economic, environmental, social and market reliability. A sustainable supply chain model referring to great economic profits, low ecological effects, and a high social acceptance rate [58], and fourth indicator is considering the reliability of market. Here it is considered five main principles to detect appropriate indicators based on sustainability assessment model. The five principles in the problem are: 1. sustainability principle, based on this principal indicator should be capable of reflecting sustainability characteristics of system in-detail, 2. Consistency principle, based on this principal indicator, should be consistent with the objective or objectives of the sustainability of the system, 3. Comparability principle, based on this principal alternative, should be comparable with respect to indicators 4. Consensuses principle, by this principle, should reach a consensus degree about appropriate alternative based on different stakeholders' opinions about the alternative based on indicators and 5. Measurable principle, based on this principle, the selected indicators should be measurable (qualitatively or quantitatively). So, we consider related following indicators based on the above-mentioned principles for our model. All of the following indicators (criteria) are selected from the literature part from different articles [81, 87–89]:

- Capital Cost:

Cost of facility, factory building, and structural construction to produce aviation fuels.

- Production Cost (per unit):

Cost related to the production of a unit of aviation fuel.

- Lifespan:

Lifespan is a significant factor of the economic aspect, and this indicator is related to the lifetime of the system.

- Availability of resources:

the availability of the necessary resources for each aviation fuel type.

- Production Capacity:

The total capacity for producing each type of aviation fuels.

- Greenhouse Gas (GHG) Emission (ppm per unit):

The total amount of GHG emission of aviation fuel.

- PM_{10} Emission (ppm per unit):

The total amount of PM_{10} emission of aviation fuel.

- $PM_{2.5}$ Emission (ppm per unit):

The total amount of $PM_{2.5}$ emission of aviation fuel.

- Water resource needed (per unit):

Total amount of water required through the entire life cycle of the aviation fuel.

- Energy Consumption to produce (per unit):

The amount of energy (all faces of that) which is consumed to produce a unit of fuel through the whole life cycle of aviation fuel.

- Social Acceptability rate:

The total point of view of society about each aviation fuel determines how much aviation fuel is socially acceptable.

- Aviation Industries acceptability rate:

The total point of view of aviation industries about each aviation fuel determines how much aviation fuel is socially acceptable.

- Innovation on technology:

Innovation of new fuels in comparison to traditional aviation fuels.

A critical step in all decision-making problems is how, in detail, the model considers all of the conditions. In addition, the method has to solve such a complicated model with many considerations and assumptions of the real world. It is considerable to mention that such a detailed problem has much vagueness around it; all the problem criteria contain uncertainty. This fact inspired us to realize that our model is more by considering it in a spherical fuzzy environment; thereby, we used the spherical fuzzy scales which is proposed in Table 2.

Another point of view to make our model more realistic is that all criteria, as mentioned earlier (indicators), are contained in some upper-level indicators, which means are sub-indicators of those upper indicators. Each level of criteria (indicators which include some sub-indicators from the previous level, together with their interactions) are self-reliant at the same time and dependent. Sub-indicators determine the indicators in a holonic way because they interact vertically and constructed indicators. The concept of Eco-holarchy is unifying the perceptions of sustainability and holonic systems together [63]; thus, an Eco-holarchy is the integrated form of a sustainable holarchy structure of a real-world case. Besides, it is considered the uncertainty of the world around the model in a spherical fuzzy environment. Also, the levels of the criteria and sub-criteria are as follows.

Therefore, it is considered all above-mentioned aviation fuels, alternative suppliers of those based on Eco-holonic structure of criteria (see Table 4) to select the sustainable aviation fuel supplier, and proposed the following Eco-holonic structure. Three decision-makers evaluate the proposed Eco-holarchy, specialists in the aviation supply chain, and fuel supply chain to handle more resilience. The decision-makers determined the uncertainties based on a spherical fuzzy environment. So, we will have the spherical fuzzy Eco-holonic structure for the sustainable supply chain of aviation fuel. Therefore, the Spherical fuzzy Eco-holonic structure of aviation fuel suppliers and Spherical Fuzzy Eco-holarchy for the sustainable supply chain of aviation fuels which shown in Fig. 6.

Table 4 Indicators (criteria) and sub layers of indicators in proposed eco-holonic structure for evaluating sustainable supply chain of aviation fuels

Main indicator (Criteria)	Sub-layer indicator (Sub-criteria)	Sub-sub-layer indicator (Sub-sub criteria)
Economic $\left(\widetilde{EH}_{s1}^{1}\right)$	Cost $\left(\widetilde{EH}_{s1}^{0}\right)$	Capital cost (u_1)
		Production cost (u_2)
	Lifespan $\left(\widetilde{EH}_{s2}^{0}\right)$	Lifespan (u_3)
Environmental $\left(\widetilde{EH}_{s2}^{1}\right)$	Ecological $\left(\widetilde{EH}_{s3}^{0}\right)$	Greenhouse gas (GHG) emission (u_4)
		PM_{10} emission (u_5)
		$PM_{2.5}$ emission (u_6)
	Supplementary resources $\left(\widetilde{EH}_{s4}^{0}\right)$	Water resource needed (u_7)
		Electricity resource needed (u_8)
	Energy efficiency $\left(\widetilde{EH}_{s5}^{0}\right)$	Energy consumption to produce (u_9)
Social $\left(\widetilde{EH}_{s3}^{1}\right)$	Acceptability rate $\left(\widetilde{EH}_{s6}^{0}\right)$	Social acceptability rate (u_{10})
		Aviation industries acceptability rate (u_{11})
	Technologic maturity related to fuel $\left(\widetilde{EH}_{s7}^{0}\right)$	Technologic maturity related to fuel (u_{12})
	Innovation on technology $\left(\widetilde{EH}_{s8}^{0}\right)$	Innovation on technology (u_{13})
Market reliability $\left(\widetilde{EH}_{s4}^{1}\right)$	Production $\left(\widetilde{EH}_{s9}^{0}\right)$	Production capacity (u_{14})
		Availability of resources (u_{15})
	Supplementary resources $\left(\widetilde{EH}_{s4}^{0}\right)$	Water resource needed (u_7)
		Electricity resource needed (u_8)
	Energy efficiency $\left(\widetilde{EH}_{s5}^{0}\right)$	Energy consumption to produce (u_9)

Presented SF Eco-holonic structure Fig. 6, consists of five suppliers, four aviation fuels, and SF Eco-holarchy for a sustainable supply chain of aviation fuels. So, up to here, steps 1 and 2 have done, and we discussed the problem (sustainable supply chain of aviation fuels), goals, alternatives to reach these goals, and criteria to evaluate the alternatives in the path of achieving goals. It has also shown the SF Eco-holonic structure of sustainable supplier of aviation fuels and defines the spherical fuzzy scales. For this problem, it is determined to have four different experts with specialties and experiences in aviation industries and supply chain management, as decision makers so we have $D = \{D_1, D_2, \ldots, D_4\}$. Then in step 3 first decision makers evaluate the alternatives based on defined atomic criteria to construct the

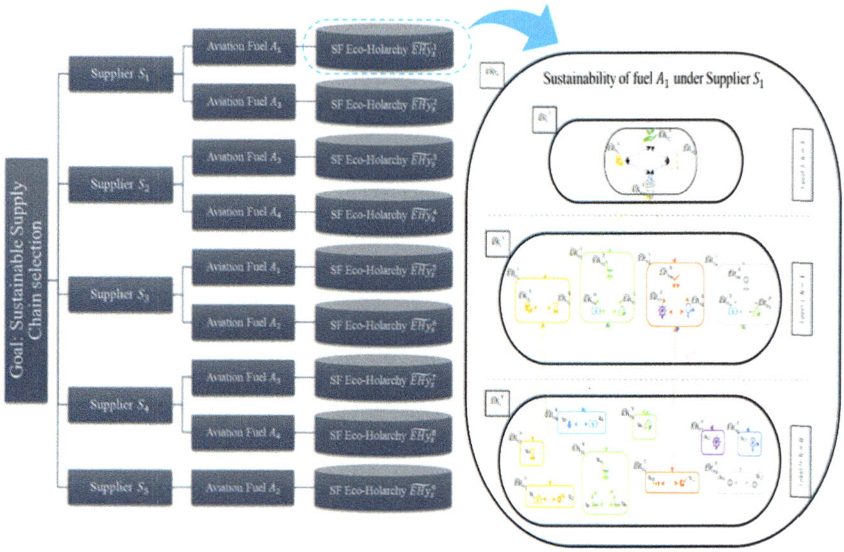

Fig. 6 Spherical fuzzy eco-holonic structure and spherical fuzzy eco-holarchy for sustainable supply chain of aviation fuels

Eco-holonic decision matrices (step 3.1) and afterward turning those matrices to spherical fuzzy Eco-holonic decision matrices (step 3.2) by using Table 2. The data are uniformly collected so there is no need to normalization of decision matrices in step 3.3.

In step 4, the weights of decision makers are determined by using the spherical fuzzy entropy method defined in Definition 9 based on [73]. The calculated entropy measures and derived weights for decision makers are shown in Table 5. In step 5, the GDM matrix for SF Eco-Holarchies is calculated by using the spherical fuzzy weighted arithmetic mean ($SFWAM$) operator, which is defined in Definition 7. Then in step 6 by using the entropy measure in Definition 9, the weights of criteria are determined and showed in Table 6. Then it is used (SFWAM) operator in Definition 7, to evaluate the weighted aggregated GDM matrix for SF Eco-Holarchies in defined SF Eco-holonic structure. The weighted aggregated GDM matrix are shown in the following Table 7.

Table 5 Evaluated spherical fuzzy entropy measure, divergence and derived weights for decision makers

Decision makers	Entropy	Divergences	Weights
DM1	0.473451852	0.526548148	0.224161963
DM2	0.417333333	0.582666667	0.248052726
DM3	0.385851852	0.614148148	0.261455016
DM4	0.3744	0.6256	0.266330295

Table 6 Evaluated spherical fuzzy entropy measure, divergence and derived weights for criteria

Criteria	Entropy	Divergences	Weights
u1	0.460893515	0.539106485	0.059297989
u2	0.394703621	0.605296379	0.06657842
u3	0.472745159	0.527254841	0.05799439
u4	0.394002738	0.605997262	0.066655512
u5	0.341449726	0.658550274	0.07243598
u6	0.334067859	0.665932141	0.073247935
u7	0.416118416	0.583881584	0.06422294
u8	0.385301924	0.614698076	0.067612542
u9	0.398803693	0.601196307	0.06612744
u10	0.430770628	0.569229372	0.062611298
u11	0.279971325	0.720028675	0.07919818
u12	0.453119218	0.546880782	0.060153108
u13	0.276823563	0.723176437	0.079544412
u14	0.496672405	0.503327595	0.055362558
u15	0.373076104	0.626923896	0.068957297

Table 7 Weighted aggregated GDM matrix for SF eco-holarchies in SF-Eco-holonic structure

Alternatives		SFNs		
Suppliers	Alternatives fuels	μ	ϑ	h
S1	A1	0.57	0.47	0.29
	A3	0.51	0.50	0.42
S2	A3	0.63	0.40	0.26
	A4	0.51	0.53	0.32
S3	A1	0.53	0.48	0.43
	A2	0.64	0.39	0.28
S4	A3	0.58	0.46	0.29
	A4	0.54	0.47	0.42
S5	A2	0.60	0.43	0.29

In step 8, the proposed MAGDM algorithm in SF Eco-holarchy, compare the spherical fuzzy values in the previous step in the framework of the weighted aggregated GDM matrix for SF Eco-Holarchies. To compare the spherical fuzzy values of weighted aggregated GDM matrix for SF Eco-Holarchies, it is proposed the score and accuracy functions, which are introduced in Definition 6, so we will have the following results for steps 8 and 9.

As shown in Table 8, alternative fuel A2, supplied by the supplier S3, stands as the first place of our sustainable supply chain for the aviation fuel problem. Alternative

Table 8 Score and accuracy functions and the ranking

Alternatives		Comparison functions		Rank
Suppliers	Alternative fuels	Score (Sc)	Accuracy (Ac)	
S1	A1	0.078647858	0.629832726	5
	A3	0.011382405	0.68615568	8
S2	*A3*	0.177588095	0.627868541	*2*
	A4	−0.013863951	0.638774398	9
S3	A1	0.03374641	0.698781097	7
	A2	0.187789746	0.636439522	*1*
S4	A3	0.085454847	0.629738742	4
	A4	0.039529045	0.687087991	6
S5	*A2*	0.126493456	0.631416982	*3*

fuel A2, which is soybean-based aviation fuel, is determined as the most optimal solution for aviation fuel; this is why it is considered the sustainability of the supplier to get the most sustainable supplier for this fuel is supplier S3. In other words, the alternative fuel A2, which is provided by Supplier S3, is the best sustainable supply chain for aviation fuels. It is evident that A2, as a biofuel, has its characteristics in Economic, Environmental, Social, and Market Reliability, which caused it to be selected. Besides the supplier S3, producing and supplying the A2 has presented significant advantages based on again Economic, Environmental, Social and Market Reliability, so altogether caused that A2 by supplier S3 stands as the best optimal solution to our problem in determining sustainable supply chain of aviation fuel. Furthermore, you could see that A3, supplied by S2, stands as the second solution for our problem, and third place among solutions belongs to A2, which is provided by S5.

6 Conclusion

In this chapter, the novel concept of spherical fuzzy Eco-holonic structure is proposed. A MAGDM method in SF Eco-holonic structure is recommended to show the advantages of SF Eco-holonic structure in providing resilience and a more reliable system. It is also demonstrated the applicability of the proposed SF Eco-holonic structure and MAGDM method based on that in the sustainable supply chain of the aviation fuel. SSC in the aviation fuel problem is one of the hardest and complicated issues, which deliberately shows SF Eco-holonic structure's advantages. Based on that, four fuel alternatives supplied by five suppliers are considered to make nine different alternatives for our proposed SSC in the aviation fuel problem. Many other criteria are then considered to catch a sustainable aviation fuel and sustainable supplier of that fuel.

Based on the proposed MAGDM method in SF Eco-holonic structures, very significant results are acquired.

A2, which is a soybean-based fuel, stands in the first place as sustainable fuel selection. So, A2 as a biofuel will give more sustainability to our SSC of aviation fuel problems. In this method, not only could determine the sustainable fuel but also at the same time could evaluate the supplier.

In this problem, S3 is determined as the best sustainable supplier for A2, which together stands in the first place.

Most important remark based on this method is that neither A2 nor S3 alone couldn't catch sustainability. It means that alternative fuel A2 by supplier S3 altogether satisfies many criteria of SSC in the aviation fuel model based on SF Eco-holonic structure.

To complete the previous remark, could refer to another alternative fuel A3, which is also supplied by S3, but stands in 7th rank among those alternatives for SSC of aviation fuels.

- Another significant remark could refer to supplier S2 that supplies two alternative fuels, A3 and A4, which stand in 2nd and 9th rank, respectively. Both alternative fuels are supplied by the same supplier, but many different Economic, Environmental, Social, and Market Reliability conditions caused such another rank.
- Another significant point could refer to supplier S2 that supplies two alternative fuels, A3 and A4, which stand in the 2nd and 9th ranks. The same supplier provides alternative fuels, but many different Economic, Environmental, Social, and Market Reliability conditions caused another rank.
- Another remark is that A3, which is not a biofuel, stands in 2nd place were supplied by S2. That is, A3 from S2 could attract more satisfactions among Economic, Environmental, Social, and Market Reliability conditions than some other biofuel-based alternatives.

For future studies, we propose various SF Eco-holonic concept applications in different fields of study like financial, banking, social, network, health care, manufacturing and transportation systems.

References

1. Nygren, E., Aleklett, K., Höök, M.: Aviation fuel and future oil production scenarios. Energy Policy (2009). https://doi.org/10.1016/j.enpol.2009.04.048
2. International Energy Agency (IEA). Available: https://www.iea.org/. Accessed 30 Oct 2020
3. U. S. E. I. Energy Information Administration: Internaltional Energy Outlook 2019 (2019)
4. Zadeh, L.A.: Fuzzy sets. Inf. Control (1965). https://doi.org/10.1016/S0019-9958(65)90241-X
5. Zadeh, L.A.: The concept of a linguistic variable and its application to approximate reasoning-I. Inf. Sci. (Ny) (1975). https://doi.org/10.1016/0020-0255(75)90036-5
6. Zimmermann, H.-J.: Fuzzy Set Theory—and Its Applications (2001)

7. Kutlu Gündoğdu, F., Kahraman, C.: A novel spherical fuzzy QFD method and its application to the linear delta robot technology development. Eng. Appl. Artif. Intell. (2020). https://doi.org/10.1016/j.engappai.2019.103348
8. Donyatalab, Y., Seyfi-Shishavan, S.A., Farrokhizadeh, E., Kutlu Gündoğdu, Y., Kahraman, C.: Spherical fuzzy linear assignment method for multiple criteria group decision-making problems. Informatica (2020). https://doi.org/10.15388/20-infor433
9. MIT PARTNER Center: U.S. Fuel Trends Analysis, Montreal, Canada
10. Mella, P.: The holonic revolution. holons, holarchies and holonic networks. The Ghost ... —Piero Mella—Google Books
11. Capra, F., March, R.: The turning point: science, society and the rising culture. Phys. Today (1982). https://doi.org/10.1063/1.2914857
12. Babiceanu, R.F., Chen, F.F., Sturges, R.H.: Real-time holonic scheduling of material handling operations in a dynamic manufacturing environment. Robot. Comput. Integr. Manuf. (2005). https://doi.org/10.1016/j.rcim.2004.11.003
13. Dani, S., Backhouse, C.J., Burns, N.D.: Application of transactional analysis in supply chain networks: a potential holonic mediating tool. Proc. Inst. Mech. Eng. Part B J. Eng. Manuf. (2004). https://doi.org/10.1177/095440540421800510
14. Shafaei, S., Aghaee, N.G.: Biological network simulation using holonic multiagent systems. In: Proceedings—UKSim 10th International Conference on Computer Modelling and Simulation, EUROSIM/UKSim2008 (2008). https://doi.org/10.1109/UKSIM.2008.23
15. Zhang, J., Gao, L., Chan, F.T.S., Li, P.: A holonic architecture of the concurrent integrated process planning system. J. Mater. Process. Technol. (2003). https://doi.org/10.1016/S0924-0136(03)00233-4
16. Ng, A.H.C., Yeung, R.W.H, Cheung, E.H.M.: HSCS—the design of a holonic shopfloor control system. In: IEEE Symposium on Emerging Technologies & Factory Automation, ETFA (1996). https://doi.org/10.1109/etfa.1996.573288
17. Chirn, D., McFarlane, J.-L.: Building holonic systems in today's factories: a migration strategy, CUED Publications database
18. Mcfarlane, S.B.D.: State of the art of holonic systems in production planning and control | Semantic Scholar
19. Akturk, M.S., Turkcan, A.: Cellular manufacturing system design using a holonistic approach. Int. J. Prod. Res. (2000). https://doi.org/10.1080/002075400050028124
20. Amiri, A.: Designing a distribution network in a supply chain system: Formulation and efficient solution procedure. Eur. J. Oper. Res. (2006). https://doi.org/10.1016/j.ejor.2004.09.018
21. Clark, K.B.: Design Rules, vol. 1. The MIT Press
22. Smith, A.P.: Worlds within worlds. The holarchy of life—Kindle edition by Smith, A.P.. Health, Fitness & Dieting Kindle eBooks @ Amazon.com (2000)
23. Wilber, K.: Messenger of the Kosmos, by Ashok, A.V. Hyderabad, India
24. Beer, S.: The Heart of Enterprise, Wiley
25. Beer, S.: Brain of the Firm, 2nd edn. Wiley
26. Mesarovic, D.T.Y., Macko, M.D.: Theory of Hierarchical, Multilevel Systems: Mesarovic, M.D., Macko, D., Takahara, Y. Amazon.com: Books
27. Shimizu, H.: A General Approach to Complex Systems in Bioholonics, pp. 204–223. Springer, Berlin, Heidelberg (1987)
28. Schilling, M.A.: Toward a general modular systems theory and its application to interfirm product modularity. Acad. Manag. Rev. (2000). https://doi.org/10.5465/AMR.2000.3312918
29. Jacak, W.: Intelligent Robotic Systems : Design, Planning, and Control, *undefined* (1999)
30. Kusumi, N., Hirasawa, K., Obayashi, M.: A holonic control system based on a universal learning network. Electr. Eng. Japan (English Transl. Denki Gakkai Ronbunshi) (1998). https://doi.org/10.1002/eej.4391240408
31. Mareschal, B.: Weight stability intervals in multicriteria decision aid. Eur. J. Oper. Res. (1988). https://doi.org/10.1016/0377-2217(88)90254-8
32. Ma, J., Fan, Z.P., Huang, L.H.: A subjective and objective integrated approach to determine attribute weights. Eur. J. Oper. Res. (1999). https://doi.org/10.1016/S0377-2217(98)00141-6

33. Roostaee, R., Izadikhah, M., Lotfi, F.H., Rostamy-Malkhalifeh, M.: A multi-criteria intuitionistic fuzzy group decision making method for supplier selection with vikor method. Int. J. Fuzzy Syst. Appl. (2012). https://doi.org/10.4018/ijfsa.2012010101
34. Tavana, M., Zareinejad, M., Di Caprio, D., Kaviani, M.A.: An integrated intuitionistic fuzzy AHP and SWOT method for outsourcing reverse logistics. Appl. Soft Comput. J. (2016). https://doi.org/10.1016/j.asoc.2015.12.005
35. Saaty, R.W.: The analytic hierarchy process-what it is and how it is used. Math. Model. (1987). https://doi.org/10.1016/0270-0255(87)90473-8
36. Hwang, C.-L., Lin, M.-J.: Group Decision Making under Multiple Criteria (1987)
37. Chu, A.T.W., Kalaba, R.E., Spingarn, K.: A comparison of two methods for determining the weights of belonging to fuzzy sets. J. Optim. Theory Appl. (1979). https://doi.org/10.1007/BF00933438
38. Pekelman, D., Sen, S.K.: Mathematical programming models for the determination of attribute weights. Manage. Sci. (1974). https://doi.org/10.1287/mnsc.20.8.1217
39. Wang, Z., Li, K.W., Xu, J.: A mathematical programming approach to multi-attribute decision making with interval-valued intuitionistic fuzzy assessment information. Expert Syst. Appl. (2011). https://doi.org/10.1016/j.eswa.2011.04.027
40. Zeleny, M.: Attribute-dynamic attitude model (ADAM). Manage. Sci. (1976). https://doi.org/10.1287/mnsc.23.1.12
41. Çalı, S., Balaman, ŞY.: A novel outranking based multi criteria group decision making methodology integrating ELECTRE and VIKOR under intuitionistic fuzzy environment. Expert Syst. Appl. (2019). https://doi.org/10.1016/j.eswa.2018.10.039
42. Hung, C.C., Chen, L.H.: A multiple criteria group decision making model with entropy weight in an intuitionistic fuzzy environment. Lect. Notes Electric. Eng. (2010). https://doi.org/10.1007/978-90-481-3517-2-2
43. Vlachos, I.K., Sergiadis, G.D.: Intuitionistic fuzzy information—applications to pattern recognition. Pattern Recognit. Lett. (2007). https://doi.org/10.1016/j.patrec.2006.07.004
44. Zhang, S.F., Liu, S.Y.: A GRA-based intuitionistic fuzzy multi-criteria group decision making method for personnel selection. Expert Syst. Appl. (2011). https://doi.org/10.1016/j.eswa.2011.03.012
45. Boran, F.E., Genç, S., Kurt, M., Akay, D.: A multi-criteria intuitionistic fuzzy group decision making for supplier selection with TOPSIS method. Expert Syst. Appl. (2009). https://doi.org/10.1016/j.eswa.2009.03.039
46. Air Transport Action Group (2020). Available: https://www.atag.org/facts-figures.html
47. Tretheway, M.W., Markhvida, K.: The aviation value chain: economic returns and policy issues. J. Air Transp. Manag. (2014). https://doi.org/10.1016/j.jairtraman.2014.06.011
48. Chen, L., Ren, J.: Multi-attribute sustainability evaluation of alternative aviation fuels based on fuzzy ANP and fuzzy grey relational analysis. J. Air Transp. Manag. (2018). https://doi.org/10.1016/j.jairtraman.2017.10.005
49. Rye, L., Blakey, S., Wilson, C.W.: Sustainability of supply or the planet: a review of potential drop-in alternative aviation fuels. Energy Environ. Sci. (2010). https://doi.org/10.1039/b918197k
50. Gudiel Pineda, P.J., Liou, J.J.H., Hsu, C.C., Chuang, Y.C.: An integrated MCDM model for improving airline operational and financial performance. J. Air Transp. Manag. (2018). https://doi.org/10.1016/j.jairtraman.2017.06.003
51. Lee, K.C., Tsai, W.H., Yang, C.H., Lin, Y.Z.: An MCDM approach for selecting green aviation fleet program management strategies under multi-resource limitations. J. Air Transp. Manag. (2018). https://doi.org/10.1016/j.jairtraman.2017.06.011
52. Feng, C.M., Wang, R.T.: Performance evaluation for airlines including the consideration of financial ratios. J. Air Transp. Manag. (2000). https://doi.org/10.1016/S0969-6997(00)00003-X
53. Chang, Y.H., Yeh, C.H.: A new airline safety index. Transp. Res. Part B Methodol. (2004). https://doi.org/10.1016/S0191-2615(03)00047-X

54. Sun, X., Gollnick, V., Stumpf, E.: Robustness consideration in multi-criteria decision making to an aircraft selection problem J. . Multi-Criteria Decis. Anal. (2011). https://doi.org/10.1002/mcda.471
55. Shojaei, P., Seyed Haeri, S.A., Mohammadi, S.: Airports evaluation and ranking model using Taguchi loss function, best-worst method and VIKOR technique. J. Air Transp. Manag. (2018). https://doi.org/10.1016/j.jairtraman.2017.05.006
56. Bongo, M.F., Alimpangog, K.M.S., Loar, J.F., Montefalcon, J.A., Ocampo, L.A.: An application of DEMATEL-ANP and PROMETHEE II approach for air traffic controllers' workload stress problem: a case of mactan civil aviation authority of the Philippines. J. Air Transp. Manag. (2018). https://doi.org/10.1016/j.jairtraman.2017.10.001
57. Shanmugam, A., Paul Robert, T.: Ranking of aircraft maintenance organization based on human factor performance. Comput. Ind. Eng. (2015). https://doi.org/10.1016/j.cie.2015.07.017
58. Rodger, J.A., George, J.A.: Triple bottom line accounting for optimizing natural gas sustainability: a statistical linear programming fuzzy ILOWA optimized sustainment model approach to reducing supply chain global cybersecurity vulnerability through information and communications. J. Clean. Prod. (2017). https://doi.org/10.1016/j.jclepro.2016.11.089
59. Cannibals with forks: the triple bottom line of 21st century business, Choice Rev. Online (1999). https://doi.org/10.5860/choice.36-3997
60. Nikolaou, I.E., Evangelinos, K.I., Allan, S.: A reverse logistics social responsibility evaluation framework based on the triple bottom line approach. J. Clean. Prod. (2013). https://doi.org/10.1016/j.jclepro.2011.12.009
61. Slaper, T., Hall, T.: The triple bottom line : what is it and how does it work? Indiana Univ. Kelley Sch. Bus. (2011)
62. Govindan, K., Khodaverdi, R., Jafarian, A.: A fuzzy multi criteria approach for measuring sustainability performance of a supplier based on triple bottom line approach. J. Clean. Prod. (2013). https://doi.org/10.1016/j.jclepro.2012.04.014
63. Ávila-Gutiérrez, M.J., Aguayo-González, F., Marcos-Bárcena, M., Lama-Ruiz, J.R., Peralta-Álvarez, M.E.: Arquitectura holónica de referencia para empresas de fabricación sostenibles distribuidas. DYNA (2017). https://doi.org/10.15446/dyna.v84n200.53095
64. Ávila-Gutiérrez, M.J., Martín-Gómez, A., Aguayo-González, F., Lama-Ruiz, J.R.: Eco-holonic 4.0 circular business model to conceptualize sustainable value chain towards digital transition. Sustainability (2020). https://doi.org/10.3390/su12051889
65. Gündoğdu, F.K., Kahraman, C.: Spherical fuzzy sets and spherical fuzzy TOPSIS method. J. Intell. Fuzzy Syst. (2019). https://doi.org/10.3233/JIFS-181401
66. Donyatalab, Y., Farrokhizadeh, E., Garmroodi, S.D.S., Shishavan, S.A.S.: Harmonic mean aggregation operators in spherical fuzzy environment and their group decision making applications. J. Mult. Log. Soft Comput. (2019)
67. Kutlu Gundogdu, F., Kahraman, C.: Extension of WASPAS with spherical fuzzy sets. Informatics (2019). https://doi.org/10.15388/Informatica.2019.206
68. Kutlu Gündoğdu, F., Kahraman, C.: A novel fuzzy TOPSIS method using emerging interval-valued spherical fuzzy sets. Eng. Appl. Artif. Intell. (2019). https://doi.org/10.1016/j.engappai.2019.06.003
69. Ashraf, S., Abdullah, S.: Spherical aggregation operators and their application in multiattribute group decision-making. Int. J. Intell. Syst. (2019). https://doi.org/10.1002/int.22062
70. Kutlu Gündoğdu, F., Kahraman, C.: A novel spherical fuzzy analytic hierarchy process and its renewable energy application. Soft Comput. (2019). https://doi.org/10.1007/s00500-019-04222-w
71. Gündoğdu, F.K., Kahraman, C.: Extension of WASPAS with spherical fuzzy sets. J. Mult. Log. Soft Comput. **30**(2), 269–292 (2019)
72. Ashraf, S., Abdullah, S., Mahmood, T.: GRA method based on spherical linguistic fuzzy Choquet integral environment and its application in multi-attribute decision-making problems. Math. Sci. (2018). https://doi.org/10.1007/s40096-018-0266-0
73. Aydoğdu, A., Gül, S.: A novel entropy proposition for spherical fuzzy sets and its application in multiple attribute decision-making. Int. J. Intell. Syst. (2020). https://doi.org/10.1002/int.22256

74. Gupta, P., Mehlawat, M.K., Grover, N.: A Generalized TOPSIS method for intuitionistic fuzzy multiple attribute group decision making considering different scenarios of attributes weight information. Int. J. Fuzzy Syst. **21**(2), 369–387 (2019). https://doi.org/10.1007/s40815-018-0563-7
75. Mardani, A., Nilashi, M., Zavadskas, E.K., Awang, S.R., Zare, H., Jamal, N.M.: Decision making methods based on fuzzy aggregation operators: three decades review from 1986 to 2017. Int. J. Inform. Technol. Decis. Making (2018). https://doi.org/10.1142/S021962201830001X
76. Solomon, S. et al.: Summary for policymakers. In: Qin, D., Manning, M., Chen, Z., Marquis, M., Averyt, K., Tignor, M., Mill. H.L.: Climate Change 2007: The Physical Science Basis. Contribution of Working Group I to the Fourth Assessment Report of the Intergovernmental Panel on Climate Change. New York Cambridge Univ. Press (2007). https://doi.org/10.1038/446727a
77. Environment and Climate Change Canada, Greenhouse Gas Sources and Sinks: Executive Summary 2019, aem (2019)
78. Metz, B., Meyer, L., Bosch, P.: Climate change 2007 mitigation of climate change (2007)
79. Annual Energy Outlook 2020. Available: https://www.eia.gov/outlooks/aeo/. Accessed 27 Oct 2020
80. U.S. Energy Information Agency: Annual Energy Outlook 2019 with projections to 2050. Annu. Energy Outlook 2019 with Proj. to 2050 (2019)
81. Ren, J., Manzardo, A., Mazzi, A., Zuliani, F., Scipioni, A.: Prioritization of bioethanol production pathways in China based on life cycle sustainability assessment and multicriteria decision-making. Int. J. Life Cycle Assess. (2015). https://doi.org/10.1007/s11367-015-0877-8
82. Zhao, S.Y., Li, W.J.: Fast asynchronous parallel stochastic gradient descent: a lock-free approach with convergence guarantee. In: 30th AAAI Conference on Artificial Intelligence, AAAI 2016 (2016)
83. Kandaramath Hari, T., Yaakob, Z., Binitha, N.N.: Aviation biofuel from renewable resources: routes, opportunities and challenges. Renew. Sustain. Energy Rev. (2015). https://doi.org/10.1016/j.rser.2014.10.095
84. Zahran, S., Iverson, T., McElmurry, S.P., Weiler, S.: The effect of leaded aviation gasoline on blood lead in children. J. Assoc. Environ. Resour. Econ. (2017). https://doi.org/10.1086/691686
85. Airport Suppliers (Aviation Fuel Suppliers). Available: https://www.airport-suppliers.com/suppliers/fuel-handling/. Accessed 31 Oct 2020
86. Aviation Fuel Market by Product and Geography—Forecast and Analysis 2020–2024 (Technavio), Feb-2020. Available: https://www.technavio.com/report/aviation-fuel-market-industry-analysis. Accessed 31 Oct 2020
87. Ren, J., Fedele, A., Mason, M., Manzardo, A., Scipioni, A.: Fuzzy multi-actor multi-criteria decision making for sustainability assessment of biomass-based technologies for hydrogen production. Int. J. Hydrogen Energy (2013). https://doi.org/10.1016/j.ijhydene.2013.05.074
88. Afgan, N.H., Carvalho, M.G.: Sustainability assessment of hydrogen energy systems. Int. J. Hydrogen Energy (2004). https://doi.org/10.1016/j.ijhydene.2004.01.005
89. Ren, J., Xu, D., Cao, H., Wei, S., Dong, L., Goodsite, M.E.: Sustainability decision support framework for industrial system prioritization. AIChE J. (2016). https://doi.org/10.1002/aic.15039

On-Board Cost Index Computation Through Fuzzy Logic

Fabio Krykhtine and Felix Mora-Camino

Abstract Air transport aircraft have turned highly complex automated systems of systems in last decades, providing safety and efficiency to the operators and the passengers of commercial flights. The Aviation 4.0 concept goes further in this direction by being concerned with the design of integrated cyber-physical systems able to increase the assistance to the human pilots, helping them to take complex decisions and to complete tasks autonomously. When considering flight management, the cost index is one of the few remaining parameters which has to be set by the pilot. Since the setting of the value of this important parameter is a difficult issue, in this chapter the automatic setting of the tactical onboard cost index through fuzzy logic techniques is developed and displayed.

1 Introduction

Over the last decades, the development of on-board computational power and data bases capacity has led to the increased automation of the main flight management and associated functions such as guidance, communication, navigation and surveillance. Aircraft flight operations optimization in the field of civil air transport is one of the many areas which has benefited greatly from this availability. In general, to tackle efficiently the management of a flight, a 3D + T trajectory optimization problem

F. Krykhtine
Industrial Engineering Department at Polytechnic School, Federal University of Rio de Janeiro, Rio de Janeiro, Brazil

Fuzzy Logic Laboratory, Federal University of Rio de Janeiro, Rio de Janeiro, Brazil

Innovation Agency, Federal University of Rio de Janeiro, Rio de Janeiro, Brazil

F. Mora-Camino (✉)
Durban University of Technology (DUT), Durban, South Africa

Ecole Nationale de L'Aviation Civile, Toulouse, France

Fluminense Federal University/PURO, Rio das Ostras, Brazil

C. Kahraman and S. Aydın (eds.), *Intelligent and Fuzzy Techniques in Aviation 4.0*, Studies in Systems, Decision and Control 372,
https://doi.org/10.1007/978-3-030-75067-1_10

is formulated and solved by a mix of exact and heuristic methods run by the Flight Management Computer [1]. A considerable amount of literature focusing on transport aircraft trajectory optimization already exists. However, airlines are faced with renewed challenges from their ever-changing operational environment which modifies the goals and constraints to be considered. Then by his capability to judge and react to complex situations, the human pilot has remained an essential actor in the flight management process.

The Cost Index-CI is a global input parameter for the flight management system which integrates two critical variables, fuel and time, in the decision process [2]. While the airlines assign, based on market considerations, a Strategic Cost Index to each flight, the pilot is in charge of regulating its value during the flights. The choice of its value, a tactical value for the pilot, depends on several factors resulting from the airline strategy: current flight delay and estimated flight delay at arrival, weather conditions and air traffic conditions in the areas crossed by the flight, aircraft condition (fully nominal, nominal under activated redundancy, downgraded with a particular attention to the engines) destination airport conditions while safety rules (Estimated Fuel On-Board - EFOB at arrival for instance) must be fulfilled. Then it appears that the on-board choice of the tactical value of the Cost Index is a very complex process. The main issue of this chapter is to display a Fuzzy-Based decision-making solution to set continuously in an automatic way the value of the Cost Index during a commercial flight, providing thus a new contribution to the automation of the flight management function.

This chapter considers first the needs for enhanced automation of flight management and displays the main functions developed by modern flight management systems for civil aircraft. Then the Cost Index is introduced with its initial definition and associated basic relations. Then the current use of the Cost Index by airlines and pilots in the flight management process is discussed, leading to the distinction between strategic and tactical cost indexes. Finally, a two-step process for the computation of the value of the Tactical Cost Index based on Fuzzy Logic [3] and Fuzzy Modelling techniques [4] is displayed. The proposed approach is illustrated by a flight case study.

2 Needs for Enhanced Automation of Flight Management

During the last decades of the previous century, on-board systems devoted to automatic flight control have been developed turning each time safer the air transportation operation. Automatic protections have been introduced to maintain the aircraft in its flight domain while fly-by-wire technology has eased the task of piloting. The introduction of a Flight Management and Guidance System-FMGS in the eighties has provided means of guiding automatically an aircraft along a flight plan computed on-board the aircraft through generic vertical and horizontal guidance modes. Initially the Flight Management Computer was mainly devoted to the computation of the necessary on-board fuel to perform a given flight and to provide some estimations such as the arrival time at destination. Then, very soon after that, the Flight Management Computer-FMC got flight plan optimization capabilities with the objective of minimizing flight operations costs, a very important issue for the airlines, especially in periods of high fuel prices such as the seventieth and eighties of the last century. The flight operations costs were considered to be composed on one side of the fuel cost and on the other side of other costs which were related more or less directly with the duration of the flight.

So, a new parameter, the cost index [5], was introduced as a relative weight in the cost function of the flight plan optimization problem. Theoretically the value of this parameter should be determined uniquely from the price composition of all the cost elements contributing to the flight operations cost. However, the exact computation was extremely complex since for example refuel could have been done at different airports with different prices while some time dependent cost elements were piecewise functions of time (the salary of the crews for instance). It was recognized that this parameter could play a strategic role for the airline with respect to a given flight by assigning a higher importance to the fuel savings or to the flight time. So, the value of the Cost Index attached to each flight was given by the airline in a scale which had no direct relation with the theoretical value of the Cost Index as a ratio between fuel cost and the other time related costs. Frequent perturbations such as delayed departures, trajectory diverting directives from the air traffic control and errors in wind predictions leaded to the necessity of in-flight optimization capability to be applied to the remaining part of a flight [6]. This required the introduction by the pilot of a tactical value for the Cost Index used in the re-optimization process. The effective choice of the value of the Tactical Cost Index, which depends of many variables and parameters, is a complex problem which has been in general tackled through the experience of the pilot with more or less success. In this communication, a fuzzy decision approach is adopted to propose a standardization and an automation of this process with the objective to ease the pilot from this task so that he can dedicate more attention to the surrounding traffic. Figure 1 displays the organization in layers with feedback loops of the main flight automation functions.

3 On-Board Flight Plan Generation

The Flight Management System (FMS) on-board modern transportation aircraft allows the pilots to manage commercial flights from the departure airport to the destination airport. It is responsible to compute or process on-board the flight plan adopted to perform a given flight mission.

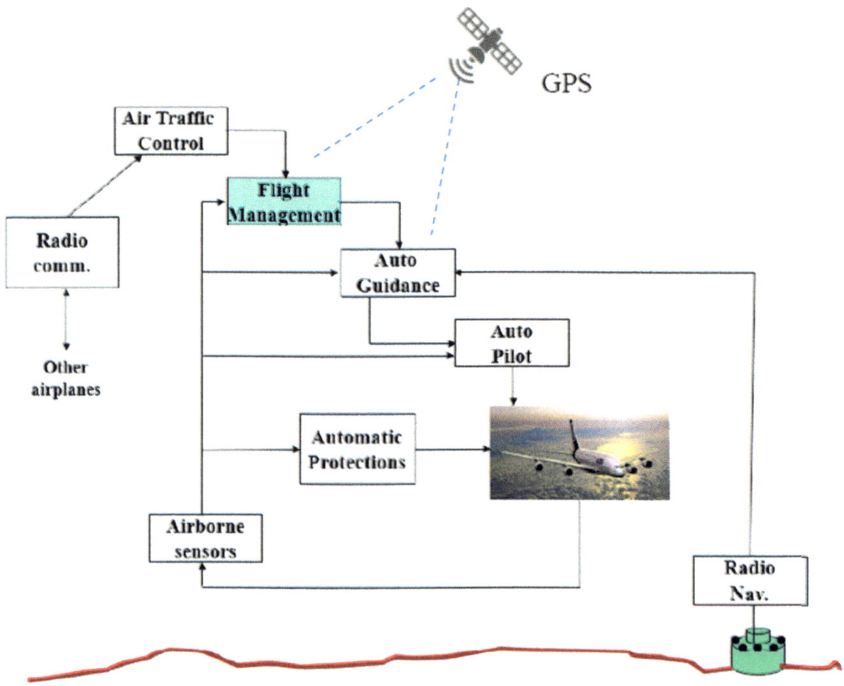

Fig. 1 The flight automation layers for modern air transport aircraft

3.1 *Main Functions of Flight Management Systems*

The main functions developed by the Flight Management System-FMS of modern transportation aircraft are:

- The navigation function, which is responsible for computing the best estimate of the current position and speed of the aircraft. This is performed using hybridization techniques to mix on-board inertial data with radio-navigation data and GPS information. In general these hybridization techniques are derived from Kalman filtering approaches.
- The flight planning function, which allows the crew to establish an efficient routing for the aircraft by optimizing the flight plan in its spatial dimensions (horizontal and vertical) and its temporal dimension;
- The trajectory prediction function, which computes flyover time predictions, time of arrival predictions and remaining fuel along the planned track based on aircraft performance data;
- The performance function, which generates performance information to the pilots such as take-off speeds, altitude capability, and profile optimization advisories;
- The guidance function, which produces reference values to the auto-guidance system so that the aircraft follows effectively computed lateral and vertical profiles

or when guidance protections are activated, produce new guidance directives to elude the threat.

These functions make use of different on-board databases: the navigation database which contains updated information about navigation aids and nominal procedures at airports, the aircraft performance database which contains aerodynamic data and engine performance data and the route company database which contains predefined routes between pairs of airports. More recently, other data such as meteorological information can also be taken into account.

Fuzzy Logic based hybridization techniques for navigation have been already introduced in the case of mobile robots and UAVs, but not yet in the case of civil aviation aircraft transporting people. Other promising uses of Fuzzy Modelling and Fuzzy Logic are related with flight planning function and the trajectory predictive function by allowing to take more easily into account weather uncertainty characteristics. Different flying devices, mainly in the military field, incorporate already in their flight control functions Fuzzy Logic elements, while Civil aviation aircraft make already use of Fuzzy Logic in their health monitoring functions such as fault detection and diagnostic for their on-board systems. In general these elements of Fuzzy Logic are imbedded in complex software and are not made explicit.

3.1.1 The Flight Planning Function

This function is of particular interest since it generates a flight plan composed of a time indexed track and a vertical profile. The flight track defines the intended horizontal flight path. It is classically composed of waypoints present in the navigation database of the FMS corresponding in general to ground navigation aids and of virtual or pseudo waypoints. The vertical profile provides the intended speed and altitude all along the track, it is composed of the take-off, the climb, the cruise, the descent, the approach and the landing. Flight plans are usually constructed from data already stored in the FMS databases. This data is mainly composed of candidate flight segments such as Standard Instrument Departures (SIDs) and Standard Terminal Arrival Routes (STARs) which are published by the civil aviation authorities and hence cannot be modified by the pilot, and by company routes which can be amended by the pilot. In the past if no acceptable company route was available in the Flight route database, the pilot had to manually construct the entire flight plan by entering via the Multipurpose Control Display Unit (MCDU) the position and other identifiers of each of the waypoint composing the flight track. Today this task can be performed automatically while integrating wind and other weather information. The flight track as well as the whole vertical profile can be optimized using mathematical programming techniques or more advanced ones which allow process more efficiently the aircraft performance data. For example in [6], Forward Dynamic Programming is associated with a Neural Network which computes local changes from one stage to the next along the flight trajectory. From the solution of this problem, many operational parameters can be derived:

- For the take-off phase, the FMS computes parameters such as the thrust needed for take-off, different speeds such as takeoff decision speed, the speed at which aircraft becomes airborne and the take-off safety speed.
- For the climb phase, the FMS computes the climb thrust and speed at the thrust reduction altitude, the acceleration to economic climb speed and the acceleration altitude.
- For the cruise phase, the FMS computes the different cruise levels and speeds or Mach numbers as well as the level change points between the top of climb point and the top of descent point.
- For the climb phase, the FMS computes the climb thrust and speed at the thrust reduction altitude, the acceleration to economic climb speed and the acceleration altitude.
- For the cruise phase, the FMS computes the different cruise levels and speeds or Mach numbers as well as the level change points between the top of climb point and the top of descent point.
- For the descent phase, the FMS defines the top of descent position while considering all the descent constraints and the successive speeds to be adopted.
- For the approach phase, the FMS defines the approach speed and the landing configuration acquisition.
- Finally, the Estimated Time of Arrival (ETA) as well as the Estimated Fuel On-Board (EFOB) at arrival are produced. From the above, it appears that the actual on-board generation techniques for flight plans do not take at all into account surrounding traffic. Then, traffic separation relies completely on the permanent work of ground-based Air Traffic Control (ATC). With the increasing trends in air traffic demand, this situation cannot be sustained anymore, and new solutions must be developed (Figs. 2 and 3).

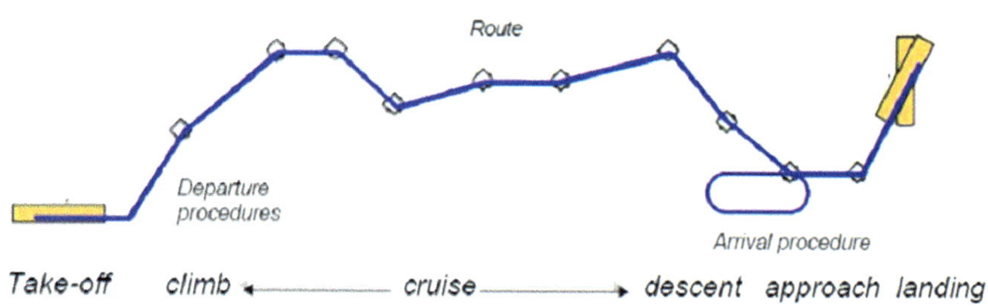

Fig. 2 Horizontal track of a flight plan

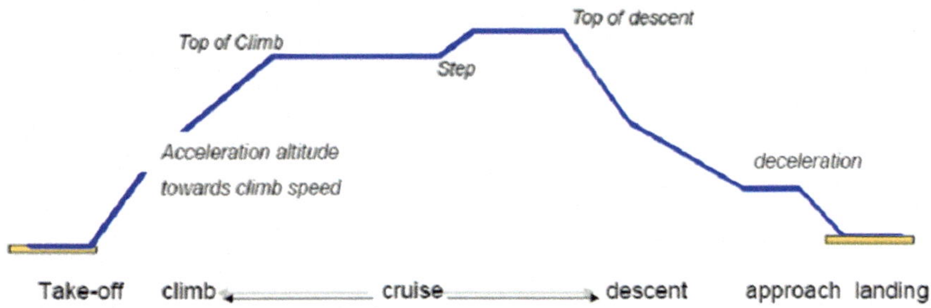

Fig. 3 Vertical profile of a flight plan

4 Cost Index

4.1 The Flight Costs

The flight costs can be divided in operating costs and time-related costs, better explained below.

4.1.1 Flight Costs Composition

The operating costs of a flight can be divided into direct operating costs and indirect operating costs. The indirect costs are related to flight background activities such as personnel training, passenger services, administration, aircraft maintenance equipment, etc. The direct operating costs are directly related to the operation of the flight. They are mainly composed of fuel costs (around 35%), crew salaries, insurance costs, navigation charges and landing fees, maintenance costs, depreciation interest on capital, and ground handling charges at airport. The total cost of a specific flight C_{TRIP} can be expressed as the sum of variable costs which depend on decisions taken for a specific flight realization and of fixed costs which are not specific of a given flight realization. This total cost can be written in a first step as:

$$C_{TRIP} = C_c + C_{Nav} + C_F \cdot F + C_T \cdot T \quad (1)$$

where C_{TRIP} is total cost of the specific flight ($), C_c represents the fixed costs ($), C_F is the unitary fuel cost ($/kg), C_T is the unitary cost of time ($/min) and C_{Nav} is the track specific navigation charges ($), T is the total flight duration (min) of the flight and F is the fuel burn (kg) during that flight. The fixed costs are common to different realizations of the same flight and they only vary with the aircraft utilization and they are often computed on an annual basis. They are composed of leasing costs, finance costs, insurance costs, maintenance labor wages, crew base salaries, administrative and operational labor wages and depreciation and planned maintenance costs according to the aircraft utilization on this flight. Navigation charges are track dependent and, therefore, they cannot be included in the fixed costs since according to traffic control and meteorological conditions the flight track can change from one realization to the next for the same commercial flight. The navigation charges are established according to regional route charges systems to recover the service expenses of Air Traffic Control (ATC) facilities in these regions. In general, they are dependent of the maximum take-off weight (*MTOW*) of the aircraft performing the flight and the distance between the entry and exit points in the traffic control region. The estimation of the cost of the fuel boarded for a particular flight can be a complex question since the acquisition price of fuel varies from place to place depending on relative supplied quantities and local market prices. Airlines practices such as fuel tankering and fuel hedging can turn this estimation even more complicated. The costs which are considered as time-related are somehow related with the time spent flying in the air or moving from and to parking places at airports. For many airlines, the effective duration of a flight is computed as the elapsed time between the aircraft door closing time at departure and the aircraft door opening time at arrival. Other ways to compute the duration of the flight can be adopted.

4.1.2 Time-Related Flight Costs

These costs are composed of hourly maintenance costs, flight crew and cabin crew costs per flight hour, marginal depreciation or leasing costs depending of the ownership of the aircraft and of flight time duration. When an aircraft arrives at destination before schedule, this can cause some problems with landing slots, parking positions and ground handling, resulting finally in a loss of time and money for the airline. When a specific flight is expected to arrive with some delay at the destination airport an additional cost related to the duration of the delay is supported by the airline [7]. It is difficult to estimate the additional cost supported by the airline since it may generate passenger related costs for the airline since in that case the passengers may have to be cared for on ground. It may have consequences on connecting flights and successive flights on the pairing assigned to the aircraft performing the delayed flight [8]. It may impede, considering the traffic situation near the arrival airport, a direct approach for landing which results in expressive additional delays and fuel burn. The cost of delay can be split up into strategic and tactical delay costs. Strategic delay costs are those accounted for in advance, they are related with buffers times introduced into the airline fleet schedule. Tactical delay costs are incurred on the day of operations and are related with passenger costs which increase as a function of the flight delay. Also, an original delay on a flight may generate delays, known as reaction delays, on the downstream airline's network. Statistics shows that a primary delay of one minute may create a reaction delay of about one minute in the network. Passenger related costs resulting from flight delays can be classified into two categories: hard costs and soft costs. Hard costs are the result of the obligation for the airlines to attend delayed passengers. Detailed regulations allow the precise estimation of this type of costs. Soft costs are difficult to estimate since they are mainly the result of passenger dissatisfaction on the deterioration of the commercial image of the airline, diverting potential demand from its flights.

4.2 Definition of the Cost Index

Different considerations have led to the introduction of the Cost Index (CI) concept. First of all, from the above time-related cost analysis, it appears quite difficult to have an a priori estimate of the unitary cost of time C_T appearing in relation (1). Also, considering that while time-related costs are made smaller by faster flights, fuel costs increases and a common basis must be established to compare these costs whose estimates are of unequal accuracy.

So, to solve this dilemma, the FMS combines both time-related and fuel costs by introducing a Cost Index parameter in the vertical flight plan optimization process presented in Eqs. (10)–(16). Then considering mean flight cost per flown nautical mile C, we get for the cruise phase:

$$C = Cost/Nm = (Time\ related\ Costs/Nm) + (Fuel\ Cost/Nm) \tag{2}$$

where:

$$\begin{aligned} Time\ related\ Cost/Nm &= (Cruise\ Duration \times (Cost\ of\ Time/hour))/Nm \\ &= (Cost\ of\ Time/hour)/(Nm)/Cruise\ duration \\ &= (Cost\ of\ Time/hour)/Mean\ Ground\ Speed \\ &= C_T/Vg \end{aligned} \tag{3}$$

and

$$\begin{aligned} Fuel\ Cost/Nm &= (Cruise\ Duration \times Hourly\ Consumption \times Fuel\ Price)/Nm \\ &= (Hourly\ Consumption \times Fuel\ Price)/Mean\ Ground\ Speed \\ &= (CH \times C_F)/Vg \end{aligned} \tag{4}$$

Here CH is the hourly consumption of fuel and Vg is the ground speed. Then:

$$C = Cost/Nm = C_F(CH + C_T/C_F)/Vg = C_F(CH + CI)/Vg \tag{5}$$

with:

$$Cost\ IndexCI = C_T/C_F = (Cost\ of\ Time/hour)/Fuel\ Price \tag{6}$$

The Cost Index can be defined by the airline either for each phase of the flight or for the whole flight and it can be computed before the flight for planning purpose and during the flight for flight managing purpose.

4.3 Current Usage of Cost Index

One of the only parameters governing the vertical flight plan optimization process which are under direct control of airline pilots is the cost index, the whole optimization process performed by the flight management computer being completely transparent to him. Here it is useful to distinguish between Strategic and Tactical Cost Indexes to explain the way pilots tackle nominal conditions as well as the frequent non-nominal ones.

4.3.1 Strategic Cost Index

A Strategic Cost Index will be adopted a priori for a route by the airline and will be a parameter of the vertical profile optimization problem which, beside the parameters of the optimal vertical profile, will produce estimates of the necessary on-board fuel as well as the arrival time for the considered flight. The value assigned to the Cost Index will not reflect in general the real value of the ratio C_T/C_f, but instead the relative importance associated by the airline to time related costs with respect to fuel costs. Many airlines adopt a scale ranging from zero to a maximum value. When the Cost Index is expressed in kg/min, it ranges from 0 to a maximum value CI_{max} taken equal to 100 for the Airbus A320 and equal to 200 for the Airbus A330. It ranges also from 0 to 9999 for the Boeing B777. The airlines assign to each flight a strategic Cost Index. A Strategic Cost Index value set to zero means that the objective is to minimize fuel costs whatever the level of the time related costs. Solving the vertical profile optimization problem will lead in that case to a flight plan with minimum fuel burn and increased flight durations. Adopting large values for the Strategic Cost Index will produce flight plans with shorter flight duration (and higher speeds) but with higher fuel burn. The choice of the value of the Strategic Cost Index can also take into account local market conditions such as competition with flights of other airlines and flight connections at the arrival airport.

The value of the strategic Cost Index can be derived from the solution of a trajectory optimization problem (see relations (10) to (16) below) where the planned time for departure (*PTD*) and the planned time for arrival (*PTA*) are imposed by the airline. In that case, let C_{flight} be the total flight cost associated to the optimal solution of this problem, then the Hidden Strategic Cost Index value can be computed from:

$$HCI = (C_{flight}/C_F - (m(PTD) - m(PTA))/(PTA - PTD)) \qquad (7)$$

The maximum value of the Hidden Cost Index, HCI_{max}, can be computed from relation (7) using the results of the optimization problem where the optimization criterion is the flight duration, while the minimum value of the Hidden Cost Index, $HCI_{\min}$, will correspond to the solution of problem (10) to (16) but with a free arrival time. Then for a given departure mass $m(PTD)$, the Strategic Cost Index can be given by:

$$SCI = f(PTA - PTD) \tag{8}$$

where f is a continuous nonlinear increasing function such as:

$$HCI_{min} \leq f(PTA - PTD) \leq HCI_{max} \tag{9}$$

4.3.2 Tactical Cost Index

The original Strategic Cost Index assigned to the flight must be revised when deviations appear during the flight between the nominal flight plan and what is being achieved online by the aircraft [9]. Deviations are the result of delayed departures at airports, errors in the prediction of en-route winds and errors in the prediction of weather conditions. An estimated time of arrival (ETA) and an estimated fuel on-board (EFOB) at arrival are produced to the pilot by the prediction function of the FMS. When these values are rather different from the nominal ones, a flight plan update is necessary. Then the pilot has to choose a Tactical Cost Index value which is submitted to the on-board flight plan optimization module of the FMS to generate online an updated flight plan which should compensate in some way the deviations from the nominal flight plan. On-board the aircraft, the pilot has no means to compute the time related additional costs resulting from these deviations from the nominal flight plan. In general, the pilot will manage these deviations by choosing arbitrarily a new value for the cost index, knowing that choosing a higher value for the Cost Index should lead to a faster flight with increased fuel consumption and choosing a lower value for the cost index should lead to a slower flight with a reduced fuel consumption. But that is far from sufficient to set efficiently this value while multiple optimization processes on-board the aircraft are not expected. Then, while the Strategic Cost Index adopted for a given flight can be computed using airlines computer facilities, there is a clear need to provide the pilot with an on-board tool to produce an efficient value of the Tactical Cost Index considering the current flight status. Since many consequences of the flight deviations cannot be assessed with high accuracy and since many factors contribute in a complex way to the variation from nominal of the overall performance of the flight, fuzzy logic appears to be a good candidate to fulfill this need. Fuzzy logic has been already proposed as a technique useful to improve the efficiency of different on-board functions for modern transportation aircraft, such as flight director operations in nominal as well as critical [10] situations.

4.4 Vertical Profile Optimization

The basic off-line vertical profile optimization problem can be formulated as [2]:

$$\min \int_{PTD}^{PTA} C_F \cdot (Q + CI)dt \text{ with } CI = C_T/C_F \tag{10}$$

with the vertical flight equations:

$$\dot{V} = (T(Q, V, z) - D(V, z))/m - g \sin \gamma \tag{11}$$

$$\dot{z} = V \sin \gamma \tag{12}$$

$$\dot{x} = V \cos \gamma + w \tag{13}$$

$$\dot{m} = -Q \tag{14}$$

with constraints with respect to the flight domain:

$$V_{min}(z) \leq V \leq V_{max}(z) \tag{15}$$

and with respect to the fuel flow:

$$Q_{min} \leq Q \leq Q_{max} \tag{16}$$

where *PTD* is the planned take-off time, *PTA* is the planned time at arrival, Q is the mass fuel flow (kg/s), V is the airspeed of the aircraft, T is the thrust of the engines, D is the aerodynamic drag, m is the total aircraft mass, γ is the ground path angle, x and z are respectively the longitudinal and vertical position of the aircraft and w is the expected longitudinal component of the wind speed. Here Eq. (11) is the longitudinal acceleration equation, (12) is the climb equation, (13) is the ground speed equation and (14) is the mass flow equation. Condition (15) characterizes the flight domain of the aircraft and condition (16) is relative to the total fuel flow constraints on the engines. Initial conditions (initial climb) are provided for V, z, x and m, final conditions (final approach) are also provided for V, z and x.

Many techniques exist to solve this problem. Direct methods as described in [11] use transcription techniques to transform the optimal control problem in a mathematical programming problem. The indirect approaches, based on the Maximum Principle, establish optimality conditions involving dual variables. In that case, when the cost index is not explicitly introduced in the formulation of the optimization problem, the dual variable associated with a flight duration constraint represents the hidden cost index of the flight.

5 Tactical Cost Index Fuzzy Computation

5.1 The Operational Framework

The computation associated with the Tactical Cost Index is divided here in two steps, both of them using fuzzy logic: first the necessity to modify the current Tactical Cost Index value is assessed, then the new value of the Tactical Cost Index is produced. In Fig. 4 is displayed the qualitative evolution of resulting effects (reduced delay and fuel reduction) of Cost Index changing versus the distance to go to final destination for a given scenario of initial aircraft mass and winds. For each distance the non-inferior solutions resulting from a cost index chosen between its maximum ($CI_{\max}$) and minimum values ($CI_{\min}$) are represented by a dotted line. For example, to achieve a delay reduction of ΔT, at distance x_1 an intermediary value (point A) of the Cost Index will be convenient, while below the limit distance $x_{\lim}$, this delay reduction will be unfeasible.

Both computational steps use data already available in the FMS without needing any new data input from the pilot or any new optimal trajectory computation by the flight management computer. However, additional information should be computed for this flight by the airline and loaded in the FMS performance database: maximum Flight duration ($ETA_{max} - EDT$) and maximum *EFOB* ($EFOB_{max}$) corresponding to the minimum value of the Strategic Cost Index (CI_{min}) and the minimum Flight duration ($ETA_{min} - EDT$) and minimum *EFOB* ($EFOB_{min}$) for the maximum value of the Strategic Cost Index (CI_{max}), where *EDT* is the effective departure time. Then, on one side the workload of the pilot is not increased by additional exchanges with the

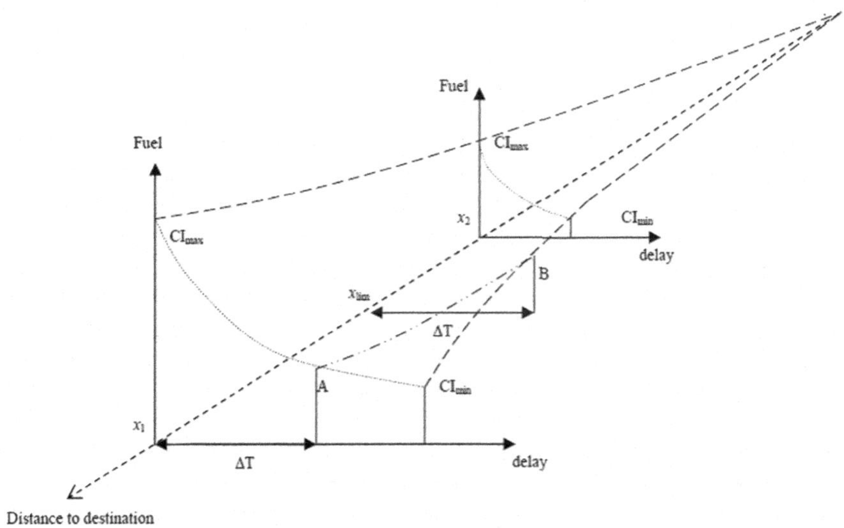

Fig. 4 Cost index impact along a flight

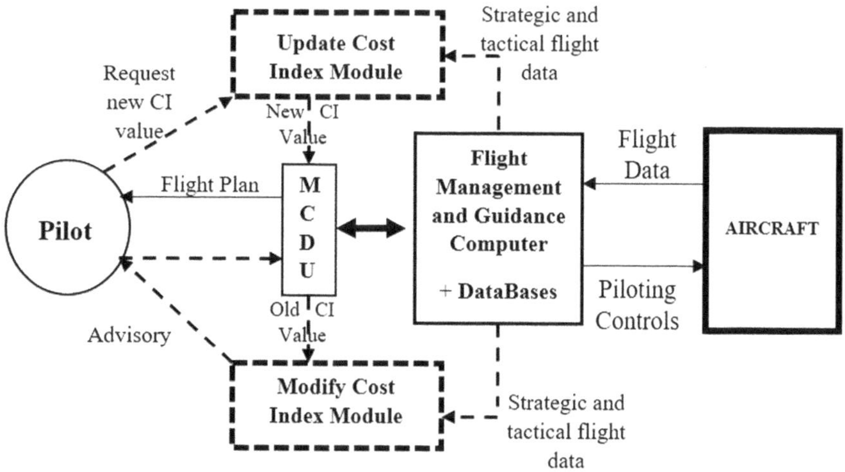

Fig. 5 Proposed cost index Management Structure

FMS and the new computational modules can be easily interfaced downstream with standard FMS functional modules. This is turned possible by the use of fuzzy expert systems [12] which are based on the activation of sets of fuzzy rules integrating aircraft performance data for this flight. Here it is considered that while the need to update the value of the tactical Cost Index is performed online and is displayed continuously to the pilot, it is the pilot who activates the effective computation of this new value at a given point of the flight and who introduces this new value in the flight planning function of the FMS which produces new *ETA* and *EFOB*. The value of the Strategic cost index chosen by the airline for the current occurrence of that flight provides the information about its strategy with respect to flight time and fuel consumption from seeking a minimum flight time to a minimum fuel consumption.

This process is represented in Fig. 5, where the MCDU is the classic interface between the pilot and the flight management function. The value of the Strategic Cost Index chosen by the airline for the current performance of the flight provides information on its strategy in relation to the flight time and fuel consumption, seeking a minimum flight time for minimum fuel consumption. In other words, the calculation establishes a balance between costs related to time and the use of fuel and adopts, through the optimization process, a value that reflects the minimum point of these two types of costs.

5.2 *Fuzzy Monitoring of the Cost Index*

The current Cost Index must be modified since it is estimated that its current value does not allow compliance with some strict restrictions associated with a particular flight, such as: the maximum arrival time or a minimum of fuel on-board. From

Fig. 4, we can also see that, the sooner the Cost Index is adjusted, the better for the fulfillment of the objectives of the current flight. The Modify CI Module, which is responsible for advising the pilot on the interest (or need) to modify the current value of the Cost Index, uses qualitative assessments of the difference between the planned ETA and the current information as inputs, the time and distance remaining for the difference between the estimated onboard fuel (EFOB) and the current, the difference between the value of the Strategic Cost Index for that flight and the current value of the Tactical Cost Index. The qualitative result of this module is chosen from a set such as: {no need, low need, moderate need, strong need, critical need, out of scope} where the orientation factor (excessive flight time or excessive fuel consumption) is informed.

The pilot must modify the value of the Cost Index when strong or critical need levels are reached, otherwise a complete review of the flight plan, including the destination, must be carried out when the system's "out of scope" opinion is displayed. This expert system report indicates that it will not be possible to meet time or fuel commitments. This type of warning could have prevented recent accidents, as was the case with the flight of the Chapecoense football team in 2016, which faced a failure of turbines due to lack of fuel associated with trying to comply with an excessive distance. Although the same aircraft previously performed the same route, with the same amount of fuel, at the time of the accident, it possibly faced weather conditions (incidence of winds) or performed an inefficient trajectory that caused greater fuel consumption. The first levels of need for change are found in many operational situations that involve a particular flight, while the extreme levels are hardly found. Then, the data to build the fuzzy rules offline according to methods like the ones described above, we have two sources: performance data from previous flights and feasibility analysis of optimal solutions of the trajectory calculation problem where PTD is replaced by current time, as well as the initial conditions are set to the current conditions. Different wind scenarios, characteristic of the chosen route, must also be considered. An update process to improve the action of the fuzzy rules adopted, must be integrated with the new data of the aircraft that operates the flight. The "Cost Index Update Module" (Update CI module) which is responsible for proposing to the pilot an updated value of the Tactical Cost Index, uses as inputs qualitative assessments of the same quantities as the "Modify CI" module, as well as information additional features such as the difference between the maximum take-off weight (MTOW) and the effective take-off weight, and additionally, the connection time restrictions with the corresponding transferred passengers. Each difference is evaluated according to a specific, non-scaled scale evaluated using trapezoidal (extreme values) and triangular (intermediate values) association functions to define a set of terms associated with the fuzzy numbers: Large Negative-LN, Medium Negative-MN, Small Negative-SN, Very Small Negative-VSN, Zero-Z, Very Small Positive-VSP, Small Positive-SP, Medium Positive-MP and Large Positive-LP (Fig. 6).

The output of the "Modify CI" module intervals over the following fuzzy sets: No Need-NNC, Very Small Need-VSNC, Small Need-SNC, Medium Need-MNC and Large Need-LNC (Fig. 7).

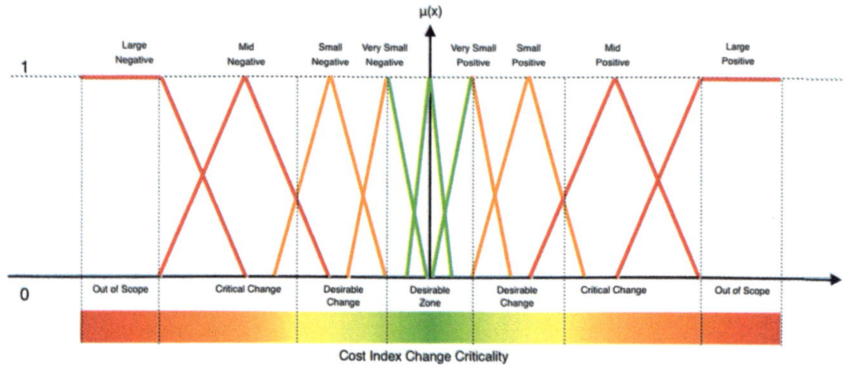

Fig. 6 The considered fuzzy scale for assessing deviations

Fig. 7 Degree of necessity to update the cost index

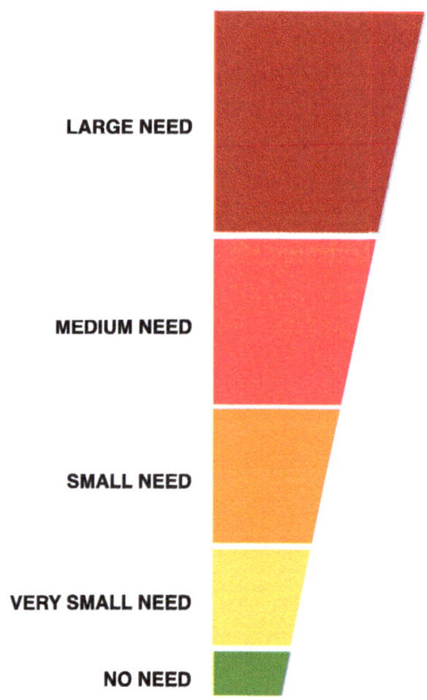

An example of a fuzzy rule for the "Modify CI" module is similar to the one shown below:

$$\textit{if}\ (\text{PETA} - \text{CETA}\ \textit{is}\ \text{VSP})\ \textit{and}\ (\text{PEFOB} - \text{CEFOB}\ \textit{is}\ \text{VSN})\ \textit{and}\ (\text{RTF}\ \textit{is}\ \text{MP})$$
$$\text{then}\ (\text{NTCCI}\ \textit{is}\ \text{SNC}) \tag{17}$$

where PETA is the planned ETA, CETA is the current ETA, PEFOB is the planned EFOB, CEFOB is the current EFOB, RTF is the remaining time for flight and NTCCI is the need to change the Cost Index. Once activated by the pilot, the module "Modify

CI" makes use of a qualitative Cost Index scale calculated from the extreme values of ETA and EFOB corresponding to the extreme values of the Cost Index, the value of the Strategic Cost Index and the current value of the Tactical Cost Index, to generate the defuzzied value of the updated and proposed Tactical Cost Index, where the defuzzification by the average of the maximums is adopted. An example of a fuzzy rule is in this case:

$$\begin{aligned} &\textit{if } ((\text{PETA} - \text{CETA } \textit{is } \text{MN}) \textit{ and } (((\text{CI}_{\text{max}} - \text{CCI}) \text{ is SP}) \text{ or } ((\text{CCI} - \text{CI}_{\text{min}}) \text{ is LP})) \\ &\qquad \textit{and } (\text{NSC } \textit{is } \text{MP})) \text{ then } (\text{Variation of CCI is MP}) \end{aligned} \tag{18}$$

where CCI is the current value of the tactical cost index and NSC is the percentage of connected passengers whose connecting flights are guaranteed. Note that this value is indicated at the beginning of the flight with the Strategic Cost Index value.

5.3 *Tactical Cost Index Calculation*

In Fig. 4 the qualitative evolution of the effects resulting (reduced delay and fuel reduction) from the variation of the cost index in relation to the distance to go to the final destination, for a given mass and initial winds scenario has been presented. For each distance, the non-inferior solutions resulting from a Cost Index chosen between the maximum (CI_{max}) and the minimum values (CI_{min}) are represented by a dotted line. For example, to obtain a delay reduction of ΔT, at distance x_1, an intermediate value (point A) of the Cost Index will be convenient, while below the limit distance x_{lim}, this delay reduction will be unfeasible. Both computational steps use data already available in the FMS without the need for new data entry by the pilot or any new computation of ideal trajectory by the FMS. However, additional information must be computed for this flight by the airline and loaded into the FMS performance database:

- maximum flight duration ($\text{ETA}_{\text{max}} - \text{EDT}$);
- maximum EFOB (EFOB max) corresponding to the minimum value of the Strategic Cost Index (CI_{min});
- minimum flight duration ($\text{ETA}_{\text{min}} - \text{EDT}$);
- minimum EFOB (EFOB_{min}) for the maximum value of the Strategic Cost Index (CI_{max}), where EDT is the effective departure time.

So, on the one hand the pilot's workload is not increased by additional pilot interactions with the FMS and the new computational modules can be easily connected downstream with the standard FMS functional modules. This is made possible by the use of specialist fuzzy systems based on the activation of diffuse rulesets that integrate with the aircraft performance data for this flight. Here, it is considered that, while the need to update the value of the Tactical Cost Index is carried out online and is continuously displayed to the pilot, it is the pilot who activates the effective calculation of this new value at a certain point in the flight, and which introduces this new value into the flight planning function of the FMS, which in turn produces new ETA and EFOB.

5.4 A Rule-Based Specialist Decision Maker in Aviation 4.0

The fuzzy controllers are extremely efficient realizing convergent approximate computations to an expected value. By these, realizing some rule-based loops, the results are obtained by approximations adjusts. On the CI computation, the result is obtained by the tradeoff parameters cross where the system suggests a Cost Index change intensity for a positive or negative value. An important concept to be understood is that the rule-based intelligent system will act over deviations at ETA and EFOB variables given by the FMC. This information is treated with linguistic values that forms very clear rules. The linguistic values can be adjusted to each aircraft model and the airlines strategies over fuel uses and delays tolerance on their daily operations. So, the limits can be adjusted and set to each case or strategy. For example, the delay tolerance limit is 30 min, each is considered moderate to a long-range flight. The Fuel tolerance is 200 kg and the linguistic values for Fuel deviations will be a set from Large Negative that means −200 kg to Large Positive that means 200 kg. Time deviations variables, as example, will be set varying from Large Positive, that means 30 min in advance to Large Negative, that means 30 min delayed. By these parameter settings, the Aviation 4.0 proposed algorithm will act to deliver a better operation in economic savings.

It is considered as assumption that the plans are well provided, enough fuel for and a capable mission time to run a flight mission. The main objective is suggesting the onboard crew better saving performance over the flights changing conditions. So, we consider that with a Low Negative Fuel situation, the aircraft still presenting a fuel buffer to complete the trip. Moreover, the undesired condition of Low Negative Fuel can be set as an alert to avoid the pilot earlier about future emergency conditions. Based on the good plan assumption, a range for cost index step limit is attributed to the linguistic variables set varying from "No Need" to "Large Need" reflecting the needs for change CI. For example, −15 to +15 value current value adjust. If the system needs to increase or decrease 30, the algorithm will suggest increase two times. If it occurs in 0 to 100 Cost Index Scale seems that initial plans are not well computed because it shows a priori a very large divergence in term of need to change. If the divergences are large, it is common to execute successive changes, the first one suggests reducing the main divergence and the second turn to minimize the divergence provide a more precise and refined value. This results in the process represented in Fig. 8.

As the ETA and the EFOB variables can be described by 9 linguistic variables each, by permuting linguistic variables a set composed by 81 rules written utilizing *if* and *then* scripts are obtained. It can be resumed in the tables below where each rule is addressed to a system behavior by crossing EFOB and ETA. Table 1 express the

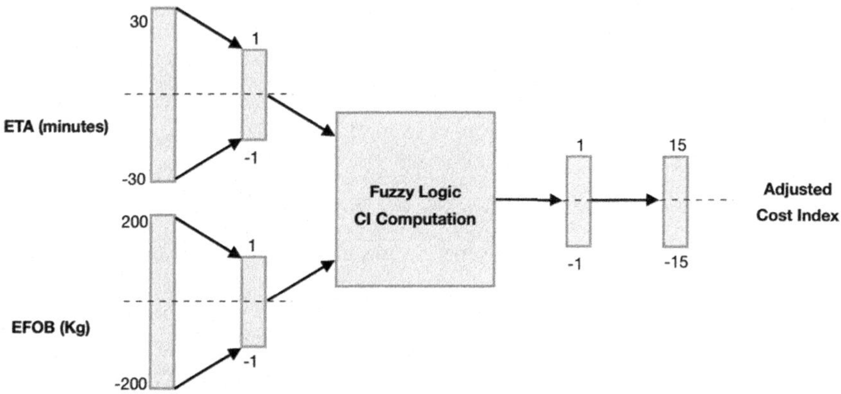

Fig. 8 Cost index computation framework

Table 1 Fuzzy rule-based specialist system

	LN	MN	SN	VSN	0	VSP	SP	MP	LP
LN	LN	LN	LN	LN	LN	LN	LN	LN	LN
MN	LN	MN	MN	MN	MN	MN	MN	MN	MN
SN	LN	SN	SN	SN	SN	VSN	VSN	VSN	VSN
VSN	LN	VSN	VSN	VSN	ZERO	ZERO	ZERO	ZERO	ZERO
0	LN	VSN	ZERO	ZERO	ZERO	ZERO	ZERO	ZERO	ZERO
VSP	LN	ZERO	VSP	VSP	VSP	VSP	VSP	VSP	VSP
SP	LN	ZERO	VSP	SP	SP	SP	SP	SP	SP
MP	LN	ZERO	SP	SP	MP	MP	MP	MP	MP
LP	LN	ZERO	MP	MP	LP	LP	LP	LP	LP

whole set of rules involving EFOB and ETA deviation linguistic terms and associating to a Cost Index Deviation Change varying from Large Negative to Large Positive. Figure 9 displays the relations obtained by crossing the EFOB and ETA deviation linguistic terms.

From an initial state or an intermediate state N (S_n), the system computes the deviations to a final state (S_f) which is the optimization objective. From this moment the rule-based system runs successive loops until the state N + 1 (S_{n+1}) presents a similarity with the final state (S_f) or zero deviation. At this moment the result is an optimal CI suggestion.

Once occurs the first adjust, the CI computation realizes loops until the parameter achieves an acceptable optimum result. In extreme deviations from the initial plans provided by the airlines flight engineers, an avoiding system can act with messages to the pilot to cope with many objectives covering from a critical fuel alert (related to fuel) to a passenger lost connection (related to time). Figure 10 displays the proposed framework.

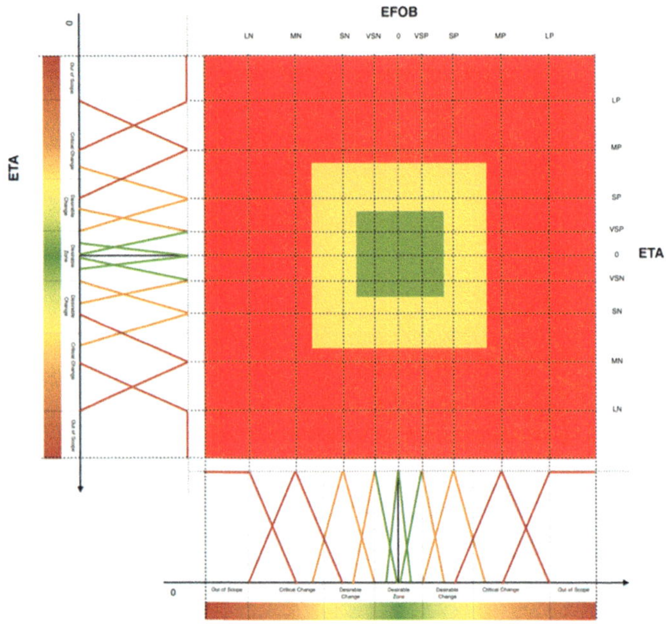

Fig. 9 Graphic view of the inference system for evaluation

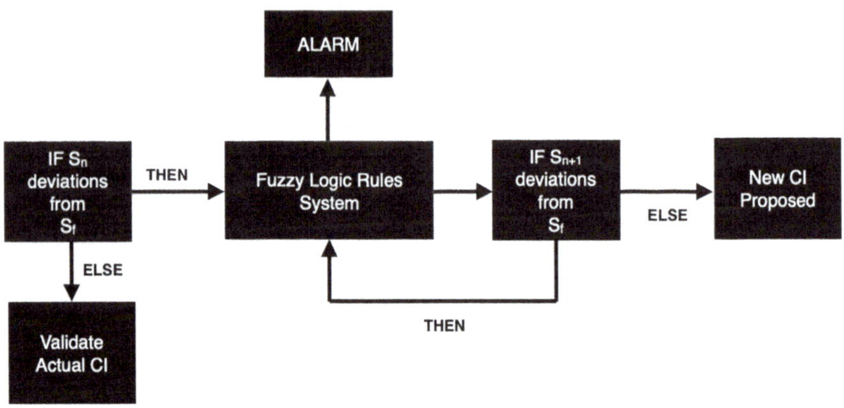

Fig. 10 Cost index monitoring and deviations computation loop and alarm framework

An additional tool to treat the uncertainty in the fuzzy cost index computation is the adoption of fuzzy dual calculus [13] to cope with the realistic scenarios of operation at airlines. Each number is composed by a real primal part and a dual part associated with the uncertainty, allowing separate computation. The fuzzy dual number limits the computation burden traditionally associated with fuzzy calculus. Using this computation method, the set of values for ETA and EFOB deviations

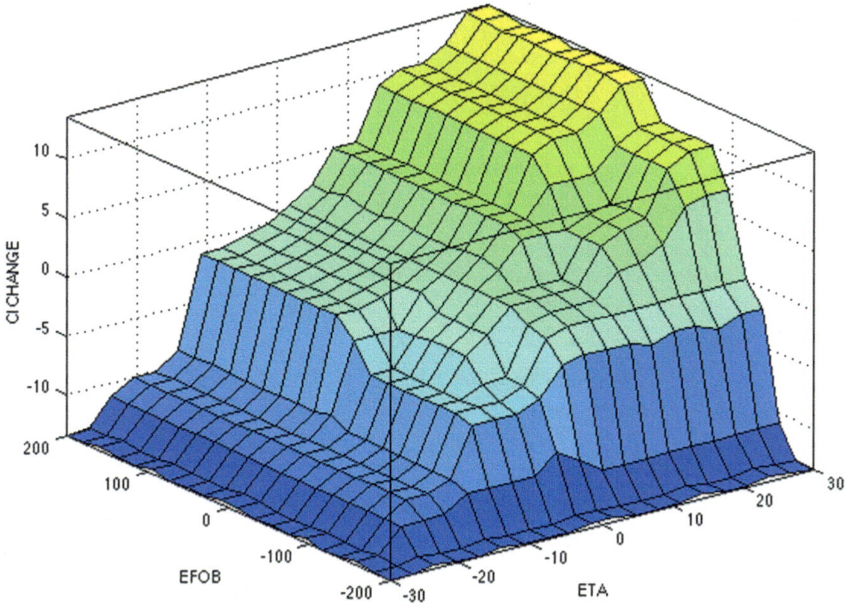

Fig. 11 Decision making surface provided by MATLAB fuzzy toolbox software

assume a new fuzzy notation where ε is the pure unit dual number. To made clearer the memberships on the fuzzy computation, the normalized operations are presented in the tables below.

The presented fuzzy dual numbers show that once the variables values are near the mission target, the uncertainty is reduced while as the variables are far, they get a larger uncertainty. This is the result of the reduction of the deviations when convergence happens. The proposed approach must be adjusted to each aircraft model and airline expectations and tolerance for delays and fuel savings (Fig. 11).

Tables 2 and 3 present the fuzzy dual numbers describing the ETA and EFOB deviations, utilized as inputs for the evaluation system followed by Table 4 which presents the fuzzy dual numbers to express the output values for the suggested Cost Index adjust. On the right side of Tables 2, 3 and 4 are presented the normalized fuzzy dual numbers associated to Fig. 8.

5.5 *Case Study*

An example of a case study considered in [14, 15] is related to a medium range flight between Madrid and Moscow using an A320 plane for a flight of about 2300 Nm. The aircraft has an MTOW of 75,000 kg and a performance range of 3300 Nm. It was assumed that the takeoff weight of the flight considered was 70,000 kg with

Table 2 Fuzzy dual representation of ETA deviations and normalized format

Fuzzy dual numbers for ETA deviations							
Value	a	b	Dual	Value	a	b	Dual
LN	−30.00	9.75	$-30 + \varepsilon 9.75$	LN	−1.00	0.33	$-1.00 + \varepsilon 0.33$
MN	−18.75	7.00	$-18.75 + \varepsilon 7.00$	MN	−0.63	0.23	$-0.63 + \varepsilon 0.23$
SN	−9.75	6.00	$-9.75 + \varepsilon 6.00$	SN	−0.33	0.20	$-0.33 + \varepsilon 0.20$
VSN	−3.75	3.75	$-3.75 + \varepsilon 3.75$	VSN	−0.13	0.13	$-0.13 + \varepsilon 0.13$
ZERO	0.00	1.88	$0 + \varepsilon 1.88$	ZERO	0.00	0.06	$0 + \varepsilon 0.06$
VSP	3.75	3.75	$3.75 + \varepsilon 3.75$	VSP	0.13	0.13	$0.13 + \varepsilon 0.13$
SP	9.75	6.00	$9.75 + \varepsilon 6.00$	SP	0.33	0.20	$0.33 + \varepsilon 0.20$
MP	18.75	7.00	$18.75 + \varepsilon 7.00$	MP	0.63	0.23	$0.63 + \varepsilon 0.23$
LP	30.00	9.75	$30.00 + \varepsilon 9.75$	LP	1.00	0.33	$1.00 + \varepsilon 0.33$

Table 3 Fuzzy dual representation of EFOB deviations and normalized format

Fuzzy dual numbers for EFOB Deviations							
Value	a	b	Dual	Value	a	b	Dual
LN	−200.00	65.00	$-200 + \varepsilon 65$	LN	−1.00	0.33	$-1 + \varepsilon 0.33$
MN	−125.00	50.00	$-125 + \varepsilon 50$	MN	−0.63	0.25	$-0.63 + \varepsilon 0.25$
SN	−65.00	45.00	$-65 + \varepsilon 45$	SN	−0.33	0.23	$-0.33 + \varepsilon 0.23$
VSN	−25.00	25.00	$-25 + \varepsilon 25$	VSN	−0.13	0.13	$-0.13 + \varepsilon 0.13$
ZERO	0.00	12.50	$0 + \varepsilon 12.50$	ZERO	0.00	0.06	$0 + \varepsilon 0.06$
VSP	25.00	25.00	$25 + \varepsilon 25$	VSP	0.13	0.13	$0.13 + \varepsilon 0.13$
SP	65.00	45.00	$65 + \varepsilon 45$	SP	0.33	0.23	$0.33 + \varepsilon 0.23$
MP	125.00	50.00	$125 + \varepsilon 50$	MP	0.63	0.25	$0.63 + \varepsilon 0.25$
LP	200.00	65.00	$200 + \varepsilon 65$	LP	1.00	0.33	$1 + \varepsilon 0.33$

13,000 kg of fuel on-board (FOB), with 145 passengers in a configuration of two classes with a flight connection at 7.10 am the next day for 30 passengers. When adopting a Strategic Cost Index of 20, the scheduled ETD was 11:10 pm (local time) and the ETA was 6:30 am (local time) with an expected flight duration of 380 min and EFOB on arrival of 3400 kg. The considered scenario assumes that the effective departure was delayed by 25 min and that, during the first part of the cruise, there was an effective wind of 10kt higher than expected. At takeoff, advising the pilot, the "Modify CI Module" produced a "LARGE POSITIVE" Cost Index and the "CI Update Module" produced a new tactical Cost Index of "60", which takes the aircraft to a cruise level higher and an ETA of 6h35 for an EFOB on arrival of 3050 kg. At the cruise level, considering the head wind, the "Modify CI Module" started to produce an increasing Cost Index, modifying the advice until the "LARGE POSITIVE" level was reached after 40 min of cruise and accepted by the pilot. A new Tactical Cost Index of 70 was produced by the "CI Update Module" and applied by the pilot to the

Table 4 Fuzzy dual representation of CI changes

Fuzzy dual numbers for CI deviations							
Value	a	b	Dual	Value	a	b	Dual
LN	−15.00	4.88	$-15 + \varepsilon 4.88$	LN	−1.00	0.33	$-1 + \varepsilon 0.33$
MN	−9.38	3.50	$-9.38 + \varepsilon 3.50$	MN	−0.63	0.23	$-0.63 + \varepsilon 0.23$
SN	−4.88	3.00	$-4.88 + \varepsilon 45$	SN	−0.33	0.20	$-0.33 + \varepsilon 0.20$
VSN	−1.88	1.88	$-1.88 + \varepsilon 1.88$	VSN	−0.13	0.13	$-0.13 + \varepsilon 0.13$
ZERO	0.00	0.94	$0 + \varepsilon 0.94$	ZERO	0.00	0.06	$0 + \varepsilon 0.06$
VSP	1.88	1.88	$1.88 + \varepsilon 1.88$	VSP	0.13	0.13	$0.13 + \varepsilon 0.13$
Table 1: Fuzzy rule-based specialist system SP	4.88	3.00	$4.88 + \varepsilon 3.00$	SP	0.33	0.20	$0.33 + \varepsilon 0.20$
MP	9.38	3.50	$9.38 + \varepsilon 3.50$	MP	0.63	0.23	$0.63 + \varepsilon 0.23$
LP	15.00	4.88	$15.00 + \varepsilon 4.88$	LP	1.00	0.33	$1 + \varepsilon 0.33$

FMS, resulting in a new 6:40 AM ETA with an EFOB arriving 2850 kg. Then, when the head wind returned to its nominal value, after 20 min, the system began again to advise the pilot by adopting a new Tactical Cost Index of "55", resulting in a final ETA of 6h35 with an EFOB on arrival of 2950 kg. The evaluation system that allows the on-line setting of the Tactical Cost Index values, composed of the two proposed modules, is based on the divergence of time and fuel, based on real data to update the pilot of relevant information for a better individual performance of the flight and mission fulfillment.

In the example, the application suggests an economy adhering to reality, considering the initial plan to reach the destination with 3400 kg as a lost goal, given the delay of 25 min before takeoff. Within the delay scenario, the fluctuation of the Cost Index at first suggests the recovery of the arrival time with an initial cost of 550 kg under wind conditions in the initial phase of the flight. With the end of the wind interference after 20 min, the change in the Cost Index allowed additional savings of over 100 kg of fuel.

6 Conclusions

The hard-competitive environment faced by airlines has led them to provide a higher quality air transport service to their customers while taking providences to manage always more efficiently their operations costs. Among the complexity of the management of operations costs in the actual context of the air transport industry, the Cost Index concept provides a strategic tool for the airlines to balance fuel and time related costs and a tactical tool for the pilots to face as efficiently as possible the common hazards of flight operations. More recently, it has been shown how the cost indexes of successive flights assigned to a given aircraft can be set globally to manage network delays and maintain the feasibility of a flights program [16].

The efficient setting of the Tactical Cost Index has become more and more complex. So, this chapter contributes to the Aviation 4.0 modern concept technology approach, by introducing a fuzzy based computation approach for the value of the Tactical Cost Index, easing the pilot in this complex task.

The proposed decision structure is composed of two modules which however remain under the authority of the pilot. This results in the use of connected fuzzy expert systems integrating performance data specific to the considered aircraft and to the considered flight. Moreover, the computational burden can be reduced using fuzzy dual numbers which are a new form to provide fast fuzzy calculations. Then a case study has been presented to illustrate a real situation faced by an airline operation, showing the effectiveness of the proposed approach based on fuzzy modelling and decision making.

References

1. Walter, R.: Flight management systems. In: Spitzer, Cary R. (ed.) The Avionics Handbook. CRC Press, Baco Raton (2001)
2. Mora-Camino, F.: Flight control systems (in French). In: Air Transportation Engineering Lecture Notes, ENAC, 393p (2014)
3. Yen, J., Langari, R., Zadeh, L.A.: Industrial Applications of Fuzzy Logic and Intelligent Systems. IEEE Press (1995)
4. Ross, T.J.: Fuzzy Logic with Engineering Applications. New Mexico, USA, Wiley (2010)
5. Airbus: Getting to grips with cost index. Flight Operations Support & Line Assistance, Issue II, May 1998
6. Hagelauer, P., Mora-Camino, F.: A soft dynamic programming approach for on-line aircraft 4D-trajectory optimization. Eur. J. Oper. Res. **107**(1), 87–95 (1998)
7. Cook, A., Tanner, G., Jovanovic, R., Lawes, A.: The cost of delay to air transport in Europe. In: 13th Air Transport Research Society, ATRS World Conference, Abu Dahbi (2009)
8. Cook, A., Tanner, G., Lawes, A.: The hidden cost of airline unpunctuality. J. Transp. Econo. Policy **46**(2), 157–173 (2012)
9. Cook, A., Tanner, G., Williams and, V., Meise, G.: Dynamic cost indexing. In: 6th EURO-CONTROL Innovative Research Workshops & Exhibition, 4–6 Dec 2007
10. Wu, H., Mora-Camino, F.: Glide control for engine-out aircraft. In: AIAA Guidance, Navigation, and Control Conference 2012, Keystone, Colorado
11. Huang, G.Q., Lu, Y.P., Nan, Y.: A survey of numerical algorithms for trajectory optimization of flight vehicles. Sci. China Technol. Sci. **55**, 2538–2560 (2012)
12. Mcneill, D., Freiberger, P.: Fuzzy logic: the revolutionary computer technology that is changing our world. Touchstone Rockfeller Center, New York (1994)
13. Mora-Camino, F., Nunes, C.A.: Fuzzy Dual Numbers: Theory and Applications, Studies in Fuzziness and Soft Computing, vol. 359. Springer (2018)
14. Krykhtine, F.L.P.: Abordagem fuzzy para otimização de planos de vôo. Ph.D. Thesis, Federal University of Rio de Janeiro (2018)
15. Sanchez-Camacho, M.G.M., Wu, H., Nunes, C.A., Krykthine, F., Mora-Camino, F.: fuzzy setting of cost index for improved flight management. In: 12th International Conference on Fuzzy Systems and Knowledge Discovery (FSKD) (2015)

16. Moyo, S., Al-Fazari, H.S., Mykoniatis, G., Mora-Camino, F.: Managing Network Flight Delays Through Collaborative Decision Making, Seminario de Transporte Aéreo—SITRAER 2019, Brasilia (2019)

Intelligent and Fuzzy Applications in Aviation 4.0 Aircraft Maintenance/Production Management

Toward Joint Application of Fuzzy Systems and Augmented Reality in Aircraft Disassembly

Samira Keivanpour

Abstract Advanced optimization models are required for the disassembly of complex products. There are stochastic variables in sequence planning, and real-time and data-driven approaches play an essential role in this context. In the case of aircraft at their end of life, the specifications and requirements of airworthiness should be followed. Hence, a collaborative and dynamic disassembly operation is required. Industry 4.0 is a promising paradigm for dealing with the uncertainties of value extracted from recovered parts and materials and the sustainability of the disassembly operation. Aviation 4.0 with its enabling technologies provides new tools and methods for addressing the challenges of the sustainability and efficiency of aircraft during the entire lifecycle. Advanced 3D simulation environment aids researchers and operators in testing and analyzing recovery options and working on different scenarios before performing a real operation. This chapter proposes a system architecture based on augmented reality and fuzzy intelligent decision-making for the disassembly of complex products. The application perspective for aircraft parts is discussed.

Keywords Aircraft at the end of life · Disassembly sequence planning · Augmented reality · Fuzzy intelligent decision making

1 Introduction

The management of aircraft at their end of life (EoL) is a new research theme in the circular economy context. With the increasing number of EoL aircraft, more initiatives and enterprises will join this business. Hence, developing industrial and managerial solutions is crucial. According to the International Air Transport Association (IATA), aircraft retirement increased annually by a rate of 4% from 1980 to 2017.

S. Keivanpour (✉)
Department of Mathematical and Industrial Engineering, Polytechnique Montreal, Montreal, Quebec, Canada
e-mail: samira.keivanpour@polymtl.ca

C. Kahraman and S. Aydın (eds.), *Intelligent and Fuzzy Techniques in Aviation 4.0*, Studies in Systems, Decision and Control 372,
https://doi.org/10.1007/978-3-030-75067-1_11

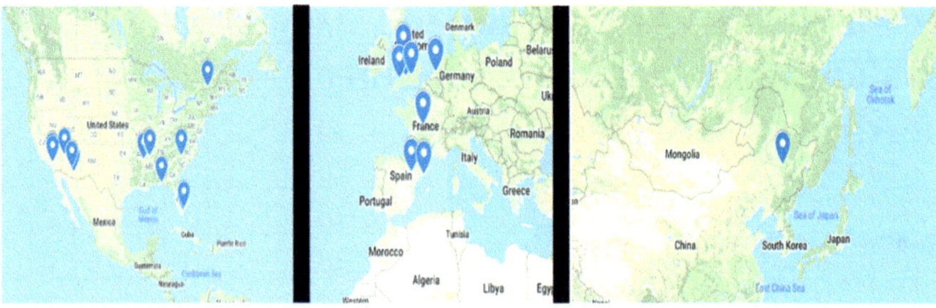

Fig. 1 The disassembly sites accredited by AFRA. *Source of Data* AFRA website, map: Google map

This retirement was affected by gross domestic product growth and fuel-price variation. According to this reference, the rate increased after the 2008 recession. Hence, we can expect that the 2020 pandemic will trigger an increase in the retirement rate in future years. A few disassembling sites around the world are accredited by the Aircraft Fleet Recycling Association (AFRA). These sites are shown in Fig. 1 according to the information provided on the AFRA website. The disassembly process should be verified by an aircraft maintenance organization to maintain a record of the airworthiness status of recovered parts. The value of the recovered parts is 80% of the aircraft value, and the recycled material is only 10% [1]. This reveals the importance of reducing the cost of disassembly for an EoL aircraft enterprise or part-out company (Fig. 2).

Different key players are contributed to the EoL aircraft problem including aircraft manufacturers, owners/operators, MRO (Maintenance, repair, and overhaul), and parting out companies. Disassembly companies play an essential role in this ecosystem. These players are actively collaborated to develop new strategies for optimizing and sustainability of parts recovery. Hence, new technologies and intelligent decision support systems could considerably improve the efficiency of the process. Smart technologies influenced product recovery operations and reverse logistics.

Guyon et al. [2] discussed the opportunities and challenges of Industry 4.0 in aeronautics. For the supply chain integration, sharing different portals between the supply chain players is one of them. Using 3D modeling and augmented reality in the construction and assembly line is the other application. For example, using the collection of smart tools for avoiding errors of operators considering several fastening points and required tools are the other implication of industry 4.0 in this industry. Junior et al. [3] studied the application of Industry 4.0 in the assembly line of aircraft manufacturing. By comparative study and using customized maturity model of technology. The authors concluded the positive impacts of this new industrial revolution on the aircraft assembly line. Barbosa and Aroca [4] also discussed the application of Industry 4.0 in fastening, inspection, sealing, manipulation, and other manufacturing processes. Like forward logistics, the new paradigm of Industry 4.0 affects dissembling, remanufacturing, and recycling [5].

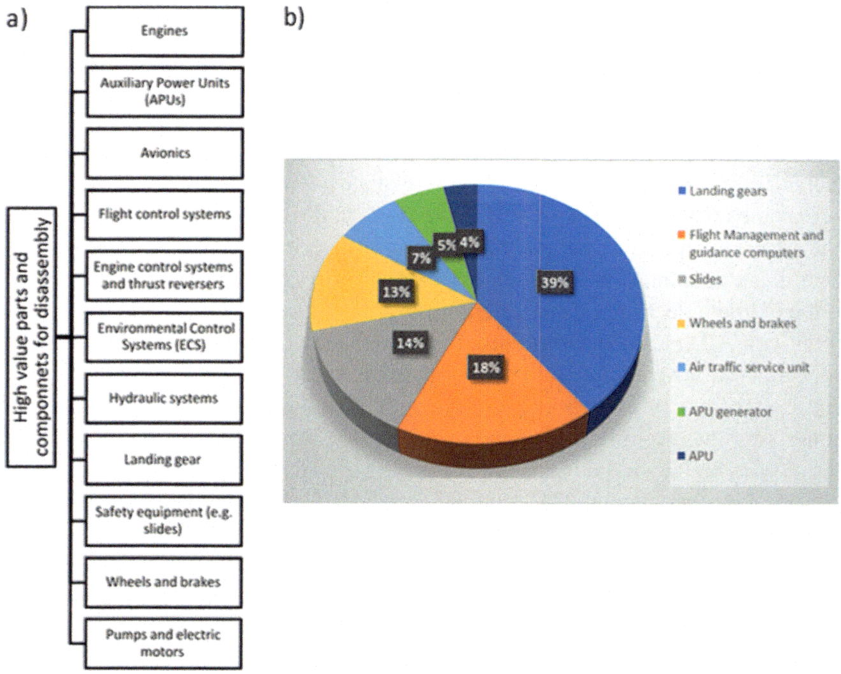

Fig. 2 **a** The high-value parts for disassembly, **b** an example of the value of each subassembly in total airframe component value. *Source of Data* IATA [1]

Different key players have contributed to the EoL aircraft problem, including aircraft manufacturers, owners/operators, maintenance, repair, and operation (MRO), and parting-out companies. Disassembly companies play an essential role in this ecosystem and have actively collaborated to develop new strategies for the optimization and sustainability of the parts recovery process. Hence, new technologies and intelligent decision support systems can considerably improve the efficiency of this process. Smart technologies influence product recovery operations and reverse logistics.

Guyon et al. [2] discussed supply-chain integration and considered the sharing of different portals between the supply-chain players as one of the opportunities and challenges of Industry 4.0 in aeronautics. Using 3D modeling and augmented reality (AR) in the construction and assembly line is another application. For example, using a collection of smart tools to avoid operator errors in the consideration of several fastening points is another implication of Industry 4.0. Junior et al. [3] studied the application of Industry 4.0 in an aircraft manufacturing assembly line by performing a comparative study and using a customized maturity model of technology. The authors demonstrated the positive impact of this new industrial revolution on the aircraft assembly line. Barbosa and Aroca [4] also discussed the application of Industry 4.0 in fastening, inspection, sealing, manipulation, and other manufacturing processes. Like forward logistics, the new paradigm of Industry 4.0 affects dissembling, remanufacturing, and recycling [5].

The learning factory construct has been used in the context of manufacturing and operation management for several years. Recently, the digital and virtual learning factory received much attention because of its evolving technologies, transformation as the result of Industry 4.0, and system complexities of real testbeds. Vijayan and Mork [6] introduced the idealab for modeling Industry 4.0, automation, and lean manufacturing. Tvenge et al. [7] explained the gap in research on the added value of digital learning activities. Mourtzis et al. [8] introduced a cloud-based platform with AR for the visualization of a customized product design. Mourtzis et al. [9] proposed a flexible manufacturing cell based on learning factory concepts to enhance the engineering skills of learners of 3D printer and computer numerical-control milling machines and a robotic arm in vitro platform. A data-driven 3D platform based on the virtual learning factory concept can support ongoing research on the circular economy in the Industry 4.0 paradigm. The two applications of Industry 4.0 in the disassembling process can be classified as modeling and simulation for virtual disassembly and smart robotic disassembly with real-time data applications [10]. As Industry 4.0 and cyber-physical systems become more popular, the role of realizing the virtual avatars of a physical system becomes increasingly significant. Digital models offer advanced analytical features and visualization. Frizziero et al. [11] proposed an AR-enabled environment to add value to computer-aided design results for designing the disassembly sequences of a gearbox. The authors used Unity and Vuforia and GameObject animation for this AR representation. Wang et al. [12] used Google Glass to disassemble a mobile phone. The authors showed how AR can be used as a potential alternative to printed manuals. Change et al. (2020) used the AR concept and 3D models to develop an interactive interface for planning disassembly sequencers. The authors developed a virtual panel in Unity using a user interface, an optimization model, and an automatic content-authoring module. Xie et al. [13] used virtual reality (VR) to simulate a turbofan engine disassembly process. The authors emphasized the role of AR/VR in MRO in the aviation industry. Rocca et al. [14] presented a laboratory application case of the disassembly of electrical and electronic equipment using VR and digital-twin models. Lee [15] also highlighted the role of AR in disassembly and improved its application via the digital-twin model. There is a gap in the literature for applying these digital models and AR approaches in the management of the complex EoL products in a systematic manner. For complex product-recovery options, comparing the different disassembly strategies with the aggregation of several dispersed databases, data analytics, and visualization dashboards is essential. This chapter discusses the recent studies on aircraft disassembly operation and provides the perspectives of fuzzy systems and AR in the disassembly of an aircraft in the new paradigm of Industry 4.0.

The rest of the chapter is organized as follows: Sect. 2 discusses the different approaches for aircraft disassembly in recent studies. Section 3 reviews the application of the fuzzy approach to disassembly planning. Section 4 discussed the application of augmented reality in optimizing the disassembly operation. Section 5 introduces the joint application of fuzzy and augmented reality and Sect. 6 discusses the application perspective. Finally, Sect. 7 concludes with some remarks and the perspectives of future research.

2 Aircraft Disassembly

There are a few studies that discussed the aircraft's disassembly. Hassan and Yoon [16] developed an Ant colony algorithm for assembly/disassembly path planning of the aircraft. The authors used a haptic-assisted environment for 3D simulation and training. Yoon [17] also used the joint application of a genetic algorithm and haptic device to develop the best sequence of assembly/disassembly for aircraft. Dayi et al. [18] discussed the lean approach in improving the part recovery process of the aircraft at the end of life. They used all the procedures of AMM (aircraft maintenance manual) for extracting the best sequence of the disassembling process. Camelot et al. [19, 20] developed a decision tool for organizing the disassembly process to optimize the disassembling of the recovered parts. The authors applied the graphical method, wave propagation approach, and Petri nets for planning the disassembling tasks. Tang and Zhong [21] also utilized the Petri nets approach for aircraft maintenance operation. Zhong et al. [22] utilized a metaheuristic approach for disassembly sequence planning of aircraft. Zhang et al. [23] applied a fuzzy clustering algorithm for disassembly planning of aircraft. Zahedi et al. [24] discussed the different alternatives for aircraft disassembly for maximizing the recovered value. Sabaghi et al. [25, 26] introduced eight different strategies for disassembling and dismantling the aircraft. One of these strategies is systematic disassembly with separating and sorting the components based on material composition. Another strategy is smart disassembling with saving time in operation by not removing rivets that are common between components. Disassembling with cutting also provides the possibility of cutting the areas with similar materials and performing dissembling for the areas with mixed materials. The authors concluded that smart disassembly is one of the most active option considering the sustainability analysis. Recently some scholars focused on the application of virtual reality in the disassembly of aircraft. Zou et al. [27] developed a virtual disassembly environment using Unity3D for landing gear. This framework includes a display module with Unity3D, a tool management module for managing the tools used during the operation, a parts module for managing the components in Bill of Material (BoM), a disassembly module for controlling and planning the disassembly sequences, and communication module for the interface between Unity3D module and the other module coded by C#. Xie et al. [13] also developed a simulation tool for the disassembly of an aircraft engine with virtual reality. The authors discussed the capability of this simulation for students' training. The synthesis in the literature highlight the following points:

- The complexity of the aircraft parts requires the advanced optimization approach for sequence planning.
- The logic of the operation should be controlled based on the maintenance manual and the specifications and the requirements of the airworthiness should be strictly followed.
- The virtual platform and visualization play an essential role in training and research.
- The smart disassembly strategy, using artificial intelligence, real-time and data-driven approach will be the next generation of the disassembly models.

3 Fuzzy Approaches to Disassembly Planning

Zhou et al. [28] performed a survey on recent developments of disassembly sequence planning. The authors conducted a taxonomy of the literature based on disassembly mode, disassembly modeling, and disassembly planning. For disassembly mode, they classified the studies into complete/partial and sequential/parallel groups. According to this survey, complete and sequential disassembly include frequent studies rather than partial and parallel. Partial disassembly in the case of aircraft disassembly considering the recovery of high-value parts and components is important. Moreover, for large complex products, parallel disassembly is more efficient as the disassembly time plays an essential role in the total costs of disassembly. For disassembly modeling, the graph-based model, Petri-net, and matrix-based method are the main categories of the model. The Petri net model considers the detailed geometrical and topological information and the number of researches on this method is considerably low in comparison to the matrixed-based methods [29]. In the case of the aircraft, the Petri net provides a reliable option for modeling. The randomness and uncertainties are a critical part of the operation that should be considered in the modeling. The other issue is considering the dynamic nature of disassembly to have the possibility to adjust the operation in real-time. In the optimization model, the costs, environmental impacts, and the revenue from recovered parts are the objective functions in the optimization models. The disassembly time is a vital variable in the cost of disassembly. Considering the uncertainty and stochastic nature of the time, fuzzy approaches are applied to consider the fuzzy time. In the modeling, fuzzy Petri net and reasoning are also used to consider the uncertainties of the operation process. Gao et al. [30] used fuzzy reasoning Petri net to provide intelligent decisions based on the real-time data during the disassembly time. Tang and Turowski [31] used a fuzzy Petri net with an iterative learning approach for designing the sequences of disassembly dynamically. Zhao et al. [32] also used the fuzzy Petri net approach to disassembly planning. The authors used a fuzzy reasoning algorithm for planning the sequences to perform decision making in real-time. Hsu [33] proposed a fuzzy knowledge-based system using a fuzzy Petri net for analyzing the disassembly operation and address the usability of the recovered parts. Tian et al. [34] considered a stochastic time, cost, and quality of the component in planning the disassembly sequence. The authors applied a fuzzy simulation and artificial colony algorithm for solving the disassembly model. Galantucci et al. [35] used fuzzy logic genetic algorithm for planning the disassembly sequences. Gonnuru [36] used a fuzzy logic model for optimizing the planning sequences. The authors applied a Bayesian method for analyzing Radio Frequency Information (RFID) data and then used the fuzzy logic for optimizing the sequence based on the input data such as product usage, component condition, etc. Xiaoyong et al. [37] used a fuzzy cognitive map for addressing the relationship among parts and provide the reasoning for planning the disassembly sequences. Feng et al. [38] used a multiobjective optimization model for designing the disassembly sequences and applied a fuzzy integral method for evaluating the results. The synthesis in the literature reveals the following points:

- The uncertainty in time and cost of disassembly operations and the quality of the components are considered in the articles with fuzzy approaches
- The dynamic disassembly operation and having the capability to adapt the operation in real-time is another reason for applying fuzzy logic and fuzzy reasoning-based methods
- Fuzzy Petri net and fuzzy logic simulation models are frequently used in the context of disassembly sequence planning
- For the case of complex products considering the stochastic nature of the variables and the value of the dynamic sequence planning, fuzzy approaches are promising.

4 Application of Augmented Reality in Disassembly

There are a few studies that addressed the applications of AR in disassembly via case studies. Here, a brief review is provided. Chang et al. [39] proposed an interactive augmented reality-based platform for designing the disassembly sequences. The authors used Unity with different modules including an optimization module for the sequences planning and a user interface based on AR for customization of the disassembly operations and creating the different scenarios. Chang et al. [40] developed an AR-guided platform for disassembly with three modules of assembly planning, content generation, and AR interface. Frizziero et al. [11] applied an AR-based model for Design for disassembly of the gearbox. Mircheski and Rizov [41] used the RFID tracking system and AR for improving the disassembly process. The architecture proposed by the authors has three phases: in phase 1, the sequences of the disassembly processes will be planned based on CAD data. Phase 2 includes the processing of RFID tags for each component and phase 3 provides an interactive user interface for visualization purposes. Osti et al. [42] used AR and CAD and an optimization model for disassembly sequence planning and a leap motion controller device. Tegeltija et al. [43] applied AR for disassembly of the heating circulation pump. The authors concluded that AR-based disassembly could reduce the disassembly time, the needed time for parts condition analysis and diagnosis as well as material flow selection for the recovered parts. The visualization and an interactive interface are the main advantages of AR-based model in disassembly. For the complex products, as the real-time feedback during the disassembly process is essential, the AR-based model could involve the users in the decision-making process, and they can customize the planning sequences based on the real-time data.

5 Fuzzy Approach and Augmented Reality in Disassembly

In this section, a system architecture is proposed based on the fuzzy approach and AR for disassembly of the complex products. The system architecture is shown in Fig. 3. This architecture has two essential modules including AR application and the fuzzy intelligent decision-making module. The details of these modules are explained in this section. AR provides an advanced simulation environment by combining virtual and real parts features for optimizing the disassembly operation. AR is used in aircraft

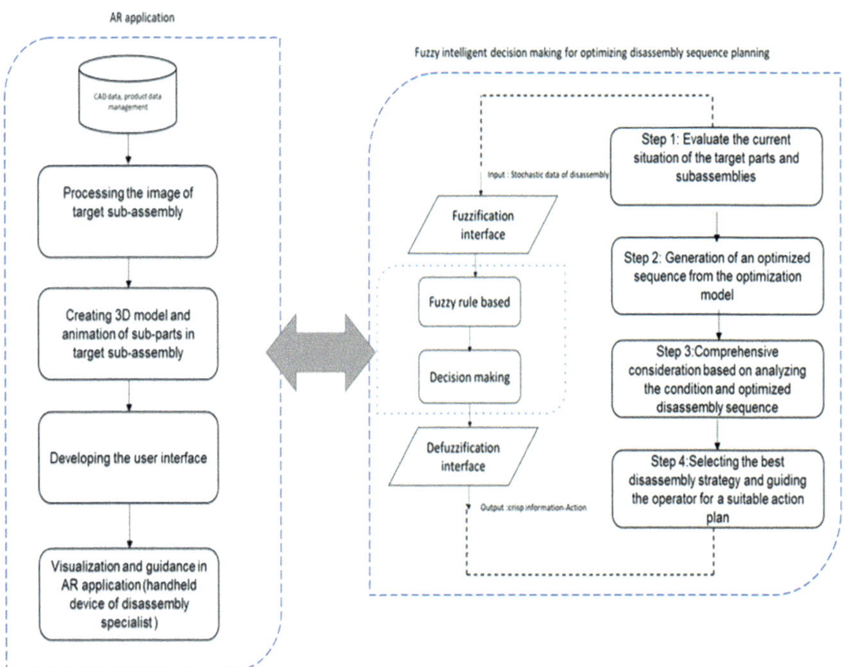

Fig. 3 The system architecture: AR-based fuzzy intelligent decision making for disassembly of a target sub-assembly

maintenance to give the position of the parts to the operator in an augmented environment [44–46]. According to Ceruti et al. [45], one of the problems that limit the application of AR is the required time for preparing the animation of the million parts. Generating a 3D maintenance manual from 2D maintenance operations is a time-consuming process for complex products with several sub-assemblies and parts. This problem has a solution in the case of disassembly of aircraft components with selecting target disassembly sequences and performing three essential processes of AR; tracking, rendering, and display when needed. For this aim, the handheld device of the operator should be connected to a remote center for creating automatic and graphic content. The operator could access the 3D models, animation, and historical data and can be guided based on an interactive decision-making module. Cetri et al. [44] proposed an AR-based user manual for the maintenance operation. The developed system includes a connected device for on-site operation, software management for providing 3D graphic content from 2D manual information, and a maintenance center for providing guidance operation by engineers and designers. The idea of collaborative disassembly operation for EoL aircraft is also proposed by Camelot et al. [19]. The advantage of a real-time interface is a dynamic operation that ensures the efficiency of the disassembly tasks. AR application includes the processing of the image of sub-assembly. The data will be inserted from the CAD database and product data management system. The second phase includes creating a 3D model and animation of sub-parts in the sub-assembly unit. The user interface will be developed by Unity by importing the animation of components and the disassembly specialist's features interface. The final step is visualization and importing the data into the operator's intelligent device. A manual disassembly is a common approach for a complex product. The disassembly time, labor costs, and skills and tooling play an essential role in disassembly operation costs. Hence, the ability to deal with vagueness and uncertainty during the operation is important in the decision-making process. The intelligent fuzzy decision-making system is an appropriate choice for sequence planning and the disassembly decision. The typical fuzzy decision-making system includes a fuzzification layer, fuzzy rule-based module, decision-making module, and defuzzification layer. The subassembly general condition, the condition of the parts, residual lifetime, the record of maintenance will be used in the first stage to evaluate the current situation of the target parts. Step 2 includes the generation of an optimized sequence from the optimization model. Step 3 includes a comprehensive consideration based on analyzing the condition and optimized disassembly sequence. Step 4 includes selecting the best disassembly strategy and guiding the operator for a suitable action plan.

6 Application Perspective

In this section, the application perspectives for the proposed architecture are discussed for the disassembly of the nose landing gear. The landing gear is one of the major components of retired aircraft with the highest recovered value. The maintenance cost of this component is also high. For the residual value of the scrap, the landing gear is also considered as a component with a high residual value [1]. The landing gear is a critical component in an aircraft and technology advancement plays an essential role in the design, development, and life cycle cost of this system [47, 4]. According to these authors, virtual tools, dynamic simulation, and advanced knowledge-based engineering tools are essential for reducing the design and development cycle time of this component. Wireless sensors and RFID technologies are used for condition monitoring and reducing operational and maintenance costs. Hence, this information could be used for the estimation of the residual lifetime of the landing gear and aiding in the EoL decision making process.

Some studies addressed the disassembly and maintenance operation of landing gear in the literature review. Table 1 shows these studies and provides highlighted points from their contributions.

The disassembly time, tooling cost, and tool options are the key variables that are considered in these studies. In developed virtual disassembly interfaces, the training and features are mainly considered. However, the stochastic nature of the disassembly time, accessibility to the tools, the conditions of the parts, and integrating the data of health monitoring and the maintenance data are not discussed. In this section,

Table 1 Disassembly of landing gear in different studies

Authors and Year	Topic of study	Approach	Results
Hao and Hongfu [48]	Disassembly sequence planning of Landing gear	Developing genetic/simulated annealing algorithm for optimization sequence planning	Increasing the efficiency of the search process in the optimization of disassembly sequences
Zhong et al. [22]	Disassembly sequence planning (case study: Landing gear)	Developing a Dijkstra's algorithm for optimizing the sequence of disassembly	Reducing the complexity of disassembly sequence planning and effectiveness in comparison to the genetic algorithm-based method
Liu et al. [49]	Virtual Disassembly for Landing Gear	using 3DMAX and Unity 3D	Reducing physical operation risk Reducing the cost of operation and improving the training
Zou et al. [27]	Virtual disassembly of landing gear	Applying Unity3D for virtual disassembly including selecting tools and disassembly operation	Increasing efficiency and decreasing the time of operation and training

using the secondary data, the implications of an AR-based fuzzy intelligent system for disassembly of an aircraft component are discussed. Figure 4 shows the graph of the component and the fastener of the nose landing gear.

The set of parts and fasteners is $\{P1, P2, P3, \ldots, P10\}$ and $\{F1, F2, F3, \ldots, F15\}$ respectively. The objective function for maximizing the disassembly profit can be defined as follows:

$$\begin{aligned} \max diss_{Profit}(Landing\ gear) = & \sum_{i=1}^{10} x_i \left(\widetilde{Vr_i} - \widetilde{Cr_i} - \widetilde{log1_i}\right) + \sum_{i=1}^{10} y_i \left(\widetilde{Rv_i} - \widetilde{log2_i}\right) \\ & - \sum_{i=1}^{10} z_i \widetilde{dis_i} - \sum_{j=1}^{15} m_j \widetilde{Fbc_j} - \sum_{j=1}^{14} m_j \widetilde{Sc_j} \end{aligned} \tag{1}$$

where

Vr	Recovered value of each part
Cr	The certification cost
$log1$	The logistics cost related to recovered parts
Rv	The residual value of each part
$log2$	The logistics cost related to residual parts
dis	The disposal cost of the part
Fbc	The cost of fasteners breaking
Sc	The cost of tooling and set-up cost
x, y, z	The binary variables, equal 1 if the part is selected to be recovered, if it is a residual part, or disposed respectively
m	The binary variable equal 1 if the fastener is selected
i, j, k	Indices for parts and fasteners respectively.

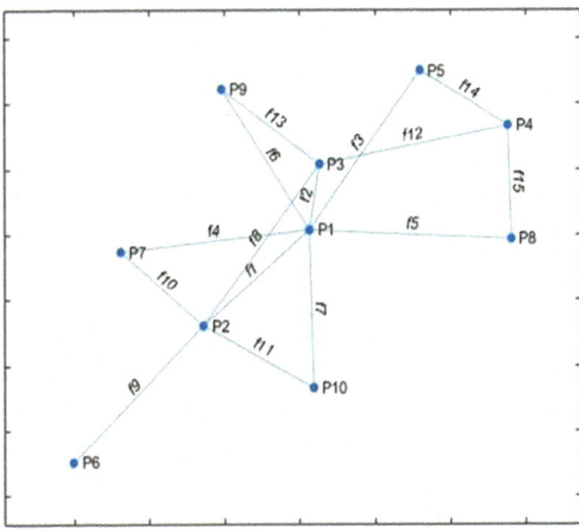

Parts	Name
P1	landing gear cabin
P2	shock strut
P3	drag strut
P4	lock stay
P5	downlock springs
P6	wheels
P7	shock strut actuator
P8	lock stay actuator
P9	front cabin door link
P10	back cabin door link

Fig. 4 Components and fastener of nose landing gear. *The Source of Data* Zhong et al. [22]

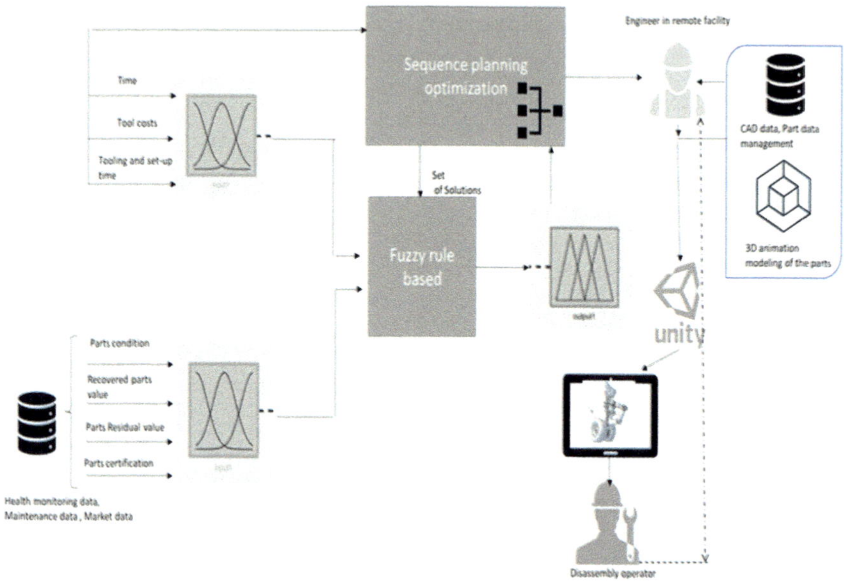

Fig. 5 Intelligent decision-making model for selecting disassembly strategy

Figure 5 shows the decision model for selecting the disassembly strategy. The fuzzy membership functions will be the interface of the fuzzy decision-making module for fuzzifications. Disassembly time and tool costs are the main variables for the disassembly sequence planning model. In addition to these variables, tooling cost and set-up time are also considered as stochastic variables. The other fuzzy module evaluates the condition of the parts and the estimation of the costs and market value. Health monitoring data, maintenance data, and the market database will be used in this section to extract the condition of the parts, the value of recovered parts, the residual value, and the certification status and costs. The set of solutions from the optimization module will be entered into a fuzzy rule-based module for integrated decision making considering the fuzzy variables and real-time value for dynamic decision making. The results will be entered into the defuzzification interface and the crisp results will be integrated into the optimization module for final optimum planning sequences. The results will be sent to engineers in a remote facility who has access to the part database and CAD data. The images of the parts during the disassembly operation via the handheld device of the operator will be sent to the engineer. These images and the result of the optimum sequence generated will be used to render 3D animation modeling of the parts. The results will be sent to the user interface developed by Unity3D for sending the AR-based view of the operation to the disassembly's operator onsite. This dynamic and data-driven decision-making process could be enhanced by integrating fuzzy variables. As the role of disassembly operators and their skills are essential, the intelligent fuzzy decision-making process allows utilizing their working experience in fuzzifying the input variables. The reason for using fuzzy membership functions is considering human-machine decision making that is essential in the disassembly operation. Considering the importance of time in disassembly, the operators should make some decisions based on their experiences. A fuzzy decision model and augmented reality could simplify and facilitates the dynamic disassembly process. It can reduce the time by avoiding unnecessary operations. The results of the optimization model could be modified accordingly based on updated information during the disassembly process.

7 Conclusion

As Industry 4.0 and cyber-physical systems become more popular, the role of realizing the virtual avatars of a physical system becomes growingly significant. The digital models offer advanced analytical features and visualization. Developing a simulated environment graphically provides the opportunity for testing the different scenarios and optimize the final model based on different parameters. For complex product recovery options, comparing the different dismantling and disassembling strategies with the aggregation of several dispersed databases, data analytics, and visualization dashboard are essential. This chapter introduced a joint application of a fuzzy decision-making model and AR for the disassembly of aircraft components. The uncertainties in the key variables of disassembly sequence planning and the conditions of the parts and disassembly situation are considered. A system architecture is proposed and the application perspective for the case of nose landing gear is discussed. Developing a digital twin model for disassembling is an interesting area for future research. Mapping the skeleton of the aircraft based on the material zones considering the complexity of the material mix is vital for optimizing the dismantling and cutting process. The different disassembling and dismantling strategies should be assessed based on the types of material, recyclability, replicability, and material scarcity. The digital twin could provide a digital mapping capability enabled with multidimensional data processing, optimization, visualization, intelligent decision making for delivering the optimum dismantling solutions. The essence of the digital twin model is comparing the different dismantling and disassembling strategies with the aggregation of several dispersed databases, data analytics, and visualization dashboard. The digital twin provides a dynamic digital profile of a physical object and integrates 3D model, historical and real-time monitoring of the object behavior for optimizing the virtual reality.

References

1. IATA Report. https://www.iata.org/contentassets/ffbed17ac843465aad778867cb23c45c/bipad.pdf (2018)

2. Guyon, I., Amine, R., Tamayo, S., Fontane, F.: Analysis of the Opportunities of Industry 4.0 in the Aeronautical Sector (2019)
3. Junior, A.L., Lemos, G.F.C., Trabasso, L.G.: Proposal of a method for the implementation of the Industry 4.0-aircraft final assembly domain. In: FT2019. Proceedings of the 10th Aerospace Technology Congress, October 8–9, 2019, Stockholm, Sweden (No. 162, pp. 199-209). Linköping University Electronic Press
4. Barbosa, G.F., Aroca, R.V.: Advances of Industry 4.0 concepts on aircraft construction: an overview of trends. J Steel Struct Constr 2472-0437 (2017)
5. Keivanpour S.: End of Life Management of Complex Products in an Industry 4.0 Driven and Customer-Centric Paradigm: A Research Agenda, accepted in MOSIM 2020 (2020)
6. Vijayan, K.K., Mork, O.J.: IdeaLab: a learning factory concept for norwegian manufacturing SME. Procedia Manuf. **45**, 411–416 (2020)
7. Tvenge, N., Olga, O., Niels, P. Ø., and Kristian, M.: Added value of a virtual approach to simulation-based learning in a manufacturing learning factory. Procedia CIRP**88**, 36–41 (2020)
8. Mourtzis, D., Angelopoulos, J., Dimitrakopoulos, G.: Design and development of a flexible manufacturing cell in the concept of learning factory paradigm for the education of generation 4.0 engineers. Procedia Manuf. **45**, 361–366 (2020)
9. Mourtzis, D., Siatras, V., Angelopoulos, J., Panopoulos, N.: An augmented reality collaborative product design cloud-based platform in the context of learning factory. Procedia Manuf. **45**, 546–551 (2020)
10. Kerin, M., Pham, D.T.: A review of emerging industry 4.0 technologies in remanufacturing. J. Cleaner Prod. 117805 (2019)
11. Frizziero, L., Liverani, A., Caligiana, G., Donnici, G., Chinaglia, L.: Design for disassembly (DfD) and augmented reality (AR): case study applied to a gearbox. Machines **7**(2), 29 (2019)
12. Wang, C.H., Tsai, N.H., Lu, J.M., Wang, M.J.J.: Usability evaluation of an instructional application based on Google Glass for mobile phone disassembly tasks. Appl. Ergon. **77**, 58–69 (2019)
13. Xie, Y., Zhang, Y., Cai, Y.: Virtual reality engine disassembly simulation with natural hand-based interaction. In: VR, Simulations and Serious Games for Education, pp. 121–128. Springer, Singapore (2019)
14. Rocca, R., Rosa, P., Sassanelli, C., Fumagalli, L., Terzi, S.: Integrating virtual reality and digital twin in circular economy practices: a laboratory application case. Sustainability **12**(6), 2286 (2020)
15. Lee, M.C.M.: Augmented Reality Applications In Product Disassembly. Doctoral Dissertation (2019)
16. Hassan, S., Yoon, J.: Haptic guided optimized aircraft maintenance assembly disassembly path planning scheme. In ICCAS 2010, pp. 1667–1672. IEEE (2010)
17. Yoon, J.: Intelligent assembly/disassembly system with a haptic device for aircraft parts maintenance. In: International Conference on Computational Science, pp. 760–767. Springer, Berlin, Heidelberg (2007)
18. Dayi, O., Afsharzadeh, A., Mascle, C.: A Lean based process planning for aircraft disassembly. IFAC-PapersOnLine **49**(2), 54–59 (2016)
19. Camelot, A., Baptiste, P., Mascle, C.: Decision support tool for the disassembly of reusable parts on an end-of-life aircraft. In: Proceedings of 2013 International Conference on Industrial Engineering and Systems Management (IESM), pp. 1–8. IEEE (2013a)
20. Camelot, A., Mascle, C., Baptiste, P.: Disassembly of spare parts on an EOL aircraft. In: 2013 IEEE International Symposium on Assembly and Manufacturing (ISAM), pp. 135–137. IEEE (2013b)
21. Tang, X.M., Zhong, S.S.: Petri nets based aircraft maintenance disassembly and assembly process planning. J. Civ. Aviat. Univ. China **5** (2006)
22. Zhong, L., Youchao, S., Gabriel, O.E., Haiqiao, W.: Disassembly sequence planning for maintenance based on metaheuristic method. Aircraft Engineering and Aerospace Technology (2011)

23. Zhang, K.F., Zhao, L., Li, Y., Shao, Y.: Effective component disassembly approach for aircraft assembly based on fuzzy-clustering algorithm. In: 2008 IEEE/ASME International Conference on Advanced Intelligent Mechatronics, pp. 850–855. IEEE (2008)
24. Zahedi, H., Mascle, C., Baptiste, P.: Advanced airframe disassembly alternatives; an attempt to increase the afterlife value. Procedia CIRP **40**, 168–173 (2016)
25. Sabaghi, M., Cai, Y., Mascle, C., Baptiste, P.: Sustainability assessment of dismantling strategies for end-of-life aircraft recycling. Resour. Conserv. Recycl. **102**, 163–169 (2015)
26. Sabaghi, M., Cai, Y., Mascle, C., Baptiste, P.: Towards a sustainable disassembly/dismantling in aerospace industry. Procedia CIRP **40**, 156–161 (2016)
27. Zou, F., Chen, Y., Wei, K.: Landing gear virtual disassembly and assembly system based on Unity3D. In: 2019 IEEE 4th Advanced Information Technology, Electronic and Automation Control Conference (IAEAC), vol. 1, pp. 2745–2748. IEEE (2019)
28. Zhou, Z., Liu, J., Pham, D.T., Xu, W., Ramirez, F.J., Ji, C., Liu, Q.: Disassembly sequence planning: recent developments and future trends. Proc. Inst. Mech. Eng. Part B J. Eng. Manuf. **233**(5), 1450–1471 (2019)
29. Zhou, Z., Jiayi, L., Pham D T., Xu, W., Javier Ramirez, F., Chunqian, Ji., and Quan, L.: Disassembly sequence planning: Recent developments and future trends. Proceedings of the Institution of Mechanical Engineers, Part B: Journal of Engineering Manufacture **233**(5) 1450-1471 (2019)
30. Gao, M., Zhou, M., Tang, Y.: Intelligent decision making in disassembly process based on fuzzy reasoning Petri nets. IEEE Trans. Syst. Man Cybern. Part B (Cybernetics) **34**(5), 2029–2034 (2004)
31. Tang, Y., Turowski, M.: Adaptive fuzzy system for disassembly process planning with uncertainty. J. Chin. Inst. Ind. Eng. **24**(1), 20–29 (2007)
32. Zhao, S.E., Li, Y.L., Fu, R., Yuan, W.: Fuzzy reasoning Petri nets and its application to disassembly sequence decision-making for the end-of-life product recycling and remanufacturing. Int. J. Comput. Integr. Manuf. **27**(5), 415–421 (2014)
33. Hsu, H.P.: A fuzzy knowledge-based disassembly process planning system based on fuzzy attributed and timed predicate/transition net. IEEE Trans. Syst. Man Cybern. Syst. **47**(8), 1800–1813 (2016)
34. Tian, G., Zhou, M., Li, P.: Disassembly sequence planning considering fuzzy component quality and varying operational cost. IEEE Trans. Autom. Sci. Eng. **15**(2), 748–760 (2017)
35. Galantucci, L.M., Percoco, G., Spina, R.: Assembly and disassembly planning by using fuzzy logic & genetic algorithms. Int. J. Adv. Robot. Syst. **1**(2), 7 (2004)
36. Gonnuru, V.K.: Disassembly planning and sequencing for end-of-life products with RFID enriched information. Robot. Comput. Integr. Manuf. **29**(3), 112–118 (2013)
37. Xiaoyong, P., Guanghong, D., Dong, X., Peng, M.: Intelligent disassembly sequence planning for EOL recycling based on hierarchical fuzzy cognitive map. In: Proceedings of the 2005 IEEE International Symposium on Electronics and the Environment, pp. 255–259. IEEE, 2005
38. Feng, Y., Zhou, M., Tian, G., Li, Z., Zhang, Z., Zhang, Q., Tan, J.: Target disassembly sequencing and scheme evaluation for CNC machine tools using improved multiobjective ant colony algorithm and fuzzy integral. IEEE Trans. Syst. Man Cybern. Syst. **49**(12), 2438–2451 (2018)
39. Chang, M.M.L., Nee, A.Y.C., Ong, S.K.: Interactive AR-assisted product disassembly sequence planning (ARDIS). Int. J. Prod. Res. 1–16 (2020)
40. Chang, M.M.L., Ong, S.K., Nee, A.Y.C.: AR-guided product disassembly for maintenance and remanufacturing. Procedia Cirp **61**, 299–304 (2017)
41. Mircheski, I., Rizov, T.: Improved nondestructive disassembly process using augmented reality and RFID product/part tracking. TEM J. (2017)
42. Osti, F., Ceruti, A., Liverani, A., Caligiana, G.: Semi-automatic design for disassembly strategy planning: an augmented reality approach. Procedia Manuf. **11**, 1481–1488 (2017)
43. Tegeltija, S.S., Lazarević, M.M., Stankovski, S.V., Ćosić, I.P., Todorović, V.V., Ostojić, G.M.: Heating circulation pump disassembly process improved with augmented reality. Therm. Sci. **20**(suppl. 2), 611–622 (2016)

44. Ceruti, A., Liverani, A., Marzocca, P.: A 3D User and Maintenance Manual for UAVs and Commercial Aircrafts Based on Augmented Reality. SAE Technical Paper 2015-01-2473 (2015)
45. Ceruti, A., Marzocca, P., Liverani, A., Bil, C.: Maintenance in aeronautics in an industry 4.0 context: the role of augmented reality and additive manufacturing. J. Comput. Des. Eng. **6**(4), 516–526 (2019)
46. Yong, S.W., Sung, A.N.: A mobile application of augmented reality for aircraft maintenance of fan cowl door opening. Int. J. Comput. Netw. Inf. Secur. **11**(6), 38 (2019)
47. Divakaran, V.N., Kumar, G.R., Rao, P.S.: Aircraft Landing Gear Design and Development (2015). https://www.infosys.com/engineering-services/white-papers/documents/landing-gear-design-and-development.pdf
48. Hao, W., Hongfu, Z.: Using genetic annealing simulated annealing algorithm to solve disassembly sequence planning. J. Syst. Eng. Electron. **20**(4), 906–912 (2009)
49. Liu, T., Chen, M., Wang, Y.: Design and research of virtual disassembly system for aircraft landing gear. In: 2nd International Conference on Computer Engineering, Information Science & Application Technology (ICCIA 2017). Atlantis Press (2016)

Some Novel Preference Relations for Picture Fuzzy Sets and Selection of 3-D Printers in Aviation 4.0

Fatma Kutlu Gündoğdu and Shahzaib Ashraf

Abstract Nowadays, one of the most significant players in the worlds' economy is the aviation sector. This sector almost supports 87.7 million people who work in the aviation industry. Thus, this sector is rapidly developing due to high demand by customers and employees. Picture fuzzy sets (PFSs) are the latest extension of the ordinary fuzzy sets (FSs). The main characteristic of the picture fuzzy sets is satisfying the condition that the sum of the positive, negative and neutral membership degrees must be at least zero and at most one. Preference relations have generally been identified as a significant strategy for expressing the interests of decision-makers over alternatives in the decision-making procedure. In this research, novel picture preference relations are established under the picture fuzzy environment. The backbone of this research, picture fuzzy preference relation, consistent picture fuzzy preference relations, incomplete picture fuzzy preference relations, consistent incomplete picture fuzzy preference relations, and acceptable incomplete picture fuzzy preference relations are established. Finally, some ranking and selection algorithms are established using these preference relations for decision-making (DM) alternatives. A few numerical examples on selection of 3D printers in aviation sector demonstrate the applicability and validity of the proposed approach.

1 Introduction

In the case of DM problems, for each criterion, comparisons between alternatives are more adaptable to DMs than allocating assessment values for all alternatives. Preference Relations (PRs) are an important and standard method for evaluating

F. Kutlu Gündoğdu (✉)
Department of Industrial Engineering, National Defence University, Turkish Air Force Academy, Istanbul, Turkey
e-mail: kgundogdu@hho.edu.tr

S. Ashraf
Department of Mathematics, Abdul Wali Khan University, Mardan, Pakistan
e-mail: shahzaibashraf@awkum.edu.pk

C. Kahraman and S. Aydın (eds.), *Intelligent and Fuzzy Techniques in Aviation 4.0*,
Studies in Systems, Decision and Control 372,
https://doi.org/10.1007/978-3-030-75067-1_12

DMPs, through which decision-makers express their preference for alternatives by pair comparisons. Furthermore, it has been observed that establishment of a PR by pairly comparisons between alternatives is much more efficient and accurate than comparisons that are not pairwise [52]. Multiplicative [34, 57] and fuzzy preference relations [23, 53] are two most common types of the preference relation. Saaty [44] has developed a classical AHP technique that focuses on the multiplicative relationship of preference, that used a linguistic scale to reflect decisions taken by paired comparisons of alternatives, where all decisions are of crisp quality. However, deterministic qualities can't express the complexity of real-life DM information. Consequently, Orlovsky [39] established a fuzzy preference relationship (FPR) where each component has assigned a degree of membership from 0 to 1. Since, negative membership can not be considered by FSs, although IFSs can compensate this weakness; therefore, Xu [54] developed intuitionistic fuzzy preference relations (IFPRs). Furthermore, Mandal and Ranadive [33] improved the IFPRs and proposed some PRs under Pythagorean fuzzy environments. Li et al. [29] strengthen the PRs to developed q-rung othopair fuzzy (q-ROF) based list of PRs to tackle the uncertainty in selection of alternatives for decision makers. Utilizing these category of methodology, DMs could characterize their ambiguous cognitions from a positive, negative, and hesitant point of view when attempting to compare alternatives. Although q-ROFPRs can communicate DM preferences more comprehensively and flexibly than intuitionistic and Pythagorean fuzzy PRs, there are still some limitations in q-ROFPRs that go beyond the capability of q-rung orthopair fuzzy numbers (q-ROFNs) to manage uncertain data that may be usual in complex decision-making circumstances.

Additive manufacturing (AM) indicates to the techniques used for creating 3D objects by adding materials. This term contrasts with the subtractive manufacturing techniques in which material is removed from a raw material block until the desired geometry is obtained. Additive manufacturing, which is also commonly designated as rapid prototyping or 3D printing, is the preferred term used in the academic community. It eliminates the notion that the producing pieces obtained using these techniques are limited to creating archetypes.

AM draws considerable attention among industrial and academic practitioners because of several possible improvements in production systems to reduce waste materials, get shorter manufacturing lead times, obtain high flexibility, feasibility of complex geometry products, and shorter product development cycle [12, 17, 22, 42]. The market demand for highly customized objects, together with the introduction of new business models and leaner supply chains, encourages AM technologies [24, 47]. According to the annual worldwide report on AM technologies' progress, the global revenue from AM production and incorporated services grew from 2.25 billion dollars in 2012 to more than 6 billion dollars in 2016. The forecast is to reach 21 billion by 2020 [50]. This enlargement is also owing to the development in technology and material development. Nowadays, AM is advancing towards becoming a proper industrialized platform that can produce the form, fit, and function of a part [15]. Hence, the selection of appropriate 3D printers based on objectives in aerospace industry is the key issue. To the best of our knowledge, there is no enough decision-making algorithms for selection of 3D printers under fuzzy environment in the literature.

In order to address the deficiencies identified above, this paper presented novel list of picture fuzzy preference relations (PFPRs). In real decision-making scenarios, PFPRs are more common and can handle more DM preference data than q-ROFPRs. For instance, assume that teams are appointed by key stakeholders to respond as DMs (reviewers) in order to overcome a massive selection of 3D printers; then, samples are collected over two shifts. Initially, stakeholders are requested to express their preference to any pair ∂_i and ∂_j of alternatives with respect to each criterion. Due to the limited information and indeterminacy inherent in the behavior of the DMs, it is much more likely that the PFN will be assessed by voting in favour, voting against or abstaining from each index test. Within that manner, the final preference relations could be achieved with respect to the criteria as allocated by PFNs [19]. The information of preferences is considerably more extensive and accurate if described in this way. The experts also conduct the survey anonymously and do not interact with each other so as not to influence one another. Using such strategy, the information found in the evaluation provided by PFNs is composed of components on $\mu(\upsilon)$, $\wp(\upsilon)$ and $\partial(\upsilon)$. We will concentrate our focus on this topic and establish novel preference relations under PFSs. In general, the proposed preference relations have a better modeling technique than the IFPRs and q-ROFPRs for evaluating uncertain information in the decision-making scenario. In this chapter the proposed preference relations have been applied to the selection of 3D printers in aviation sector.

To achieve the list of goals the structure of the chapter is arranged as: Sect. 2 presents the literature review of previous studies regarding 3D printing technology. In Sect. 3, some basic concepts of fuzzy sets are introduced. In Sect. 4, gave the main contribution of the chapter, introduced the picture fuzzy preference relations. Section 5, proposed the ranking and selection algorithms based on the proposed preference relations and numerical example are demonstrated to shows the applicability and effectiveness of the proposed technique. Concluded remarks are discussed in Sect. 6.

2 Literature Review: Aviation 4.0

Nowadays, the aerospace industry has been one of the most important sectors in the adopting of AM technologies. Figure 1 shows the application areas of additive manufacturing around the world. As you seen, industrial machines, aerospace and motor vehicles are the top leading sectors to adopt AM technologies [36]. Aerospace companies were some of the first to adopt additive manufacturing. Some of the most rigid industry performance standards exist in this area, requiring parts to hold up in harsh conditions. Engineers designing and manufacturing for commercial and military aerospace platforms need flight-worthy components made from high-performance materials.

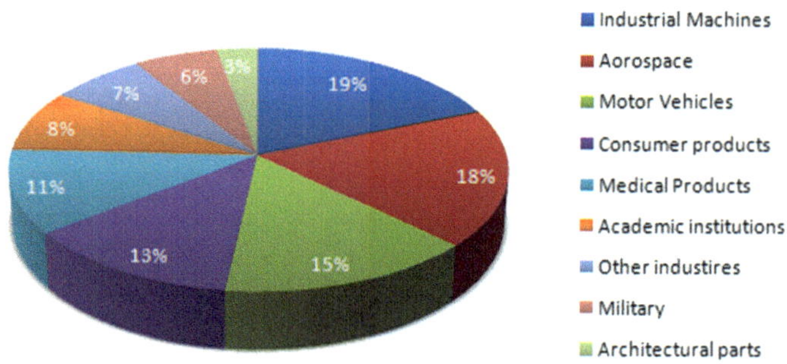

Fig. 1 Application areas of additive manufacturing

Another important application area of 3D printers is the medical industry. The rapidly innovating medical industry utilizes additive manufacturing solutions to deliver breakthroughs to doctors, patients, and research institutions. Medical manufacturers employ a wide range of high-strength and bio-compatible 3D printing materials, from inflexible to flexible and opaque to transparent, to customize designs like never before.

The transportation industry is one of the top sectors in terms of the application of additive manufacturing. To avoid unnecessary drag, it needs parts that stand up to harsh testing and are lightweight enough. With a wide array of strong, high-temperature materials and the ability to build very complex geometries, transportation businesses are just scratching the surface of what can be manufactured for their vehicles.

Success in the energy sector relates to developing customized, mission-critical components that can withstand extreme conditions. Additive manufacturing's advancements in producing efficient, lightweight components and environmentally friendly materials provide answers for various requirements and field functions.

The time it takes to form an idea and deliver it to the market is everything for designers and marketing teams. Simulating the final product's look and feel during design reviews to demonstrate ideas to key stakeholders quickly. Consumer product producers have adopted 3D printing to help develop iterations and quickly adjust the design.

Indeed, one of the main drivers of the aircraft industry's improvement is the production of complex design and weight reduction of flying components. Aviation experts perceived the cost-saving reason as the most important after complexity-for-free and reduction weight. The cost-saving item does not have this priority on the other sectors except the 3D printing platforms. In manufacturing cost saving, it is worth noting that the aircraft industry's structural materials must be highly performant—such as Titanium alloys or composite materials—and thus quite costly. An exciting opportunity is to increase creativity and produce rapid and economic prototyping. Figure 2 indicates the ranking of main reasons for adopting AM technologies in aviation industry.

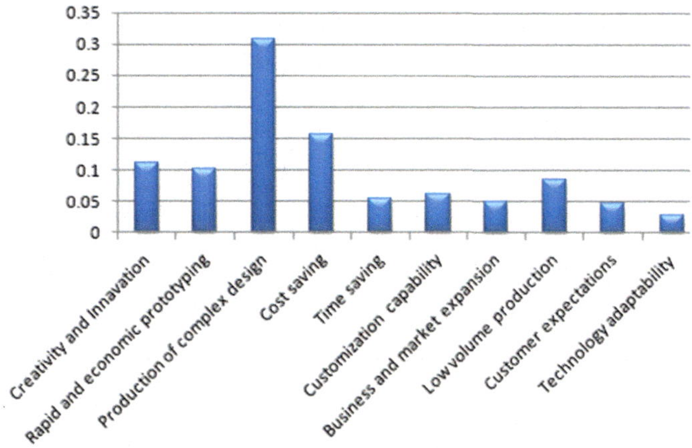

Fig. 2 The ranking of main reasons for adopting AM technologies in aviation sector

Additive manufacturing techniques generally can be classified into seven categories. The categories are vat photopolymerization, material jetting, binder jetting, material extrusion, powder bed fusion, sheet lamination, and directed energy deposition. The short definitions of these techniques are as follows:

Vat photopolymerization: In this process, a light beam in the ultraviolet (UV) spectrum is used to reticulate materials with high molecular weight (polymers) from monomer molecules or oligomers. Thus, it transforms liquid photopolymers into solid objects by applying light beams of a specific wavelength [25, 28]. It is the base technology for several commercial photopolymer technologies and is currently used in various sectors such as arts, optical adhesion, microelectronics, and the dental, medical, and biomedical fields [21]. Stereolithographic (SLA) printers were developed in the USA [21] based on vat photopolymerization in 1980 and causes controlled solidification of the liquid resins via photopolymerization by directly using the light beams from a laser or through a projection obtained using an optical configuration.

Material jetting: The material jetting technology was developed in 1984 for building 3D objects using a mobile inkjet printhead that directly injects a photopolymer in a drop form into the platform. The drops are selectively injected in a specific shape using one or more inkjet print heads [6]. The main commercial applications for inkjet printing remain in graphics, product marking, coding, and dating. In recent years, there has been considerable interest in technological areas, such as aerospace and defense, architecture, commercial products, consumer products, automotive, medical. Over the last decade, aerospace innovators have pushed 3D printing beyond the realm of prototyping into advanced tooling and production applications.

Binder jetting: Massachusetts Institute of Technology developed binder jetting in the 1990s, binder jetting facilitates the fabrication of pieces using powder materials such as sand, glass, and metals. The principal parts of the binder jetting process include the building platform, material feed platform, roll, dry unity, and printhead

[5, 11]. The main drawback of binder jet printing (BJP) process is known to produce parts not suitable for most structural applications [11].

Fused deposition modeling: The material extrusion techniques lead a material to selectively pass through an extrusion die due to the applied pressure, resulting in an extruded material having a uniform cross section and the desired stroke diameter. The first mechanism developed using this technique was fused deposition modeling (FDM), which was created, patented, and commercialized by Stratasys in 1992 [10, 35]. FDM involves placing a thin thread of a thermoplastic material on a printing platform. The process is initiated by heating this material until it reaches the melting point. Subsequently, the material is poured on a printing platform with a predefined pattern until the first layer of the desired object is completed. The most important benefits of this method are to prove low cost, high speed and simplicity. This method can only work with limited materials such as thermoplastics. It has also week mechanical properties. Its' resolution range is between 50 and 200 μ [49].

Powder bed fusion: This technique begins by transferring a layer of powder material from the delivery system to the building platform. An energy source is programmed to dose energy directly to the platform surface for fusing and sintering the material with a desired shape. Further, the process is repeated until the desired piece is obtained [12]. Selective laser melting (SLM) [4] selective laser sintering (SLS), and electron beam melting (EBM) [43] are based on the powder bed fusion technique and represent a large part of the 3D printer market [8]. This method presents us fine resolution and high quality. When we select this technique, we should know that we will face some problems such as slow printing, expensive steps, high porosity in the binder method. Its resolution range is greater than 80 and less than 250 μm. [16].

Sheet lamination: The sheet lamination technique can be used to create 3D objects from successive cuts and lamination of the two dimensional sections [41]. The sheet lamination machines can be categorized as rolled and unrolled systems connected by a sheet of material guided by the rollers. This system feeds and stores materials simultaneously [20]. In addition, it has a build platform (that can perform vertical movements), a laser, and a roller [30]. The main benefits of this technique are reducing tooling and manufacturing time, a huge range of materials, low cost, and excellent for manufacturing of large structures. The main drawback of it is the limitation in manufacturing complex shapes. Resolution range is depend on the thickness of the laminates.

Directed energy deposition: The directed energy deposition technique forms the basis for technologies such as laser engineered net shaping, directed light fabrication, ion fusion formation, direct metal deposition, and 3D laser cladding [45]. These technologies are used to fix and add material to the already fabricated pieces; however, they can also be used to fabricate entire pieces from scratch [13]. This technique' advantages are reducing manufacturing time and cost, controlling micro structures, and being excellent repair and retrofitting. Nevertheless, this technique can cause the low accuracy and low surface quality. It has also limited to print complex shapes with fine details [14]. Additive Manufacturing techniques and application areas are briefly given in Table 1.

Table 1 Additive manufacturing methods and application areas [37]

Methods	Materials	Applications
Fused deposition modeling	Continuous filaments of thermoplastic polymers, continuous fiber-reinforced polymers	Rapid prototyping, toys, advanced composite parts
Powder bed fusion	Compacted fine powders, metals, alloys, limited polymers (SLS or SLM), ceramic, polymers	Biomedical, aerospace, lightweight structures
Material Jetting (Inkjet printing)	A concentrated dispersion of particles in a liquid, ceramic, concrete, and soil	Biomedical, large structures, buildings
Direct energy deposition	Metals and alloys in the form of powder, ceramics, and polymers	Aerospace, retrofitting, repair, cladding, biomedical
Sheet Lamination	Polymer composites, ceramics, paper, metal-filled tapes, and metal rolls	Paper manufacturing, foundry industries, electronics, smart structures

One of the advantages of AM technology is that it has more degree of freedom in design and utility when we compared with traditional techniques in aviation. Hence, complexity can be added to without any additional cost resulting in a range of design possibilities [31].

Design flexibility leads to reduce weight during the production stage. In the aviation industry, light weighting results in real and tangible benefits. It enables fuel savings and allows the greater weight to be apportioned for mission-critical cargo [7]. Lightweight means fewer materials are being used; thus, this technology enables financial savings associated with reducing waste across processes.

Prototyping and agility are other capabilities provided by additive manufacturing technology [48]. General Electric (GE) noted that these abilities are the main reasons AM was selected for the Federal Aviation Administration since they achieved quickly, inexpensively, and iteratively [38].

Current Additive manufacturing costs are higher than traditional manufacturing costs considering the spare parts' availability. The limitations of adopting additive manufacturing; in other words, 3D printing, are the operating and acquisition costs in the aviation sector [46]. The critically of the component, manufacturing feasibility, regulations, including initial and continuing airworthiness, are critical factors. All factors should be considered before changing the techniques from a traditional to an AM production process [32].

The use of 3D printing is a multi-criteria decision-making problem. Human evaluations about preferences are always very imprecise and subjective. Since the information may be insufficient, the main factors influencing 3D printing adoption are challenging to identify. The relative difference in importance between any two objectives is difficult to confirm. Moreover, the usability of 3D printers has intolerable costs in the aviation industry. The selection of right additive manufacturing techniques, raw materials, and the process has a critical impact on reducing mentioned costs. Thus, this chapter aims to develop the picture fuzzy preference relations to handle uncertainty and propose the decision-making algorithm based on these relations. The methodology is applied to an illustrative example that is the selection of 3D printing under impreciseness information.

3 Preliminaries

Let us briefly recall in this segment the rudiments of FSs and picture FSs. For the following review these definitions will be included in the here.

Definition 1 [55] A fuzzy set (FS) F in a universe set U is an object having the form

$$F = \{\langle \upsilon, \mu(\upsilon)\rangle \mid \upsilon \in U\},$$

where $\mu(\upsilon) \in [0, 1]$ is represented by the positive membership grade.

Definition 2 [3] A intuitionistic FS F_s in a universe set U is an object having the form

$$F_s = \{\langle \upsilon, \mu(\upsilon), \partial(\upsilon)\rangle \mid \upsilon \in U\},$$

where $\mu(\upsilon) \in [0, 1]$ positive and $\partial(\upsilon) \in [0, 1]$ negative membership grades repectivity. In addition $0 \leq \mu(\upsilon) + \partial(\upsilon) \leq 1, \forall\ \upsilon \in U$.

Definition 3 [9] A picture FS F_s in a universe set U is an object having the form

$$F_s = \{\langle \upsilon, \mu(\upsilon), \wp(\upsilon), \partial(\upsilon)\rangle \mid \upsilon \in U\},$$

where $\mu(\upsilon) \in [0, 1]$ positive, $\wp(\upsilon) \in [0, 1]$ neutral and $\partial(\upsilon) \in [0, 1]$ negative membership grades respectively. In addition $0 \leq \mu(\upsilon) + \wp(\upsilon) + \partial(\upsilon) \leq 1, \forall\ \upsilon \in U$.

In what follows, we signify by $PFS(U)$ the family of all picture FSs. We shall signify the picture fuzzy number (PFN) with the triplet $F_{ts} = (\mu(\upsilon), \wp(\upsilon), \partial(\upsilon))$ for simplicity.

Definition 4 [1] Suppose that for any $F_{ts_{(1)}}, F_{ts_{(2)}} \in PFS(U)$.
(1) $F_{ts_{(1)}} \subseteq F_{ts_{(2)}}$ if and only if $\mu_1 \leq \mu_2, \wp_1 \leq \wp_2$ and $\partial_1 \geq \partial_2$. Clearly $F_{ts_{(1)}} = F_{ts_{(2)}}$ if $F_{ts_{(1)}} \subseteq F_{ts_{(2)}}$ and $F_{ts_{(2)}} \subseteq F_{ts_{(1)}}$.
(2) $F_{ts_{(1)}} \cap F_{ts_{(2)}} = \{\min(\mu_1, \mu_2), \min(\wp_1, \wp_2), \max(\partial_1, \partial_2)\}$,
(3) $F_{ts_{(1)}} \cup F_{ts_{(2)}} = \{\max(\mu_1, \mu_2), \min(\wp_1, \wp_2), \min(\partial_1, \partial_2)\}$,
(4) $F^c_{ts_{(1)}} = \{\partial_1, \wp_1, \mu_1\}$.

Definition 5 [1] Let $F_{ts_{(1)}}, F_{ts_{(2)}} \in PFS(U)$ with $\varrho > 0$. The operating laws are as follows:
(1) $F_{ts_{(1)}} \otimes F_{ts_{(2)}} = \{\mu_1\mu_2, \wp_1\wp_2, \partial_1 + \partial_2 - \partial_1\partial_2\}$;
(2) $F_{ts_{(1)}} \oplus F_{ts_{(2)}} = \{\mu_1 + \mu_2 - \mu_1\mu_2, \wp_1\wp_2, \partial_1\partial_2\}$;
(3) $F_{ts_{(1)}}^{\varrho} = \left\{\mu_1^{\varrho}, \wp_1^{\varrho}, 1 - (1 - \partial_1)^{\varrho}\right\}$;
(4) $\varrho \cdot F_{ts_{(1)}} = \left\{1 - (1 - \mu_1)^{\varrho}, \wp_1^{\varrho}, \partial_1^{\varrho}\right\}$.

Definition 6 [1] Let $F_{ts_{(1)}} = (\mu_1, \wp_1, \partial_1) \in PFS(U)$. Then, the score values $\hat{S}c$ is described as

$$\hat{S}c\left(F_{ts_{(1)}}\right) = \mu_1 - \wp_1 - \partial_1$$

where $\hat{S}c\left(F_{ts_{(1)}}\right) \in [0, 1]$.

Definition 7 [1] Let $F_{ts_{(1)}} = (\mu_1, \wp_1, \partial_1)$ & $F_{ts_{(2)}} = (\mu_2, \wp_2, \partial_2) \in PFS(U)$. $\hat{S}c\left(F_{ts_{(1)}}\right)$ and $\hat{S}c\left(F_{ts_{(2)}}\right)$ are the score value of $F_{ts_{(1)}}$ and $F_{ts_{(2)}}$, and $\hat{A}c\left(F_{ts_{(1)}}\right)$ and $\hat{A}c\left(F_{ts_{(2)}}\right)$ are the accuracy value of $F_{ts_{(1)}}$ and $F_{ts_{(2)}}$ respectively.
(1) If $\hat{S}c\left(F_{ts_{(1)}}\right) < \hat{S}c\left(F_{ts_{(2)}}\right)$, then $F_{ts_{(1)}} < F_{ts_{(2)}}$.
(2) If $\hat{S}c\left(F_{ts_{(1)}}\right) = \hat{S}c\left(F_{ts_{(2)}}\right)$, then
(a) if $\hat{A}c\left(F_{ts_{(1)}}\right) < \hat{A}c\left(F_{ts_{(2)}}\right)$, then $F_{ts_{(1)}} < F_{ts_{(2)}}$
(b) $\hat{A}c\left(F_{ts_{(1)}}\right) = \hat{A}c\left(F_{ts_{(2)}}\right)$, then $F_{ts_{(1)}} = F_{ts_{(2)}}$.

4 Picture Fuzzy Preference Relations and Incomplete Picture Fuzzy preference Relations

In order to illustrate the expert preferences, Orlovsky [39] established a fuzzy preference relationship using a unit interval [0, 1]. Although it is nearly impossible for experts to provide appropriate value due to the complex nature of the DMPs and the limited expertise of the experts, which reduces the application. Fuzzy preference relations are a reasonable choice to identify the uncertain and ambiguous decisions of the experts. In this section, picture fuzzy preference relation, the consistent picture fuzzy preference relation, the incomplete picture fuzzy preference relation, the consistent incomplete picture fuzzy preference relation and the acceptable incomplete picture fuzzy preference relation are proposed.

4.1 Picture Fuzzy Preference Relations

Definition 8 Let $V = \{\nu_1, \nu_2, ...\nu_n\}$ be the collection of the alternatives. A matrix $M = (m_{pq})_{n\times n} \subset V \times V$ is known to be the picture fuzzy preference relation (PFPR) on V, $m_{pq} = \left(\mu_{pq}, \wp_{pq}, \partial_{pq}\right)$ is the PFN with $p, q \in \mathbb{N}$. In addition, the following condition is satisfy:

(a) $\mu_{pq} \in [0, 1]$, $\wp_{pq} \in [0, 1]$ and $\partial_{pq} \in [0, 1]$,
(b) $\mu_{pq} + \wp_{pq} + \partial_{pq} \leq 1$,
(c) $\mu_{qp} = \partial_{pq}$, $\partial_{qp} = \mu_{pq}$, $\wp_{pq} = \wp_{qp}$ and $\mu_{pp} = \partial_{pp} = 0.5$, $\wp_{pp} = 0$
where μ_{pq}, $\wp_{pq}$ and ∂_{pq} known to be preferred, neutral and non prefered grades of the alternative ν_p over ν_q.

Theorem 1 *Let $M = (m_{pq})_{n\times n}$ to be a PFPR. If the ith row and the ith column are removed from M, the prefernce relation consisting of the remaining $n - 1$ rows and $n - 1$ columns of M is also PFPR.*

Proof This directly addresses from Definition 8.

Definition 9 Let $M = (m_{pq})_{n\times n}$ to be an PFPR. If the arithmetic mean is met

$$m_{pq} = \frac{1}{2}\left(m_{pr} \oplus m_{rq}\right),$$

where p, q and $r \in \mathbb{N}$,then, M is said to be additive consistent PFPR.

Example 1 Suppose M_1 is a PFPR as follows:

$$M_1 = \begin{pmatrix} (0.50, 0.00, 0.50) & (0.30, 0.50, 0.20) & (0.20, 0.40, 0.40) \\ (0.20, 0.50, 0.30) & (0.50, 0.00, 0.50) & (0.60, 0.20, 0.20) \\ (0.40, 0.40, 0.20) & (0.20, 0.20, 0.60) & (0.50, 0.00, 0.50) \end{pmatrix}$$

then, M_1 is additive consistent PFPR by Definition 9.

Definition 10 Let $M = (m_{pq})_{n\times n}$ to be an PFPR. If the geometric mean is met

$$m_{pq} = \left(m_{pr} \oplus m_{rq}\right)^{\frac{1}{2}},$$

where p, q and $r \in \mathbb{N}$,then, M is said to be multiplicative consistent PFPR.

Example 2 Suppose M_1 is a PFPR as follows:

$$M_1 = \begin{pmatrix} (0.50, 0.00, 0.50) & (0.30, 0.50, 0.20) & (0.20, 0.40, 0.40) \\ (0.20, 0.50, 0.30) & (0.50, 0.00, 0.50) & (0.40, 0.40, 0.20) \\ (0.40, 0.40, 0.20) & (0.20, 0.40, 0.40) & (0.50, 0.00, 0.50) \end{pmatrix}$$

then, M_1 is multiplicative consistent PFPR by Definition 10.

Let $M = (m_{pq})_{n\times n}$ be a PFPR, where $m_{pq} = \left(\mu_{pq}, \wp_{pq}, \partial_{pq}\right)$, $p, q \in \mathbb{N}$. The different properties of PFPR are examined according to Definition 8.

(1) If $m_{pr} \oplus m_{rq} \geq m_{pq}$, $\forall p, q$ and $r \in \mathbb{N}$, then we conclude that M validates the axiom of the triangle.

(2) If $m_{pr} \geq (0.50, 0.00, 0.50)$, $m_{rq} \geq (0.50, 0.00, 0.50) \Rightarrow m_{pq} \geq (0.50, 0.00, 0.50)$, $\forall p, q$ and $r \in \mathbb{N}$, then we conclude that M validates the axiom of weak transitive.

(3) If $m_{pq} \geq \min\left\{m_{pr}, m_{rq}\right\}$, $\forall p, q$ and $r \in \mathbb{N}$, then we conclude that M validates the axiom of max-min transitive.

(4) If $m_{pq} \geq \max\left\{m_{pr}, m_{rq}\right\}$, $\forall p, q$ and $r \in \mathbb{N}$, then we conclude that M validates the axiom of max-max transitive.

(5) If $m_{pr} \geq (0.50, 0.00, 0.50)$, $m_{rq} \geq (0.50, 0.00, 0.50) \Rightarrow m_{pq} \geq \min\left\{m_{pr}, m_{rq}\right\}$, $\forall p, q$ and $r \in \mathbb{N}$, then we conclude that M validates the axiom of restricted max-min transitive.

(6) If $m_{pr} \geq (0.50, 0.00, 0.50)$, $m_{rq} \geq (0.50, 0.00, 0.50) \Rightarrow m_{pq} \geq \max\left\{m_{pr}, m_{rq}\right\}$, $\forall p, q$ and $r \in \mathbb{N}$, then we conclude that M validates the axiom of restricted max-max transitive.

4.2 Incomplete Picture Fuzzy Preference Relations

Definition 11 Let $M = (m_{pq})_{n \times n}$ be a PFPR, where $m_{pq} = \left(\mu_{pq}, \wp_{pq}, \partial_{pq}\right)$, $p, q \in \mathbb{N}$. Then, M is known to be the incomplete picture fuzzy prefernce relation, if some of its components can not be provided by the decision maker, which we identify as an unknown variable "ξ". Also other components provided by the decision maker, that satisfy

(a) $\mu_{pq} \in [0, 1]$, $\wp_{pq} \in [0, 1]$ and $\partial_{pq} \in [0, 1]$,
(b) $\mu_{pq} + \wp_{pq} + \partial_{pq} \leq 1$,
(c) $\mu_{qp} = \partial_{pq}$, $\partial_{qp} = \mu_{pq}$, $\wp_{pq} = \wp_{qp}$ and $\mu_{pp} = \partial_{pp} = 0.5$, $\wp_{pp} = 0$.

Definition 12 Let $M = (m_{pq})_{n \times n}$ to be an incomplete PFPR. If the arithmetic mean is met

$$m_{pq} = \frac{1}{2}\left(m_{pr} \oplus m_{rq}\right),$$

where p, q and $r \in \mathbb{N}$,then, M is said to be additive incomplete consistent PFPR.

Example 3 Suppose M_1 is an incomplete PFPR as follows:

$$M_1 = \begin{pmatrix} (0.50, 0.00, 0.50) & (0.30, 0.20, 0.50) & \xi \\ (0.50, 0.20, 0.30) & (0.50, 0.00, 0.50) & (0.40, 0.20, 0.40) \\ \xi & (0.40, 0.20, 0.40) & (0.50, 0.00, 0.50) \end{pmatrix}$$

then, M_1 is additive incomplete consistent PFPR by Definition 12.

Definition 13 Let $M = (m_{pq})_{n \times n}$ to be an incomplete PFPR. If the geometric mean is met

$$m_{pq} = \left(m_{pr} \oplus m_{rq}\right)^{\frac{1}{2}},$$

where p, q and $r \in \mathbb{N}$,then, M is said to be multiplicative incomplete consistent PFPR.

Let $M = (m_{pq})_{n\times n}$ be an incomplete PFPR, where $m_{pq} = \left(\mu_{pq}, \wp_{pq}, \partial_{pq}\right)$, $p, q \in \mathbb{N}$ and φ be the set of all known elements, then different properties of an incomplete PFPR are examined according to Definition 11.

(1) If $m_{pr} \oplus m_{rq} \geq m_{pq}$, $\forall m_{pr}, m_{rq}$ and $m_{pq} \in \varphi$, then we said M validates the triangle axiom.
(2) If $m_{pr} \geq (0.50, 0.00, 0.50)$, $m_{rq} \geq (0.50, 0.00, 0.50) \Rightarrow m_{pq} \geq (0.50, 0.00, 0.50)$, $\forall m_{pr}, m_{rq}$ and $m_{pq} \in \varphi$, then we said M validates the weak transitive axiom.
(3) If $m_{pq} \geq \min\left\{m_{pr}, m_{rq}\right\}$, $\forall m_{pr}, m_{rq}$ and $m_{pq} \in \varphi$, then we said M validates the max-min transitive axiom.
(4) If $m_{pq} \geq \max\left\{m_{pr}, m_{rq}\right\}$, $\forall m_{pr}, m_{rq}$ and $m_{pq} \in \varphi$, then we said M validates the max-max transitive axiom.
(5) If $m_{pr} \geq (0.50, 0.00, 0.50)$, $m_{rq} \geq (0.50, 0.00, 0.50) \Rightarrow m_{pq} \geq \min\left\{m_{pr}, m_{rq}\right\}$, $\forall m_{pr}, m_{rq}$ and $m_{pq} \in \varphi$, then we said M validates the restricted max-min transitive axiom.
(6) If $m_{pr} \geq (0.50, 0.00, 0.50)$, $m_{rq} \geq (0.50, 0.00, 0.50) \Rightarrow m_{pq} \geq \max\left\{m_{pr}, m_{rq}\right\}$, $\forall m_{pr}, m_{rq}$ and $m_{pq} \in \varphi$, then we said M validates the restricted max-max transitive axiom.

Definition 14 Let $M = (m_{pq})_{n\times n}$ be an incomplete PFPR. If $(p, q) \cap (r, l) \neq \phi$, then the elements m_{pq} and m_{rl} are said to be adjoining. For the unknown element m_{pq}, suppose $\exists$ two adjoining elements m_{pr} and m_{rq}, then m_{pq} is said to be available.

In this situation, the element m_{pq} can be acquired indirectly by the known element values m_{pr} and m_{rq}.

Definition 15 Let $M = (m_{pq})_{n\times n}$ to be an incomplete PFPR. If any unknown element can be identified indirectly using any known function (e.g. Definition 12 or 13) according to the known elements adjoining to it), then M is named PFPR as acceptable incomplete, otherwise M is declared as unacceptable incomplete.

Let $M = (m_{pq})_{n\times n}$ be an acceptable incomplete PFPR. Then, using Definition 12 and 13, every unknown element m_{pq} can be indirectly computed by

$$\widetilde{m}_{pq} = \frac{1}{2\Psi_{pq}}\left(\sum_{r\in I_{pq}}\left(m_{pr} \oplus m_{rq}\right)\right) \tag{1}$$

or

$$\widetilde{m}_{pq} = \left(\prod_{r\in I_{pq}}\left(m_{pr} \oplus m_{rq}\right)\right)^{\frac{1}{2\Psi_{pq}}}, \tag{2}$$

where $I_{pq} = \{r | m_{pr}, m_{rq} \in \varphi\}$, Ψ_{pq} represents the total number of elements in I_{pq}. Hence, we obtained a complete PFPR $\overline{M} = (\widehat{m}_{pq})_{n \times n}$, where

$$\widehat{m}_{pq} = \begin{cases} \widetilde{m}_{pq} & m_{pq} \notin \varphi, \\ m_{pq} & m_{pq} \in \varphi. \end{cases}$$

5 Some Alternative Ranking Algorithms

Some ranking and alternative selection algorithms are established in this section based on the preference relations described in Sect. 4.

5.1 *PFPR Based Algorithm for Rank the Alternatives*

We build a novel algorithm in this subsection to rank the alternatives based on PFPRs. Following are the algorithm steps:

Step-1 For any decision support issue, let $\bar{O}_i$ $(i \in \mathbb{N})$ be the non-empty collection of alternatives. Decision makers equate one alternative to each other and builds PFPR $M = (m_{pq})_{n \times n} \subset V \times V$, with $m_{pq} = (\mu_{pq}, \wp_{pq}, \partial_{pq})$ is the PFN with $p, q \in \mathbb{N}$. Where $\mu_{pq} \in [0, 1]$, $\wp_{pq} \in [0, 1]$, $\partial_{pq} \in [0, 1]$, $\mu_{pq} + \wp_{pq} + \partial_{pq} \leq 1$ and

$$\mu_{qp} = \partial_{pq}, \partial_{qp} = \mu_{pq}, \wp_{pq} = \wp_{qp} \;\&\; \mu_{pp} = \partial_{pp} = 0.5, \wp_{pp} = 0.$$

Step-2 Aggregated information ∂_i of every alternative $\bar{O}_i$ is calculated using picture fuzzy arithmetic averaging operator

$$\partial_i = \frac{1}{n} \left(m_{p1} \oplus m_{p2} \oplus m_{p3} \oplus \cdots \oplus m_{pn} \right), \; p \in \mathbb{N} \tag{3}$$

or the picture fuzzy geometric averaging operator

$$\partial_i = \left(m_{p1} \otimes m_{p2} \otimes m_{p3} \otimes \cdots \otimes m_{pn} \right)^{\frac{1}{n}}, \; p \in \mathbb{N}. \tag{4}$$

Step-3 Rank all the ∂_i $(i \in \mathbb{N})$ using the novel score function defined in Definition 6 & 7 and next ranked the $\bar{O}_i$ $(i \in \mathbb{N})$ alternatives and choose the finest one as per the ∂_i $(i \in \mathbb{N})$ values.

Table 2 PFN information

$$M = \begin{pmatrix} (0.5, 0.0, 0.5) & (0.3, 0.4, 0.3) & (0.7, 0.2, 0.1) & (0.5, 0.2, 0.3) & (0.4, 0.4, 0.2) \\ (0.3, 0.4, 0.3) & (0.5, 0.0, 0.5) & (0.8, 0.1, 0.1) & (0.6, 0.3, 0.1) & (0.5, 0.2, 0.3) \\ (0.1, 0.2, 0.7) & (0.1, 0.1, 0.8) & (0.5, 0.0, 0.5) & (0.3, 0.6, 0.1) & (0.4, 0.4, 0.2) \\ (0.3, 0.2, 0.5) & (0.1, 0.3, 0.6) & (0.1, 0.6, 0.3) & (0.5, 0.0, 0.5) & (0.3, 0.3, 0.4) \\ (0.2, 0.4, 0.4) & (0.3, 0.2, 0.5) & (0.2, 0.4, 0.4) & (0.4, 0.3, 0.3) & (0.5, 0.0, 0.5) \end{pmatrix}$$

5.2 *Case Study*

First of all, 3D printers is widely employed by architects to manufacture aesthetic and functional prototypes due to its rapid production and low cost prototyping capabilities used. AM technologies are increasingly integrated in aerospace and missile applications for both military and civil purposes. Many companies in the aviation sector have started to produce plastic parts by taking advantage of the 3D printers. For instance, Boeing has produced various thermoplastic parts using commercially using laser sintering techniques. Our proposed algorithm is implemented to a descriptive example based on a 3D printer selection problem in Aviation sector. 3D printer selection decision criteria can vary depending on the number of qualitative and quantitative elements [18]. Printer selection decision criteria can vary depending on the number of qualitative and quantitative elements. Based on a comprehensive literature review, there are five dimensions of the criteria to select 3D printers. These dimensions are technology, organization, environment, quality and cost [51]. Suppose ∂_i $(i = 1, 2, \ldots, 5)$ be the non-empty collection of alternatives. The decision-maker need to evaluated the preference information on ∂_i by PFNs and describe the PFN in Table 2 as follows:

By Eq. 3, we have

$$\partial_1 = (0.499203, 0., 0.245951)\ \partial_2 = (0.57418, 0., 0.214113)\ \partial_3 = (0.298317, 0., 0.354517)$$
$$\partial_4 = (0.276347, 0., 0.447769)\ \partial_5 = (0.330609, 0., 0.412892)$$

and then by Definition 6, we get

$$\hat{S}c\,(\partial_1) = (0.253252) \quad \hat{S}c\,(\partial_2) = (0.360068) \quad \hat{S}c\,(\partial_3) = (-0.0562003)$$
$$\hat{S}c\,(\partial_4) = (-0.171422)\ \hat{S}c\,(\partial_5) = (-0.0822828)$$

Thus,

$$\hat{S}c\,(\partial_2) > \hat{S}c\,(\partial_1) > \hat{S}c\,(\partial_3) > \hat{S}c\,(\partial_5) > \hat{S}c\,(\partial_4)$$

That is

$$\partial_2 > \partial_1 > \partial_3 > \partial_5 > \partial_4$$

By using Eq. 4, we get

$\partial_1 = (0.461789, 0., 0.293195)$ $\partial_2 = (0.514352, 0., 0.276347)$ $\partial_3 = (0.226793, 0., 0.535602)$
$\partial_4 = (0.214113, 0., 0.469543)$ $\partial_5 = (0.299256, 0., 0.424735)$

and then by Definition 6, we get

$$\hat{S}c(\partial_1) = (0.168594) \quad \hat{S}c(\partial_2) = (0.238005) \quad \hat{S}c(\partial_3) = (-0.308808)$$
$$\hat{S}c(\partial_4) = (-0.255431) \; \hat{S}c(\partial_5) = (-0.125479)$$

Thus,

$$\hat{S}c(\partial_2) > \hat{S}c(\partial_1) > \hat{S}c(\partial_5) > \hat{S}c(\partial_4) > \hat{S}c(\partial_3)$$

That is

$$\partial_2 > \partial_1 > \partial_5 > \partial_4 > \partial_3.$$

5.3 *Incomplete PFPR Based Algorithm for Rank the Alternatives*

We build a novel algorithm in this subsection to rank the alternatives depending on incomplete PFPRs. The algorithm steps are as follows:

Step-1 For any decision support issue, let $\bar{O}_i$ $(i \in \mathbb{N})$ be the non-empty collection of alternatives. Decision makers equate one alternative to each other and builds an acceptable incomplete PFPR $M = (m_{pq})_{n\times n} \subset V \times V$, with $m_{pq} = (\mu_{pq}, \wp_{pq}, \partial_{pq})$ is the PFN with $p, q \in \mathbb{N}$. Where $\mu_{pq} \in [0, 1]$, $\wp_{pq} \in [0, 1]$, $\partial_{pq} \in [0, 1]$, $\mu^{\kappa}_{pq} + \wp^{\kappa}_{pq} + \partial^{\kappa}_{pq} \leq 1$, $\kappa > 2$ and

$$\mu_{qp} = \partial_{pq}, \partial_{qp} = \mu_{pq}, \wp_{pq} = \wp_{qp} \;\&\; \mu_{pp} = \partial_{pp} = 0.5, \wp_{pp} = 0.$$

Step-2 To build the complete PFPR $\overline{M} = (\widehat{m}_{pq})_{n\times n}$, using Eq. 1 or 2.

Step-3 Aggregated information ∂_i of every alternative is calculated using picture fuzzy arithmetic averaging operator

$$\partial_i = \frac{1}{n}\left(m_{p1} \oplus m_{p2} \oplus m_{p3} \oplus \cdots \oplus m_{pn}\right), \; p \in \mathbb{N} \tag{5}$$

or the picture fuzzy geometric averaging operator

$$\partial_i = \left(m_{p1} \otimes m_{p2} \otimes m_{p3} \otimes \cdots \otimes m_{pn}\right)^{\frac{1}{n}}, \; p \in \mathbb{N}. \tag{6}$$

Step-4 Rank all the ∂_i $(i \in \mathbb{N})$ using the novel score function defined in Definition 6 and 7 and then ranked the $\bar{O}_i$ $(i \in \mathbb{N})$ alternatives and choose the finest one as per the ∂_i $(i \in \mathbb{N})$ values.

Table 3 Acceptable incomplete PFPR information

$$M=\begin{pmatrix}(0.5,0.0,0.5) & (0.3,0.4,0.3) & (0.7,0.2,0.1) & (0.5,0.2,0.3) & (0.4,0.4,0.2)\\ (0.3,0.4,0.3) & (0.5,0.0,0.5) & \xi & (0.6,0.3,0.1) & (0.5,0.2,0.3)\\ (0.1,0.2,0.7) & \xi & (0.5,0.0,0.5) & \xi & (0.4,0.4,0.2)\\ (0.3,0.2,0.5) & (0.1,0.3,0.6) & \xi & (0.5,0.0,0.5) & \xi\\ (0.2,0.4,0.4) & (0.3,0.2,0.5) & (0.2,0.4,0.4) & \xi & (0.5,0.0,0.5)\end{pmatrix}$$

5.4 Numerical Example

For the same problem mentioned above, suppose ∂_i $(i=1,2,\ldots,5)$ be the non-empty family of alternatives. The decision maker need to evaluated the preference information on ∂_i by PFNs and describe an acceptable incomplete PFPR in Table 3 as follows:

Firstly, we find out the incomplete elements. For this, utilizing Eq. 1, we have

$$\begin{aligned}\widetilde{m}_{23} &= \frac{1}{4}\left((m_{21}\oplus m_{13})\oplus(m_{25}\oplus m_{53})\right)\\ &= \frac{1}{4}\left(((0.3,0.4,0.3)\oplus(0.7,0.2,0.1))\oplus((0.5,0.2,0.3)\oplus(0.2,0.4,0.4))\right)\\ &= (0.978304, 0.0282843, 0.000901218)\end{aligned}$$

$$\begin{aligned}\widetilde{m}_{34} &= \frac{1}{2}\left(m_{31}\oplus m_{14}\right)\\ &= \frac{1}{2}\left((0.1,0.2,0.7)\oplus(0.5,0.2,0.3)\right)\\ &= (0.64162, 0.2, 0.111181)\end{aligned}$$

and

$$\begin{aligned}\widetilde{m}_{45} &= \frac{1}{4}\left((m_{41}\oplus m_{15})\oplus(m_{42}\oplus m_{25})\right)\\ &= \frac{1}{4}\left(((0.3,0.2,0.5)\oplus(0.4,0.4,0.2))\oplus((0.1,0.3,0.6)\oplus(0.5,0.2,0.3))\right)\\ &= (0.948976, 0.0263215, 0.0045307)\,.\end{aligned}$$

Then, using above calculation of incomplete information and Eq. 2, we have complete PFPR $\overline{M}=(\widehat{m}_{pq})_{5\times 5}$ as given in Table 4.

By Eq. 3, we have

$\partial_1=(0.499203, 0., 0.245951)$ $\partial_2=(0.726155, 0., 0.0834624)$ $\partial_3=(0.373078, 0., 0.376836)$
$\partial_4=(0.570816, 0., 0.207492)$ $\partial_5=(0.259199, 0., 0.519725)$

Table 4 Complete PFPR information

(0.5, 0.0, 0.5)	(0.3, 0.4, 0.3)	(0.7, 0.2, 0.1)	(0.5, 0.2, 0.3)	(0.4, 0.4, 0.2)
(0.3, 0.4, 0.3)	(0.5, 0.0, 0.5)	(0.978, 0.028, 0.0009)	(0.6, 0.3, 0.1)	(0.5, 0.2, 0.3)
(0.1, 0.2, 0.7)	(0.0009, 0.028, 0.978)	(0.5, 0.0, 0.5)	(0.641, 0.2, 0.111)	(0.4, 0.4, 0.2)
(0.3, 0.2, 0.5)	(0.1, 0.3, 0.6)	(0.111, 0.2, 0.641)	(0.5, 0.0, 0.5)	(0.948, 0.026, 0.004)
(0.2, 0.4, 0.4)	(0.3, 0.2, 0.5)	(0.2, 0.4, 0.4)	(0.004, 0.0263, 0.948)	(0.5, 0.0, 0.5)

and then by Definition 6, we get

$$\hat{S}c\,(\partial_1) = 0.253252 \quad \hat{S}c\,(\partial_2) = 0.642693 \quad \hat{S}c\,(\partial_3) = -0.00375753$$
$$\hat{S}c\,(\partial_4) = 0.363324 \quad \hat{S}c\,(\partial_5) = -0.260526$$

Thus,

$$\hat{S}c\,(\partial_2) > \hat{S}c\,(\partial_4) > \hat{S}c\,(\partial_1) > \hat{S}c\,(\partial_3) > \hat{S}c\,(\partial_5)$$

That is

$$\partial_2 > \partial_4 > \partial_1 > \partial_3 > \partial_5$$

By using Eq. 4, we get

$$\partial_1 = (0.461789, 0., 0.293195) \quad \partial_2 = (0.535439, 0., 0.26107) \quad \partial_3 = (0.102903, 0., 0.702079)$$
$$\partial_4 = (0.275198, 0., 0.486346) \quad \partial_5 = (0.119136, 0., 0.657982)$$

and then by Definition 6, we have

$$\hat{S}c\,(\partial_1) = 0.168594 \quad \hat{S}c\,(\partial_2) = 0.27437 \quad \hat{S}c\,(\partial_3) = -0.599176$$
$$\hat{S}c\,(\partial_4) = -0.211148 \quad \hat{S}c\,(\partial_5) = -0.538846$$

Thus,

$$\hat{S}c\,(\partial_2) > \hat{S}c\,(\partial_1) > \hat{S}c\,(\partial_4) > \hat{S}c\,(\partial_5) > \hat{S}c\,(\partial_3)$$

That is

$$\partial_2 > \partial_1 > \partial_4 > \partial_5 > \partial_3.$$

6 Conclusion

The aviation industry requires stronger, lighter and and durable parts. To meet these requirements, today's 3D printers provides new possibilities. The concept of picture fuzzy sets help to handle uncertainty in multiple criteria decision-making problems by considering membership functions that are membership, non-membership, and hesitancy degrees independently for each element. This chapter consists of two main parts. One of the parts is what additive manufacturing and application areas are used in aviation and the critical factors when choosing this technology. The second part of the study is to develop new fuzzy relations under the picture fuzzy environment and decision-making process based on these relations under imperfect data. We proposed novel preference relations under PFSs. Based on developing the picture fuzzy preference relations, we proposed a novel decision-making method for solving multiple criteria decision-making problems. In general, the proposed preference relations have result in a better modeling technique than the IFPRs and q-ROFPRs for evaluating uncertain information in the decision-making scenario. The proposed preference relations applied to the illustrative example of selecting 3D printers in the aviation sector.

We will continue to explore various preferential relations to address ambiguous information in the future and apply them to real decision-making problems in aviation industry.

References

1. Ashraf, S., Mahmood, T., Abdullah, S., Khan, Q.: Different approaches to multi-criteria group decision making problems for picture fuzzy environment. Bull. Braz. Math. Soc. New Ser. **50**(2), 373–397 (2019)
2. Ashraf, S., Abdullah, S., Mahmood, T., Aslam, M.: Cleaner production evaluation in gold mines using novel distance measure method with cubic picture fuzzy numbers. Int. J. Fuzzy Syst. **21**(8), 2448–2461 (2019)
3. Attanassov, K.: Intuitionistic fuzzy sets. Fuzzy Sets Syst. **20**, 87–96 (1986)
4. Attar, H., Calin, M., Zhang, L.C., Scudino, S., Eckert, J.: Manufacture by selective laser melting and mechanical behavior of commercially pure titanium. Mater. Sci. Eng. A **593**, 170–177 (2014)
5. Bai, Y., Williams, C.B.: Binder jetting additive manufacturing with a particle-free metal ink as a binder precursor. Mater. Des. **147**, 146–156 (2018)
6. Calignano, F., Manfredi, D., Ambrosio, E.P., Biamino, S., Lombardi, M., Atzeni, E., Salmi, A., Minetola, P., Iuliano, L., Fino, P.: Overview on additive manufacturing technologies. Proc. IEEE **105**(4), 593–612 (2017)
7. Carter, W. T., Erno, D. J., Abbott, D. H., Bruck, C. E., Wilson, G. H., Wolfe, J. B., Finkhousen, D.M., Tepper, A. and Stevens, R.G.: The GE aircraft engine bracket challenge: an experiment in crowdsourcing for mechanical design concepts. In: Solid Freeform Fabrication Symposium, pp. 1402–1411 (2014)
8. Casalino, G., Campanelli, S.L., Contuzzi, N., Ludovico, A.D.: Experimental investigation and statistical optimisation of the selective laser melting process of a maraging steel. Opt. Laser Technol. **65**, 151–158 (2005)
9. Cuong, B.C., Kreinovich, V.: Picture fuzzy sets-a new concept for computational intelligence problems. In: Proceedings of 3rd World Congress on Information and Communication Technologies (WICT), pp. 1–6 (2013)
10. Crump, S.S.: U.S. Patent No. 5,121,329. U.S. Patent and Trademark Office, Washington, DC (1992)
11. Do, T., Kwon, P., Shin, C.S.: Process development toward full-density stainless steel parts with binder jetting printing. Int. J. Mach. Tools Manuf. **121**, 50–60 (2017)
12. Frazier, W.E.: Metal additive manufacturing: a review. J. Mater. Eng. Perf. **23**(6), 1917–1928 (2014)
13. Gardan, J.: Additive manufacturing technologies: state of the art and trends. Int. J. Prod. Res. **54**(10), 3118–3132 (2016)

14. Gibson, I.G.: Additive Manufacturing Technologies 3D Printing, Rapid Prototyping, and Direct Digital Manufacturing (2015)
15. Gisario, A., Kazarian, M., Martina, F., Mehrpouya, M.: Metal additive manufacturing in the commercial aviation industry: a review. J. Manuf. Syst. **53**, 124–149 (2019)
16. González-Henríquez, C.M., Sarabia-Vallejos, M.A., Rodriguez-Hernandez, J.: Polymers for additive manufacturing and 4D-printing: materials, methodologies, and biomedical applications. Progr. Polym. Sci. **94**, 57–116 (2019)
17. Guo, N., Leu, M.C.: Additive manufacturing: technology, applications and research needs. Front. Mech. Eng. **8**(3), 215–243 (2013)
18. Gündoğdu, F.K., Kahraman, C.: A novel fuzzy TOPSIS method using emerging interval-valued spherical fuzzy sets. Eng. Appl. Artif. Intell. **85**, 307–323 (2019)
19. Gundogdu, F.K.: picture fuzzy linear assignment method and its application to selection of pest house location. In International Conference on Intelligent and Fuzzy Systems, pp. 101–109. Springer, Cham (2020)
20. Hehr, A., Dapino, M.J.: Interfacial shear strength estimates of NiTi-Al matrix composites fabricated via ultrasonic additive manufacturing. Compos. Part B Eng. **77**, 199–208 (2015)
21. Hull, C.W.: Apparatus for Production of Three-Dimensional Objects by Stereolithography. United States Patent, Appl., No. 638905, Filed (1984)
22. Huang, S.H., Liu, P., Mokasdar, A., Hou, L.: Additive manufacturing and its societal impact: a literature review. Int. J. Adv. Manuf. Technol. **67**(5–8), 1191–1203 (2013)
23. Jin, F., Ni, Z., Chen, H., Li, Y.: Approaches to group decision making with intuitionistic fuzzy preference relations based on multiplicative consistency. Knowl. Based Syst. **97**, 48–59 (2016)
24. Khajavi, S.H., Partanen, J., Holmström, J.: Additive manufacturing in the spare parts supply chain. Comput. Ind. **65**(1), 50–63 (2014)
25. Khudyakov, I.V.: Fast photopolymerization of acrylate coatings: achievements and problems. Progr. Org. Coat. **121**, 151–159 (2018)
26. Khan, S., Abdullah, S., Ashraf, S.: Picture fuzzy aggregation information based on Einstein operations and their application in decision making. Math. Sci. **1–17** (2019)
27. Khan, S., Abdullah, S., Abdullah, L., Ashraf, S.: Logarithmic aggregation operators of picture fuzzy numbers for multi-attribute decision making problems. Mathematics **7**(7), 608 (2019)
28. Lee, H., Lim, C.H.J., Low, M.J., Tham, N., Murukeshan, V.M., Kim, Y.J.: Lasers in additive manufacturing: a review. Int. J. Precis. Eng. Manuf. Green Technol. **4**(3), 307–322 (2017)
29. Li, H., Yin, S., Yang, Y.: Some preference relations based on q-rung orthopair fuzzy sets. Int. J. Intell. Syst. **34**(11), 2920–2936 (2019)
30. Li, J., Monaghan, T., Nguyen, T.T., Kay, R.W., Friel, R.J., Harris, R.A.: Multifunctional metal matrix composites with embedded printed electrical materials fabricated by ultrasonic additive manufacturing. Compos. Part B Eng. **113**, 342–354 (2017)
31. Lindemann, C., Jahnke, U., Moi, M., Koch, R.: Analyzing product lifecycle costs for a better understanding of cost drivers in additive manufacturing. In: 23th Annual International Solid Freeform Fabrication Symposium, An Additive Manufacturing Conference, Austin Texas USA, 6–8 Aug 2012
32. Manco, P., Macchiaroli, R., Maresca, P., Fera, M.: The additive manufacturing operations management maturity: a closed or an open issue? Procedia Manuf. **41**, 98–105 (2019)
33. Mandal, P., Ranadive, A.S.: Pythagorean fuzzy preference relations and their applications in group decision-making systems. Int. J. Intell. Syst. **34**(7), 1700–1717 (2019)
34. Meng, F., Chen, X.: An approach to incomplete multiplicative preference relations and its application in group decision making. Inf. Sci. **309**, 119–137 (2015)
35. Mohamed, O.A., Masood, S.H., Bhowmik, J.L.: Optimization of fused deposition modeling process parameters: a review of current research and future prospects. Adv. Manuf. **3**(1), 42–53 (2015)
36. Najmon, J.C., Raeisi, S., Tovar, A.: Review of additive manufacturing technologies and applications in the aerospace industry. In: Additive Manufacturing for the Aerospace Industry, pp. 7–31 (2019)

37. Ngo, T.D., Kashani, A., Imbalzano, G., Nguyen, K.T., Hui, D.: Additive manufacturing (3D printing): a review of materials, methods, applications and challenges. Compos. Part B Eng. **143**, 172–196 (2018)
38. Norris, G.: Higher altitudes cleared for GE-powered 787, 747-8 in icing. In: Aviation Week & Space Technology (2015)
39. Orlovsky, S.A.: Decision-making with a fuzzy preference relation. Fuzzy Sets Syst. **1**(3), 155–167 (1978)
40. Qiyas, M., Abdullah, S., Ashraf, S., Aslam, M.: Utilizing linguistic picture fuzzy aggregation operators for multiple-attribute decision-making problems. Int. J. Fuzzy Syst. **22**(1), 310–320 (2020)
41. Parandoush, P., Lin, D.: A review on additive manufacturing of polymer-fiber composites. Compos. Struct. **182**, 36–53 (2017)
42. Rejeski, D., Zhao, F., Huang, Y.: Research needs and recommendations on environmental implications of additive manufacturing. Addit. Manuf. **19**, 21–28 (2018)
43. Rivera, F.J.M., Arciniegas, A.J.R.: Additive manufacturing methods: techniques, materials, and closed-loop control applications. Int. J. Adv. Manuf. Technol. **109**(1), 17–31 (2020)
44. Saaty, T.L.: Axiomatic foundation of the analytic hierarchy process. Manage. Sci. **32**(7), 841–855 (1986)
45. Tan, X., Kok, Y., Tan, Y.J., Descoins, M., Mangelinck, D., Tor, S.B., Leong, K.F., Chua, C.K.: Graded microstructure and mechanical properties of additive manufactured Ti-6Al-4V via electron beam melting. Acta Mater. **97**, 1–16 (2015)
46. Thomas, D.S., Gilbert, S.W.: Costs and cost effectiveness of additive manufacturing. NIST Special Publication **1176**, 12 (2014)
47. Thompson, M.K., Moroni, G., Vaneker, T., Fadel, G., Campbell, R.I., Gibson, I., Bernard, A., Schulz, J., Graf, P., Ahuja, B., Martina, F.: Design for additive manufacturing: trends, opportunities, considerations, and constraints. CIRP Ann. **65**(2), 737–60 (2016)
48. Tuck, C., Hague, R., Burns, N.: Rapid manufacturing: impact on supply chain methodologies and practice. Int. J. Serv. Oper. Manage. **3**(1), 1–22 (2007)
49. Wang, X., Jiang, M., Zhou, Z., Gou, J., Hui, D.: 3D printing of polymer matrix composites: a review and prospective. Compos. Part B Eng. **110**, 442–458 (2017)
50. Wohlers, T.: Additive manufacturing and 3D printing state of the industry. Wohlers Assoc. Fort Collins, CO **6**, 269–375 (2013)
51. Yeh, C.C., Chen, Y.F.: Critical success factors for adoption of 3D printing. Technol. Forecast. Soc. Change **132**, 209–216 (2018)
52. Xu, G.L., Wan, S.P., Wang, F., Dong, J.Y., Zeng, Y.F.: Mathematical programming methods for consistency and consensus in group decision making with intuitionistic fuzzy preference relations. Knowl.-Based Syst. **98**, 30–43 (2016)
53. Xu, Z., Liao, H.: A survey of approaches to decision making with intuitionistic fuzzy preference relations. Knowl.-Based Syst. **80**, 131–142 (2015)
54. Xu, Z.: Intuitionistic preference relations and their application in group decision making. Inf. Sci. **177**(11), 2363–2379 (2007)
55. Zadeh, L.A.: Fuzzy sets. Inf. Control **8**(3), 338–353 (1965)
56. Zeng, S., Asharf, S., Arif, M., Abdullah, S.: Application of exponential jensen picture fuzzy divergence measure in multi-criteria group decision making. Mathematics **7**(2), 191 (2019)
57. Zhang, Y., Li, K.W., Wang, Z.J.: Prioritization and aggregation of intuitionistic preference relations: a multiplicative-transitivity-based transformation from intuitionistic judgment data to priority weights. Group Decis. Negot. **26**(2), 409–436 (2017)

A Conceptual Framework for Estimating the Remaining Operational Lifetime of the Recovered Components from End of Life Aircraft Using Fuzzy Simulation and Digital Twin

Samira Keivanpour

Abstract Industry 4.0 promotes product lifecycle management and promises to provide digital models and integrating product lifecycle data. Aviation 4.0 addresses the implications of the fourth industrial revolution in the operation and management of aircraft during their lifecycles. The management of the end of life (EoL) phase of aircraft is essential and contains uncertainty data. The revenue from recovered parts plays an important role in the sustainability of the EoL aircraft business. Estimating the operational life of high-value parts is critical in EoL decision-making. This chapter proposes a framework including use of fuzzy simulation and digital twins to estimate the remaining useful life (RUL) of recovered parts. The proposed framework contains three digital models including the digital twins of parts health management, disassembly process, and recovered parts. A fuzzy intelligent decision-making model generates the rules based on sensor measurements and operational settings for the estimation of the RUL of a part. The application of the model related to aircraft engines is discussed.

Keywords Remaining useful life · End of life aircraft · Recovered parts · Digital twin · Fuzzy simulation

1 Introduction

The different technologies in the context of Industry 4.0 include cyber-physical systems, big-data analytics, the Internet of Things, additive manufacturing, and advanced simulation. The advantages of Industry 4.0 are its vertical and horizontal integration and the possibility for integrated and real-time management of a product's lifecycle data in a supply chain [36]. These value creations highlight the great advantages of this digital transformation in the aviation industry. Aviation 4.0 is a novel term used to address the implications of the Industry 4.0 revolution in the aviation

S. Keivanpour (✉)
Department of Mathematical and Industrial Engineering, Polytechnique Montreal, Montreal, Quebec, Canada
e-mail: samira.keivanpour@polymtl.ca

C. Kahraman and S. Aydın (eds.), *Intelligent and Fuzzy Techniques in Aviation 4.0*,
Studies in Systems, Decision and Control 372,
https://doi.org/10.1007/978-3-030-75067-1_13

industry. Some recent studies discussed the role of Industry 4.0 in safety, manufacturing, construction, maintenance and spare parts management in the aviation industry (e.g., [6, 42]; Guyon et al. [16, 20]). EoL aircraft recycling is a new challenge in the aviation industry, and the application of Industry 4.0 technologies promotes decision-making to assure the sustainability of the treatment process. Economic recessions affect the retirement rates of aircraft. The 2020 pandemic increased the number of EoL aircraft. The sustainability of this business depends on the efficiency and safety of eco-friendly practices, the collaboration of key players, and the consideration of the circular economy in this industrial sector.

The decision to decommission an aircraft depends on its operational lifetime and maintenance status [18]. The RUL of components plays an essential role in the viability of the decommissioning and the revenue flow of the EoL enterprise and depends on maintenance operations. Maintenance operations include C-checks, structural checks, and the maintenance of major components [18]. The interval of these operational tasks is determined by the total number of flight hours and aircraft cycles. Different tasks must be performed for maintenance checks of high-value parts such as landing gear, engines, and auxiliary power unit. The value of recovered parts depends on their maintenance and fault regimes. For life-limited parts, the history of the parts is essential as they should be disposed of after a specific time period or number of flight cycles. Time-control components or on-condition parts can be certified for reuse if their conditions are aligned based on airworthiness regulations. Factors that impact the trade of a recovered part include the actual status of the part, the supply of the component, its availability in spare-parts stocks, the certification cost, and its maintenance history. For high-value parts such as engines, the health condition of life-limited parts and time-controlled parts is essential [14, 18]. The scheduled maintenance of 89% of aircraft components is ineffective because of failure rate patterns ([34], cited in [19]). The challenges in estimating the RUL of recovered parts include consideration of the complexities of aircraft parts, management of the uncertainties of the operational environment, consideration of the role of operators on-site, and validation and verification concerns based on regulatory agencies. Fuzzy decision-making is a promising approach for considering uncertainties and a lack of data and information. The intuitive decisions of operators and engineers can be integrated into the fuzzy system. Simulating a digital model provides the opportunity to adjust the model with the real-time behavior of parts. The data of a part's lifecycle can be integrated and updated in a digital model.

Industry 4.0 is promising in the provision of digital models and the integration of product lifecycle data. Digital twins as an enabling technology in the Industry 4.0 context provide a virtual profile of a product and integrate real-time data behavior for optimizing virtual reality. Several applications of digital twins have been discussed in the literature including additive manufacturing [32], system failure detection [17], and defect prediction [40]. This chapter introduces a framework for the application of digital twins and fuzzy simulation for estimating the RUL of the recovered parts of an EoL aircraft. The contributions of this research:

- considers the context of EoL aircraft decommissioning and the role of estimating the RUL in part-recovery decision-making, and
- develops a conceptual framework for the joint application of fuzzy simulation and digital twins in estimating the RUL of recovered parts.

The rest of his chapter is organized as follows: Sect. 2 discusses the application of Industry 4.0 in the aircraft maintenance operation, Sect. 3 provides a short review of the application of fuzzy approaches to aircraft maintenance. Section 4 discusses the estimating RUL of complex products. Section 5 addresses the advantages of digital models in the health monitoring of complex products. Section 6 proposes the framework and application perspective and finally, Sect. 7 concludes with some remarks and the future research direction.

2 Application of Industry 4.0 in Aircraft Maintenance

MRO (Maintenance, Repair, and Overhaul) operation includes inspection and replacement of parts, refilling the oil and gas, and the related logistics activities. There are different applications of Industry 4.0 in the context of MRO. The main technologies that impact MRO in the future include big data analytics, additive manufacturing, advanced simulation, digital twin, Internet of things, and blockchain. In this section, some recent studies that addressed these applications are reviewed. Chemweno and Pintelon [8] discussed the application of data-driven approaches to aircraft maintenance. The authors explained the challenges of e-maintenance in the aviation industry. According to this work, sometimes the maintenance data from technicians and maintenance crew are recorded in text format. Hence, there are no decision support tools to extract valuable information from these records, and the operations mainly are based on a standardized maintenance manual. The authors proposed the text mining and rule-based approach to integrating the results of statistical analysis from aircraft sensors with data during maintenance operations for developing a maintenance decision support tool. Deng et al. [10] proposed an advanced optimization model for scheduling aircraft maintenance. This optimization model could reduce the checking time for the aircraft to 15 min.

Bouarfa et al. [5] developed a deep learning approach for an automated aircraft maintenance inspection. The authors used a neural network architecture with a small dataset size of training and showed the benefits of this approach in detecting the different types of damage. Dinis et al. [11] used big data and Bayesian networks in the capacity planning of MRO. Liu et al. [26] proposed an advanced simulation model based on agent-based modeling for developing an autonomous system for supporting aircraft maintenance. Utzig et al. [41] introduced a conceptual framework for the application of augmented reality and digital twin for assessing the damage of aircraft components. Ceruti et al. [7] proposed an augmented reality approach to aircraft maintenance operation. This system comprises a connected device for an on-site maintenance operation, 3D graphic content generation from the maintenance manual, and a maintenance center for guiding the operation. The authors explained that big data, smart sensors, connectivity, and visualization capability are the advantages of Industry 4.0 in the aviation industry. The new technologies also improve the health monitoring systems by gathering large data from sensors. The authors also explained the role of additive manufacturing in the maintenance operation. Additive manufacturing could be integrated into the spare parts supply and can reduce the inventory of the spare parts. The collaboration between OEM (Original Equipment manufacturer) and MRO and Airlines could facilitate the supply of spare parts too. Ramudhin et al. [37] discussed the application of RFID (Radio-frequency identification technology) in tracking the control system. They applied the concept to a service center for aircraft engines for MRO operations. The synthesis in the literature reveals that the implications of Industry 4.0 focused on maintenance decision support tools, health monitoring, capacity planning, and spare part management. Figure 1 shows a summary of the implications of Industry 4.0 technologies in MRO.

3 Fuzzy Models in Aircraft Maintenance

In this section, the applications of fuzzy models in aircraft maintenance and operation are reviewed. Atli and Kahraman [2] applied fuzzy critical path analysis for reducing the maintenance time planning. The authors explained the uncertainty of the task time in the operation as the result of the information uncertainty and different

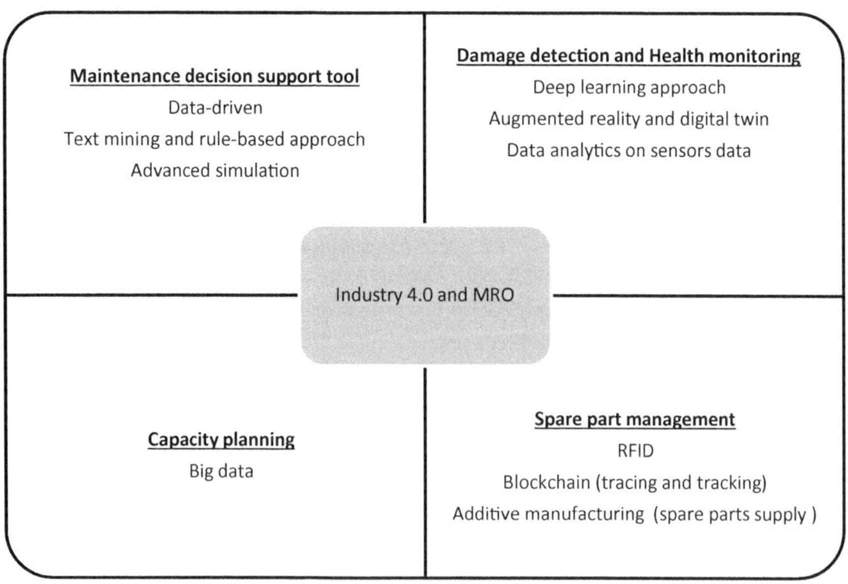

Fig. 1 The implications of Industry 4.0 in MRO

management approach. Hence, they considered fuzzy logic in analyzing the nodes, and activity time in the critical path. Azadeh et al. [3] used fuzzy simulation for maintenance planning. Considering the uncertainty of time between failures and maintenance activity time, a joint application of data envelopment analysis and fuzzy simulation is applied. Demirci et al. [9] applied fuzzy logic for health monitoring of commercial aircraft. The authors explained the role of the fuzzy approach in the identification of faulty engines. The authors used a user interface for entering the data from gas turbine engines and using the fuzzy rule-based model for engine health monitoring in the Turkish airline fleet. Gholizadeh et al. [15] proposed a fuzzy approach for analyzing aircraft wing reliability. Lu and Sy [29] addressed the fault diagnosis of aircraft maintenance without enough data. The authors used an agent-based modeling approach and Fuzzy-Rough Nearest Neighbour. They used a fault classification method considering service strategy, supply chain, and the flight requirement. Lu and Sun [30] applied a fuzzy decision-making model for evaluating the maintainability of the aircraft systems. The authors proposed a virtual simulation model for assessing the maintainability of nose landing gear based on accessibility, simplicity of tasks, testability, and ergonomic aspect [31].

The synthesis in the literature shows that uncertainty in the maintenance time and the time between failures and the lack of data are the main motivation for considering the fuzzy approaches to aircraft maintenance.

4 Estimating RUL of the Complex Products

The success of reusing or remanufacturing the recovered parts depends on the reliability of the parts, ease of disassembling, and required maintenance operations. According to Li and Kuo [25], the key factors in the effective management of spare parts includes part factors, demand, time, and associated parts. Part factors include the mean time between failures (MTBF), the probability of damage, the number of recovered parts, and the existence of substitutes [22]. MTBF is essential in estimating the residual lifetime of the parts. There are different approaches in estimating the RUL of the parts based on data availability, uncertainties factors, the role of experts, etc. These estimation methods include the deterministic method, engineering approaches, statistical methods, condition monitoring, and hybrid methods. Elsayed [12] discussed the role of experts' judgment in experimental methods. The standards, industrial methods, and the estimation of parts behavior are the other approaches. Condition monitoring data also plays an essential role in this context. This data includes operational data that usually are collected via different sensors such as oil, pressure, temperature, vibration, etc., performance data of parts, and environmental information. The uncertainties in this data are inevitable and should be considered in the estimation of RUL.

Si et al. [39] performed a survey on RUL estimation methods. The authors focused on statistical data-driven approaches. The challenges of estimation include the availability of data, data fusion, considering the external and environmental factors, and the presence of multiple failure modes. Stochastic Filtering models with the advantage of using the linear and Gaussian assumptions are also used in the health monitoring of aircraft engines and power generators. According to the authors, the needs for failure event data and the possible changes as the results of the preventive maintenance and operating environment are the challenges of applying this method.

Okoh et al. [35] also conducted a study on the different approaches to estimating RUL that decrease the uncertainties. The authors classified these methods into analytical based, model-based, knowledge-based, and hybrid-based methods. The knowledge-based approach uses computational intelligence integrating into the experts' experience. Generating the set of rules could improve the management of assets during their life cycle. The expert judgment could facilitate the maintenance decisions based on the experience and the observed evidence. The authors also discussed the fusion method that integrates different databases. The fuzzy methods could be applied here for data mining purposes.

Figure 2 shows the RUL for a recovered part. Equations 1–5 shows the formulation of RUL based on probability density and reliability function. Different factors contribute to the failure rate of the parts. The nature of parts, the operating context, and the definition of failure in that specific context should be considered in RUL estimation.

$$RLT(t) = E(LT - t|LT \geq t) \quad t \geq 0 \tag{1}$$

$$f_{LT|LT\geq t} = \frac{f(\tau)}{R(t)} \tag{2}$$

$$E(LT|LT \geq t) = \int_t^\infty \tau f_{LT|LT\geq t}(\tau)d\tau = \int_t^\infty \tau \frac{f(\tau)}{R(t)}d\tau \tag{3}$$

$$\mathrm{R}UL(t) = \int_t^\infty \tau \frac{f(\tau)}{R(t)}d\tau - t \tag{4}$$

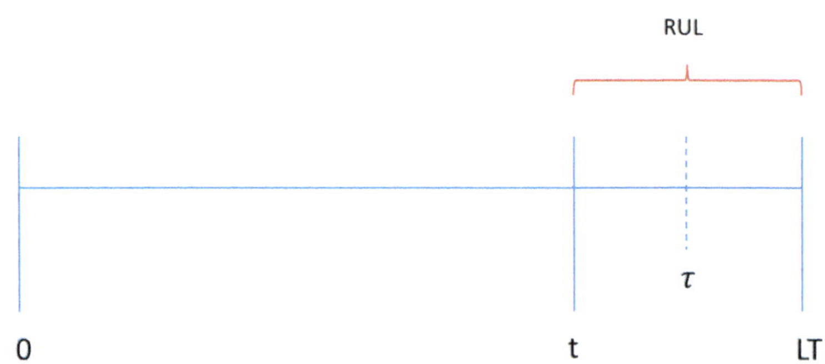

Fig. 2 Remaining useful life of a recovered part

$$RUL(t) = \frac{1}{R(t)} \int_{t}^{\infty} \tau f(\tau) d\tau - t \quad (5)$$

Fornlöf [14] performed a study for improving the estimation of RUL of on-condition parts in an aircraft engine. The author concluded that for an effective estimation, integrating data-driven, context awareness and physics-based techniques are essential.

Xiongzi et al. [44] performed a survey on methods for estimating RUL for aircraft systems. The authors divided the health degradation of the parts into four stages of health, caution, repair, and failure regime. The authors explained that RUL is a random variable and there are different uncertainties in its estimation based on the current age of parts, operational condition, and monitoring. The authors addressed the gaps and proposed the research agenda. The key gaps in the existing approaches and the advantages of digital models and fuzzy systems for addressing these gaps are shown in Table 1.

5 Digital Models and the Advantages in Maintenance and Complex Products Health Monitoring

The digital twin provides a dynamic digital profile of a physical object and integrates 3D model, historical and real-time monitoring of the object behavior for optimizing the virtual reality. Recently, it has received much attention in design, production, operation, and maintenance [21]. In this section, a brief review of applications in maintenance is provided. Lu et al. [28] reviewed digital twin models in manufacturing. The authors discussed that the application in operation and maintenance and its integration to smart manufacturing is limited. They mentioned two types of architecture patterns or digital twin models; severed based and edge-based. In the server-based architecture, the data from the target device is backed to a central server for performing data analysis and forming a digital twin. In the case of maintenance operation and having a remote center, this type of architecture could manage the needed computing resources and operate effectively. Khan et al. [23] discussed the role of the digital twin in autonomous maintenance. The authors proposed a framework including physical blocks, data processing, performance, and autonomous block, product aspect, and construction aspect. Errandonea et al. [13] performed a literature review on applications of the digital twin in maintenance. The authors discussed that the applications in the manufacturing, energy industry, and aerospace are the top among the other contexts. They addressed the applications in reactive maintenance, preventive, condition-based, predictive, and perspective maintenance. They concluded that the application in predictive maintenance is essential as it could combine and analyze the current condition, the behavior of the system, and degradation. If the information about the physical characteristics of the system and the operation process will be integrated, the capability of digital models will be improved. The historical data is also critical for validation purposes. Aivaliotis et al. [1] proposed a methodology for

Table 1 The gaps in existing approaches for estimating RUL and the advantages of fuzzy decision making and digital model

Existing challenges in estimating RUL for aircraft parts	The advantages of fuzzy decision models for addressing these challenges	The advantage of digital models in addressing these challenges
• The need for novel hybrid models • Considering the different types of failure modes • Considering the complexity of aircraft sub-assemblies and components	• The fuzzy systems could integrate the different optimization models with generating the rule-based knowledge management platform	• The digital twin could be updated with the maintenance records and health monitoring data via the life cycle of parts • The integration of digital twin and advanced simulation tools provide the solution for the complexity of aircraft configuration
• Managing the uncertainties of the environment and the work conditions • The role of operators • Quantifying the uncertainties	• Fuzzy decision making is promising in lack of data and information and uncertainties • The intuitive decisions of operators and engineers could be integrated into the fuzzy system	• Simulating the digital model provide the opportunity for adjusting the model with the real-time behavior of parts • The data during the life cycle of the parts could be integrated and updated in the digital model
• The need for validation and verification of the developed approach considering security and safety concerns	• Fuzzy methods could be used for the fusion of multiple databased and sustaining the validity	• Digital twin provides an experimental testbed for assessing different scenarios
• The need for huge computing resources and applying the complicated algorithms	• Connectivity to a remote maintenance center and using the fuzzy decision-making module in the center could balance the needed computing resources	• The connectivity and selecting target sub-assembly for creating digital twin and using the remote center could address this gap

using digital twins in predictive maintenance. They developed a framework based on three phases: modeling of dynamic behavior, modeling the virtual sensors for monitoring and gathering data, and online updating via real-time data. Mi et al. [33] developed a prediction maintenance decision tool using a digital twin. The authors discussed that predictive maintenance requires gathering data during the life cycle of the complex products and there are different players involved in the problem. Moreover, the limitation of fault diagnosis methods and considering the operating environment and performance could limit the expected accuracy.

Figure 3 shows a digital twin driven framework for disassembly and parts recovery from the complex products. The dynamic data stored inside the digital model is the core of the whole system architecture. Hence, the first step is developing a data model considering the main categories of data attributes and the involved entities in the EoL products treatment process. For the successful operation of re-used and

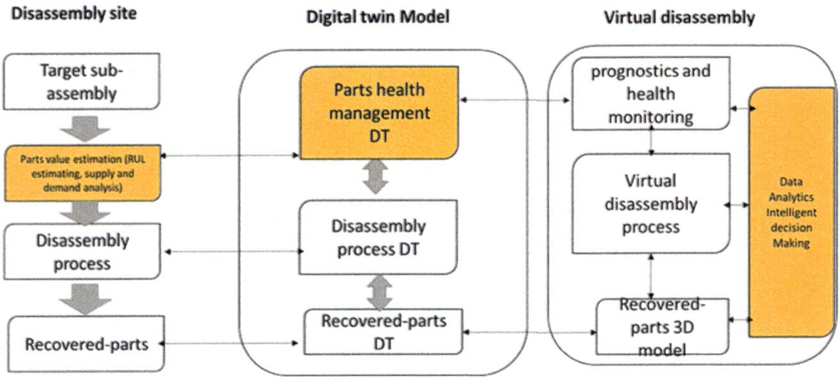

Fig. 3 Digital twin driven framework for decommissioning and parts recovery

remanufactured parts and components, the reliability of the components, ease of disassembling, inspection, cleaning, and maintenance data should be considered.

The essence of the digital twin model is comparing the different disassembling scenarios with the aggregation of several dispersed databases, data analytics, and visualization dashboard. These databases include OEM data, maintenance, and operation historical data, and the parts distributors databases. A configuration tool based on virtual reality for disassembling process design should be considered. 3D models (e.g. CAD, digital mockup, or open-source web-based) will be captured and parts selection and manipulation features for disassembling operation will be prepared. After designing the 3D models, the related assembly/maintenance and disassembly information can be added. A multi-purpose graphic user interface will be designed for the data management system, front-end application, and visualized interactive decision-making dashboard. The optimization model and data analytics module will be able to perform prognostics and health monitoring and estimation of the RUL. The proper disassembling scenario based on the retrieved detailed data also will be provided. This architecture provides the flexibility of the data processing and integrates the massive data for dealing with uncertainties of the disassembly and recovery parts trading. Latorella and Prabhu [24] reviewed the human errors in the aviation maintenance context. The majority of these errors could happen in the case of aircraft decommissioning and the EoL part recovery. For example, the incomplete and incorrect disassembly instructions, misinterpretation of the disassembly steps, nonavailability, and non-appropriateness of the required equipment for disassembly are some of these human errors. Hence, the digital model and the visual human–machine interface could decrease the number of errors during operation. The expert systems and intelligent data-driven decision making could improve the quality of diagnosis and performing the related tasks.

6 A Conceptual Framework Using the Digital Model and Fuzzy Simulation for Estimating RUL

In this section, the framework of applying digital twin and fuzzy simulation in aircraft disassembly operation is provided (Fig. 4). The digital twin (DT) module includes recovered parts DT, disassembly process DT, and RUL DT. RUL DT includes a data mining layer and a fuzzy intelligent decision-making model.

For the explanation of the represented block diagram, some details about the application in the case of an aircraft engine are provided. The database of PHM08 prognostics data challenge [38] is used for illustration and visualization. The data set includes operational settings and degradation data based on 21 sensor measurements. The first 14% of time series data is considered as a healthy regime and the last 14% is the failure regime.

Figures 5 and 6 show the difference between the healthy stage and the last stage of the aircraft engine lifetime based on eight sensor measures. As shown in these two figures, the average difference in sensor measures is increased considerably in the later stage of the engine operation. Hence, extracting the rules from the health behavior of engines could aid in estimating RUL.

The data mining layer analyzes the data from different sensors and operational settings. For the case of the engine, based on the PHM08 dataset, the data from 21 sensors and operation settings could be used for health monitoring purposes.

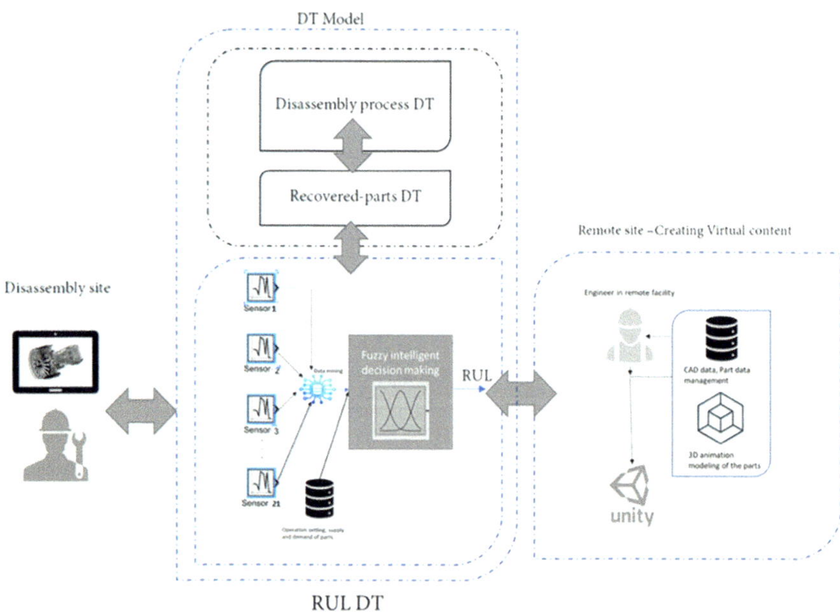

Fig. 4 The conceptual framework for using digital twin and fuzzy intelligent decision making in aircraft decommissioning

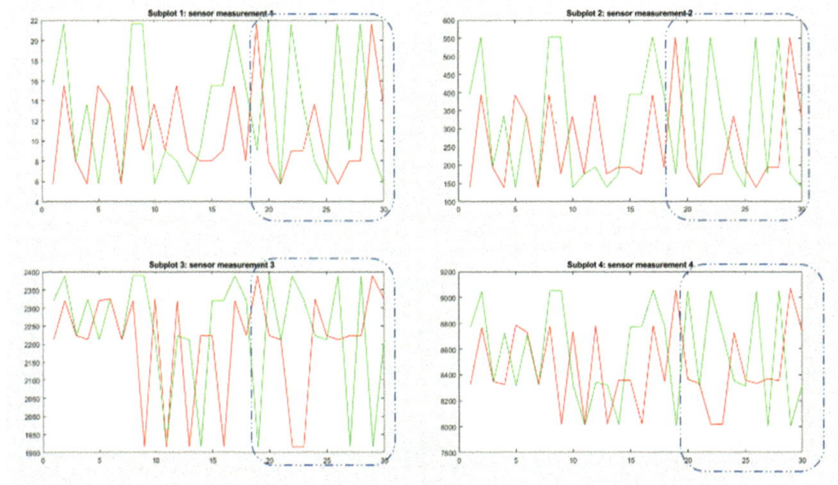

Fig. 5 The difference between sensor measures (sensors 1, 2, 3, 4) of the healthy and failure regime of an aircraft engine (green: health period, red: the last 14% of the operation time series)

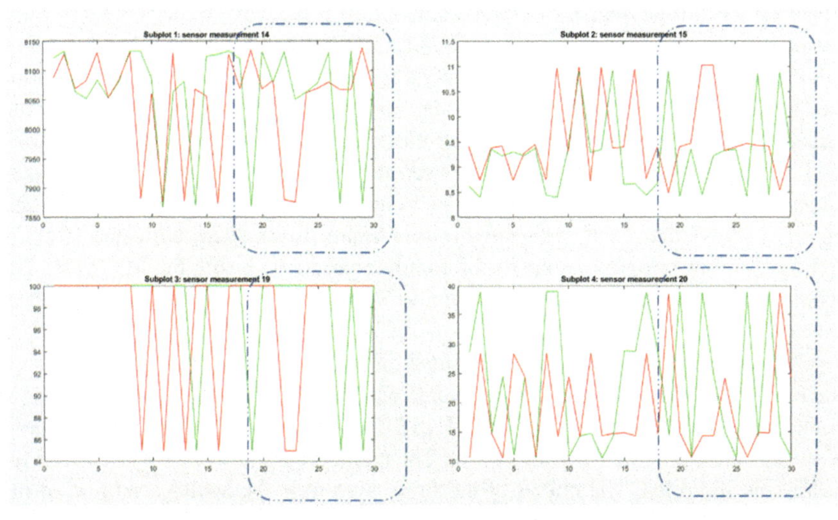

Fig. 6 The difference between sensor measures (sensors, 14, 15, 19, and 20) of healthy and failure regime of an aircraft engine (green: health period, red: the last 14% of the operation time series)

Different clustering approaches could be used for analyzing the time series of the engine units for extracting the degradation pattern. For example, the fuzzy clustering method could be used to consider the ambiguity in assigning the data to each cluster in the complex product failure behavior. Fuzzy c-means clustering is a data mining approach for classifying the data into certain clusters that each data in the database

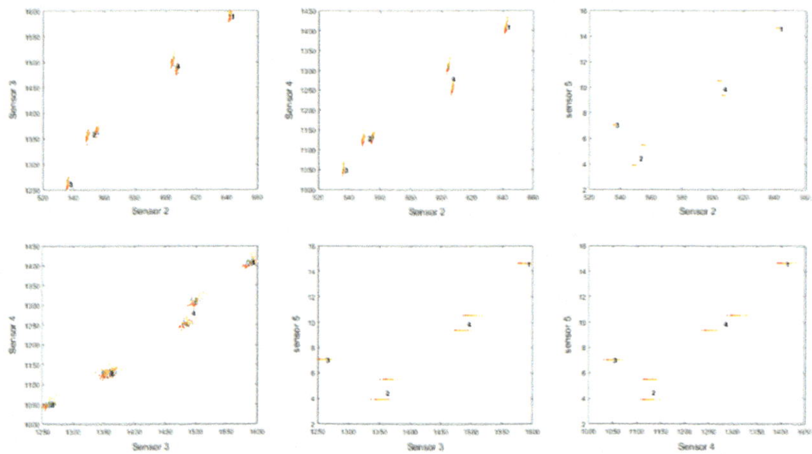

Fig. 7 Fuzzy C-means clustering for the sensors measures of three units

belongs to each cluster with a certain degree. If the data is close to the center of the cluster, it has a high degree of membership and if it is far from the centers, it has a low degree. The algorithm includes randomly initialize the cluster membership value, calculating the center of each cluster, updating the membership value, and repeating the steps to minimize the difference of data assigned to each cluster and the cluster centers [4]. In this part, a fuzzy c-mean clustering approach is used for clustering the time series data of three engine units based on the PHM08 dataset. Matlab fuzzy c-means clustering is used for forming four clusters from the four sensors measures. Figure 7 shows the result. As shown in this figure, fuzzy c-mean clustering could be used for analyzing the behavior of complex products during the life cycle. The healthy, caution, repair, and failure regime could be detected based on the result of clusters.

The other data that should be integrated into sensors data is operational settings and the supply and demand of the recovered parts. This data will be used as the input of the fuzzy logic model. For example, Fig. 8 shows the fuzzy membership functions of sensors based on the PHM08 dataset. The membership functions will be used for addressing the rules. The output of the fuzzy system is the health condition of the unit and leads to the estimation of RUL. When the operator on the disassembly site selects the target disassembly unit, the virtual content of sub-assembly will be created in a remote site. This virtual content will be created based on CAD data and parts database. A 3D model of the parts will be developed and will be accessible for the operator on the disassembly site. The sequence of disassembly and the performance of the operation will be controlled via disassembly DT. Hence, the operator has access to real-time and dynamic disassembly sequence and could decide based on the RUL of the target part.

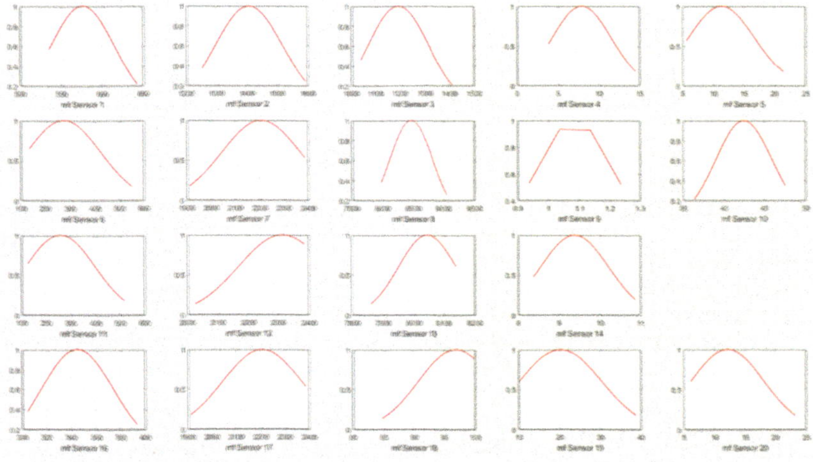

Fig. 8 Fuzzy membership functions of engine sensor measures

7 Conclusion

Estimating RUL plays an essential role in the EoL decision of the recovered parts from EoL aircraft. Digital twin provides a real-time digital profile of the target parts and integrates the 3D models and historical data for health monitoring of the parts. The fuzzy simulation also provides a knowledge-based decision-making module for integrating multi-layers data and dealing with uncertainty and ambiguity in the context of the operational environment and failure regime of the parts. This chapter introduced a joint application of digital twin and fuzzy decision making for aircraft decommissioning decision making. The advantages and application perspectives of the proposed framework are discussed. In this chapter, the health monitoring of the engine unit is addressed. However, the complexity of applying the DT model for all levels in engine configuration including replacement unit level, shop replacement level, and parts should be addressed in future research. Wang et al.[43] considered three modules for the digital twin model including digital model, data analytics, and knowledge base. The proposed architecture in this study also considered these three essential elements. Digital model for simulation of the target subassembly and parts behavior, data analytics for clustering and data mining of multidimensional data and knowledge part enabling by fuzzy decision making for integrating the experts and operators knowledge into the decision-making process. In this study, a conceptual framework is proposed. An experimental analysis is required to evaluate this framework. A pilot digital twin model is proposed as future research.

References

1. Aivaliotis, P., Georgoulias, K., Arkouli, Z., Makris, S.: Methodology for enabling digital twin using advanced physics-based modelling in predictive maintenance. Proc. CIRP **81**, 417–422 (2019)
2. Atli, O., Kahraman, C.: Aircraft maintenance planning using fuzzy critical path analysis. Int. J. Comput. Intell. Syst. **5**(3), 553–567 (2012)
3. Azadeh, A., Sheikhalishahi, M., Khalili, S.M., Firoozi, M.: An integrated fuzzy simulation–fuzzy data envelopment analysis approach for optimum maintenance planning. Int. J. Comput. Integr. Manuf. **27**(2), 181–199 (2014)
4. Bezdec, J.C.: Pattern Recognition with Fuzzy Objective Function Algorithms. Plenum Press, New York (1981)
5. Bouarfa, S., Doğru, A., Arizar, R., Aydoğan, R., Serafico, J.: Towards automated aircraft maintenance inspection. A use case of detecting aircraft dents using Mask R-CNN. In: AIAA Scitech 2020 Forum, p. 0389 (2020)
6. Büyüközkan, G., Feyzioğlu, O., Havle, C.A.: Analysis of success factors in aviation 4.0 using integrated intuitionistic fuzzy MCDM methods. In: International Conference on Intelligent and Fuzzy Systems, pp. 598–606. Springer, Cham (2019)
7. Ceruti, A., Marzocca, P., Liverani, A., Bil, C.: Maintenance in aeronautics in an Industry 4.0 context: the role of augmented reality and additive manufacturing. J. Comput. Des. Eng. **6**(4), 516–526 (2019)
8. Chemweno, P.K., Pintelon, L.: Towards e-maintenance: an exploration approach for aircraft maintenance data. In: Applications and Challenges of Maintenance and Safety Engineering in Industry 4.0, pp. 189–212. IGI Global (2020)
9. Demirci, S., Hajiyev, C., Schwenke, A.: Fuzzy logic-based automated engine health monitoring for commercial aircraft. Aircr. Eng. Aerosp. Technol. (2008)
10. Deng, Q., Santos, B.F., Curran, R.: A practical dynamic programming based methodology for aircraft maintenance check scheduling optimization. Eur. J. Oper. Res. **281**(2), 256–273 (2020)
11. Dinis, D., Barbosa-Póvoa, A., Teixeira, A.P.: Valuing data in aircraft maintenance through big data analytics: a probabilistic approach for capacity planning using Bayesian networks. Comput. Ind. Eng. **128**, 920–936 (2019)
12. Elsayed, E.A.: Mean residual life and optimal operating conditions for industrial furnace tubes. In: Case Studies in Reliability and Maintenance, pp. 497–515 (2003)
13. Errandonea, I., Beltrán, S., Arrizabalaga, S.: Digital twin for maintenance: a literature review. Comput. Ind. **123**, 103316 (2020)
14. Fornlöf, V.: Improved remaining useful life estimations for on-condition parts in aircraft engines. Doctoral dissertation, University of Skövde (2016)
15. Gholizadeh, H., Javadian, N., Fazlollahtabar, H.: An integrated fuzzy-genetic failure mode and effect analysis for aircraft wing reliability. Soft Comput. 1–12 (2020)
16. Guyon, I., Amine, R., Tamayo, S., Fontane, F.: Analysis of the opportunities of industry 4.0 in the aeronautical sector. In: 10th International conference on complexity, Informatics and Cybernetics: IMCIC2019 (2019)
17. Hazbon, O., Gutierrez, L., Bil, C., Napolitano, M., Fravolini, M.: Digital twin concept for aircraft system failure detection and correction. In: AIAA Aviation 2019 Forum, p. 2887 (2019)
18. IATA Report. https://www.iata.org/contentassets/ffbed17ac843465aad778867cb23c45c/bipad.pdf (2018)
19. Keivanpour, S., Kadi, D.A.: A sustainable approach to aircraft engine maintenance. IFAC-PapersOnLine **48**(3), 977–982 (2015)
20. Keivanpour, S., Kadi, D.A.: The effect of "internet of things" on aircraft spare parts inventory management. IFAC-PapersOnLine **52**(13), 2343–2347 (2019)
21. Keivanpour, S.: End of Life Management of Complex Products in an Industry 4.0 Driven and Customer-Centric Paradigm: A Research Agenda. Accepted in MOSIM 2020 (2020)

22. Keivanpour, S., Kadi, D.A., Mascle, Ch.: The role of estimation of residual life time. In: End of Life Aircraft Components Decision Making. Sustainability 2015, Sponsored by AHS International's Montréal-Ottawa Chapter, September 22–24, Hotel Bonaventure, Montréal, Québec, Canada (September 2015)
23. Khan, S., Farnsworth, M., McWilliam, R., Erkoyuncu, J.: On the requirements of digital twin-driven autonomous maintenance. Annu. Rev. Control (2020)
24. Latorella, K.A., Prabhu, P.V.: A review of human error in aviation maintenance and inspection. Int. J. Ind. Ergon. **26**(2), 133–161 (2000)
25. Li, S.G., Kuo, X.: The inventory management system for automobile spare parts in a central warehouse. Expert Syst. Appl. **34**(2), 1144–1153 (2008)
26. Liu, Y., Wang, T., Zhang, H., Cheutet, V., Shen, G.: The design and simulation of an autonomous system for aircraft maintenance scheduling. Comput. Ind. Eng. **137**, 106041 (2019)
27. Liu, Y., Wang, T., Zhang, H., Cheutet, V.: Simulation-based fuzzy-rough nearest neighbour fault classification and prediction for aircraft maintenance. J. Simul. 1–15 (2019)
28. Lu, Y., Liu, C., Kevin, I., Wang, K., Huang, H., Xu, X.: Digital Twin-driven smart manufacturing: connotation, reference model, applications and research issues. Robot. Comput.-Integr. Manuf. **61**, 101837 (2020)
29. Lu, K.Y., Sy, C.C.: A real-time decision-making of maintenance using fuzzy agent. Expert Syst. Appl. **36**(2), 2691–2698 (2009)
30. Lu, Z., Sun, Y.C.: Maintainability virtual evaluation method based on fuzzy multiple attribute decision making theory for civil aircraft system. In: 2009 8th International Conference on Reliability, Maintainability and Safety, pp. 684–689. IEEE, New York (July 2009)
31. Lu, Z., Zhou, J., Li, N.: Maintainability fuzzy evaluation based on maintenance task virtual simulation for aircraft system. Eksploatacja i Niezawodność **17**(4) (2015)
32. Mandolla, C., Petruzzelli, A.M., Percoco, G., Urbinati, A.: Building a digital twin for additive manufacturing through the exploitation of blockchain: a case analysis of the aircraft industry. Comput. Ind. **109**, 134–152 (2019)
33. Mi, S., Feng, Y., Zheng, H., Wang, Y., Gao, Y., Tan, J.: Prediction maintenance integrated decision-making approach supported by digital twin-driven cooperative awareness and interconnection framework. J. Manuf. Syst. (2020)
34. Nowlan, F.S., Heap, H.F.: Reliability-Centered Maintenance. United Air Lines Inc., San Francisco, CA (1978)
35. Okoh, C., Roy, R., Mehnen, J., Redding, L.: Overview of remaining useful life prediction techniques in through-life engineering services. Proc. CIRP **16**, 158–163 (2014)
36. Rahman, S.M., Perry, N., Müller, J.M., Kim, J., Laratte, B.: End-of-life in industry 4.0: ignored as before? Resour. Conserv. Recycl. **154**, 104539 (2020)
37. Ramudhin, A., Paquet, M., Artiba, A., Dupré, P., Varvaro, D., Thomson, V.: A generic framework to support the selection of an RFID-based control system with application to the MRO activities of an aircraft engine manufacturer. Prod. Plann. Control **19**, 183–196 (2008)
38. Saxena, A., Goebel, K.: PHM08 Challenge Data Set. NASA Ames Prognostics Data Repository (https://ti.arc.nasa.gov/project/prognostic-data-repository), NASA Ames Research Center, Moffett Field, CA (2008)
39. Si, X.S., Wang, W., Hu, C.H., Zhou, D.H.: Remaining useful life estimation—a review on the statistical data driven approaches. Eur. J. Oper. Res. **213**(1), 1–14 (2011)
40. Tyncherov, T., Rozkova, L.: Aircraft lifecycle digital twin for defects prediction accuracy improvement. In: International Conference on Reliability and Statistics in Transportation and Communication, pp. 54–63, Springer, Cham (October 2019)
41. Utzig, S., Kaps, R., Azeem, S.M., Gerndt, A.: Augmented reality for remote collaboration in aircraft maintenance tasks. In: 2019 IEEE Aerospace Conference, pp. 1–10. IEEE, New York (March 2019)
42. Valdés, R.A., Comendador, V.F.G., Sanz, A.R., Castán, J.P.: Aviation 4.0: more safety through automation and digitization. In: Aircraft Technology. IntechOpen (2018)

43. Wang, J., Ye, L., Gao, R.X., Li, C., Zhang, L.: Digital twin for rotating machinery fault diagnosis in smart manufacturing. Int. J. Prod. Res. **57**(12), 3920–3934 (2019)
44. Xiongzi, C., Jinsong, Y., Diyin, T. Yingxun, W.: Remaining useful life prognostic estimation for aircraft subsystems or components: a review. In: 2011 10th International Conference on Electronic Measurement & Instruments (ICEMI), pp. 94–98. IEEE, New York (2011)

Designing a System Architecture for the Management of the Recovered Parts from End-of-Life Aircraft Using Fuzzy Decision Making and Blockchain

Samira Keivanpour

Abstract The recovery of high-value parts from end-of-life (EoL) aircraft as a business model is a main source of revenue. Following airworthiness regulations, performing parts inspections and certifications and tracing documentation are critical for the safety and efficiency of parts trading. In this chapter, an architecture is proposed based on fuzzy intelligent decision making and blockchain-enabled network for real-time data accessibility of multi-actors involved in the ecosystem of EoL aircraft treatment and the effective management of the recovered parts after disassembly. This architecture has three layers: the management layer contains fuzzy inference system models for revenue, supply and demand management as well as data mining module, blockchain layer for tracing parts, certifications, and following airworthiness regulations supported by a connection layer. The application perspective for EoL aircraft is also discussed.

Keywords Blockchain technology · Fuzzy decision making · End of life aircraft · Recovered parts trading

1 Introduction

The number of aircraft being retired each year is increasing. The efficiency, safety, and sustainability of the treatment process are essential. Time plays an essential role in the decommissioning process, because it can guarantee the value of recovered parts and recycled materials and the sustainability of the entire process [10]. This research area is a new field, and the studies and publications have been increasing since 2010. A few studies have addressed and systematically analyzed the EoL aircraft via case studies (e.g., [13, 22, 29]). Considering the limited experience in managing an EoL aircraft, developing intelligent decision tools and testing and evaluating prototypes are critical. The Industry 4.0 revolution has influenced the dismantling, recycling,

S. Keivanpour (✉)
Department of Mathematical and Industrial Engineering, Polytechnique Montreal, Montreal, Quebec, Canada
e-mail: samira.keivanpour@polymtl.ca

C. Kahraman and S. Aydın (eds.), *Intelligent and Fuzzy Techniques in Aviation 4.0*, Studies in Systems, Decision and Control 372,
https://doi.org/10.1007/978-3-030-75067-1_14

remanufacturing, and logistics of recovered parts and materials. Hence, exploring the opportunities and challenges of the different technologies of Industry 4.0 in aircraft decommissioning provides great advantages.

The application of blockchain for tractability in a supply chain is a fresh research topic. The perspective of application in the food supply chain, pharmaceutical, and spare-part inventory management have received much more attention recently [5, 18, 24, 25, 26]. Blockchain technology provides real-time and secure accessibility of data in a multi-actor collaboration context. The quick and wide-ranging access to information accelerates the processes and avoids bottlenecks [7]. Rajkov [21] performed research on the application of blockchain for aircraft spare-part management. The author interviewed several maintenance, repair, and overhaul (MRO) experts regarding the application of blockchain in the traceability of parts and the communication between MRO and other partners.

One of the main challenges regarding the management of aircraft at EoL is developing a robust business model for recovered parts and materials. Selling the parts in the second-hand parts market with traceability features considering the criticality of tracking parts in the aviation industry is essential for EoL aircraft recycling initiatives. Operational experience is essential in the treatment process. The optimization tools and methods should be integrated with operational experience and industry best practices to maximize the efficiency and sustainability of the process. Fuzzy interactive decision-making facilitates the problem solving of uncertain, unknown, or incomplete information [12]. Randomness and uncertainties are a critical part of a parting-out operation that should be considered.

The disassembly time is an important factor in the cost of disassembly. An operator's skills, the availability of tools, and set-up time are other uncertain variables that influence the supply of parts. The value of the parts in the market and the required directives and regulations are other stochastic parameters that impact the supply and demand of recovered parts. Considering the uncertainty and stochastic nature of these factors, the fuzzy decision model is promising.

To the best of my knowledge, the application of fuzzy intelligent decision-making and blockchain has not been addressed in the management of recovered parts from an EoL aircraft. The contribution of this chapter is two-fold. It discusses the opportunities and challenges of the smart management of recovered parts and then proposes a system architecture for revenue management, visualization of supply and demand, clustering of high-value parts, and logistics management. The blockchain layer provides an opportunity to follow regulations, to trace parts in terms of certification, documentation, and inspection and to trade parts safely and efficiently.

The rest of this chapter is organized as follows: Sect. 2 provides a literature review on blockchain technology and its application in spare parts management and the aviation industry, Sect. 3 discusses the management of the recovered parts at the EoL, Sect. 4 proposes a system architecture, Sect. 5 discusses the application perspective and finally Sect. 6 concludes with some remarks and the direction of future research.

2 Literature Review

In this section, a brief review is provided based on the application of blockchain in supply chain and the product life cycle. The implication of this technology in the spare part management and some perspectives in the aviation industry are discussed. Wang et al. [28] performed a literature review in the application of blockchain in the supply chain context [28]. The authors discussed the extended visibility and traceability, digitalization of supply chain, improved data security, and information sharing and smart contracts as the advantages of blockchain technology in the supply chain. Casino et al. [4] provided a taxonomy of the application of blockchain in different contexts. For the business and industrial applications, the authors discussed the role of this technology in decentralized business management processes, workflows and automating the intra-organizational process, and reducing the costs. In logistics, it leads to robust contract management between the logistics parties and improving inventory management.

Fraga-Lamas and Fernández-Caramés [6] provided a literature review on the application of blockchain in the automotive industry. The authors discussed inefficient stock management, the market of counterfeit parts, lack of transparency and warranty monitoring, low margins as the existing challenges in the automotive industry that can be addressed with blockchain technology. Hasan et al. [8] also proposed a blockchain-enabled smart contract system for spare part management in the automotive industry considering the details from upstream to downstream of the supply chain.

Keivanpour et al. [15] proposed a software architecture for blockchain adoption in the management of the complexity in the offshore wind energy supply chain. The authors analyzed the procurement process for turbine manufacturing. Keivanpour et al. [17] also addressed the adoption of blockchain technology in upstream, midstream, and downstream of the supply chain for offshore wind energy. In another study, Keivanpour and Ait-Kadi [14] studied the implication of the Internet of things (IoT) and RFID (Radio-frequency identification) technology in four generic practices of spare part management in the aviation industry. These technologies have challenges of energy consumption, security, and synchronization. Hence, designing an inter-organizational distributed network based on blockchain technology is promising. In the inventory pooling model, there is not a centralized agent and the parties agree on sharing the sources based on a certain arrangement. This model is a better candidate for adopting blockchain technology.

Wang and Li [27] discussed the application of IoT and blockchain in the aviation industry. According to these authors, RFID technology is frequently used by Airbus and Boeing in the supply chain for quality management and cost reduction purposes. However, this technology cannot remove fraud possibility. Moreover, the interoperability of IoT considering involving several actors in the aviation supply chain is challenging. The authors suggested an alliance chain with pre-selected nodes for certain actors for the aviation industry. They concluded that blockchain technology will maximize the airworthiness of the equipment during the life cycle. According to IATA blockchain white paper [10], analyzing the blockchain use cases in the aviation industry is essential as it can provide a timely assess the opportunities and challenges of this technology. The relevant prototypes in this context offer low-risk solutions for different applications of blockchain in distribution and payment, spare part management, passenger and crew management, and across the value chain.

Mandolla et al. [20] applied blockchain for developing the digital twin for additive manufacturing in the aircraft industry. The authors designed the different blocks for scanning and designing, building and monitoring, testing, and delivery. The different blocks (phases) will be added to a private blockchain and are accessible by different actors to enter the data into the blocks. The hash function could provide security of the network and guarantee compliances in terms of quality and safety.

Aleshi [2] and Aleshi et al. [3] proposed a secure blockchain for aircraft maintenance records. The author explained the three essential elements of the maintenance logbooks including the total maintenance history, the parts, and airframe records, and the certification based on airworthiness regulations. The author suggested a secure blockchain ledger for tracing the maintenance records. The author emphasized the need for a different data model, use case, and consensus algorithm in comparison to the bitcoin context. Storing maintenance records is more complex rather than transactions. In this case, blockchain is private and will be accessible to the actors. Madhwal and Panfilov [18] discussed the application of blockchain in the management of aircraft's spare parts. The authors explained that maintenance features, the life expectancy of different parts, the number of parts and related parameters, and the number of actors involved in the aviation supply chain leads to analyzing the large data. Rajkov [21] studies the implications of blockchain from a business model perspective. The author also discussed the complexity of the supply chain in the aviation industry due to aircraft configuration, different tiers in the supply chain, spare part management challenges, and communication [16].

Schyga et al. [23] studied the strengths and weaknesses of blockchain technology in aircraft maintenance operations. The authors mentioned the complexity of maintenance operations, involving several players, and the need for efficient communication and transparency as the applicability of blockchain technology in aircraft maintenance.

The synthesis of the literature reveals the following points:

- Blockchain technology is a reliable mechanism for traceability and ensuring the airworthiness of parts or equipment in the aviation industry
- The distributed ledger is an appropriate mechanism for data sharing and data security
- The application of the blockchain for the management of EoL components are not studied in a systematic way
- Developing the use case and designing prototypes are the consistent solution for analyzing the opportunities and challenges of this technology in the context of recovered parts distribution.

3 Aircraft at the End of Life and Parts Management

The life expectancy of the parts, the commonality of parts among aircraft types, and regulatory requirements are critical parameters in addressing the demand and supply of the recovered parts. According to the International Air Transport Association [9], the parts will be classified into three groups of life-limited parts, time-controlled components, and other parts. The life-limited parts should be replaced after a certain period determined by the manufacturer's maintenance manual (mainly based on flight hours or calendar days). Time-controlled components require periodic replacement and inspection based on the scheduled maintenance program. The other category of the parts does not need detailed maintenance records and is called "on condition" parts. The main revenues from parts will be obtained for 1–2 years and the other parts will be sold slowly during the rest period. Figure 1 shows the Pareto diagram of sales revenue of the recovered parts from narrow-body aircraft according to the data of IATA. Different online platforms provide market analysis. They aid search and finding the parts in different vendors or regions and MRO resources (ILS website [11]).

Hence, market visibility can help to arrange the sale process of the recovered parts. The other concern is parts certifications. The recovered parts should be certified by airworthiness to be sold in the market. Besides, the agreements between regions and countries should be followed in the context of international trade. Hence, the recovered parts should be inspected and certified by an authorized organization before

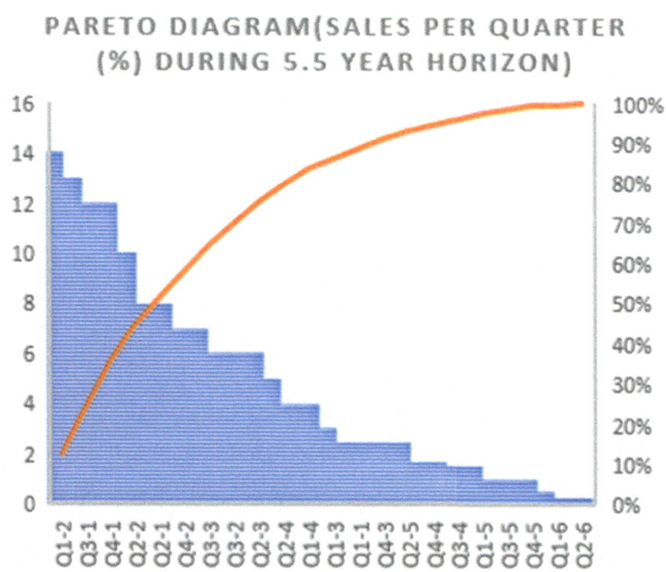

Fig. 1 Pareto diagram of selling the recovered parts during the 5.5-year horizon (source of data IATA: [9])

returning to fleet operation. Multiple actors are involved in ensuring the safety aspects of the parts. All the records should be tracked to avoid entering an unapproved part in the market. Hence, the aircraft owner, dismantling, and disassembly initiative, MRO should be involved in tracking all steps. Some different inter-related data and records should be traced and tracked for ensuring the safety of parts and the successful trading of the recovered parts. These data are in different separate databases with different owners and actors who are involved in this business. This increases the complexity of tracing data and reliable trading operations. Figure 2 shows the complexity based on the required data and records that will be connected in a multi-actor setting. Table 1 shows the flow of information that should be shared among different actors participating in the parts distribution value chain.

4 Proposed System Architecture Using Fuzzy Logic and Blockchain

Figure 3 shows the proposed system architecture for application of fuzzy systems and block chain in the EoL aircraft recovered parts management. This architecture has three main layers. The management layer includes a fuzzy intelligent decision-making module for revenue management and visibility of supply and demand of the parts market. The classification of the parts based on the value, market demand, the remaining life per airworthiness directives, the required inspections or repairments, and the mandatory records and documentation will be performed. The health monitoring data, maintenance data, and market data will be entered into the fuzzy model to analyze the supply and demand, revenue flow, and the safety of the parts. This fuzzy decision-making module controls the sustainability of the part's re-certification and distribution. The second layer, the blockchain layer includes the transaction layer, censuses algorithm, and smart contracts development and application. The connection layer encompasses the connection between different objects including operators, smart tools, disassembly sites, detecting equipment, parts, and key actors involved in the parts recovery and distribution and the required communication channels.

5 Application Perspectives

In this section, more details of the fuzzy intelligent system and the blockchain use case are provided. According to IATA's best industry practices for aircraft decommissioning [10], identification of high-value parts is performed based on market analysis and previous experience. The experience of key players such as MRO, OEMs (original equipment manufacturers), and EoL enterprise translate into the main decision for the treatment of aircraft at EoL and should lead to efficiency and effectiveness of the different sub-processes. One of these key decisions is classifying

Fig. 2 The complexity of interconnected data and records for tracing and tracking the recovered parts

Table 1 The information flow and relationship between actors

Information flows	Disassembler	MRO	Parts trader	OEM	Regulatory agencies
Tracing HAZMAT material	√		√	√	
Tracing contaminated parts	√		√	√	
Tracking parts for ensuring continued airworthiness	√	√	√	√	√
Tracing the payment method	√		√		
Tracing the available quantity of parts	√		√		
Accessibility of the documentations	√	√	√	√	√
Accessibility of parts historical records	√	√	√	√	√
Tracing delivery time of parts	√		√		
Tracing the price of parts	√		√		
Tracing suspected parts	√	√	√		√
Tracing counterfeit parts	√	√	√		√
Following the applicable airworthiness regulations	√	√	√	√	√
Verifying the arrangements between countries	√	√	√		√
Tracking inspection and certifications records	√	√	√	√	√
Following airworthiness certification	√	√	√	√	√
Following FAA (federal aviation administration) export airworthiness	√	√	√		√
Identifying high-value parts	√		√		

the parts into three categories of slow movers, mid movers, and fast movers parts based on the speed of selling [10]. The priority of the disassembly process should be performed based on this classification. The systematic way of using this experience represents the characteristics of human learning and are in categories of artificial intelligence [19]. According to Mamdani and Assilian [19], many engineering decisions should be made intuitively. What they perform during strategic tasks is the translation of the optimization model, data analytic, and AI of engineers that will be

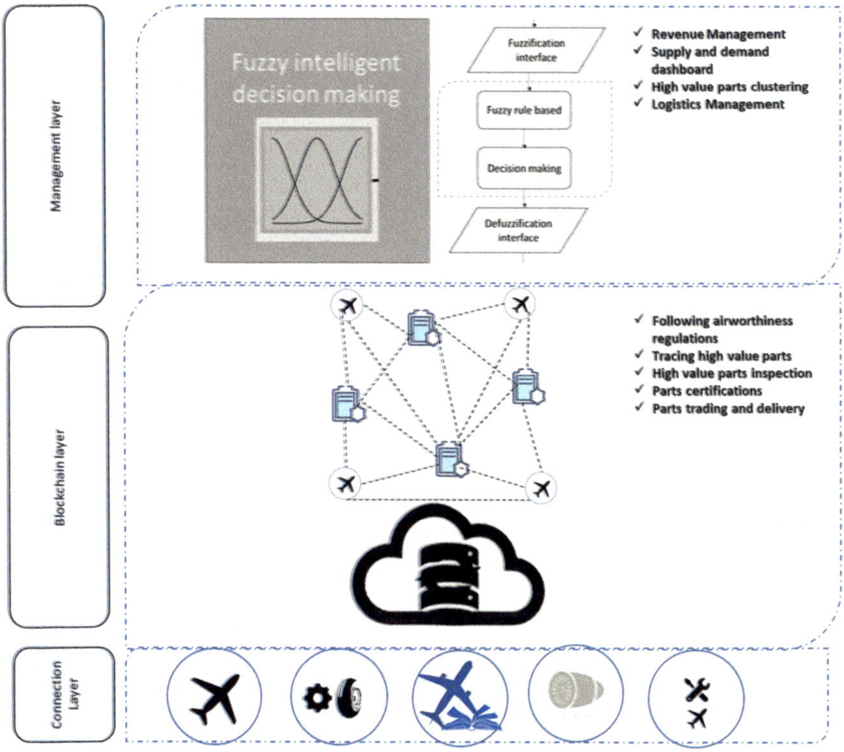

Fig. 3 The system architecture for fuzzy intelligent and blockchain-enabled recovered parts trading

interpreted into certain decisions and actions. Hence, supporting the explanation of vaguer and non-numeric states as the results of different variables are essential. The fuzzy decision-making model proposed by Mamdani is applied here to provide a semi-quantitative approach based on market data and the experience of the decision-makers. Figure 4 shows the fuzzy logic model for controlling revenue flow based on the supply and demand of the parts and market data. The sources of input data are health monitoring data, maintenance database, and market data. Different trading portals provide information related to recovered parts. The part condition and life expectancy, the commonality of the parts, and related directives control the demand of the target parts. The operational environment, disassembly, and logistics costs are key factors in determining the supply of the parts. It should be noted that, here the influential factors for the part-out company and disassembly operation are considered. Time of disassembly, tooling and set-up cost can determine the disassembly cost of operation. Other macroeconomic factors also impact the aircraft retirement rate such as fuel price, GDP, pandemic. These parameters are not considered in the model. However, another fuzzy logic model could be designed for addressing these factors. Figures 5 and 6 shows the Fuzzy logic designer view and rule view in MATLAB. A Gaussian membership function is considered in this example. For the

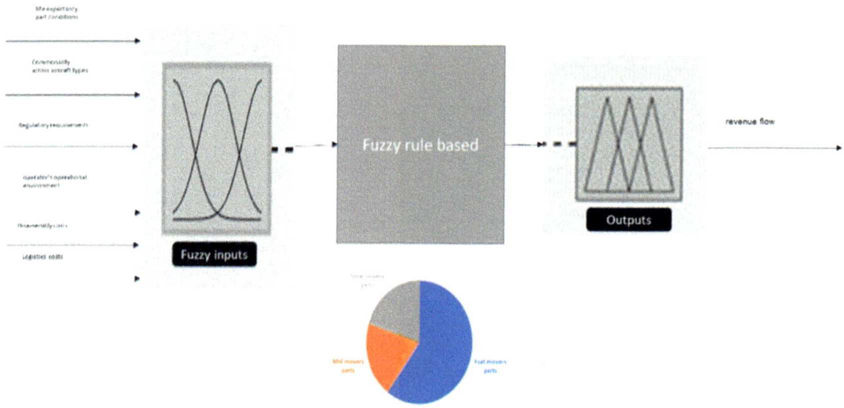

Fig. 4 The fuzzy decision-making model for controlling the revenue flows of recovered parts

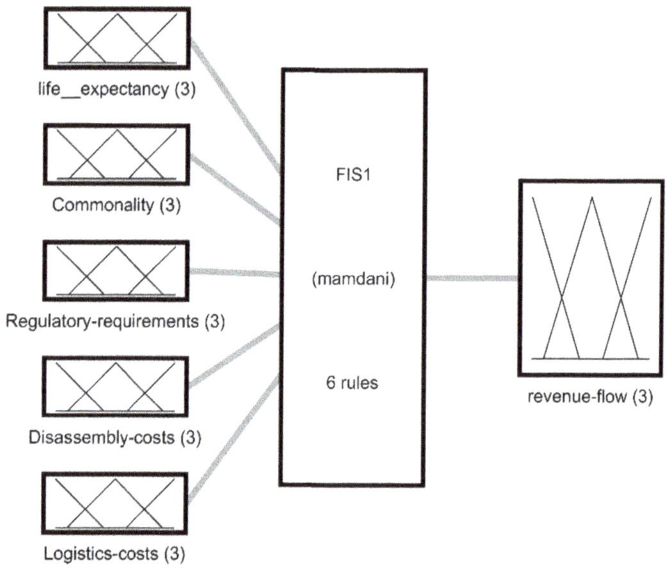

Fig. 5 Fuzzy logic designer view for controlling the revenue flows of recovered parts

rule view, for example, the first rule shows if the life expectancy of the part is low, the commonality is low, the regulatory requirements are strict, the disassembly cost and logistics cost are high, the revenue of the parts is low.

The revenue flow in Fig. 4 is related to the high-value parts. The other revenue stream is related to recovered materials. The negative value of the safe disposal of hazardous waste also should be considered. The efficiency of recycling depends on the level of sorting the material before dismantling. Smart sorting is an appropriate

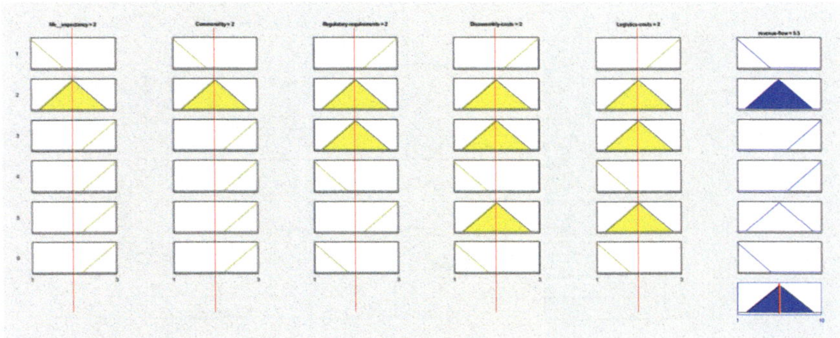

Fig. 6 The example of rule view for fuzzy logic model

solution to manage the revenue of the recovered materials. Hence, the total revenue could be addressed as follows:

$$\tilde{V}_{Total} = \tilde{V}_{parts} + \tilde{V}_{material} - \tilde{C}_{Hazardous\_w} \tag{1}$$

where

$\tilde{V}_{Total}$ is the total value of the EoL project
$\tilde{V}_{parts}$ is the value of recovered parts
$\tilde{V}_{material}$ is the value of recovered materials
$\tilde{C}_{Hazardous\_w}$ is the cost of the safe disposal of the hazardous material.

According to the IATA report, the value from high-value parts contributes to 90% of the total value of the aircraft at EoL while it is only about 10–15% of the aircraft's empty weight. In this study, the value chain of the parts is considered. The second and third element of the system architecture presented in Fig. 3 is the blockchain and connection layer. A use case is presented here to illustrates the application of smart contracts for trading the recovered parts from EoL aircraft. The class diagram of the players and a section of the UML (Unified Modeling Language) class diagram are shown in Figs. 7 and 8. The recovered high-value parts class is defined in the diagram. The inspection, certification, trading, and records, and documentation are

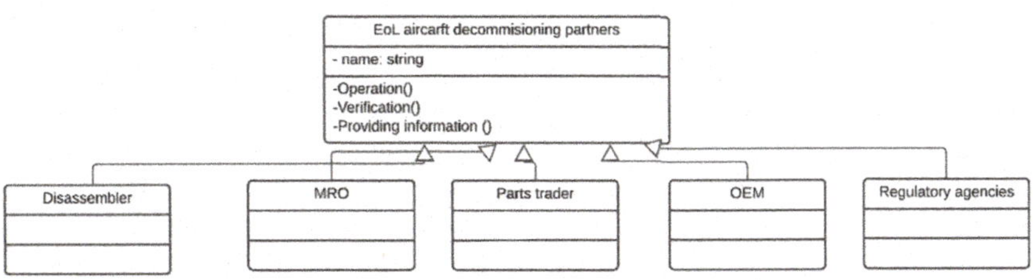

Fig. 7 Class diagram of players (Created in Lucidchart, https://www.lucidchart.com) [1]

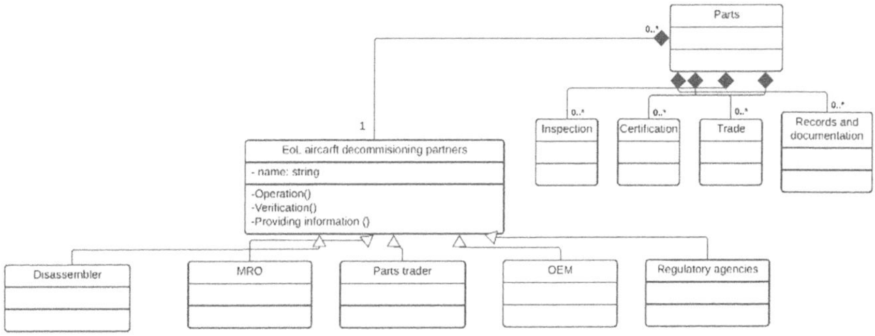

Fig. 8 A section of the UML class diagram (Created in Lucidchart, https://www.lucidchart.com)

assigned classes in the data structure that enabling the partners to modify and change the statuses. A private blockchain network is considered and the service is trading the recovered parts with ensuring the safety and profitability of the key actors involved in the problem. The first phase is identifying the actors involved in the context and planning the interviews with them for explaining the blockchain technology, the opportunities and challenges, and determining the roles and responsibilities of each actor in the network. The EoL enterprise or aircraft owner and the buyers (reliable distributor) are considered as external owned accounts. The accounts should be controlled by private keys. These actors can provide smart contracts and manage transactions. For the smart contract, there should be a related account to receive the messages and executes the codes. The smart contracts are between parts owners (EoL enterprise or disassembly company or aircraft owner) and the reliable distributors, original equipment manufacturers, airworthiest, and authorized maintenance organizations. The purpose of these smart contracts is to make sure that the applicable airworthiness regulations are followed, the related arrangement between countries (for the case of the international distributor) are considered, high-value parts are appropriately inspected by authorized MRO, the parts are tracked, the value of the part is paid on time and the target part will be delivered safely to the reliable distributor. Figure 9 shows different smart contracts between the actors involved in the blockchain.

6 Conclusion

The efficiency and sustainability of the aircraft decommissioning require the application of the appropriate decision tools with integrating the industrial best practices and operational experience. Moreover, traceability and following the airworthiness directives are essential for entering the recovered parts in the market and optimizing the revenue stream. The complexity of process treatment, different actors participated in parts trading and the need for secure data exchange among players made

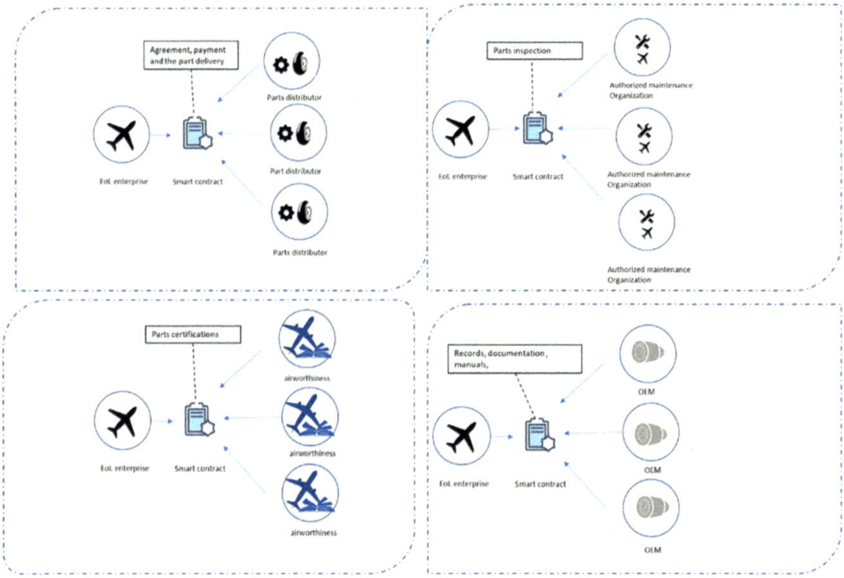

Fig. 9 The main processes of smart contracts between the actors involved in trading the recovered parts

blockchain as a promising technology for the management of high-value parts from EoL aircraft. This chapter proposed a system architecture for integrating the operational experience via a fuzzy decision model and the blockchain for security and safety of the recovered parts tracing. A preliminary use case and model architecture proposed in this chapter. The detailed use cases and a SWOT (strength, weakness, opportunities, and threats) analysis are proposed as future research. A discrete event simulation could be developed for assessing information sharing scenarios too.

References

1. https://www.lucidchart.com
2. Aleshi, A.: Secure aircraft maintenance records using blockchain (SAMR) (2018)
3. Aleshi, A., Seker, R., Babiceanu, R.F.: Blockchain model for enhancing aircraft maintenance records security. In: 2019 IEEE International Symposium on Technologies for Homeland Security (HST), pp. 1–7. IEEE, New York (November 2019)
4. Casino, F., Dasaklis, T.K., Patsakis, C.: A systematic literature review of blockchain-based applications: current status, classification, and open issues. Telematics Inform. **36**, 55–81 (2019)
5. Chong, M., Perez, E., Castilla, J., Rosario, H.: Blockchain technology applied to the cocoa export supply chain: a Latin America case. In: Handbook of Research on Emerging Technologies for Effective Project Management, pp. 323–339. IGI Global (2020)

6. Fraga-Lamas, P., Fernández-Caramés, T.M.: A review on blockchain technologies for an advanced and cyber-resilient automotive industry. IEEE Access **7**, 17578–17598 (2019)
7. Francisconi, M.: An explorative study on blockchain technology in application to port logistics. Master Thesis, Delft University of Technology (2017)
8. Hasan, H.R., Salah, K., Jayaraman, R., Ahmad, R.W., Yaqoob, I., Omar, M.: Blockchain-based solution for the traceability of spare parts in manufacturing. IEEE Access (2020)
9. IATA Report. https://www.iata.org/contentassets/ffbed17ac843465aad778867cb23c45c/bipad.pdf (2018)
10. IATA Blockchain: International Air Transport Association: Blockchain in aviation: Exploring the fundamentals, use cases, and industry initiatives. Available at: https://www.iata.org/contentassets/965ffb686aa54cfaba93fce9f35edb80/blockchain-in-aviation-white-paper.pdf (2018)
11. ILS. https://www.ilsmart.com/
12. Jimenez, M., Arenas, M., Bilbao, A.: Linear programming with fuzzy parameters: an interactive method resolution. Eur. J. Oper. Res. **177**(3), 1599–1609 (2007)
13. Keivanpour, S., Ait Kadi, D., Mascle, C.: End-of-life aircraft treatment in the context of sustainable development, lean management, and global business. Int. J. Sustain. Transp. **11**(5), 357–380 (2017)
14. Keivanpour, S., Kadi, D.A.: The effect of the "internet of things" on aircraft spare parts inventory management. IFAC-PapersOnLine **52**(13), 2343–2347 (2019)
15. Keivanpour, S., Ramudhin, A., Ait Kadi, D.: An empirical analysis of complexity management for offshore wind energy supply chains and the benefits of blockchain adoption. Civ. Eng. Environ. Syst. **37**(3), 117–142 (2020)
16. Keivanpour, S.: End of life management of complex products in an industry 4.0 driven and customer-centric paradigm: a research agenda. In 13ème conference internationale de modelisation, optimisation et simulation (mosim 2020), 12–14 Nov 2020, agadir, maroc. (2020)
17. Keivanpour, S., Ramudhin, A., Kadi, D.A.: Towards the blockchain-enabled offshore wind energy supply chain. In: Proceedings of the Future Technologies Conference, pp. 904–913. Springer, Cham (November 2018)
18. Madhwal, Y., Panfilov, P.B.: Blockchain and supply chain management: aircrafts 'parts' business case. In: Annals of DAAAM & Proceedings, vol. 28 (2017)
19. Mamdani, E.H., Assilian, S.: An experiment in linguistic synthesis with a fuzzy logic controller. Int. J. Hum. Comput. Stud. **51**(2), 135–147 (1999)
20. Mandolla, C., Petruzzelli, A.M., Percoco, G., Urbinati, A.: Building a digital twin for additive manufacturing through the exploitation of blockchain: a case analysis of the aircraft industry. Comput. Ind. **109**, 134–152 (2019)
21. Rajkov, D.: Blockchain for aircraft spare part management: evaluating the robustness of the maintenance, repair and overhaul business model (2018)
22. Sabaghi, M., Cai, Y., Mascle, C., Baptiste, P.: Sustainability assessment of dismantling strategies for end-of-life aircraft recycling. Resour. Conserv. Recycl. **102**, 163–169 (2015)
23. Schyga, J., Hinckeldeyn, J., Kreutzfeldt, J.: Prototype for a permissioned blockchain in aircraft MRO. In: Hamburg International Conference of Logistics (HICL) 2019, pp. 469–505. epubli GmbH (September 2019)
24. Tian, F.: An agri-food supply chain traceability system for China based on RFID & blockchain technology. In: 2016 13th International Conference on Service Systems and Service Management (ICSSSM), pp. 1–6. IEEE, New York (June 2016)
25. Tian, F.: A supply chain traceability system for food safety based on HACCP, blockchain & internet of things. In: 2017 International Conference on Service Systems and Service Management, pp. 1–6. IEEE, New York (June 2017)
26. Wang, Z., Wang, T., Hu, H., Gong, J., Ren, X., Xiao, Q.: Blockchain-based framework for improving supply chain traceability and information sharing in precast construction. Autom. Constr. **111**, 103063 (2020)
27. Wang, C., Li, B.: Research on traceability model of aircraft equipment based on blockchain technology. In: 2019 IEEE 1st International Conference on Civil Aviation Safety and Information Technology (ICCASIT), pp. 88–94. IEEE, New York (October 2019)

28. Wang, Y., Han, J.H., Beynon-Davies, P.: Understanding blockchain technology for future supply chains: a systematic literature review and research agenda. Supply Chain Manage.: Int. J. (2019)
29. Zahedi, H.: End-of-life efficient disassembly of complex structures using product and process focused approach. Doctoral Dissertation, École Polytechnique de Montréal (2016)

Intelligent and Fuzzy Applications in Aviation 4.0 Transportation and Cargo Management

A Hybrid Model Based on FAHP and WASPAS for Evaluation of Explosive and Narcotics Trace Detection Devices

Eren Ozceylan, Baris Ozkan, and Cihan Cetinkaya

Abstract As a result of Industry 4.0, the aviation industry is started to be re-designed with cyber–physical systems, Internet of Things, cloud computing and similar technologies. In Aviation 4.0, where cyber–physical systems are used to assist humans' unkind work and to complete tasks autonomously, security applications are one of the promising and tight operations that need to be considered. The detection of explosives and narcotics material for the purposes of aviation security is an important area for preventing terrorism and smuggling. For this purpose, there are some novel technologies (devices) in the framework of Aviation 4.0. Evaluation and selection the best device is a multi-criteria decision making (MCDM) problem which includes ambiguities and vagueness. To do so, fuzzy analytic hierarchy process (FAHP) is applied for assigning weights of the attributes and weighted aggregated sum product assessment (WASPAS) method is used to determine the most suitable alternative device for explosive and narcotics trace detection. As a result, three well-known devices in the market are evaluated and the best alternative is suggested.

Keywords Aviation 4.0 · Evaluation · Explosive trace detection · Fuzzy AHP · Security · WASPAS

E. Ozceylan (✉)
Industrial Engineering Department, Gaziantep University, Gaziantep, Turkey
e-mail: eozceylan@gantep.edu.tr

B. Ozkan
Industrial Engineering Department, Ondokuz Mayıs University, Samsun, Turkey
e-mail: baris.ozkan@omu.edu.tr

C. Cetinkaya
Management Information Systems Department, Adana Alparslan Turkes Science and Technology University, Adana, Turkey
e-mail: ccetinkaya@atu.edu.tr

C. Kahraman and S. Aydın (eds.), *Intelligent and Fuzzy Techniques in Aviation 4.0*,
Studies in Systems, Decision and Control 372,
https://doi.org/10.1007/978-3-030-75067-1_15

1 Introduction

After the start of fourth industrial revolution, all other industries are trying to keep up with the necessities of the new world. Even if there is still a long way to go before the first fully automated airplane is produced, all the aviation firms are investing in technology [1]. Revolution of Aviation 1.0 accounts for the beginning of the aviation that the flights were conducted by visual flight rules, following visuals signals and there were no help to the pilots for flying. These times were ruled by the technological challenges for building or flying an aircraft [2]. There were very simple instruments for the virtualization of the environment and some basic indicators such as the first anemometers/altimeters to determine the altitude and speed; basic mechanical autopilots; pneumatic and electric gyroscopes to measure the attitude.

Aviation 2.0 contained the involvement of electric devices. Technological advances were driven by two important challenges; How to fly an aircraft in bad weather conditions? How to control multiple aircrafts flying in dense traffic? Very high frequency omnidirectional range and instrument landing systems allowed the pilots to follow safe tracks. Also airborne weather radars, electric autopilots, flight directors, navigation instruments etc. resulted in better safety improvements [3].

Aviation 3.0, involved the giant incorporation of electronics in aircraft cockpit, driven by the availability of data communication technology. Technological solutions were progressively designed to support the operators informed decisions, with the help of aggregated, visualized, understandable information. Operations on board and outside of the aircraft shifted from tactical to strategic; and assistance systems and safety nets became crucial elements to increase the level of safety in commercial aviation [4].

Aviation 4.0 is concerned with the design of cyber-physical systems which help people by aiding them to give decisions or to complete tasks autonomously. Thus the cyber-physical systems will form a digital and smart airplane [5]. In Aviation 4.0, big data acquisition and processing networks make possible automation and integration with information technology (IT) systems. Examples of cyber components can be; weather/traffic forecast, aircraft digital communications, flight planning/optimization algorithms, decision support software, situation awareness and etc. While examples of physical components can be; mobile aircraft; dynamic airspace traffic, weather, pollution, noise; pilots, air traffic controllers, airlines crew, etc. [3]. Some of the current and emerging applications in Aviation 4.0 can be found in Table 1.

The aviation industry is influenced very highly by the coronavirus pandemic, so the adoption of artificial intelligence (AI) became more important. The AI can help to transform the aviation industry, improve operational efficiency and it can provide better decision-making tools. In addition, in case of a second wave of the pandemic, using AI data sets will provide a better crisis response.

It seems inevitable that the Aviation 4.0 will transform the airports too. Especially the security issues will be much tighter by the use of data sharing systems [7]. Security checks at airports are crucial to ensuring the flight safety. The most common technique

Table 1 Emerging applications of aerospace and defense 4.0 [6]

Technology	Current applications	Emerging applications
Additive manufacturing	Prototyping, tooling and functional end-use parts manufacturing	Combining new types of novel materials to improve aerospace parts and accessories
Advanced analytics	Monitoring real-time aircraft health; identifying system/component failures in advance; making intelligent scheduling and forecasting models	Developing large-scale digitation of plane maintenance data and schedules; creating synergies across business and functional areas by enabling the "connected plane"
Advanced robotics and cognitive automation	Using robotics in aircraft manufacturing for more efficient production, with fewer errors and quality issues	Simplifying simulations of aircraft, weapons, and satellite performance, avoiding time-consuming analysis and tests of algorithms, software, and hardware
Artificial intelligence	Applying AI to robotics, automatic programming to tasks and processes in industrial settings, and enabling predictive maintenance	Leveraging AI and computer vision technologies to augment advanced safety features in aircraft; incorporating advanced AI into drones; and replacing human copilots in new, autonomous aircraft
Block-chain	Enabling greater transparency of information between different parties; improving just-in-time logistics; reducing erroneous orders; improving inventory turnover	Improving tracking in supply chains and procurement using a shared database with suppliers and partners; improving validation of supplier performance and reputation; and improve supply chain security
Digital reality (augmented reality/virtual reality/mixed reality)	Replacing assembly manuals with smart-glasses displays, which substantially reduces wiring production time	Using VR to optimize and design factories, and simulate an entire factory or warehouse to train workers to use equipment more safely and efficiently
Internet of Things	Monitoring aircraft engine health and optimizing engine performance based on data collected from sensors	Managing material costs and demand fluctuations by analyzing big data, enabling integrated smart connected assets and operations, and, eventually, an autonomous production environment

for checking passengers at the airport is scanning them for drugs, explosives, weapons etc. by using explosives and narcotic detection systems.

There are many different kinds of these detection systems in the market, having different advantages and disadvantages. In addition, a cost-effective solution should be found for the rapid detection of explosives and narcotics. Choosing the most appropriate detection system is a complex problem that contains different selection criteria. The MCDM techniques are generally used for these types of decision problems. However, to cope with the uncertainty in Aviation 4.0 and decision making, traditional approaches are not sufficient. The fuzzy logic unlike conventional logic system is able to model inaccurate or imprecise models. The fuzzy logic approach offers a simpler, quicker and more reliable solution that is clear advantages over conventional techniques. Integrating the fuzzy logic into MCDM approaches overcomes the vagueness in decision making process.

Thus, this paper aims to determine the most appropriate explosives and narcotics detection device in the light of Aviation 4.0 for securing the airports. To do so, firstly nine criteria are determined for evaluating the detection device alternatives. Then the FAHP method is used for determining the criteria weights because the fuzzy nature minimizes the negative effects of uncertainty in expert judgments. Later, the alternative explosives and narcotics detection devices are ranked using WASPAS method.

The outline of the paper is as follows; the first section gives information about the Aviation 4.0 then gives examples about its applications. The second section gives a detailed literature review about the MCDM applications in aviation field. The third section presents the applied methodology including FAHP and WASPAS. The fourth section defines the case study and discusses the results. Last section sums up the conclusion and future recommendations.

2 Literature Review

In this part of the study, related literature is given about the MCDM techniques in aviation sector and especially about the papers that utilize FAHP and WASPAS methods in different fields. To the best of author's knowledge, this is the first study in the literature that uses FAHP and WASPAS methods for the evaluation of explosive and narcotics trace detection devices. When the literature review is examined, the proposed methodology (FAHP and WASPAS) is applied especially in recent years.

Liou and Tzeng [8] developed a non-additive model for evaluating the service quality of airline companies and compared the results by using a conventional additive method. They made use of factor analysis to extract independent common factors and the fuzzy integral was used for integrating the performance ratings of attributes in each factor. In addition they used AHP to determine the factor weights. Finally, they used grey relation analysis and simple additive weighting method to find the airline company service quality ranks.

Nejati et al. [9] examined the quality service factors in Iranian airline industry by using a questionnaire based on SERVQUAL model and fuzzy TOPSIS methodology. They used 22 evaluation criteria for their problem. The findings showed that; flight safety and good appearance of the crew were determined as the most important service quality factors according to the Iranian respondents.

Yeh and Chang [10] presented a fuzzy MCDM approach for aircraft selection problem. A hierarchical weighting method was developed for determining the weights of each selection criteria by the help of triangular fuzzy numbers. In addition, a new fuzzy MCDM algorithm was developed to determine the final rank of decision alternatives.

Wang et al. [11] examined the aero-engine health assessment problem by using a three-step model. They used; FAHP, TOPSIS and fuzzy preference programming (FPP). FPP and FAHP methods were used for determining the weights of criteria and obtain the ratings of aero-engines. Then TOPSIS was applied to obtain an overall performance value for alternatives. They also conducted a real case study for Air China Company evaluating 11 selection criteria and 10 alternative aero-engines.

Kuo [12] presented a paper that combined VIKOR, grey relational analysis and interval valued fuzzy sets for evaluating the service quality of five different Chinese airlines. They used 14 evaluation criteria and five alternatives (airline companies) in their study. Also they conducted an empirical study to prove the proposed approach and in addition a sensitivity analysis was done to understand the stability of the model.

Aydogan [13] evaluated the performance of aviation firms in Turkey by using rough AHP and fuzzy TOPSIS. The AHP improved by rough sets theory (Rough-AHP) was used for determining the selection criteria weights and fuzzy TOPSIS method was used to obtain final ranks of the alternatives.

Chen et al. [14] proposed a FAHP, and quality function deployment (QFD) method to evaluate the importance levels of the technical criteria in aviation safety field, as defined by the International Air Transport Association. They made use of AHP to determine the criteria weights, and they used QFD to categorize the relationship between the main criteria and sub-criteria. They used four main and 12 sub-criteria in the study and the most important criteria were determined as "Extensive engine failure, uncontained engine fire".

Bruno et al. [15] proposed a model for aircraft evaluation, depending on the airlines' needs. They used a hybrid model that combines AHP and the fuzzy set theory. Four main and eight sub-criteria were used for the evaluation process.

Li et al. [16] evaluated the service quality levels of airline companies by using FAHP and 2-tuple fuzzy linguistic method. First, they used SERVQUAL method to determine the hierarchy of the evaluation criteria then they used FAHP to determine the main and sub-criteria weights by using linguistic variables. Second, the 2-tuple linguistic arithmetic mean operator was used to determine the final rank of criteria.

Karaman and Akman [17] investigated the corporate social responsibility (CSR) program assessment problem in airline industry. They used AHP to determine the weights of three main and 15 sub-criteria. They used a questionnaire—which was answered by 31 airline company experts—to determine the weights of criteria. Then, fuzzy linguistic variables were used to rank the CSR programs of airline firms.

Mahtani and Garg [18] evaluated the factors that are used to assess the financial performance of airline companies in India. These factors are analyzed and ranked by using FAHP. They determined the criticality of the criteria as; financial factors, operational factors, economic/government factors, performance factors, market related factors and external factors. They also conducted a sensitivity analysis to see how the solution changed in case of different criteria weights.

Percin [19] evaluated the airline service quality levels in Turkey by using a combined fuzzy decision making approach. Five main and 16 sub-criteria were defined for the evaluation of airline companies. The fuzzy DEMATEL and fuzzy ANP methods were used for calculating the importance of each criterion. Finally fuzzy VIKOR method was used for ranking the airlines in Turkey according to their service quality performance.

Chen [20] proposed a remoteness index based Pythagorean fuzzy VIKOR methodology for evaluating the quality service levels of domestic airlines. In addition, they compared four different algorithm performances on different case studies. In addition they also conducted different case studies such as investment stock decisions and R&D project investments.

Kiraci and Bakir [21] examined the aircraft selection problem by using the AHP, COPRAS and MOORA methods. They evaluated the most demanded aircrafts in the light of seven selection criteria and four aircraft alternatives. According to the results of AHP weighting, the most important criteria were found as fuel consumption.

Buyukozkan et al. [22] proposed a new digital service quality model for evaluating the airline service qualities of Turkish airline companies. The model contained; digital interaction, digital tangibles, digital trust, reliability, and customer-centricity dimensions and also 35 related criteria. The weights of each criterion were determined by AHP technique. The validity of the model was verified by conducting a real case of Turkish airline industry.

Until now, the papers that made use of MCDM techniques in aviation industry were summarized. Below, the papers that use FAHP and WASPAS method can be seen.

Turskis et al. [23] examined the site selection problem for a shopping center construction site in Vilnius. They proposed a FAHP and WASPAS methodology to solve the problem. They used FAHP to determine the weights of the selection criteria and they used WASPAS method to determine the most suitable site.

Alam et al. [24] investigated the public cloud service selection problem. They used FAHP and WASPAS methods to solve the problem. They used nine main and 30 sub-criteria in the study. FAHP technique was used to determine the criteria weights and WASPAS was used to rank the candidate public cloud services.

Ustali and Tosun [25] examined the new product selection problem of a technology firm which is concentrated on the market needs. For the solution process, FAHP and fuzzy WASPAS methods were used together. The criteria were; legal and political regulations, economic conditions of the market, prices of competitors, market situation, target market, cost structure and marketing strategies.

Turskis et al. [26] proposed a method to solve the problem of ensuring the sustainable development of EU countries in terms of identifying critical information infrastructures. They made use of fuzzy WASPAS and AHP methods to solve the problem. Six main criteria were used for the study. They determined "The danger to life or health" criteria as the most important criteria as a result of AHP calculations. Later, they used fuzzy WASPAS to rank the alternative information infrastructures.

Ayyildiz and Gumus [27] examined the petrol station location selection problem. The novel spherical FAHP-integrated spherical WASPAS methodology was applied for the solution. 13 different alternative locations and 16 criteria were used for evaluation. A real application was to show the applicability of the methodology.

Finally, Otay et al. [28] took into account the outsource manufacturer evaluation and selection problem. They made use of spherical FAHP and WASPAS methods as the first study in the literature. They used FAHP for determining the selection criteria weights and they used WASPAS to determine the final ranks of different manufacturers.

As seen in the literature review, there is a still lack of studies which use MCDM approaches on security applications in Aviation 4.0. To fill this gap, an approach which combines FAHP and WASPAS is applied for evaluation of explosive and narcotics trace detection devices in this study.

3 Methodology

This section aims to introduce the basic theory of FAHP to prioritize the weights and WASPAS to rank the alternatives.

3.1 FAHP

The fuzzy logic theory was firstly presented by Zadeh [29]. In classical set theory, an element is either the element of the set or not. An object cannot partially belong to a set. Fuzzy set theory performs graded data modelling using linguistic variables such as low, frequent, medium, low, high and many. Fuzzy set is a useful technique for identifying and solving uncertainty and uncertain real life issues.

Fuzzy logic sets are described by membership functions. The membership function of the fuzzy set A, represented by $\mu_{\tilde{A}}(x)$, takes values in the range [0, 1]. If $\mu_{\tilde{A}}(x) = 1$, the number x is definitely the element of the set, and if $\mu_{\tilde{A}}(x) = 0$, the number x is not the element of the set. In this case, Hsieh et al. [30] indicated that fuzzy numbers are a fuzzy subset of real numbers, representing the expansion of the view of the dependence interval. Triangular fuzzy numbers (TFNs) are a special class of fuzzy numbers. In TFNs $\tilde{A}$ expressed by $(l/m,\ m/u)$ or $(l,\ m,\ u)$, l and u stand for the lower and upper limits, respectively, and m for the modal value. Triangular type membership function is defined in Eq. (1) [31]. The operational laws are used to calculate with TFNs [30].

$$\mu_{\tilde{A}}(x) = \begin{cases} (x-l)/(m-l), l \le x \le m, \\ (u-x)/(u-m), m \le x \le u, \\ 0, \text{otherwise} \end{cases} \quad (1)$$

AHP is one of the most frequently used methods in MCDM. AHP is a mathematical approach that allows a combination of qualitative and quantitative variables to be used in decision making, giving priority to a group or individual. However, the method is deficient, especially in the case of pairwise comparisons where there is no certainty [32]. In addition, the AHP method may not fully reflect people's thoughts, even if it contains expert knowledge. The FAHP method has been proposed as a means of overcoming these deficiencies. The method has been used for different problems, including prioritization of dimensions of visual merchandising [33], bioprinter selection [34], flood risk assessment [35], optimum fuel blend selection [36], land suitability analysis [37] and intellectual property rights managers' evaluation [38]. The process for determining the weightings of the assessment criteria with FAHP can be summed up briefly as follows [39].

Step 1. *Construct the pairwise comparison matrix.*

Pairwise comparison matrices are constructed between all criteria in the hierarchical structure. Linguistic terms are allocated to the pairwise comparisons by asking "which criterion is more important" as given in Eq. (2).

$$\tilde{A} = \begin{bmatrix} 1 & \tilde{a}_{12} & \cdots & \tilde{a}_{1n} \\ \tilde{a}_{21} & 1 & \cdots & \tilde{a}_{2n} \\ \vdots & \vdots & \ddots & \vdots \\ \tilde{a}_{n1} & \tilde{a}_{n2} & \cdots & 1 \end{bmatrix} = \begin{bmatrix} 1 & \tilde{a}_{12} & \cdots & \tilde{a}_{1n} \\ 1/\tilde{a}_{12} & 1 & \cdots & \tilde{a}_{2n} \\ \vdots & \vdots & \ddots & \vdots \\ 1/\tilde{a}_{1n} & \tilde{a}_{n2} & \cdots & 1 \end{bmatrix} \quad (2)$$

where $\tilde{a}_{ij}$ measure denotes, let $\tilde{1}$ be *(1,1,1)* when i equal j (i.e. $i = j$); if criterion i has one of the above linguistic variables assigned to it when compared with criterion j, then j has the reciprocal value when compared with i. If criterion i has one of the above linguistic variables assigned to it when compared with criterion j, then j has the reciprocal value when compared with i. If criterion i is relatively important to criterion j, then that criterion j is relatively important to criterion i. If $\tilde{1}, \tilde{2}, \tilde{3}, \tilde{4}, \tilde{5}, \tilde{6}, \tilde{7}, \tilde{8}, \tilde{9}$ measure that criterion i is relatively important to criterion j and then $\tilde{1}^{-1}, \tilde{2}^{-1}, \tilde{3}^{-1}, \tilde{4}^{-1}, \tilde{5}^{-1}, 6^{-1}, \tilde{7}^{-1}, \tilde{8}^{-1}, \tilde{9}^{-1}$ measure that criterion j is relatively important to criterion i. The linguistic comparison terms and their equivalent TFNs used in this study are given in Table 2.

Step 2. *Compute the fuzzy geometric mean for each criterion.*

The use of the geometric mean technique to define the fuzzy geometric mean and the fuzzy weighting of each criterion was described by Buckley [41]. The geometric

Table 2 Linguistic variables for importance of criteria [40]

Fuzzy number	Linguistic	Scale of triangular fuzzy number	Scale of triangular reciprocal fuzzy number
1	Equal	(1, 1, 1)	(1, 1, 1)
2	Weak advantage	(1, 2, 3)	(1/3, 1/2, 1)
3	Not bad	(2, 3, 4)	(1/4, 1/3, 1/2)
4	Preferable	(3, 4, 5)	(1/5, 1/4, 1/3)
5	Good	(4, 5, 6)	(1/6, 1/5, 1/4)
6	Fairly good	(5, 6, 7)	(1/7, 1/6, 1/5)
7	Very good	(6, 7, 8)	(1/8, 1/7, 1/6)
8	Absolute	(7, 8, 9)	(1/9, 1/8, 1/7)
9	Perfect	(8, 9, 10)	(1/10, 1/9, 1/8)

mean of each row of $\tilde{A} = \left[\tilde{a}_{ij}\right]$ is calculated as in Eq. (3).

$$\tilde{r}_i = \left(\tilde{a}_{i1} \otimes \tilde{a}_{i2} \otimes \cdots \otimes \tilde{a}_{in}\right)^{1/n} \tag{3}$$

where $\tilde{a}_{in}$ is the fuzzy value of criterion i compared to criterion n; thus, $\tilde{r}_i$ is the geometric mean of the fuzzy comparison values of criterion i for each criterion.

Step 3. *Compute the fuzzy weight for each criterion.*

The fuzzy weight of the ith criterion is calculated as in Eq. (4).

$$\tilde{w}_i = \tilde{r}_i \otimes \left(\tilde{r}_1 \otimes \cdots \otimes \tilde{r}_n\right)^{-1} \tag{4}$$

where $\tilde{w}_i$ is the fuzzy weighting of the ith criterion and can be indicated by a TFN, $\tilde{w}_i = (lw_i, mw_i, uw_i)$, with lw_i, mw_i and uw_i representing the lower, middle and upper values, respectively, of the fuzzy weighting of the ith criterion.

Step 4. *Convert fuzzy weight to crisp value.*

The result of the fuzzy synthetic decision reached for each alternative is a fuzzy number. Defuzzification is a mathematical process performed to convert fuzzy output into a crisp value. The center of area (COA) method is a simple and practical method and there is no need to introduce the preferences of any evaluators [39]. The best non-fuzzy performance (BNP) value of the fuzzy number, $\tilde{w}_i$, which is equal to (lw_i, mw_i, uw_i), can be determined as in Eq. (5).

$$BNP_i = lw_i + \frac{(uw_i - lw_i) + (mw_i - lw_i)}{3}, \forall i. \tag{5}$$

Step 5. *The crisp weights of criteria are normalized.*

The crisp weights are normalized so that their sum is 1.

3.2 WASPAS

The WASPAS (Weighted Aggregated Sum Product Assessment) method is based on the combination of Weighted Sum (WS) and Weighted Product Model (WP) [42]. The method is described by Zavadskas et al. [43]. The WASPAS method uses the performance values and criterion weights of alternatives. Its steps are given below [44, 45]:

Step 1. *Construct the decision matrix.*

The decision matrix is shown in Eq. (6), which shows the performance of different alternatives under different criteria. In the decision matrix X, the performance of the ith alternative under jth criteria is shown by x_{ij}.

$$X = \left[x_{ij}\right]_{m\times n} = \begin{bmatrix} x_{11} & x_{12} & \cdots & x_{1n} \\ x_{21} & x_{22} & \cdots & x_{2n} \\ \vdots & \vdots & & \vdots \\ x_{m1} & x_{m2} & \cdots & x_{mn} \end{bmatrix}, i = 1, 2, ..., m; \ j = 1, 2, ..., n. \tag{6}$$

Step 2. *Construct the normalized decision matrix.*

Linear normalization is applied to the decision matrix by using maximization and minimization as in Eqs. (7) and (8), respectively.

$$\overline{x}_{ij} = \frac{x_{ij}}{\max\limits_{i} x_{ij}} \tag{7}$$

$$\overline{x}_{ij} = \frac{\min\limits_{i} x_{ij}}{x_{ij}} \tag{8}$$

In the equations above, normalized performance value of the ith alternative under the jth criterion is shown with $\overline{x}_{ij}$.

Step 3. *Calculate the relative importance of* ith *alternative, based on WS method.*

The relative importance of ith alternative, based on WS method, is calculated as in Eq. (9).

$$Q_i^{(1)} = \sum_{j=1}^{n} \overline{x}_{ij} w_j \tag{9}$$

In Eq. (9), $Q_i^{(1)}$ denotes the relative importance of ith alternative in relation to the jth criterion, based on WS method.

Step 4. *Calculate the relative importance of* ith *alternative, based on WP method.*

The relative importance of ith alternative, based on WP method, is calculated as in Eq. (10).

$$Q_i^{(2)} = \prod_{j=1}^{n} (\overline{x}_{ij})^{w_j} \tag{10}$$

In Eq. (10), $Q_i^{(2)}$ denotes the relative importance of ith alternative in relation to the jth criterion, based on WP method.

Step 5. *Calculate total relative importance, for each alternative.*

The total relative importance is calculated by using Eq. (11). The alternatives are ranked according to the size of the Q_i value, the highest Q_i shows the best alternative.

$$Q_i = 0.5Q_i^{(1)} + 0.5Q_i^{(2)} \tag{11}$$

In order to have increased ranking accuracy and effectiveness of the decision making process, a more generalized equation which is given in Eq. (12) for determining the total relative importance of ith alternative is developed [44, 45].

$$Q_i = \lambda Q_i^{(1)} + (1 - \lambda) Q_i^{(2)}, \quad \lambda = 1, \ldots, 0. \tag{12}$$

4 Case Study

In this section, three best-selling explosive and narcotics trace detection devices are considered. These devices are IONSCAN 500DT, ITEMISER 4DX and DEtector FLEX. These innovative devices have capabilities that far exceed the regular devices and they are very appropriate for Aviation 4.0 concept. Since they use a non-radioactive ionization source and simultaneous, dual-mode detection, they can detect a broad range of explosives and narcotics without the use of a radioactive source, eliminating the need for annual wipe tests. Among these three alternatives, the best one will be determined for the decision makers. The devices are shown in Fig. 1.

The selection criteria are formed from the literature and also a decision-making team of three people was formed. To evaluate the alternatives, totally nine criteria are determined. The considered criteria and associated value for each device are given in Table 3. All data for each device are collected from [46].

The first step of the applied methodology is determining the weights of criteria. By using the scale in Table 2, a pairwise comparison was provided to each decision maker. The pair wise comparison matrix of decision makers is given in Table 4.

IONSCAN 500DT ITEMISER 4DX DE-tector FLEX

Fig. 1 Three alternative explosive and narcotics trace detection devices

Table 3 Criteria and associated values

Criteria	Unit	IONSCAN 500DT	ITEMISER 4DX	DE-tector FLEX
C1. Analysis time	s	6.50	8.00	7.00
C2. Warm-up time	min	30.00	30.00	10.00
C3. Weight	kg	19.00	12.99	10.30
C4. Volume (screen down)	m^3	0.0496	0.0397	0.0424
C5. Volume (screen up)	m^3	0.0912	0.0912	0.0424
C6. Data display	cm	26.40	26.40	30.73
C7. Price	$	48,886	48,200	30,000
C8. Hard drive	GB	40	120	100
C9. Technology readiness	9 is best, 7 is worst	9	9	8

The linguistic terms obtained after pairwise comparisons were transformed into TFNs (Table 5) using the scale given in Table 2.

Computing the elements of synthetic pairwise comparison matrix by using the geometric mean method suggested by Buckley [41] in Eq. (3), Table 6 is created. For $\tilde{a}_2$ as example:

$$\begin{aligned}\tilde{a}_2 &= (1, 2, 3)^{1/3} \otimes (1, 1, 1)^{1/3} \otimes (1, 1, 1)^{1/3} \\ &= \left((1 \times 1 \times 1)^{1/3}, (2 \times 1 \times 1)^{1/3}, (3 \times 1 \times 1)^{1/3}\right) \\ &= (1,\ 1.26,\ 1.44)\end{aligned}$$

After the pairwise comparison matrix was transformed into TFNs, firstly $\tilde{r}_i$ value was calculated by using Eq. (3) to calculate the fuzzy weights of criteria. For $\tilde{r}_1$ as example:

Table 4 Pairwise comparisons matrix

DM	Criteria	C1	C2	C3	C4	C5	C6	C7	C8	C9
DM1	C1	–	2	6	9	7	5	1/3	6	6
	C2	1/2	–	4	5	5	5	1/6	5	4
	C3	1/6	1/4	–	7	6	4	1/5	3	3
	C4	1/9	1/5	1/7	–	2	1/2	1/6	2	1
	C5	1/7	1/5	1/6	1/2	–	1/2	1/6	2	1
	C6	1/5	1/5	1/4	2	2	–	1/6	2	1
	C7	3	6	5	6	6	6	–	6	5
	C8	1/6	1/5	1/3	1/2	1/2	1/2	1/6	–	7
	C9	1/6	1/4	1/3	1	1	1	1/5	1/7	–
DM2	C1	–	1	5	7	6	4	1/5	5	6
	C2	1	–	4	4	4	4	1/7	5	4
	C3	1/5	1/4	–	7	6	3	1/6	3	3
	C4	1/7	1/4	1/7	–	2	1/3	1/8	1	1/3
	C5	1/6	1/4	1/6	1/2	–	1/3	1/8	1	1/3
	C6	1/4	1/4	1/3	3	3	–	1/7	1	1/3
	C7	5	7	6	8	8	7	–	5	5
	C8	1/5	1/5	1/3	1	1	1	1/5	–	3
	C9	7	1/4	1/3	3	3	3	1/5	1/3	–
DM3	C1	–	1	4	6	5	3	1/6	5	5
	C2	1	–	1	3	3	4	1/7	3	4
	C3	1/4	1	–	6	5	2	1/7	3	4
	C4	1/6	1/3	1/6	–	1/2	1/4	1/8	1/2	1/3
	C5	1/5	1/3	1/5	2	–	1/4	1/8	1/2	1/3
	C6	1/3	1/4	1/2	4	4	–	1/8	1/2	1/3
	C7	6	7	7	8	8	8	–	3	4
	C8	1/5	1/3	1/3	2	2	2	1/3	–	2
	C9	1/5	1/4	1/4	3	3	3	1/4	1/2	–

$$\begin{aligned}\tilde{r}_1 = {} & ((1 \times 1 \times 3.91 \times 5.94 \times 4.93 \times 2.88 \times 0.18 \times 4.31 \times 4.64)^{1/9},\\ & (1 \times 1.26 \times 4.93 \times 6.95 \times 5.94 \times 3.91 \times 0.22 \times 5.31 \times 5.65)^{1/9},\\ & (1 \times 1.44 \times 5.94 \times 7.96 \times 6.95 \times 4.93 \times 0.29 \times 6.32 \times 6.65)^{1/9})\\ = {} & (2.199, 2.663, 3.129)\end{aligned}$$

Similarly, the remaining $\tilde{r}_t$ values were calculated:

$\tilde{r}_2 = (1.542,\ 1.908,\ 2.293), \tilde{r}_3 = (0.999,\ 1.260,\ 1.528), \tilde{r}_4 = (0.300,\ 0.379,\ 0.493),$
$\tilde{r}_5 = (0.297,\ 0.373,\ 0.494), \tilde{r}_6 = (0.470,\ 0.611,\ 0.801), \tilde{r}_7 = (4.022,\ 4.800,\ 5.549),$

Table 5 Pairwise comparison matrix with TFNs

DM	Criteria	C1	C2	C3	C4	C5	C6	C7	C8	C9
DM1	C1	–	(1, 2, 3)	(5, 6, 7)	(8, 9, 10)	(6, 7, 8)	(4, 5, 6)	(1/4, 1/3, 1/2)	(5, 6, 7)	(5, 6, 7)
	C2	(1/3, 1/2, 1)	–	(3, 4, 5)	(4, 5, 6)	(4, 5, 6)	(4, 5, 6)	(1/7, 1/6, 1/5)	(4, 5, 6)	(3, 4, 5)
	C3	(1/7, 1/6, 1/5)	(1/5, 1/4, 1/3)	–	(6, 7, 8)	(5, 6, 7)	(3, 4, 5)	(1/6, 1/5, 1/4)	(2, 3, 4)	(2, 3, 4)
	C4	(1/10, 1/9, 1/8)	(1/6, 1/5, 1/4)	(1/8, 1/7, 1/6)	–	(1, 2, 3)	(1/3, 1/2, 1)	(1/7, 1/6, 1/5)	(1, 2, 3)	(1, 1, 1)
	C5	(1/8, 1/7, 1/6)	(1/6, 1/5, 1/4)	(1/7, 1/6, 1/5)	(1/3, 1/2, 1)	–	(1/3, 1/2, 1)	(1/7, 1/6, 1/5)	(1, 2, 3)	(1, 1, 1)
	C6	(1/6, 1/5, 1/4)	(1/6, 1/5, 1/4)	(1/5, 1/4, 1/3)	(1, 2, 3)	(1, 2, 3)	–	(1/7, 1/6, 1/5)	(1, 2, 3)	(1, 1, 1)
	C7	(2, 3, 4)	(5, 6, 7)	(4, 5, 6)	(5, 6, 7)	(5, 6, 7)	(5, 6, 7)	–	(5, 6, 7)	(4, 5, 6)
	C8	(1/7, 1/6, 1/5)	(1/6, 1/5, 1/4)	(1/4, 1/3, 1/2)	(1/3, 1/2, 1)	(1/3, 1/2, 1)	(1/3, 1/2, 1)	(1/7, 1/6, 1/5)	–	(6, 7, 8)
	C9	(1/7, 1/6, 1/5)	(1/5, 1/4, 1/3)	(1/4, 1/3, 1/2)	(1, 1, 1)	(1, 1, 1)	(1, 1, 1)	(1/6, 1/5, 1/4)	(1/8, 1/7, 1/6)	–
DM2	C1	–	(1, 1, 1)	(4, 5, 6)	(6, 7, 8)	(5, 6, 7)	(3, 4, 5)	(1/6, 1/5, 1/4)	(4, 5, 6)	(5, 6, 7)
	C2	(1, 1, 1)	–	(3, 4, 5)	(3, 4, 5)	(3, 4, 5)	(3, 4, 5)	(1/8, 1/7, 1/6)	(4, 5, 6)	(3, 4, 5)
	C3	(1/6, 1/5, 1/4)	(1/5, 1/4, 1/3)	–	(6, 7, 8)	(5, 6, 7)	(2, 3, 4)	(1/7, 1/6, 1/5)	(2, 3, 4)	(2, 3, 4)
	C4	(1/8, 1/7, 1/6)	(1/5, 1/4, 1/3)	(1/8, 1/7, 1/6)	–	(1, 2, 3)	(1/4, 1/3, 1/2)	(1/9, 1/8, 1/7)	(1, 1, 1)	(1/4, 1/3, 1/2)
	C5	(1/7, 1/6, 1/5)	(1/5, 1/4, 1/3)	(1/7, 1/6, 1/5)	(1/3, 1/2, 1)	–	(1/4, 1/3, 1/2)	(1/9, 1/8, 1/7)	(1, 1, 1)	(1/4, 1/3, 1/2)
	C6	(1/5, 1/4, 1/3)	(1/5, 1/4, 1/3)	(1/4, 1/3, 1/2)	(2, 3, 4)	(2, 3, 4)	–	(1/8, 1/7, 1/6)	(1, 1, 1)	(1/4, 1/3, 1/2)
	C7	(4, 5, 6)	(6, 7, 8)	(5, 6, 7)	(7, 8, 9)	(7, 8, 9)	(6, 7, 8)	-	(4, 5, 6)	(4, 5, 6)
	C8	(1/6, 1/5, 1/4)	(1/6, 1/5, 1/4)	(1/4, 1/3, 1/2)	(1, 1, 1)	(1, 1, 1)	(1, 1, 1)	(1/6, 1/5, 1/4)	–	(2, 3, 4)
	C9	(6, 7, 8)	(1/5, 1/4, 1/3)	(1/4, 1/3, 1/2)	(2, 3, 4)	(2, 3, 4)	(2, 3, 4)	(1/6, 1/5, 1/4)	(1/4, 1/3, 1/2)	–
DM3	C1	–	(1, 1, 1)	(3, 4, 5)	(5, 6, 7)	(4, 5, 6)	(2, 3, 4)	(1/7, 1/6, 1/5)	(4, 5, 6)	(4, 5, 6)
	C2	(1, 1, 1)	–	(1, 1, 1)	(2, 3, 4)	(2, 3, 4)	(3, 4, 5)	(1/8, 1/7, 1/6)	(2, 3, 4)	(3, 4, 5)
	C3	(1/5, 1/4, 1/3)	(1, 1, 1)	–	(5, 6, 7)	(4, 5, 6)	(1, 2, 3)	(1/8, 1/7, 1/6)	(2, 3, 4)	(3, 4, 5)

(continued)

Table 5 (continued)

DM	Criteria	C1	C2	C3	C4	C5	C6	C7	C8	C9
	C4	(1/7, 1/6, 1/5)	(1/4, 1/3, 1/2)	(1/7, 1/6, 1/5)	–	(1/3, 1/2, 1)	(1/5, 1/4, 1/3)	(1/9, 1/8, 1/7)	(1/3, 1/2, 1)	(1/4, 1/3, 1/2)
	C5	(1/6, 1/5, 1/4)	(1/4, 1/3, 1/2)	(1/6, 1/5, 1/4)	(1, 2, 3)	–	(1/5, 1/4, 1/3)	(1/9, 1/8, 1/7)	(1/3, 1/2, 1)	(1/4, 1/3, 1/2)
	C6	(1/4, 1/3, 1/2)	(1/5, 1/4, 1/3)	(1/3, 1/2, 1)	(3, 4, 5)	(3, 4, 5)	–	(1/9, 1/8, 1/7)	(1/3, 1/2, 1)	(1/4, 1/3, 1/2)
	C7	(5, 6, 7)	(6, 7, 8)	(6, 7, 8)	(7, 8, 9)	(7, 8, 9)	(7, 8, 9)	–	(2, 3, 4)	(3, 4, 5)
	C8	(1/6, 1/5, 1/4)	(1/4, 1/3, 1/2)	(1/4, 1/3, 1/2)	(1, 2, 3)	(1, 2, 3)	(1, 2, 3)	(1/4, 1/3, 1/2)	–	(1, 2, 3)
	C9	(1/6, 1/5, 1/4)	(1/5, 1/4, 1/3)	(1/5, 1/4, 1/3)	(2, 3, 4)	(2, 3, 4)	(2, 3, 4)	(1/5, 1/4, 1/3)	(1/3, 1/2, 1)	–

Table 6 Synthetic pairwise comparison matrix

Criteria	C1	C2	C3	C4	C5	C6	C7	C8	C9
C1	(1, 1, 1)	(1, 1.26, 1.44)	(3.91, 4.93, 5.94)	(5.94, 6.95, 7.96)	(4.93, 5.94, 6.95)	(2.88, 3.91, 4.93)	(0.18, 0.22, 0.29)	(4.31, 5.31, 6.32)	(4.64, 5.65, 6.65)
C2	(0.69, 0.79, 1)	(1, 1, 1)	(2.08, 2.52, 2.92)	(2.88, 3.91, 4.93)	(2.88, 3.91, 4.93)	(3.3, 4.31, 5.31)	(0.13, 0.15, 0.18)	(3.17, 4.22, 5.24)	(3, 4, 5)
C3	(0.17, 0.2, 0.26)	(0.34, 0.4, 0.48)	(1, 1, 1)	(5.65, 6.65, 7.65)	(4.64, 5.65, 6.65)	(1.82, 2.88, 3.91)	(0.14, 0.17, 0.2)	(1.59, 2.08, 2.52)	(1.59, 2.62, 3.63)
C4	(0.13, 0.14, 0.17)	(0.2, 0.26, 0.35)	(0.13, 0.15, 0.18)	(1, 1, 1)	(0.69, 1.26, 2.08)	(0.26, 0.35, 0.55)	(0.12, 0.14, 0.16)	(0.69, 1, 1.44)	(0.4, 0.48, 0.63)
C5	(0.14, 0.17, 0.2)	(0.2, 0.26, 0.35)	(0.15, 0.18, 0.22)	(0.48, 0.79, 1.44)	(1, 1, 1)	(0.26, 0.35, 0.55)	(0.12, 0.14, 0.16)	(0.69, 1, 1.44)	(0.4, 0.48, 0.63)
C6	(0.2, 0.26, 0.35)	(0.19, 0.23, 0.3)	(0.26, 0.35, 0.55)	(1.82, 2.88, 3.91)	(1.82, 2.88, 3.91)	(1, 1, 1)	(0.13, 0.14, 0.17)	(0.69, 1, 1.44)	(0.4, 0.48, 0.63)
C7	(3.42, 4.48, 5.52)	(5.65, 6.65, 7.65)	(4.93, 5.94, 6.95)	(6.26, 7.27, 8.28)	(6.26, 7.27, 8.28)	(5.94, 6.95, 7.96)	(1, 1, 1)	(3.42, 4.48, 5.52)	(3.63, 4.64, 5.65)
C8	(0.16, 0.19, 0.23)	(0.19, 0.24, 0.31)	(0.4, 0.48, 0.63)	(0.69, 1, 1.44)	(0.69, 1, 1.44)	(0.69, 1, 1.44)	(0.18, 0.22, 0.29)	(1, 1, 1)	(2.29, 3.48, 4.58)
C9	(0.15, 0.18, 0.22)	(0.2, 0.25, 0.33)	(0.28, 0.38, 0.63)	(1.59, 2.08, 2.52)	(1.59, 2.08, 2.52)	(1.59, 2.08, 2.52)	(0.18, 0,22, 0.28)	(0.22, 0.29, 0.44)	(1, 1, 1)

$\tilde{r}_8 = (0.491,\ 0.634,\ 0.829),\ \tilde{r}_9 = (0.477,\ 0.596,\ 0.762).$

Then, $\tilde{w}_i$ values were calculated using Eq. (4). For $\tilde{w}_1$ as example:

$$\begin{aligned}\tilde{w}_1 &= (2.199, 2.663, 3.129) \otimes (1/(3.129 + 2.293 + 1.528 + 0.493 + 0.494\\&\quad + 0.801 + 5.549 + 0.829 + 0.762), 1/(2.663 + 1.908 + 1.260 + 0.379 + 0.373\\&\quad + 0.611 + 4.800 + 0.634 + 0.596), 1/(2.199 + 1.542 + 0.999 + 0.300 + 0.297\\&\quad + 0.470 + 4.022 + 0.491 + 0.477))\\&= (0.138, 0.201, 0.290)\end{aligned}$$

Similarly, the remaining $\tilde{w}_i$ values were calculated:

$$\tilde{w}_2 = (0.097,\ 0.144,\ 0.212), \tilde{w}_3 = (0.063,\ 0.095,\ 0.142), \tilde{w}_4 = (0.019,\ 0.029,\ 0.046),$$
$$\tilde{w}_5 = (0.019,\ 0.028,\ 0.046),\ \tilde{w}_6 = (0.030,\ 0.046,\ 0.074),\ \tilde{w}_7 = (0.253,\ 0.363,\ 0.514),$$
$$\tilde{w}_8 = (0.031,\ 0.048,\ 0.077), \tilde{w}_9 = (0.030,\ 0.045,\ 0.071).$$

The COA defuzzification method was used to calculate BNP weights of criteria. For BNP_1 as example; $BNP_1 = 0.138+[(0.290 - 0.138) + (0.201 - 0.138)]/3 = 0.210$. Similarly, the remaining BNP weights of criteria were calculated (Table 7). The weights representing the importance levels of the criteria were found as in Table 7 by normalizing the BNP values. According to Table 7, the most important criteria are price and analysis time, respectively, while the least important criteria appear as volume (screen down) and volume (screen up).

After determining the weights of evaluation criteria, alternatives are ranked by WASPAS method. The types of the criteria to be used in WASPAS method are given in Table 8. In Table 8, the symbol "–" indicates that the relevant criterion is cost, and the "+" symbol indicates that the relevant criterion is the benefit criterion. As can be seen in Table 8, while three criteria (data display, hard drive, technology readiness level) are tried to be maximized, six criteria are minimized.

Table 7 Criteria weights

Criteria	BNP	Weight	Importance rank
C1. Analysis time	0.210	0.200	2
C2. Warm-up time	0.151	0.144	3
C3. Weight	0.100	0.095	4
C4. Volume (screen down)	0.031	0.030	8
C5. Volume (screen up)	0.031	0.029	9
C6. Data display	0.050	0.048	6
C7. Price	0.377	0.359	1
C8. Hard drive	0.052	0.049	5
C9. Tech. readiness level	0.049	0.046	7

Table 8 Types of criteria

Criteria	C1	C2	C3	C4	C5	C6	C7	C8	C9
Types of criteria	−	−	−	−	−	+	−	+	+

Table 9 Initial matrix

	C1	C2	C3	C4	C5	C6	C7	C8	C9
Max	8.000	30	19.000	0.050	0.091	30.730	48,886	120	9
Min	6.500	10	10.300	0.040	0.042	26.400	30,000	40	8

The max and min catalog values of nine criteria are given in Table 9.

For the WASPAS method, firstly, the criteria are normalized separately with Eqs. (7) and (8). The results of these calculations are given in Table 10 as a normalized decision matrix.

The total relative importance of the alternatives was calculated as to the WS method by using Eq. (9) in Table 11 and according to the WP method by using Eq. (10) in Table 12. During these calculations, the criterion weights obtained from the FAHP method were used in the previous section (Table 7).

Table 10 Normalized matrix

Alternatives	C1	C2	C3	C4	C5	C6	C7	C8	C9
A1 (IONSCAN 500DT)	1.000	0.333	0.542	0.801	0.465	0.859	0.614	0.333	1.000
A2 (ITEMISER 4DX)	0.813	0.333	0.793	1.000	0.465	0.859	0.622	1.000	1.000
A3 (DE-tector FLEX)	0.929	1.000	1.000	0.937	1.000	1.000	1.000	0.833	0.889

Table 11 Weighted normalized matrix

Alternatives	C1	C2	C3	C4	C5	C6	C7	C8	C9
A1	0.200	0.048	0.052	0.024	0.014	0.041	0.220	0.016	0.046
A2	0.162	0.048	0.075	0.030	0.014	0.041	0.223	0.049	0.046
A3	0.186	0.144	0.095	0.028	0.029	0.048	0.359	0.041	0.041

Table 12 Exponentially weighted matrix

Alternatives	C1	C2	C3	C4	C5	C6	C7	C8	C9
A1	1.000	0.854	0.943	0.993	0.978	0.993	0.839	0.947	1.000
A2	0.959	0.854	0.978	1.000	0.978	0.993	0.844	1.000	1.000
A3	0.985	1.000	1.000	0.998	1.000	1.000	1.000	0.991	0.995

Table 13 Ranking of alternatives according to WASPAS method

Alternatives	Q_{i1}	Q_{i2}	Q_i	Ranking
A1 (IONSCAN 500DT)	0.661	0.617	0.639	3
A2 (ITEMISER 4DX)	0.689	0.656	0.672	2
A3 (DE-tector FLEX)	0.970	0.969	0.970	1

The final total relative importance of each alternative was calculated with Eq. (12) by assuming the λ value as 0.5 and given in Table 13. As can be seen in Table 13, A3 (DE-tector FLEX) is the best alternative among other devices, while the worst alternative is A1 (IONSCAN 500DT).

4.1 Scenario Analysis

In this sub-section, two different sensitivity analyses are conducted. In the first scenario analysis (SA1), the effects of different λ values are investigated. When $\lambda = 0$, WASPAS is transformed to WP; and when $\lambda = 1$, WASPAS is transformed to WS [42]. Depending on the change in λ in SA1, the changes in ranking is examined. As can be seen in Table 14, it was observed that the ranking is not changed according to the different λ values. The results show that the initial ranking is robust and not affected by change of λ.

A common approach is to change input factors (weights of criteria) to see what effect this produces on the output. In the second scenario analysis (SA2), sensitivity analysis was done where weight values of criteria were changed to evaluate the differences in the final ranking. To assess the sensitivity, the weight (w_i) of a criterion at a certain percent change (PC) level can be calculated as follows [47]:

$$w_i = w_{i0} \mp w_{io} \times PC \tag{13}$$

where w_{i0} is the weight of the main changing criterion at the base run. The weights of the other criterion w_j are adjusted proportionally in accordance with w_i derived in the equation:

$$w_j = (1 - w_i) \times \frac{(w_{j0})}{(1 - w_{i0})} \tag{14}$$

where w_j is the new weight value assigned to the criterion *j* and w_i is the weight of criterion *i* at a certain PC level. w_{i0} and w_{j0} are weight values of criteria *i* and *j* at the base run. For instance, when the weight value (0.359) of price criterion is decreased by 20%, the new weight values of price (w_i) and analysis time (w_j) criteria can be calculated using Eqs. (13) and (14) as follows:

Table 14 Results of SA1

Alternatives	Relative significances (ranks) of alternatives Q_i										
	$\lambda = 0$	$\lambda = 0.1$	$\lambda = 0.2$	$\lambda = 0.3$	$\lambda = 0.4$	$\lambda = 0.5$	$\lambda = 0.6$	$\lambda = 0.7$	$\lambda = 0.8$	$\lambda = 0.9$	$\lambda = 1$
A1	0.617 (3)	0.622 (3)	0.626 (3)	0.630 (3)	0.635 (3)	0.639 (3)	0.643 (3)	0.648 (3)	0.652 (3)	0.656 (3)	0.661 (3)
A2	0.656 (2)	0.659 (2)	0.663 (2)	0.666 (2)	0.669 (2)	0.672 (2)	0.676 (2)	0.679 (2)	0.682 (2)	0.686 (2)	0.689 (2)
A3	0.969 (1)	0.969 (1)	0.970 (1)	0.970 (1)	0.970 (1)	0.970 (1)	0.970 (1)	0.970 (1)	0.970 (1)	0.970 (1)	0.970 (1)

$$w_i^{price} = 0.359 - 0.359 \times 0.2 = 0.287$$

$$w_j^{analysistime} = (1 - 0.287) \times \frac{0.200}{(1 - 0.359)} = 0.222$$

In the analysis, the price criterion (C1) is selected as the main changing criterion. The weight of C1 criterion is decreased by 5% in 10 times. It must be noted that the sensitivity analysis is done just for the price criterion. Due to having the highest weight among other criteria, price criterion is selected. Updated weights of other criteria according to price criterion are shown in Table 15. The sum of all weights at any percent change level equals to 1.0. The effect of ranking was examined by applying the WASPAS method according to the change in criterion weights. As can be seen in Table 16, it has been observed that the rank of all analyses is fixed.

Two scenario analyses which measure the effects of λ and criteria weights change on the ranking are generated. Results show that the initial rank (A3 $\rightarrow$ A2 $\rightarrow$ A1) is remained same in two scenarios. It can be said that the rank is robust.

5 Conclusion

Aviation security is a combination of human and material measures and resources designed to protect passengers, staff, aircraft, and airport property against acts of unlawful interference, such as terrorism, crime, and other threats. The modern technologies and approaches driven by the Industry 4.0 increase the security options in Aviation 4.0. Especially, the big data analytics can provide precise data for operational control, and the Internet of Things might improve the equipment safety through a better maintenance. Also cyber-physical systems can assist the humans which will result in safer flights. However, determining the most suitable option among alternative security applications is a challenging task for the decision-makers.

In this study, the evaluation problem of explosive and narcotics trace detection devices is considered. For the problem, different alternatives (three devices) and selection criteria (nine criteria) are determined. The device alternatives were there different devices which are being sold frequently in this market. The specifications of these devices were generally close to each other. To reflect the ambiguity of decision-makers, FAHP approach is applied. After the determination of the criteria weights, the three devices are evaluated and ranked by using WASPAS approach. According to the FAHP, price and analyzing time are found as first two most important criteria. Finally, DE-tector FLEX device is ranked first among three alternative devices. The superior specifications of the selected device were; its price, its warm-up time (which was determined as the third most important criteria) and its analysis time. The warm-up time expresses the amount of time it takes for the sensor to attain its highest performance. This time is also very important in security detections because generally there are generally long queues in important airports.

Table 15 Sensitivity analysis in SA2

Scenario	% Change	Criteria weight values									
		C1	C2	C3	C4	C5	C6	C7	C8	C9	C10
Current	0	0.200	0.144	0.095	0.030	0.029	0.048	0.359	0.049	0.046	0.200
S1	−5	0.206	0.148	0.098	0.031	0.030	0.049	0.341	0.050	0.047	0.206
S2	−10	0.211	0.152	0.100	0.032	0.031	0.051	0.323	0.052	0.049	0.211
S3	−15	0.217	0.156	0.103	0.033	0.031	0.052	0.305	0.053	0.050	0.217
S4	−20	0.222	0.160	0.106	0.033	0.032	0.053	0.287	0.054	0.051	0.222
S5	−25	0.228	0.164	0.108	0.034	0.033	0.055	0.269	0.056	0.052	0.228
S6	−30	0.234	0.168	0.111	0.035	0.034	0.056	0.251	0.057	0.054	0.234
S7	−35	0.239	0.172	0.114	0.036	0.035	0.057	0.233	0.059	0.055	0.239
S8	−40	0.245	0.176	0.116	0.037	0.035	0.059	0.215	0.060	0.056	0.245
S9	−45	0.250	0.180	0.119	0.038	0.036	0.060	0.197	0.061	0.058	0.250
S10	−50	0.256	0.184	0.122	0.038	0.037	0.061	0.180	0.063	0.059	0.256

Table 16 Results of SA2

Alternatives	Relative significances (ranks) of alternatives Q_i									
	S1	S2	S3	S4	S5	S6	S7	S8	S9	S10
A1	0.662 (3)	0.664 (3)	0.665 (3)	0.666 (3)	0.667 (3)	0.669 (3)	0.670 (3)	0.671 (3)	0.673 (3)	0.674 (3)
A2	0.691 (2)	0.693 (2)	0.695 (2)	0.696 (2)	0.698 (2)	0.700 (2)	0.702 (2)	0.704 (2)	0.706 (2)	0.708 (2)
A3	0.970 (1)	0.969 (1)	0.968 (1)	0.967 (1)	0.966 (1)	0.966 (1)	0.965 (1)	0.964 (1)	0.963 (1)	0.962 (1)

For the future studies, the number of criteria can be increased such as electricity consumption, cloud services and also the device alternatives can be widened. Other well-known MCDM approaches such as DEMATEL, ARAS, and TOPSIS can be applied. Finally, other cargo screening technology equipment such as carbon dioxide monitors can be evaluated.

References

1. Atalik, O., Akan, S., Bakir, M.: Aviation 4.0: Current practices of industry 4.0 in the airline and airport industries. In: Proceedings of 2nd International Conference on Empirical Economics and Social Science, pp. 894–906 (2019)
2. Barbosa, G.F., Aroca, R.V.: Advances of industry 4.0 concepts on aircraft construction: an overview of trends. J. Steel Struc. Const. **3**(1), 1–8 (2017)
3. Valdes, R.A., Comendador, V.F.G., Sanz, A.R., Castan, J.P.: Aviation 4.0: more safety through automation and digitization. Intech Open (2018). https://doi.org/10.5772/intechopen.73688
4. Mosterman, P.J., Zander, J.: Cyber-physical systems challenges: a needs analysis for collaborating embedded software systems. Softw. Syst. Model. **15**, 5–16 (2016)
5. Sahin, D.R., Haitmurodov, U., Turan, P.: Industry 4.0: opportunities, challenges of airport and airline management practices. In: Proceedings of European Social & Behavioral Sciences, pp. 568–577 (2018)
6. Rodriguez, M.D., Libbey, R., Mondal, S., Carbeck, J., Michalik, J.: Exponential technologies in manufacturing, technical report. https://www2.deloitte.com/us/en/pages/manufacturing/articles/advanced-manufacturing-technologies-report.html. Access date 1 Nov 2020 (2018)
7. Nowacki, G., Paszukow, B.: Security requirements for new threats at international airports. Int. J. Mar. Navig. Saf. Sea Transp. **12**(1), 187–192 (2008)
8. Liou, J.J.H., Tzeng, G.H.: A non-additive model for evaluating airline service quality. J. Air Transp. Manag. **13**, 131–138 (2007)
9. Nejati, M., Nejati, M., Shafaei, A.: Ranking airlines' service quality factors using a fuzzy approach: study of the Iranian society. Int. J. Qual. Reliab. Manag. **26**(3), 247–260 (2008)
10. Yeh, C.H., Chang, Y.H.: Modeling subjective evaluation for fuzzy group multi-criteria decision making. Eur. J. Oper. Res. **194**, 464–473 (2009)
11. Wang, J., Fan, K., Wang, W.: Integration of fuzzy AHP and FPP with TOPSIS methodology for aeroengine health assessment. Expert Syst. Appl. **37**, 8516–8526 (2010)
12. Kuo, M.S.: A novel interval-valued fuzzy MCDM method for improving airlines' service quality in Chinese cross-strait airlines. Transp. Res. Part E Logistics Transp. Rev. **47**, 1177–1193 (2011)
13. Aydogan, E.K.: Performance measurement model for Turkish aviation firms using the rough-AHP and TOPSIS methods under fuzzy environment. Expert Syst. Appl. **38**, 3992–3998 (2011)
14. Chen, C.J., Yang, S.M., Chang, S.C.: A model integrating fuzzy AHP with QFD for assessing technical factors in aviation safety. Int. J. Mach. Learn. Cybern. **5**, 761–774 (2014)
15. Bruno, G., Esposito, E., Genovese, A.: A model for aircraft evaluation to support strategic decisions. Expert Syst. Appl. **42**, 5580–5590 (2015)
16. Li, W., Yu, S., Pei, H., Zhao, C., Tian, B.: A hybrid approach based on fuzzy AHP and 2-tuple fuzzy linguistic method for evaluation in-flight service quality. J. Air Transp. Manag. **60**, 49–64 (2017)
17. Karaman, A.S., Akman, E.: Taking-off corporate social responsibility programs: an AHP application in airline industry. J. Air Transp. Manag. **68**, 187–197 (2018)
18. Mahtani, U.S., Garg, C.P.: An analysis of key factors of financial distress in airline companies in India using fuzzy AHP framework. Transp. Res. Part A Policy Pract. **117**, 87–102 (2018)

19. Percin, S.: Evaluating airline service quality using a combined fuzzy decision-making approach. J. Air Transp. Manag. **68**, 48–60 (2018)
20. Chen, T.Y.: Remoteness index-based Pythagorean fuzzy VIKOR methods with a generalized distance measure for multiple criteria decision analysis. Inf. Fusion **41**, 129–150 (2018)
21. Kiraci, K., Bakir, M.: Application of commercial aircraft selection in aviation industry through multi-criteria decision making methods. Manisa Celal Bayar Univ. J. Soc. Sci. **16**(4), 307–331 (2018)
22. Buyukozkan, G., Havle, C.A., Feyzioglu, O.: A new digital service quality model and its strategic analysis in aviation industry using interval-valued intuitionistic fuzzy AHP. J. Air Transp. Manag. **86**, 101817 (2020)
23. Turskis, Z., Zavadskas, E.K., Antucheviciene, J., Kosareva, N.: A hybrid model based on fuzzy AHP and fuzzy WASPAS for construction site selection. Int. J. Comput. Commun. Control **10**(6), 113–128 (2015)
24. Alam, K.A., Ahmed, R., Butt, F.S., Kim, S., Ko, K.: An uncertainty-aware integrated fuzzy AHP-WASPAS model to evaluate public cloud computing services. Proc. Comput. Sci. **130**, 504–509 (2018)
25. Ustali, N.K., Tosun, N.: New product selection with fuzzy AHP and fuzzy WASPAS methods. J. Mark. Res. **3**(2), 25–34 (2019)
26. Turskis, Z., Goranin, N., Nurusheva, A., Boranbayev, S.: A fuzzy WASPAS-based approach to determine critical information infrastructures of EU sustainable development. Sustainability **11**(2), 424 (2019)
27. Ayyildiz, E., Gumus, A.T.: A novel spherical fuzzy AHP-integrated spherical WASPAS methodology for petrol station location selection problem: a real case study for İstanbul. Environ. Sci. Pollut. Res. **27**, 36109–36120 (2020)
28. Otay, I., Kahraman, C., Oztaysi, B., Onar, S.C.: A novel single-valued spherical fuzzy AHP-WASPAS methodology. In: Developments of Artificial Intelligence Technologies in Computation and Robotics, pp. 190–198 (2020)
29. Zadeh, L.: Fuzzy sets. Inf. Control **8**(3), 338–353 (1965)
30. Hsieh, T.Y., Lu, S.T., Tzeng, G.H.: Fuzzy MCDM approach for planning and design tenders selection in public office buildings. Int. J. Proj. Manag. **22**(7), 573–584 (2004)
31. Laarhoven, P.J.M., Pedrycz, W.: A fuzzy extension of Saaty's priority theory. Fuzzy Sets Syst. **11**(3), 229–241 (1983)
32. Huang, C.C., Chu, P.Y., Chiang, Y.H.: A fuzzy AHP application in government-sponsored R&D project selection. Omega **36**(6), 1038–1052 (2008)
33. Jakhar, R., Verma, D., Rathore, A.P.S., Kumar, D.: Prioritization of dimensions of visual merchandising for apparel retailers using FAHP. Benchmarking **27**(10), 2759–2784 (2020)
34. Chen, T.C.T., Lin, Y.C.: A FAHP-FTOPSIS approach for bio-printer selection. Health Technol. (2020). https://doi.org/10.1007/s12553-020-00469-8
35. Lyu, H.M., Shen, S.L., Zhou, A.N., Zhou, W.H.: Flood risk assessment of metro systems in a subsiding environment using the interval FAHP-FCA approach. Sustain. Cities Soc. **50**, 101682 (2019)
36. Ramesh, W.V., Sakthivel, G.: Selection of optimum fuel blend in an Ic engine using FAHP-PROMETHEE. IOP Conf. Ser., Mater. Sci. Eng. **390**(1), 1–12 (2018)
37. Noorollahi, E., Fadai, D., Akbarpour Shirazi, M., Ghodsipour, S.H.: Land suitability analysis for solar farms exploitation using GIS and fuzzy analytic hierarchy process (FAHP)—a case study of Iran. Energies **9**(8), 643 (2016)
38. Wang, M.K., Hwang, K.P.: Using FAHP methods evaluation and screening of intellectual property rights managers in Taiwan. Asia Pac J. Oper. Res. **31**(6), 1450048 (2014)
39. Sun, C.C.: A performance evaluation model by integrating fuzzy AHP and fuzzy TOPSIS methods. Expert Syst. Appl. **37**(12), 7745–7754 (2010)
40. Gumus, A.T.: Evaluation of hazardous waste transportation firms by using a two-step fuzzy-AHP and TOPSIS methodology. Expert Syst. Appl. **36**(2), 4067–4074 (2009)
41. Buckley, J.J.: Fuzzy hierarchical analysis. Fuzzy Sets Syst. **17**(3), 233–247 (1985)

42. Zolfani, S.H., Aghdaie, M.H., Derakhti, A., Zavadskas, E.K., Varzandeh, M.H.M.: Decision making on business issues with foresight perspective: an application of new hybrid MCDM model in shopping mall locating. Expert Syst. Appl. **40**(17), 7111–7121 (2013)
43. Zavadskas, E.K., Turskis, Z., Antucheviciene, J., Zakarevicius, A.: Optimization of weighted aggregated sum product assessment. Elektron Elektrotech. **122**(6), 3–6 (2012)
44. Zavadskas, E.K., Antucheviciene, J., Šaparauskas, J., Turskis, Z.: Multi-criteria assessment of facades' alternatives: peculiarities of ranking methodology. Proc. Eng. **57**, 107–112 (2013)
45. Karabašević, D., Stanujkić, D., Urošević, S., Maksimović, M.: An approach to personnel selection based on SWARA and WASPAS methods. J. Econ. Manag. Strategy **7**(1), 1–11 (2016)
46. URL-1 (https://www.cbrnetechindex.com/)
47. Chen, Y., Yu, J., Khan, S.: Spatial sensitivity analysis of multi-criteria weights in GIS based land suitability evaluation. Environ. Model. Softw. **25**, 1582–1591 (2010)

Selection of the Best Face Recognition System for Check in and Boarding Services

Dorota Kuchta, Duygu Sergi, and Irem Ucal Sari

Abstract Check-in and boarding services are one of the most human oriented pre-flight services in aviation industry. The process of using face recognition systems increase with the aviation 4.0 concept, decreases need for manpower and increases the efficiency of the processes. Therefore, problems, developments and challenges of face recognition in terms of aviation 4.0 are discussed in this chapter to determine the best face recognition system for check in and boarding systems. Analytic hierarchy process and grey relational analysis are used to analyze current system providers. To handle the ambiguity in the linguistic evaluations, fuzzy Z- numbers are used. 10 face recognition system providers are evaluated according to five criteria with the proposed methodology and the results are discussed.

Keywords Face recognition · Intelligent check-in · Intelligent boarding · Fuzzy Z-AHP · Fuzzy Z-grey relational analysis

1 Introduction

Aviation 4.0 emerged with the use of Industry 4.0 tools in the aviation industry. The number of automated systems used in aviation is increasing day by day and intelligent automated systems are beginning to be used even in jobs that require human judgment. Face recognition systems are one of the intelligent systems used

D. Kuchta (✉)
Wroclaw University of Science and Technology, Wyb. Wyspianskiego 27, 50-370 Wroclaw, Poland
e-mail: dorota.kuchta@pwr.edu.pl

D. Sergi
Department of Industrial Engineering, MEF University, Ayazağa Cad. No. 4, 34396 Maslak, Sarıyer, Istanbul, Turkey
e-mail: sergid@mef.edu.tr

I. Ucal Sari
Industrial Engineering Department, Istanbul Technical University, Macka, Istanbul, Turkey
e-mail: ucal@itu.edu.tr

C. Kahraman and S. Aydın (eds.), *Intelligent and Fuzzy Techniques in Aviation 4.0*, Studies in Systems, Decision and Control 372,
https://doi.org/10.1007/978-3-030-75067-1_16

in aviation, capable of identifying human faces and matching them with identity information. The use of face recognition systems in pre-flight processes brings many benefits such as increased security and transit speed and decreased documentation.

Check-in and boarding services are one of the most manpower demanding pre-flight services in the aviation industry. The main processes in check-in and boarding services can be summarized as passenger identity validation, baggage registration and handling, and seating assignment. In the passenger identity validation process, valid ID, passport, visa and ticket of each passenger are checked. In baggage registration and handling, luggage is checked for compliance with weight limits, tagged and sent to the baggage handling system. In the last process, a suitable seat is assigned to the passenger and boarding pass card is prepared. Although online check-in, which has been used in check-in processes for a long time, shortens the process, it does not prevent check-in queues as it requires additional processing for identity control. Automating identity control with face recognition systems significantly reduces check-in queues, requiring manpower only in case of special circumstances or when additional control is required. Similarly, the use of face recognition systems in boarding services creates benefits such as shortening processes, reducing documentation, and preventing troubles caused by human errors. For all these reasons, the number of airline companies using face recognition systems is gradually increasing.

Existing face recognition technologies differ, particularly in terms of face recognition speeds, false acceptance rates, and false reject rates. Besides that, in order not to affect flight safety, many factors such as system reliability, compliance with airline regulations, technical requirements play an important role in the selection of the system provider. These systems have also high initial investment costs, which makes the choice of facial recognition system provider more critical. Therefore, the purpose of this study is determined as a selection of the best face recognition system provider for check-in and boarding services for airline companies. In the determination of the criteria used in the analysis, first, the main differences of the systems that are face detection speed and accuracy rate are considered. It is also found that most of them haven't been used in the aviation industry. Therefore, the experience with air industry applications is taken as another criterion which could increase the success of the implementation. Due to the security restrictions of the flights, experience with border control applications is also determined as an important criterion. Fifth and last criteria is determined as the presence on the market to reduce the risk of failure. Since these criteria can be evaluated with verbal expressions, fuzzy logic is used in digitizing linguistic expressions in the study. Fuzzy Z-numbers [1] which is an extension of fuzzy numbers that additionally considers the reliability of the linguistic statement are used in the analysis to handle the vagueness of linguistic judgements. Fuzzy Z-AHP [2] is used in the determination of the criteria weights, and for the ranking of the vendors fuzzy Z-GRA method is proposed and used.

The rest of the chapter is organized as follows: In Sect. 2, problems, developments and challenges of face recognition in the context of Aviation 4.0 are discussed. Then a literature review on AHP and GRA are given in Sect. 3. Preliminaries for the calculations and the steps of the proposed methods are determined in Sect. 4. In Sect. 5, the proposed methods are applied to select the best face recognition system for check-in and boarding services. Finally in Sect. 6, the chapter terminates with the conclusion and future research suggestions.

2 Face Recognition—State of Art, Developments and Challenges in the Context of Aviation 4.0

Recognition of objects is an easy task for humans because our brain does not see faces as a set of isolated pieces but as a whole. In automatic face recognition systems, it is necessary to isolate significant features from an image, transform them into a useful presentation and produce an answer or a classification. We encounter two types of face recognition problems: identification problems, in which the input to the system is an unknown face, and the system seeks this face in a database, and verification problems, where the system needs to confirm or reject the claimed identity of the input face. The respective algorithm is being developed and searched for which would satisfy the requirements of real-world applications of face recognition systems [3, 4].

Biometric face recognition systems can be or are implemented in ABC (Automated Border Control) electronic gates. An ABC [5] is a fully automated system that performs border checks. It is used in Europe, in some Asian airports (e.g. Singapore [6] and Tokyo [7]) and in the USA [7, 8]. The basic functions of the system are to authenticate the travel document, establish that the traveller is the rightful holder of the document, check border control records, and on this basis, automatically verify the entry conditions for authorised citizens [9]. One of the components of ABC is an ABC- or e-gate, consisting among others of a face recognition system, which should be able to compare the face of the traveller with the passport picture (verification problem). The passport pictures requirements for second-generation passport pictures are very specific [9], so that an effective comparison is possible.

Requirements set to face recognitions systems include the following ones: they should be invariant to external factors, such as lighting or humidity. Also, they should be invariant to non-idealities, such as changes in pose and expression, skin problems, aging [10], changes in hairstyle, wearing or not wearing glasses, inhomogeneous background and object occlusion, as well as to the fact of having at our disposal only a single sample per person [9]. The false accept rate (FAR) for facial capture and verification should be less than 0, 1% and the false rejection rate (FRR) less than 5% [9].

In order to achieve the desired FAR and FRR values, both hardware and software part of the face recognition system have to be designed appropriately. As far as hardware is concerned, the different components of the system (cameras, document reader, display, biometric modules etc.), must be strategically located in the most ergonomic position [9], taking into account such factors as light conditions or the angle at which individuals will be positioned considering the cameras [6].

For the software, various technologies are available and are being developed. In order to avoid sensitivity to illumination, poses, expressions and occlusions, advanced technologies for face feature extraction should be used, e.g. methods based on neural networks, methods based on linear discriminant analysis, elastic graph method, methods based on hidden Markov models, on flexible contour models of the face, method of comparison of standards, optical flux method; methods based on lines of the same intensity, Karunen-Loeve decomposition; fuzzy logic, Gabor filters or sinusoidal projection [3, 11, 12]. A high performance and robust face recognition system has been tested with respect to differences in pose and low image quality and the results were close to optimal [13, 14]. Promising, although rather old studies focused on software invariant to non-idealities of different expressions, are presented in [15, 16]. More recent results regarding this problem can be found in [4]. The aspect of illumination problems has been considered in the study described in [15]. Face recognition systems invariant to the non-ideality of age was presented in [17, 18]. Solutions aiming at solving the problem of the availability of an only single sample per person are discussed in [19, 20].

In [9] we can find a comparative analysis of various technologies (not limited to facial recognition ones) described in the literature (state of art for the year 2016), which might be suitable for the use in ABC systems. The following criteria were applied:

- **Maturity**—this criterion refers to the level of plenitude (advancement, variety) of technologies used;
- **Applicability** is the measure of the suitability of the respective technology for the ABC environment;
- **Acceptance**—this criterion refers to the users' attitude to dealing with the respective technology, to the question of whether individuals consider it an intrusive or a neutral technique;
- **Performance**—this criterion was measured in a different way and in different terms according to the nature of the respective technology. For facial recognition systems values like FAR or FRR were used;
- **Costs** refer to the cost of technology implementation and maintenance;
- **Constraints** refer to the limitations that affect the performance of the technologies, such as limits to the number of cases analysed per time unit or problems of interoperability between systems;
- **Deployment** refers to the degree of difficulty in implementing the respective technology.

On the basis of the values of the above criteria, a rating was performed. Three systems were ranked at the highest position, both including facial recognition. The first two are described in [5] and have been implemented in Portugal and Austria airports, respectively. The third one is presented in [21] and has been implemented at Shiphol airport.

Coming back to the facial recognition systems performance, it has to be underlined that it has been improving rapidly. In 2012 the following performance values attainted by the face recognition algorithms were reported [6]: FAR remained at the level of 0.001, and FRR decreased from 0.2 in 2002 to 0.003 in 2010. However, it has to be underlined that these values were attained in laboratory conditions, whilst the real world conditions are different. The quality of pictures taken in laboratories is obviously higher than that of the pictures taken by cameras at the airports. That is why some research has been conducted taking into account the difference between the lab and the reality. For example, the recognition of faces in pictures taken 8 years apart was checked and the number of errors per 100 cases was between 5 and 17. Similar experiments were performed for pictures taken from the 250 angles and here the number of errors per 100 cases was in the range of [4, 13, 19]. Also, low-quality pictures were tested in laboratories and an on the spot experiment was performed at an airport in 2010. The top-performing vendor participating in the test was able to achieve a real-world FRR rate of 1.1% and the worst 5.6%. This is significantly a better result than can be obtained by a human who checks passport pictures against the passport holders [6]. There have been no published studies of the FAR and FRR achieved by humans, but it is generally accepted that face recognition systems operate at a higher level of accuracy, especially when a human border guard has to work continuously for two hours and deals with multiple races of passengers. Most e-gates in Europe today achieved in 2010 an FRR of approximately 6% and FAR 0.1% [6].

As already mentioned above, facial recognition systems are not used uniquely for border control purposes. Security and surveillance issues as well as facilitation of airport management, seamless passenger travel and the gathering of valuable statistical information pertaining to passenger movements [6] are among the reasons for the implementation of such systems. For example, MFlow Journey system of facial recognition is installed at Edinburgh Airport in order to conduct a performance review of passenger traffic through security screening and support Edinburgh Airport in reducing queuing times at peak hours [22]. The Berlin airports purchased a few years ago a face recognition system [23] in order to make the administration of access rights to restricted zones more efficient: facial characteristics of authorized personnel are stored and the access limits are easier to control. An important application of face recognition systems is real-time watch lists alerts [6]: matching faces captured from cameras against photographic databases in search for wanted individuals and raising an alarm in case of positive verification. The objective of these systems is to minimize the false alerts to a reasonable level so that people do not start to ignore them while detecting the highest possible percentage of people moving past the cameras who actually are in the watchlists.

Researchers are working intensively on technologies assuring insensibility to combined non-idealities and real-world conditions in which the facial recognition systems are used [24, 25]. Main challenges include recognizing young children and impaired individuals, monitoring travellers' behaviour, recognising non-frontal images of the subject, capturing facial gestures and expressions. The objectives of the research include the possibility to infer the individual attributes, such as gender or age, and detect possibly dangerous situations [9]. Experiments are conducted with non-frontal photographs (where the subject's image is captured at an angle) [6] and with video-based face recognition. Both the still image and the video have certain advantages and disadvantages, depending on the objective of facial recognition, thus both approaches have to be explored [4]. In the future cloud-based services will enable the transfer of expensive computing power out of the airport into shared server facilities. Internet of Things will enable more efficient control of humans present in public places [26]. Also, face recognition will assist passengers in their journey, for example by guiding individual passengers to their specific gate [6].

3 Literature Review

The analytic hierarchy process (AHP) is a methodology developed by Saaty [27] for structuring, measuring and synthesizing multi-criteria problems. AHP method is applied to the problems of choosing among the alternatives, especially in complex situations with many criteria. Fuzzy analytic hierarchy process (FAHP) is an improvement of the classical AHP method proposed by Saaty. It has the ability to include verbal evaluations into the problem, and paired comparisons are made with verbal expressions that can be expressed by fuzzy numbers.

The earliest study on FAHP was proposed by Van Laarhoven and Pedrycz [28] and triangular fuzzy numbers were used. Later, new approaches have been developed in the literature regarding the application of this method. Geometric mean method [29], fuzzy logarithmic least squares method [30], logarithmic fuzzy preference programming [31], and extended analysis method [32] are some of these FAHP extensions.

FAHP method is used in the selection of suppliers [33], in the management of distribution channels by combining with fuzzy TOPSIS [34], in the selection of computer-aided maintenance management systems [35], in evaluating the quality of the website [36], in measuring the relative efficiency of R&D performance by integrating with data envelopment analysis [37], in the selection of medical decision support systems [38], in making bank loan decisions in the selection of small and medium enterprises [39], in the selection of machine bench in flexible manufacturing systems integrated with PROMETHEE [40], in the route selection [41] and many other decision-making problems.

The grey system theory, which is an interdisciplinary approach, was introduced by Deng [42] first in the early 1980s. Grey system theory is an alternative method for quantifying uncertainty and its one of the most widely used model is grey relational analysis [43]. The basic idea in its emergence is to predict the behaviour of uncertain systems that cannot be overcome by stochastic or fuzzy methods with the help of limited data [44].

Grey Relational Analysis (GRA) is a suitable solution method for problems involving complex relationships between factors [45]. Due to this feature, it is seen that it is frequently used alone or in combination with other methods to solve Multi-Criteria Decision Making (MCDM) problems in the literature.

The GRA method has been widely used in many areas such as the selection of the wastewater treatment process [46], weapon systems [47], host country for refugees [48], and the evaluation of software projects [49]. The TOPSIS and OCRA methods were compared with the GRA method by Wu [50] and it was recommended to use more than one method instead of sticking to a single method for solving MCDM problems. Song and Jamalipour [51] used AHP method together with GRA in the selection of the best mobile network in terms of service quality in a dynamic environment. Jadidi and others [52] used GRA together with TOPSIS method in supplier selection. Chen and Tzeng [48] combined GRA method with TOPSIS concept to select an expatriate host country. Zhang and Liu [53], Pramanik and Mukhopadhyaya [54] proposed a heuristic fuzzy logic-based integrated approach with GRA for the personnel selection problem, while Goyal and Grover [55] proposed a fuzzy logic-based GRA for the selection of advanced production systems. Wu and Luyan [56] benefited from GRA with fuzzy AHP in making investment decisions.

4 Methodology

In this chapter fuzzy extensions of Analytic Hierarchy Process and Fuzzy Grey Relational Analysis using fuzzy z-numbers are used for the evaluation. Fuzzy Z-AHP is proposed by Otay and Kahraman [2] to incorporate vagueness in the evaluations and reliabilities to these evaluations into the AHP. Fuzzy Z-Grey Relational Analysis (Fuzzy Z-GRA) is proposed in this chapter for the first time. In this section, first preliminaries on fuzzy Z-numbers are given. Then the steps of the Fuzzy Z-AHP and proposed Fuzzy Z-GRA are determined as follows.

4.1 Fuzzy Z-Numbers

Zadeh [1] introduced fuzzy Z-numbers that is an ordered pair of fuzzy sets; $Z\left(\tilde{A}, \tilde{B}\right)$ in which the first component $\left(\tilde{A}\right)$ is a fuzzy restriction of the values of X variable, and the second component $\left(\tilde{B}\right)$ is the certainty of the fuzzy restriction.

Figure 1 represents a simple fuzzy Z-number; $Z = \left(\widetilde{A}, \tilde{B}\right)$ which has triangular membership functions both for fuzzy restriction and fuzzy reliability.

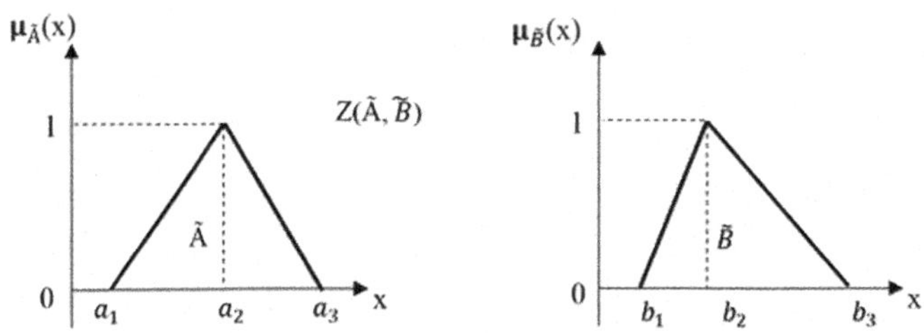

Fig. 1 A simple fuzzy Z number with triangular membership functions

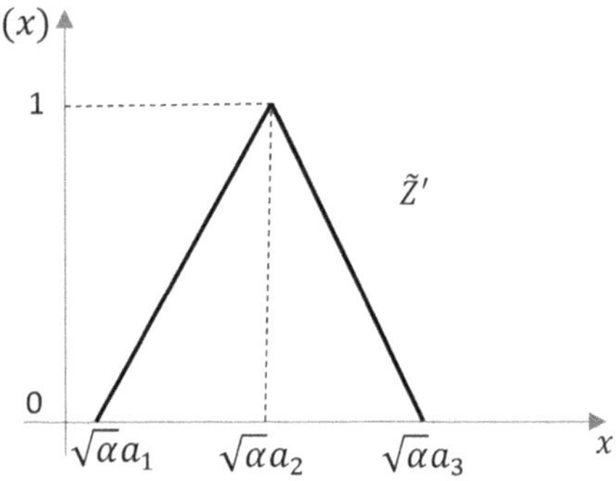

Fig. 2 Ordinary triangular fuzzy number transformed from Z-number

Let $\tilde{A} = \{\langle x, \mu_{\tilde{A}}(x)\rangle | \mu(x) \in [0, 1]\}$ and $\tilde{B} = \{\langle x, \mu_{\tilde{B}}(x)\rangle | \mu(x) \in [0, 1]\}$ where $\mu_{\tilde{A}}(x)$ $\mu_{\tilde{B}}(x)$ are a triangular membership functions. Equations (1–3) could be used to convert a fuzzy Z-number into an ordinary fuzzy number [57]. Equation (1) converts the reliability into a crisp number:

$$\alpha = \frac{b_l + 2b_m + b_u}{4} \tag{1}$$

Then, the weighted Z-number that is denoted by $\tilde{Z}^\alpha$ can be calculated by adding the weight of the reliability to the restriction as given in Eq. (2):

$$\tilde{Z}^\alpha = \{\langle x, \mu_{\tilde{A}^\alpha}(x)\rangle | \mu_{\tilde{A}^\alpha}(x) = \alpha\mu_{\tilde{A}}(x), \mu(x) \in [0, 1]\} \tag{2}$$

At the end, Eq. (3) converts the weighted Z-number to an ordinary fuzzy number $\tilde{Z}'$ which is shown in Fig. 2:

$$\tilde{Z}' = \left\{\langle x, \mu_{\tilde{Z}'}(x)\rangle | \mu_{\tilde{Z}'}(x) = \mu_{\tilde{A}}\left(\frac{x}{\sqrt{\alpha}}\right), \mu(x) \in [0, 1]\right\} \tag{3}$$

4.2 Fuzzy Z-AHP

Fuzzy Z-AHP steps are determined as follows [2]:

Step 1. Determination of the hierarchical structure and the linguistic scales for restriction and reliability functions. Triangular fuzzy Z scales for the linguistic restrictions and linguistic reliabilities are defined in Tables 1 and 2, respectively.

Step 2. Construction of the pairwise comparison matrices and conversion with fuzzy Z-numbers using the linguistic terms defined in Step 1. Each element $(\tilde{a}_{ij})$ of the fuzzy restriction matrix $\tilde{A}$ of pairwise comparison that is given

Table 1 Triangular Z-fuzzy restriction scale for the pairwise comparisons of criteria

Linguistic restriction	Triangular fuzzy Z restriction scale
Equally important (EI)	(1, 1, 1; 1)
Weakly more important (WI)	(1, 3, 5; 1)
Moderately more important (MI)	(3, 5, 7; 1)
Greatly more important (GI)	(5, 7, 9; 1)
Absolutely more important (AI)	(7, 9, 9; 1)
If factor *i* has one of the above fuzzy numbers assigned to it when compared with factor *j*, then *j* has the reciprocal value when compared with *i*	Reciprocals of above

Table 2 Reliability scale with its corresponding triangular Z-fuzzy numbers

Linguistic reliability	Triangular Z-fuzzy reliability scale
Absolutely Reliable (AR)	(1, 1, 1; 1)
Strongly Reliable (SR)	(0.7, 0.8, 0.9; 1)
Very Highly Reliable (VHR)	(0.6, 0.7, 0.8; 1)
Highly Reliable (HR)	(0.5, 0.6, 0.7; 1)
Fairly Reliable (FR)	(0.4, 0.5, 0.6; 1)
Weakly Reliable (WR)	(0.3, 0.4, 0.5; 1)
Very Weakly Reliable (VWR)	(0.2,0.3,0.4;1)
Strongly Unreliable (SU)	(0.1, 0.2, 0.3; 1)
Absolutely Unreliable (AU)	(0, 0.1, 0.2; 1)

in Eq. (4) is a linguistic term presenting fuzzy restriction for which is the more important of two criteria. Each element $\left(\tilde{b}_{\text{ij}}\right)$ of the fuzzy reliability matrix $\tilde{B}$ of pairwise comparison is a linguistic term presenting the fuzzy reliability of the respective fuzzy restriction.

$$\tilde{A} = \begin{vmatrix} 1 & \tilde{a}_{12} & \dots & \tilde{a}_{1n} \\ \tilde{a}_{21} & 1 & \dots & \tilde{a}_{2n} \\ \vdots & \vdots & \vdots\vdots\vdots & \vdots \\ \tilde{a}_{m1} & \tilde{a}_{m2} & \dots & 1 \end{vmatrix} \quad \tilde{B} = \begin{vmatrix} 1 & \tilde{b}_{12} & \dots & \tilde{b}_{1n} \\ \tilde{b}_{21} & 1 & \dots & \tilde{b}_{2n} \\ \vdots & \vdots & \vdots\vdots\vdots & \vdots \\ \tilde{b}_{m1} & \tilde{b}_{m2} & \dots & 1 \end{vmatrix} \tag{4}$$

Step 3. Conversion of Z-numbers to corresponding equivalent ordinary fuzzy numbers using Eqs. (1–3).

Step 4. Applying Buckley's ordinary fuzzy AHP method [29]. The steps of this method are summarized as follows:

Step 4.1. Checking the consistency of each fuzzy pairwise comparison matrix. Assume $A = \left[a_{ij}\right]$ is a positive reciprocal matrix and

$\tilde{A} = [\tilde{a}_{ij}]$ is a fuzzy positive reciprocal matrix. If the result of the comparisons of $A = [a_{ij}]$ is consistent, then it can imply that the result of the comparisons of $\tilde{A} = [\tilde{a}_{ij}]$ is also consistent. In [2], it is suggested to ignore reliability functions in the consistency measurement, because using transformed z-numbers could make consistent pairwise matrix inconsistent.

Step 4.2. Calculation of the geometric mean for each parameter of $\tilde{a}_{ij}$ in the n dimensional pairwise comparison matrix. If $(\tilde{r}_i)$ is geometric mean of each row of $\tilde{A} = [\tilde{a}_{ij}]$ then it is calculated as below;

$$\tilde{r}_i = [\tilde{a}_{i1} \otimes \cdots \otimes \tilde{a}_{in}]^{1/n} \tag{5}$$

Step 4.3. Computation of the fuzzy weights by normalization. The fuzzy weight $(\tilde{w}_i)$ of the ith criterion is calculated as below;

$$\tilde{w}_i = \tilde{r}_i \otimes [\tilde{r}_1 \oplus \cdots \oplus \tilde{r}_i \oplus \cdots \oplus \tilde{r}_n]^{-1} \tag{6}$$

Step 4.4. Defuzzification of the normalized weights vector using the center of gravity method given by Eq. (1).

4.3 Fuzzy Z-Grey Relational Analysis

Steps of the proposed fuzzy Z-GRA are determined as follows:

Step 1. Determination of the decision matrices for fuzzy restriction and fuzzy reliability of the comparisons. Linguistic statements of the decision makers are converted into fuzzy Z-numbers using the scale given in Tables 2 and 3. Fuzzy restriction of the decision matrix is denoted by $\tilde{A}_D$ and fuzzy reliability of the decision matrix is denoted by $\tilde{B}_D$ for the multi-attribute decision making problem with m alternatives and n criteria in Eq. (7).

Table 3 Triangular Z-fuzzy restriction Scale for rating the performance of alternatives. Adapted from [58]

Linguistic restriction	Triangular fuzzy Z restriction scale
Very weak (VW)	(0, 0, 0.15)
Weak (W)	(0, 0.15, 0.30)
Medium weak (MW)	(0.15, 0.30, 0.50)
Medium (M)	(0.30, 0.50, 0.65)
Medium High (MH)	(0.50, 0.65, 0.80)
High (H)	(0.65, 0.80, 1)
Very High (VH)	(0.80, 1, 1)

$$\tilde{A}_D = \begin{vmatrix} 1 & \tilde{a}_{D12} & \dots & \tilde{a}_{D1n} \\ \tilde{a}_{D21} & 1 & \dots & \tilde{a}_{D2n} \\ \vdots & \vdots & \vdots & \vdots \\ \tilde{a}_{Dm1} & \tilde{a}_{Dm2} & \dots & 1 \end{vmatrix} \quad \tilde{B}_D = \begin{vmatrix} 1 & \tilde{b}_{D12} & \dots & \tilde{b}_{D1n} \\ \tilde{b}_{D21} & 1 & \dots & \tilde{b}_{D2n} \\ \vdots & \vdots & \vdots & \vdots \\ \tilde{b}_{Dm1} & \tilde{b}_{Dm2} & \dots & 1 \end{vmatrix} \tag{7}$$

where $\tilde{a}_{\mathrm{Dij}}$ represents fuzzy restriction of the performance of the respective alternative under the respective criteria and $\tilde{b}_{\mathrm{Dij}}$ represents fuzzy reliability of the respective fuzzy restriction.

Step 2. Constructing ordinary fuzzy decision matrix. Z-numbers for each element of the decision matrix are converted to corresponding equivalent ordinary fuzzy numbers using Eqs. (1–3). Ordinary fuzzy decision matrix is presented as in Eq. (8):

$$\tilde{A}'_D = \begin{vmatrix} 1 & \tilde{a}'_{\mathrm{D12}} & \dots & \tilde{a}'_{\mathrm{D1n}} \\ \tilde{a}'_{\mathrm{D21}} & 1 & \dots & \tilde{a}'_{\mathrm{D2n}} \\ \vdots & \vdots & \vdots & \vdots \\ \tilde{a}'_{\mathrm{Dm1}} & \tilde{a}'_{\mathrm{Dm2}} & \dots & 1 \end{vmatrix} \tag{8}$$

Step 3. Applying Chen and Ren's [59] ordinary fuzzy GRA method that is an extension of [60]. The steps of this method are summarized as follows:

Step 3.1. Standardization. Equations (9) and (10) can be used to standardize ordinary fuzzy decision matrix transformed from fuzzy Z-decision matrix for the benefit and cost type of indicators, respectively.

$$\tilde{d}_{ij} = \left(d_{ijl}, d_{ijm}, d_{iju}\right) = \begin{cases} d_{ijl} = \dfrac{\tilde{a}'_{\mathrm{Dijl}}}{\sqrt{\sum_{i=1}^{m}\left(\tilde{a}'_{\mathrm{Diju}}\right)^2}} \\ d_{ij\mathrm{m}} = \dfrac{\tilde{a}'_{\mathrm{Dijm}}}{\sqrt{\sum_{i=1}^{m}\left(\tilde{a}'_{\mathrm{Dijm}}\right)^2}} \\ d_{ij\mathrm{u}} = \dfrac{\tilde{a}'_{\mathrm{Diju}}}{\sqrt{\sum_{i=1}^{m}\left(\tilde{a}'_{\mathrm{Dijl}}\right)^2}} \end{cases} \tag{9}$$

$$\tilde{d}_{ij} = \left(d_{ijl}, d_{ij\mathrm{m}}, d_{iju}\right) = \begin{cases} d_{ijl} = \dfrac{\frac{1}{\tilde{a}'_{\mathrm{Diju}}}}{\sqrt{\sum_{i=1}^{m}\left(\frac{1}{\tilde{a}'_{\mathrm{Dijl}}}\right)^2}} \\ d_{ij\mathrm{m}} = \dfrac{\frac{1}{\tilde{a}'_{\mathrm{Dijm}}}}{\sqrt{\sum_{i=1}^{m}\left(\frac{1}{\tilde{a}'_{\mathrm{Dijm}}}\right)^2}} \\ d_{iju} = \dfrac{\frac{1}{\tilde{a}'_{\mathrm{Dijl}}}}{\sqrt{\sum_{i=1}^{m}\left(\frac{1}{\tilde{a}'_{\mathrm{Diju}}}\right)^2}} \end{cases} \tag{10}$$

Step 3.2. Generating the reference series. The maximum standardized value of the jth criterion that is denoted by $\tilde{d}_j^*(j = 1, 2, \ldots, n)$ can be calculated using Eq. (11).

$$\tilde{d}_j^* = \left(\max_{i=1,2,\ldots,m} d_{ijl}, \max_{i=1,2,\ldots,m} d_{ijm}, \max_{i=1,2,\ldots,m} d_{iju}\right) = \left(d_{jl}^*, d_{jm}^*, d_{ju}^*\right) \tag{11}$$

Then, reference series consisting of ideal solutions relating to all indicators that is denoted by D^* can be represented as in Eq. (12):

$$D^* = \left[\ \tilde{d}_1^*\ \tilde{d}_2^*\ \ldots\ \tilde{d}_n^*\ \right] \tag{12}$$

Step 3.3. Calculating the grey relational coefficients. Equations (13–14) can be used to calculate grey relational coefficients:

$$\xi_{ij} = \frac{\min_i \min_j \Delta_{ij} + \rho \max_i \max_j \Delta_{ij}}{\Delta_{ij} + \rho \max_i \max_j \Delta_{ij}} \tag{13}$$

$$\Delta_{ij} = \sqrt{\frac{1}{3}\left[\left(d_{ijl} - d_{jl}^*\right)^2 + \left(d_{ijm} - d_{jm}^*\right)^2 + \left(d_{iju} - d_{ju}^*\right)^2\right]} \tag{14}$$

where $\rho \in [0, 1]$ is the distinguishing coefficient that is most taken as 0.5 in the literature [59].

Step 3.4. Determining the integrated superiority of each alternative for ranking. The integrated superiority of each alternative, which is denoted by S_i for the ith alternative, can be determined by Eq. (15):

$$S_i = \sum_{j=1}^{n} \xi_{ij} w_j i = 1, 2, \ldots m \tag{15}$$

where w_j is the weight of the jth indicator that is calculated using Fuzzy Z-AHP. Finally, alternatives are ranked according to their integrated superiority. The greatest superiority determines the best alternative.

5 Application

The results of a recent study [61], states that the top facial recognition vendors include Accenture, Aware, BioID, Certibio, Fulcrum Biometrics, Thales, HYPR, Idemia, Leidos, NEC. This study focuses on these ten facial recognition vendors that are chosen for the application, and each vendor has different characteristics within different levels of detail.

Accenture offers unique identity service platform to set up a breakthrough biometric system that can manage facial recognition, fingerprints, iris scans and other data to solve problems faced by people without official identities. It aims to combine the power of biometrics and blockchain to provide a global identity solution.

Aware uses five different products for facial recognition which are Knomi, PreFace, NexalFace, AwareABIS. Knomi is a mobile authentication with facial matching and liveness detection, whereas the PreFace is a face image auto-capture and processing on mobile device or desktop. NexalFace is used as a face-matching algorithm SDK. AwareABIS is used as an automated biometric identification system while FaceWorkbench used as forensic examiner workstation application.

BioID develops sensor-independent solutions that are compatible with any standard camera with facial recognition and face liveness detection software. BioID, which develops special software for face liveness detection and recognition, also includes PhotoVerify, a software that biometrically verifies document ownership for remote authentication. All technologies are combined in BioID Web Service and offered as cloud service via APIs. BioID implementations extent from counterfeiting prevention for existing face recognition systems to verification of identity document ownership, e-signing and multi-factor authentication for sign-in scenarios.

Certibio offers user authentication solutions using face, fingerprint, voice and iris biometry. OnePassID that offers identification and authentication for mobile platforms; OnIdentity, which biometrically authenticates users using information obtained from existing databases; Biometric Hub, which ensures sharing; and Snap-hit, which provides facial biometrics through a web platform, constitutes the company's biometric authentication portfolio.

Fulcrum Biometrics offers advanced biometric hardware and software solutions that support fingerprint, face, iris, palm vein and voice biometric modalities to improve identity management, improve logical and physical security, and increase operational efficiency. Technologies that Fulcrum offers include fast biometric software activation tools, mobile biometric systems, fingerprint scanners and other biometric sensors, LiveScan solutions and more. Fulcrum provides custom software application development and integration for multiple platforms and also offers Cloud-based and hosted biometric solutions.

Thales offers biometric technology and solutions by fingerprint, palm print, face and iris recognition for identification, verification and registration solutions. LFIS, a state-of-the-art live biometric face recognition solution that provides efficiency and accuracy for face detection, tracking and recognition, has been developed based on deep neural networks. By providing an industrial-grade solution, LFIS benefits from facial recognition in a wide variety of use cases, such as improving the customer/passenger experience, increased operational efficiency or enhanced security.

HYPR has dozens of pre-integrated authenticators, including fingerprint, voice, face, eye, palm, decentralized PIN and more. Some face authentication solutions are designed in a decentralized model using FIDO standards that allow a consumer or employee face template to be secured on the user's mobile device. Here, a user's face scan is locally verified against him/her, a token is sent to the service provider and access is granted. The biometer itself is not stored at the service provider, it is a real secret.

Idemia offers a reliable and accurate solution to capture facial biometrics with its MFace technology. MFace provides real-time tracking in a specified area using IDEMIA's independently proven face recognition technology and efficient 3D video tracking.

MFace performs multiple face tracking and capturing for a wide variety of applications where high efficiency and safety are key. Typical applications include a self-service kiosk, bag self-drop, border control, and self-boarding. It can be deployed as a stand-alone unit, desktop unit or as an OEM kit ready to be integrated.

IDHaystack, released by Leidos, provides pay-as-you-go identification and verification services, all of which reduce costs and increase flexibility. Currently, available modalities include voice, DNA, finger, palm-planned, face, and iris. With fast, low-cost startup and fast, accurate operational biometric matching, IDHaystack creates an opportunity to easily adding biometric matching to business services.

NEC's biometric facial recognition technology is used worldwide to fighting crime, prevent fraud, ensure public safety, and improve the customer experience in a wide variety of locations and industries. NEC's NeoFace recognition and matching capability have been verified by all benchmark tests and independent NIST tests as the world's fastest and most accurate in demanding conditions. Multimode technologies including the face, iris, voice recognition, finger and palm identification and ear acoustic authentication were developed by NEC, and they were supported by AI and data analytics.

The five criteria used in the study are determined as follows:

C1 length of presence on the market
C2 experience with border control applications
C3 experience with air industry applications
C4 accuracy of identification rate
C5 speed of facial recognition.

The selection of criteria C1, C2, C3, C4, C5 is a subjective choice of the authors of the present paper. The criteria were selected considering the constraints of the decision-making process. The experts had no access to hard and detailed data about the companies' implemented projects and their products. The only source of information were freely accessible sites in Internet. In real world situations the purchase department of a potential buyer would be able to ask detailed product descriptions and demonstrations, in which case criteria C4 and C5 should be expanded and made more detailed. Also, in a real world situation decision makers might replace criterion C1 with "company's rentability" or some other financial criteria, and criteria C2 and C3 could be adapted to the profile of the potential buyer.

In order to determine the values of the criteria, the experts were free to consult websites of the companies as well as other sources available in Internet (like Wikipedia information and the bibliography listed there). Of course, this kind of information is not always reliable, as it is biased by the endeavour of the companies to present their products and competences in the best possible light and by the largely uncontrolled process of entering information into Wikipedia. Z-fuzzy numbers, thanks to the reliability scale, allow to model the degree to which the experts judge the respective information to be reliable. Thus, the information about the length of the company's presence on the market (criterion C1) can usually be judged as reliable (although it may also be distorted through various mergers and acquisitions in the company's history). The information about "success stories" and customers (criteria C2 and C3) and the parameters of the software (criteria C4 and C5) cannot always be assessed as fully reliable. Experts expressed by means of Z-fuzzy sets their belief degree.

The company names are not given in the application part because the information that can be obtained about the companies is subjective and based on often incomplete information on their websites and the results are not binding for any company. Companies were named in random order (as A1, A2, A3, …, A10). After the criteria and alternatives were determined, the Fuzzy Z-AHP and Fuzzy Z-GRA methods were applied as follows.

In order to determine the importance weight of the criteria to be used in face recognition vendor selection, a pairwise comparison matrix is constructed in Table 4 (using the terminology from Table 1).

For example, the expert considered criterion C3 (experience with air industry application) to be greatly more important (GI) than criterion C1 (length of presence on the market). The cell expressing to which extent criterion C1 was seen as more important than C3 was then filled in automatically with the value reciprocal to GI (denoted as RGI). By using the linguistic terms, pairwise comparison matrix for reliability function is constructed in Table 5 (using the terminology from Table 2).

Here the expert was able to express his belief degree as to the values from Table 4. Basically, low-reliability values are linked here to the fact that the expert was

Table 4 Pairwise comparison matrix for restriction function

	C1	C2	C3	C4	C5
C1	EI	RMI	RGI	RWI	WI
C2	MI	EI	RWI	WI	GI
C3	GI	WI	EI	MI	AI
C4	WI	RWI	RMI	EI	MI
C5	RWI	RGI	RAI	RMI	EI

Table 5 Pairwise comparison matrix for reliability function

	C1	C2	C3	C4	C5
C1	AR	WR	SR	FR	FR
C2	WR	AR	WR	FR	HR
C3	SR	WR	AR	VWR	HR
C4	FR	FR	VWR	AR	WR
C5	FR	VWR	HR	WR	AR

not acting in a real-world situation, having an actual buyer as a partner in the decision-making process. Various buyers will have different preferences as to the importance of the criteria and the expert expressed his doubts in the form of reliability values. For example, the relation between criterion C3 and C4 (C3 was evaluated as being moderately more important than C4, see Table 4) is presented as very weakly reliable (VWR), which means that the expert was rather hesitant about it.

In order to check the validity of the solution, the consistency ratio (CR) of the fuzzy pairwise comparison matrix is calculated as 0.003. If the calculated CR value is less than 0.10, it means that the comparisons are consistent. After converting triangular z-fuzzy numbers into equivalent ordinary fuzzy numbers, normalized and fuzzy weights of each criterion are calculated as shown in Table 6 using Eqs. (5–6).

Thus, we can conclude from the application of fuzzy Z-AHP the following: criterion C3 (experience with air industry applications) is considered to be most important (weight 0.488), being followed by criterion C2 (experience with border control applications, weight 0.274) and criterion C4 (accuracy of identification, weight 0.132). The other two criteria (length of presence on the market and speed of facial recognition) turned out to be less important.

Table 6 Fuzzy and defuzzified weights of decision criteria

	Fuzzy weights	Defuzzified weights	Normalized defuzzified weights
C1: length of presence on the market	(0.30, 0.067, 0.171)	0.084	0.072
C2: experience with border control applications	(0.115, 0.262, 0.638)	0.319	0.274
C3: experience with air industry applications	(0.236, 0.516, 1.008)	0.569	0.488
C4: accuracy of identification rate	(0.054, 0.123, 0.314)	0.154	0.132
C5: speed of facial recognition	(0.018, 0.032, 0.077)	0.040	0.034

Table 7 Decision matrix including linguistic evaluations

	Evaluation for restriction function						Evaluation for reliability function				
	Criterion						Criterion				
Alternatives		C1	C2	C3	C4	C5	C1	C2	C3	C4	C5
	A1	MW	H	H	H	VH	SR	FR	HR	SU	WR
	A2	H	H	MW	H	VH	SR	VHR	HR	SU	WR
	A3	MW	W	W	H	VH	SR	HR	VHR	SU	VWR
	A4	VW	H	MW	H	VH	SR	VHR	VHR	HR	VWR
	A5	MW	MW	H	H	VH	VHR	FR	HR	HR	VWR
	A6	MW	W	W	H	VH	SR	FR	HR	HR	FR
	A7	W	W	W	H	VH	VHR	VHR	HR	VWR	WR
	A8	H	MW	MW	H	VH	SR	HR	VHR	AR	HR
	A9	VH	H	H	H	VH	SR	VHR	VHR	AR	VHR
	A10	VH	H	VH	H	VH	VHR	FR	HR	FR	HR

In the next part, the alternatives are evaluated using the Fuzzy GRA method, considering the criteria weights obtained. First, the decision matrix is constructed in Table 7 by linguistic terms depending on the decision maker's assessment.

Here the expert was asked to evaluate each company according to each criterion (using the terminology from Table 3). For example, company A1 was assessed as fulfilling almost all the criteria to a high degree. However, as mentioned before, the information the expert had access to be not always highly reliable. The expert was able to express his belief degree as to the assessments from the left-hand side of Table 7 in the right-hand side of the same table. For example, the very positive evaluation of company A1 with respect to criterion C4 (accuracy of identification rate) was given by the expert with a rather high consciousness of its low reliability, because this assessment was probably based on the publicity type data presented on the website of the company.

Linguistic statements of the decision-maker are converted into fuzzy Z-numbers using the scales, and initial decision matrices for fuzzy restriction and reliability functions are constructed. Then as shown in Table 8, triangular fuzzy Z-numbers are transformed to corresponding equivalent ordinary fuzzy numbers using Eqs. (1–3).

The ordinary fuzzy decision matrix is standardized as in Table 9, considering the indicators are all benefit type using Eq. (9).

Then, the maximum standardized values of each criterion are calculated using Eq. (11). Reference series consisting of ideal solutions for all indicators are denoted by D^* and given in Table 10.

By using Eqs. (13–14), grey relational coefficients are calculated and matrix is constructed in Table 11 where the distinguishing coefficient (ρ) is taken as 0.5.

At the end, after determining the integrated superiority of each alternative (S_i) alternatives are ranked and the results are given in Table 12.

Table 8 Ordinary fuzzy decision matrix

	Criterion					
Alternatives		C1	C2	C3	C4	C5
	A1	(0.13, 0.27, 0.45)	(0.46, 0.57, 0.71)	(0.50, 0.62, 0.77)	(0.29, 0.36, 0.45)	(0.51, 0.63, 0.63)
	A2	(0.58, 0.72, 0.89)	(0.54, 0.67, 0.84)	(0.12, 0.23, 0.39)	(0.29, 0.36, 0.45)	(0.51, 0.63, 0.63)
	A3	(0.13, 0.27, 0.45)	(0, 0.12, 0.23)	(0, 0.13, 0.25)	(0.29, 0.36, 0.45)	(0.44, 0.55, 0.55)
	A4	(0, 0, 0.13)	(0.54, 0.67, 0.84)	(0.13, 0.25, 0.42)	(0.50, 0.62, 0.77)	(0.44, 0.55, 0.55)
	A5	(0.13, 0.25, 0.42)	(0.11, 0.21, 0.35)	(0.50, 0.62, 0.77)	(0.50, 0.62, 0.77)	(0.44, 0.55, 0.55)
	A6	(0.13, 0.27, 0.45)	(0, 0.11, 0.21)	(0, 0.12, 0.23)	(0.50, 0.62, 0.77)	(0.57, 0.71, 0.71)
	A7	(0, 0.13, 0.25)	(0, 0.13, 0.25)	(0, 0.12, 0.23)	(0.36, 0.44, 0.55)	(0.51, 0.63, 0.63)
	A8	(0.58, 0.72, 0.89)	(0.12, 0.23, 0.39)	(0.13, 0.25, 0.42)	(0.65, 0.80, 1.00)	(0.62, 0.77, 0.77)
	A9	(0.72, 0.89, 0.89)	(0.54, 0.67, 0.84)	(0.54, 0.67, 0.84)	(0.65, 0.80, 1.00)	(0.67, 0.84, 0.84)
	A10	(0.67, 0.84, 0.84)	(0.46, 0.57, 0.71)	(0.62, 0.77, 0.77)	(0.46, 0.57, 0.71)	(0.62, 0.77, 0.77)

As it is seen in Table 12 A9 with a S_i value of 0.98 was determined as the first company to be selected as a face recognition vendor.

It is important to underline that the ranking given in Table 12 is highly sensitive to the expert opinions and, above all, to the circumstances in which the decision about selecting a vendor is taken. The example described here is merely an illustration of how the decision-making process might look like. In each real-world case, the buyer will deliver a detailed description of the product sought for and its future application requirements as well as a description of the criteria that should be applied. It is also the buyer who would be responsible to determine the relative importance of the criteria and to assess to which extent the product offered by each vendor suits his preferences. On the other hand, the vendor will have the possibility to demonstrate in detail several products, as well as to offer product modification or adjustment possibilities. A negotiation process will take place, which may lead to several decision-making processes, similar to the one described here.

Table 9 Standardized ordinary fuzzy decision matrix

	Criterion					
Alternatives		C1	C2	C3	C4	C5
	A1	(0.07, 0.16, 0.34)	(0.24, 0.39, 0.62)	(0.28, 0.43, 0.70)	(0.13,0.20, 0.30)	(0.24, 0.30, 0.37)
	A2	(0.29, 0.43, 0.68)	(0.29, 0.46, 0.73)	(0.07, 0.16, 0.35)	(0.13, 0.20, 0.30)	(0.24, 0.30, 0.37)
	A3	(0.07, 0.16, 0.34)	(0, 0.08, 0.20)	(0, 0.09, 0.23)	(0.13, 0.20, 0.30)	(0.21, 0.26, 0.32)
	A4	(0, 0, 0.10)	(0.29, 0.46, 0.73)	(0.07, 0.18, 0.38)	(0.22, 0.34, 0.52)	(0.21, 0.26, 0.32)
	A5	(0.07, 0.15, 0.32)	(0.06, 0.14, 0.30)	(0.28, 0.43, 0.70)	(0.22, 0.34, 0.52)	(0.21, 0.26, 0.32)
	A6	(0.07, 0.16, 0.34)	(0, 0.08, 0.18)	(0, 0.08, 0.21)	(0.22, 0.34, 0.52)	(0.27, 0.33, 0.42)
	A7	(0, 0.08, 0.19)	(0, 0.09, 0.22)	(0, 0.08, 0.21)	(0.16, 0.24, 0.37)	(0.24, 0.30, 0.37)
	A8	(0.29, 0.43, 0.68)	(0.06, 0.16, 0.34)	(0.07, 0.18, 0.38)	(0.29, 0.44, 0.68)	(0.29, 0.36, 0.45)
	A9	(0.36, 0.53, 0.68)	(0.29, 0.46, 0.73)	(0.30, 0.47, 0.76)	(0.29, 0.44, 0.68)	(0.32, 0.40, 0.49)
	A10	(0.34, 0.50, 0.64)	(0.24, 0.39, 0.62)	(0.35, 0.54, 0.70)	(0.20, 0.31, 0.48)	(0.29, 0.36, 0.45)

Table 10 Reference series values

Criterion					
D^*	C1	C2	C3	C4	C5
	(0.36, 0.53, 0.68)	(0.29, 0.46, 0.73)	(0.35, 0.54, 0.76)	(0.29, 0.44, 0.68)	(0.32, 0.40, 0.49)

Table 11 Fuzzy grey relational coefficient matrix

ξ_{ij}	C1	C2	C3	C4	C5
A1	0.43	0.72	0.86	0.33	0.41
A2	0.78	1.00	0.45	0.33	0.41
A3	0.43	0.34	0.39	0.33	0.33
A4	0.33	1.00	0.46	0.55	0.33
A5	0.42	0.39	0.86	0.55	0.33
A6	0.43	0.33	0.39	0.55	0.53
A7	0.36	0.34	0.39	0.38	0.41
A8	0.78	0.40	0.46	1.00	0.68
A9	1.00	1.00	0.96	1.00	1.00
A10	0.89	0.72	1.00	0.49	0.68

Table 12 S_i values and ranking of alternatives

Alternatives	S_i	Ranking
A1	0.71	3
A2	0.61	5
A3	0.37	10
A4	0.61	6
A5	0.64	4
A6	0.40	8
A7	0.37	9
A8	0.55	7
A9	0.98	1
A10	0.84	2

6 Conclusion

The check-in and border control are processes which are crucial in passenger air traffic. In the framework of Aviation 4.0, it is thus extremely important to search for efficient, possibly failure-free and user-friendly automatization of these processes, which can be attained thanks to the application of intelligent facial recognition software.

Numerous conditions have to be fulfilled by this type of software. It seems to be obvious that still some time is needed till the market will offer "close to ideal" products. For the nearest future airports will repeatedly face the problem of selecting one product among many different ones, none of which will fully satisfy all the requirements. In formal terms, this means that airports will have to solve multi-criteria selection problems.

In this paper, we propose a possible way of solving such problems. In order to be able to formulate and solve a multi-criteria selection problem, we basically need the following elements:

- alternatives
- evaluation criteria
- a procedure and model supporting the decision maker to select one alternative, taking into account the criteria importance
- expert opinions on the model parameters and on the outcomes of each step of the procedure.

In individual sections of this paper, we give some indications concerning each of the above aspects.

As for the alternatives, we presented the leaders on the market of facial recognition systems. The products to be purchased in a real-world case can be searched for in their offer. As far as evaluation criteria are concerned, we discussed the evaluation criteria used for facial recognition systems in the literature and proposed our own proposal. Also, we discussed in detail the challenges set to facial recognition systems in the framework of Aviation 4.0 as well as recent developments in this area. We also proposed a procedure for selecting one alternative from a given set of alternatives, based on fuzzy AHP and fuzzy Grey Relational Analysis methods. With respect to expert opinions, it has to be underlined that in the fuzzy AHP and fuzzy Grey Relational Analysis methods we propose to use a relatively recent modification of fuzzy numbers, called Z-fuzzy numbers. They make it possible for the experts to express their opinion not only using linguistic expressions, which do not require any mathematical preparation but also to give the information about the reliability degree of their judgment. This is especially important in the discussed case, where for example information contained in publicity-type documents has to be taken into account, which is not always completely reliable.

From the point of view of the theory of multi-criteria decision making, the fuzzy Z-Grey Relational Analysis is an original proposal, not presented in the literature before.

Further research will go in two directions. The first one is related to Aviation 4.0 specific applications: real-world case studies of decision-making situations, where the emphasis will be put on the criteria used to evaluate various facial recognition products and on the dialogue with experts in order to adapt the procedure to the needs of practitioners. It has to be emphasized that the evaluation criteria and their importance will change, not only because each potential buyer will have other needs, but also because the development of intelligent software will offer new perspectives in facial recognition and consequently will whet the appetite of potential users. Better and better product parameters and functionalities will be expected.

The other research direction is the improvement of multi-criteria decision-making methods. For example, an open problem in the fuzzy Z-AHP method is the global consistency measure—both of the criteria relative importance and the reliability of the opinions about relative importance. Also, other multi-criteria methods should be tried out or elaborated, so that they suit the needs of Aviation 4.0 practitioners to the highest possible degree.

References

1. Omarov, B., Omarov, B., Shekerbekova, S., Gusmanova, F., Oshanova, N., Sarbasova, A., Sultan, D.: Applying face recognition in video surveillance security systems. In: International Conference on Objects, Components, Models and Patterns, pp. 271–280. Springer, Cham (2019)
2. Ye, J., Zhu, J., Jiang, A., Li, H., Zuo, J.: Overview of facial expression recognition: a survey, data acquisition and processing. **1**, 21–34 (2020). https://doi.org/10.16337/j.1004-9037.2020.01.002
3. Frontex, B.I.: Automated biometric border crossing systems based on electronic passports and facial recognition: RAPID and SmartGate. Tech. rep., agency, European cooperation, operational borders, external states, Member Union, European, www. frontex. europa. eu. (2010)
4. Gohringer, C.: The application of face recognition in airports. Biometric Technol. Today **2012**(7), 5–9 (2012)

5. Ring, T.: Los Angeles and Tokyo airports trial face systems. Biometric Technol. Today **1**, 3–12 (2018). https://doi.org/10.1016/s0969-4765(18)30005-5
6. Anonymous: Facial recognition tech at JFK airport as DHS stalls on biometric exit system. Biometric Technol. Today (2016). https://doi.org/10.1016/s0969-4765(16)30023-6
7. del Rio, J.S., Moctezuma, D., Conde, C., de Diego, I.M., Cabello, E.: Automated border control e-gates and facial recognition systems. Comput. Secur. **62**, 49–72 (2016)
8. Hwang, W., Huang, X., Li, S.Z., Kim, J.: Face recognition using extended curvature Gabor classifier bunch. Pattern Recogn. **48**(4), 1247–1260 (2015)
9. Oh, B.S., Toh, K.A.: Face identity verification based on sinusoidal projection. In: 2013 IEEE Symposium on Computational Intelligence in Biometrics and Identity Management (CIBIM), pp. 71–76. IEEE, New York (2013)
10. Wu, C.H., Su, J., Chen, Y.: Age invariant face recognition. Tien Tzu Hsueh Pao/Acta Electronica Sinica **46**(7), 1593–1600 (2018). https://doi.org/10.3969/j.issn.0372-2112.2018.07.008
11. Ding, C., Tao, D.: A comprehensive survey on pose-invariant face recognition. ACM Trans. Intel. Syst. Technol. (TIST) **7**(3), 1–42 (2016)
12. Dutta, A., Veldhuis, R., Spreeuwers, L.: The impact of image quality on the performance of face recognition. In: 33rd WIC Symposium on Information Theory in the Benelux, Boekelo, The Netherlands, pp. 141–148 (2012)
13. Er, M.J., Chen, W., Wu, S.: High-speed face recognition based on discrete cosine transform and RBF neural networks. IEEE Trans. Neural Networks **16**(3), 679–691 (2005)
14. Zhang, X., Gao, Y.: Face recognition across pose: a review. Pattern Recogn. **42**(11), 2876–2896 (2009)
15. Hwang, W., Wang, H., Kim, H., Kee, S.C., Kim, J.: Face recognition system using multiple face model of hybrid Fourier feature under uncontrolled illumination variation. IEEE Trans. Image Process. **20**(4), 1152–1165 (2010)
16. Park, U., Tong, Y., Jain, A.K.: Age-invariant face recognition. IEEE Trans. Pattern Anal. Mach. Intell. **32**(5), 947–954 (2010)
17. Hu, C., Ye, M., Ji, S., Zeng, W., Lu, X.: A new face recognition method based on image decomposition for single sample per person problem. Neurocomputing **160**, 287–299 (2015)
18. Hu, X., Peng, S., Wang, L., Yang, Z., Li, Z.: Surveillance video face recognition with single sample per person based on 3D modeling and blurring. Neurocomputing **235**, 46–58 (2017)
19. Spreeuwers, L.J., Hendrikse, A.J., Gerritsen, K.J.: Evaluation of automatic face recognition for automatic border control on actual data recorded of travellers at Schiphol Airport. In: 2012 BIOSIG-Proceedings of the International Conference of Biometrics Special Interest Group (BIOSIG), pp. 1–6. IEEE, New York (2012)
20. Anonymous: MFlow journey facial recognition at Edinburgh airport. Biometric Technol. Today (2013). https://doi.org/10.1016/s0969-4765(13)70105-x
21. Anonymous: Berlin airports purchase facial recognition system. Biometric Technol. Today (2002). https://doi.org/10.1016/s0969-4765(02)00909-8
22. Ahmed, H., Umair, M., Murtaza, A., Bajwa, U.I., Vardasca, R. Multi-spectral face recognition system. In: European Congress on Computational Methods in Applied Sciences and Engineering, pp. 983–997. Springer, Cham (2017)
23. Kwon, B.M., Lee, K.H.: An introduction to face-recognition methods and its implementation in software applications. Int. J. Inf. Technol. Manage. **17**(1–2), 33–43 (2018)
24. Balla, P.B., Jadhao, K.T.: IoT based facial recognition security system. In: 2018 International Conference on Smart City and Emerging Technology (ICSCET), pp. 1–4. IEEE, New York (2018)
25. Saaty, T.L.: The Analytic Hierarchy Process: Planning, Priority Setting, Resource Allocation. McGraw-Hill, New York (1980)
26. Van Laarhoven, P.J., Pedrycz, W.: A fuzzy extension of Saaty's priority theory. Fuzzy Sets Syst. **11**(1–3), 229–241 (1983)
27. Buckley, J.J.: Fuzzy hierarchical analysis. Fuzzy Sets Syst. **17**(3), 233–247 (1985)
28. Wang, Y.M., Elhag, T.M., Hua, Z.: A modified fuzzy logarithmic least squares method for fuzzy analytic hierarchy process. Fuzzy Sets Syst. **157**(23), 3055–3071 (2006)

29. Wang, Y.M., Chin, K.S.: Fuzzy analytic hierarchy process: a logarithmic fuzzy preference programming methodology. Int. J. Approx. Reason. **52**(4), 541–553 (2011)
30. Chang, D.Y.: Applications of the extent analysis method on fuzzy AHP. Eur. J. Oper. Res. **95**(3), 649–655 (1996)
31. Shaw, K., Shankar, R., Yadav, S.S., Thakur, L.S.: Supplier selection using fuzzy AHP and fuzzy multi-objective linear programming for developing low carbon supply chain. Expert Syst. Appl. **39**(9), 8182–8192 (2012)
32. Paksoy, T., Pehlivan, N.Y., Kahraman, C.: Organizational strategy development in distribution channel management using fuzzy AHP and hierarchical fuzzy TOPSIS. Expert Syst. Appl. **39**(3), 2822–2841 (2012)
33. Durán, O.: Computer-aided maintenance management systems selection based on a fuzzy AHP approach. Adv. Eng. Softw. **42**(10), 821–829 (2011)
34. Lin, H.F.: An application of fuzzy AHP for evaluating course website quality. Comput. Educ. **54**(4), 877–888 (2010)
35. Lee, S.K., Mogi, G., Lee, S.K., Hui, K.S., Kim, J.W.: Econometric analysis of the R&D performance in the national hydrogen energy technology development for measuring relative efficiency: the fuzzy AHP/DEA integrated model approach. Int. J. Hydrogen Energy **35**(6), 2236–2246 (2010)
36. Uzoka, F.M.E., Obot, O., Barker, K., Osuji, J.: An experimental comparison of fuzzy logic and analytic hierarchy process for medical decision support systems. Comput. Methods Programs Biomed. **103**(1), 10–27 (2011)
37. Che, Z., Wang, H.S., Chuang, C.L.: A fuzzy AHP and DEA approach for making bank loan decisions for small and medium enterprises in Taiwan. Expert Syst. Appl. **37**(10), 7189–7199 (2010)
38. Taha, Z., Rostam, S.: A hybrid fuzzy AHP-PROMETHEE decision support system for machine tool selection in flexible manufacturing cell. J. Intell. Manuf. **23**(6), 2137–2149 (2012)
39. Arslan, T., Khisty, J.: A rational approach to handling fuzzy perceptions in route choice. Eur. J. Oper. Res. **168**(2), 571–583 (2006)
40. Ju-Long, D.: Control problems of grey systems. Syst. Control Lett. **1**(5), 288–294 (1982)
41. Liu, S., Yang, Y., Forrest, J.: Grey Data Analysis. Springer, Singapore (2017). https://doi.org/10.1007/978-981-10-1841-1
42. Liu, S., Lin, Y.: Grey Information: Theory and Practical Applications. Springer Science & Business Media, Berlin (2006)
43. Köse, E., Aplak, H.S., Kabak, M.: Personel Seçimi için Gri Sistem Teori Tabanli Bütünlesik Bir Yaklasim/An integrated approach based on grey system theory for personnel selection. Ege Akademik Bakis **13**(4), 461 (2013)
44. Karimi, A., Ahmadpour, B., Marjani, M.R.: Using the fuzzy grey relational analysis method in wastewater treatment process selection. Iran. J. Health Saf. Environ. **5**(3), 1041–1050 (2018)
45. Yeh, M.F., Lu, H.C.: Evaluating weapon systems based on grey relational analysis and fuzzy arithmetic operations. J. Chin. Inst. Eng. **23**(2), 211–221 (2000)
46. Chen, M.F., Tzeng, G.H.: Combining grey relation and TOPSIS concepts for selecting an expatriate host country. Math. Comput. Model. **40**(13), 1473–1490 (2004)
47. Song, Q., Shepperd, M., Mair, C.: Using grey relational analysis to predict software effort with small data sets. In: 11th IEEE International Software Metrics Symposium (METRICS'05), p. 10. IEEE, New York (2005)
48. Wu, H.H.: A comparative study of using grey relational analysis in multiple attribute decision making problems. Qual. Eng. **15**(2), 209–217 (2002)
49. Song, Q., Jamalipour, A.: An adaptive quality-of-service network selection mechanism for heterogeneous mobile networks. Wireless Commun. Mobile Comput. **5**(6), 697–708 (2005)
50. Jadidi, O., Sai Hong, T., Firouzi, F., Yusuff, R.M.: An optimal grey based approach based on TOPSIS concepts for supplier selection problem. Int. J. Manage. Sci. Eng. Manage. **4**(2), 104–117 (2009)
51. Zhang, S.F., Liu, S.Y.: A GRA-based intuitionistic fuzzy multi-criteria group decision making method for personnel selection. Expert Syst. Appl. **38**(9), 11401–11405 (2011)

52. Pramanik, S., Mukhopadhyaya, D.: Grey relational analysis based intuitionistic fuzzy multi-criteria group decision-making approach for teacher selection in higher education. Int. J. Comput. Appl. **34**(10), 21–29 (2011)
53. Goyal, S., Grover, S.: Applying fuzzy grey relational analysis for ranking the advanced manufacturing systems. Grey Syst. Theory Appl. (2012)
54. Wu, Q., ve Luyan, L.: Research on investment decision-making of construction engineering projects based on the grey relation grade. Adv. Sci. Lett. **15**(1), 407–409 (2012)
55. Otay, I., Kahraman, C.: Solar PV power plant location selection using a Z-fuzzy number based AHP. Int. J. Anal. Hierarchy Process **10**(3) (2018)
56. Zadeh, L.A.: A note on Z-numbers. Inf. Sci. **181**(14), 2923–2932 (2011)
57. Yildiz, N., Kahraman, C.: Evaluation of social sustainable development factors using Buckley's fuzzy AHP based on Z-numbers. In: International Conference on Intelligent and Fuzzy Systems, pp. 770–778. Springer, Cham (2019)
58. Elevli, B.: Logistics freight center locations decision by using Fuzzy-PROMETHEE. Transport **29**(4), 412–418 (2014)
59. Chen, L., Ren, J.: Multi-attribute sustainability evaluation of alternative aviation fuels based on fuzzy ANP and fuzzy grey relational analysis. J. Air Transp. Manage. **68**, 176–186 (2018)
60. Wei, G.W.: Theory and Methods of Multiple Attribute Decision Making Based on Fuzzy Information. China Economic Publishing House, Beijing (in Chinese) (2010)
61. Markets and Markets Research Private Ltd.: Facial Recognition Market by Component (Software Tools (2D Recognition, 3D Recognition, and Facial Analytics) and Services), Application Area (Emotion Recognition, Access Control, and Law Enforcement), Vertical, and Region-Global Forecast to 2024. REPORT CODE TC 3421. https://www.marketsandmarkets.com/Market-Reports/facial-recognition-market-995.html(2019)

Intelligent and Fuzzy Approaches in Aviation 4.0 Transportation and Cargo Applications

Ecem Cem and Selcuk Cebi

Abstract Airline transportation presents many advantages in terms of transit time, security, handling, technology, and coverage. Containers are indispensable for air cargo transportation as in maritime, railway, and highway transportation. The history of the container goes back many years and it is still in use today as it was first used. Today, innovations based on the Internet of Things technology are effective in the transportation sector as well as in the other sectors. This technology change has also affected container designs and the definition of smart containers has come to the fore. The smart container is a new term and since the design process of a new product is difficult and complex, the design requirements of a smart container and the design parameters to meet these requirements must be well defined. Therefore, within the scope of the proposed study, it is aimed to define the functional requirements (FRs) that a smart container design to be used in air transportation should meet and the design parameters (DPs) that will meet these requirements. For this aim, the fuzzy independence axiom method, which is the first axiom of the axiomatic design method, is used to define FRs and the DPs of a smart container. In the application section, a total of 23 FRs and 23 DPs under 4 main factors are defined for the smart container design. The limitation of the proposed study is that the design parameters regarding smart and intelligent technologies except for the physical design features of the containers were only taken into consideration for the smart container design except for the physical design features of the containers.

S. Cebi (✉)
Department of Industrial Engineering, Yildiz Technical University, Besiktas, Istanbul, Turkey
e-mail: scebi@yildiz.edu.tr

E. Cem
Department of Industrial Engineering, Aydin University, Buyukcekmece, Istanbul, Turkey
e-mail: ecemcem@aydin.edu.tr

C. Kahraman and S. Aydın (eds.), *Intelligent and Fuzzy Techniques in Aviation 4.0*, Studies in Systems, Decision and Control 372,
https://doi.org/10.1007/978-3-030-75067-1_17

1 Introduction

While the importance of the transportation industry is becoming more vital in the world, the number of studies for the solutions of sectoral problems are increasing day by day. Innovations based on Industry 4.0 show their effects in the air transportation sector as well as other sectors. One of the leading problems in the transportation sector is that the products are out of use due to various factors during transportation and this cannot be tracked before it happened. With the application of Industry 4.0 tools in the transportation sector, great progress has been made in solving this problem. Industry 4.0 tools such as the Internet of Things and Cloud Technologies are leading of these. Also, the difficulties and confusion in tracking the products have been resolved with the use of Industry 4.0 tools. Besides, one of the biggest signs of progress is experienced in the security field. However, airline transportation offers many advantages in terms of transit time, security, handling, technology, and coverage. Since the airline is the fastest vehicle of transportation today, it offers a great opportunity in terms of transportation speed in cargo. Accident rates in air transport are very low compared to road transport and there are high-level security measures at airports. This makes it safe to transport cargo. Since every step of the handling is monitored, the operations are done more meticulously and safely. Moreover, since airports are always the areas where high-level technology is used, the latest technologies are used in cargo operations. Another advantage of air transportation is that there is at least one airline in almost every city which creates many delivery points around the world. All these advantages make air transport superior to other vehicles. The biggest disadvantage of airline transportation is its high cost, but when the benefit and cost of the cargo are compared, it is a great advantage to provide important cargoes by air transportation. Another disadvantage is that the restrictions on cargo size and cargo quality are stricter than other alternatives.

The indispensable part of cargo transportation is the container. A container is a tank that is usually made of metal, used to transport, store, or pack a product. Containers are widely used in cargo transportation because they shorten the handling (loading and unloading) time of the products, allow regular loading, and better protect the materials they contain against damage and theft. Although containers are a traditional tool, they are a product that keeps up with the development of technology. Today, the definition of smart containers has entered the literature [1]. In the Business Requirements Features Report published by the United Nations Economic Commission for Europe in 2019 [2], Smart Container is defined as a container equipped with the Internet of Things Device. Internet of Things Device is defined as an electronic device that can be connected to the Internet to exchange information with an IoT platform. In a transportation operation using a smart container, real-time geographic monitoring of the container or measurement of humidity, temperature, oxygen content, shock, light, pressure, and even ethylene concentration and air pollution within it becomes possible. All these data provide a great advantage in preventing the loss of goods and time in the supply chain process. Furthermore, smart devices should be used not only for monitoring but for an instant intelligent decision-making system in the smart cargo concept to be able to intervene in logistics events immediately without delay. By this way, a significant improvement in communication costs and transportation costs may be provided [3]. This is an actual requirement to achieve various optimizations such as flexibility, compatibility, and time in logistics and procurement in order to create a common denominator and to use resources sustainably [4]. For this, the design characteristics and design requirements of a smart container should be well defined.

The smart container is a new term and the design requirements of a smart container should be well defined since the design process of a new product is difficult and complex. For this, in the scope of the proposed study, it is aimed to define the design characteristics of a smart container to be used in air transportation by using the fuzzy independence axiom method. The independence axiom, which is the first axiom of the Axiomatic Design method is proposed by Suh [5] to provide a scientific basis to define the design specifications that will meet the functional requirements of the design.

In the first part of the study, general information about smart containers and aviation 4.0 is given. The rest of the study is organized as follows; In Sect. 2, a literature review study on smart container applications is presented. Section 3 presents the fundamentals of the axiomatic design method and its fuzzy extensions. Then, in Sect. 4, the basic components of a smart container design are defined using the fuzzy axiomatic design method which is explained in the third part of the study. Finally, concluding remarks are given in Sect. 5.

2 Literature Review

In this section, a literature review study is presented to present hot topics and trends in smart transportation systems and aviation transportation applications in aviation. Despite the limited number of studies in the literature, the prominent publications among the studies are as follows:

Büyüközkan et al. [6] present a new digital service quality model on aviation transportation. The model consists of 6 dimensions and 35 criteria. These dimensions are; digital tangible elements, reliability, digital interaction, digital trust, and customer focus. The importance of the criteria was calculated using the Interval-Valued Intuitionistic Fuzzy Analytic Hierarchy Process (IVIF AHP) method. As a result of the calculation, it was revealed that the most important dimension is digital trust and the three most important criteria are proactive customer service, cybersecurity, and customer insight. The validity of this model has been verified by Turkish Airlines.

Valdés et al. [7] define the concept of Aviation 4.0, which emerged with the application of Industry 4.0 to the aviation industry and outline the issue of increasing aviation security. The study aims to provide new improvements with Industry 4.0 applications in the aviation industry operating at high-security levels.

Malygin et al. [8] designed the 3rd Generation Multimodal Intelligent Transportation System (MITS 3.0) for Russia using information, network, and artificial intelligence technologies. The proposed MITS 3.0 design is a new information and telecommunications system. The proposed system is an in-mode and multi-mode transport flow project, as well as making it possible to control and manage their security. By the implementation of the project, it is aimed to achieve outputs such as shortening the time of transportation of people and goods, improving the quality, safety, and environmental friendliness of transportation services, effectively controlling and managing intra-modal and multi-modal transport flows, and increasing the capacity of the national transportation system.

May et al. [9] aim to facilitate the optimization of key performance metrics used in the demand management area of Virgin Atlantic Air Cargo's air cargo operations by using the fuzzy group decision-making method. At the end of the study, they listed 10 main performance criteria based on results. It has been demonstrated that Network Optimization is the most important criterion in ranking. The study also aims to contribute to the Revenue Management literature.

In the study conducted by May et al. [10], smart fuzzy multi-criteria decision-making methods were used in ordering result-based key performance indicators. Actual operational data of Virgin Atlantic Cargo were used in the study. It is aimed to optimize the performance measurement of air cargo capacity demand management by using the best indicators among the key performance indicators (KPI). As a result of the study, the order is as follows, network optimization, optimizing density, PB/Free-sale mix, overbooking, flight plan audit, dynamic bid price, PB utilization, empty moves, maintaining station budgets, and flight plan released.

Zhou and Lou [11] develop Cargo Information Modeling and Smart Cargo Tracking System in their study. The proposed tracking system is an internet and mobile communication-based tracking system based on the Internet of Things. The smart cargo tracking system has been successfully presented as an exemplary system application in the field of cotton transportation. JADE, a framework implemented in Java language, was used in the integration. It is stated that the proposed Smart Cargo Tracking System can be applied in all transportation and cargo types.

Dalmolen et al. [12] present a proposal on how to decide on a sustainable logistics process for the European Union iCargo project through cooperation and integration and information sharing among logistics supply chain stakeholders. The aim is stated to be reducing the carbon footprint and they try to achieve this with optimal use of resources and infrastructure. For this, the optimal use of logistic resources and physical infrastructure should be aimed. Information accessibility, traceability, and transparency among the stakeholders in the logistics value chain and the presence of an infrastructure that provides them are taken as a prerequisite in this process. In the iCargo project, this infrastructure and this project are provided by using Information and Communication Technologies.

Forcolin et al. [13] reveal the technical details of ICT applications and solutions developed within the scope of the EURIDICE project, presented for viewing, tracking and transmitting, and processing the information of the cargo throughout the supply chain, as well as in-field applications. In the article, the solutions provided to the logistics industry by the Internet of Things applications, which are one of the leading Industry 4.0 tools, are presented. The purpose of all these applications is to be able to react quickly to events developing in the supply chain and thus increase efficiency. With the established EURIDICE platform, secure information exchange between all stakeholders and rapid response to events that disrupt the chain is aimed.

Sternberg et al. [14] use smart cargo to monitor, control, and reduce the environmental impacts of freight transportation. Within the scope of the study, a large literature on logistics, green logistics, and information science was presented and empirical data were collected from a binary case study. The dual case study was conducted through interviews and observation. As a result of the study, it was revealed that smart cargo is a sustainable method in environmental and financial fields and contributed to the literature by offering a new perspective to logistics models for use in future studies.

Schumacher et al. [3] develop a platform to intervene in logistics events immediately without delay. By this way, a significant improvement in communication costs and transportation loss is obtained. The designed platform is based on accessible standards such as SOA (Service Oriented Architecture) and Multiple Agent Systems (MAS). By using information and communication technologies together, a fast and effective smart cargo design that can decide on its own to the events defined in the system has been created.

Matthews et al. [15] propose a new model for driver behavior analysis in next-generation intelligent transportation systems. It is stated that the proposed model can be used as a basis for understanding the positive or negative impact that new generation intelligent transportation systems can have on driver performance. It is stated that the model can provide a roadmap to determine how the next generation intelligent transportation system supports or interferes with the situational awareness (SA) required to achieve a driving goal.

Although the studies in the literature are limited, these studies mostly focused on cargo transportation. In cargo transportation, subjects such as Industry 4.0 applications, performance analysis, load, and demand optimization were discussed. However, there is not any study considering the design requirements of a smart container product. Although the demand for smart container designs is on the agenda in the industrial market, none of the studies in the literature addresses the smart container design needs. There is a lack of information about the smart container since the design characteristics of the system are unclear and customer expectations are not defined before is the main shortcoming in the literature. In order to cope with this research gap in the literature, it is aimed to determine the design characteristics and functional requirements of a smart container by using the fuzzy axiomatic design method.

3 Axiomatic Design

The Axiomatic Design (AD) method developed by Suh [5] provides a scientific and systematic basis for the design process. The main purpose of the method is to make designers more creative, to reduce random scanning processes, to minimize trial and error processes, and to help choose the best design [5, 16, 17]. The philosophy underlying the method is "WHAT does the designer or client want to do?—HOW can he achieve what he wants? "It is based on the continuous interaction between the questions [5].

The axiomatic design approach has two axioms, namely "*independence axiom*" and "*information axiom*". The *Independence Axiom* which is the first axiom of the method requires ensuring independence between functional requirements. The second axiom, the axiom of knowledge, requires choosing the best alternative among the alternatives that satisfy the first axiom. In other words, after the designs that provide the independence axiom are determined, the second axiom is used to determine the best design. The flow diagram of the design process given in Fig. 1 presents the relationship between design axioms [18].

Fig. 1 The relationship between independence and information axioms

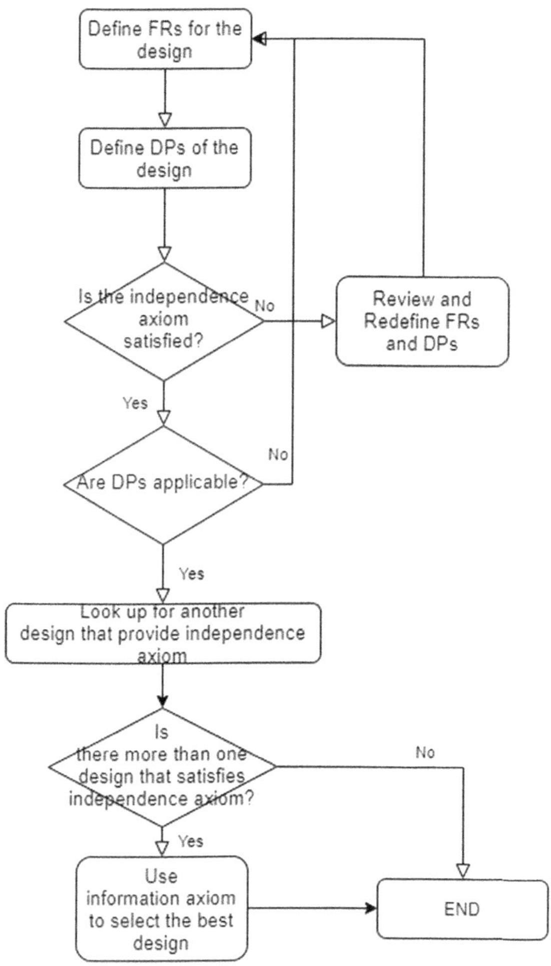

3.1 Independence Axiom

The first axiom of the axiomatic design method is the *Independence Axiom* and it requires ensuring independence among functional requirements of a design. If there is more than one functional requirement in the design stage, these requirements should not affect each other. This means that when selecting design parameters, the designer should ensure that functional requirements are met and the independence of the functional requirements are satisfied. The relationship between functional requirements (FRs) and design parameters (DPs) is defined by the matrix format shown in Eq. 1 [5, 19, 20]

$$\{FR\} = [A]\{DP\}, \quad A = \left[A_{ij}\right]_{mxm} \tag{1}$$

where A is the relation matrix that defines the relationship between FR and DP. Equation 1 is also called as the design equation [17]. A relationship matrix for the design is a matrix of 1 and 0 elements that indicates whether there is a relationship between FR-DP. However, the relationship matrix does not show the degree and direction of the relationship. The relationship matrix is given in Eq. 2.

$$\mathrm{A} = \begin{bmatrix} \mathrm{A}_{11} & \mathrm{A}_{12} & \ldots & \mathrm{A}_{1\mathrm{n}} \\ \mathrm{A}_{21} & \mathrm{A}_{22} & \ldots & \mathrm{A}_{2\mathrm{n}} \\ \ldots & \ldots & \ldots & \ldots \\ \mathrm{A}_{\mathrm{m}1} & \mathrm{A}_{\mathrm{m}2} & \mathrm{A}_{\mathrm{m}3} & \mathrm{A}_{\mathrm{mn}} \end{bmatrix} \tag{2}$$

where *m* and *n* show the number of functional requirements design parameters, respectively. The differential form of Eq. 2 is given in Eq. 3.

$$A_{ij} = \frac{\partial FR_i}{\partial DP_j} \tag{3}$$

The design relationship matrix is defined in two different ways according to the values of the elements and the shape of the design matrix. One of them is linear or nonlinear design according to the values of the elements in the relationship matrix. The other is uncoupled, decoupled, and uncoupled design definitions according to the shape formed by the elements in the relationship matrix. Linear design or nonlinear design is defined as follows [16].

- If the relationship matrix, A_{ij} consists of fixed elements, the design is called linear design.
- If any of the elements of the relationship matrix, A_{ij} consists of a function that depends on design parameters, the design is called nonlinear design.

According to the shape formed by the elements of the relationship matrix, the type of the design is one of uncoupled, decoupled, and coupled design. If the type of design is uncoupled, the created design is the ideal design. If the design is decoupled, the proposed design is acceptable and the design works. If the design is coupled design, the design is unacceptable and the design does not work well. In this case, the design process should be reconsidered with different design parameters. The type of design is defined as follows [5, 16].

- If the number of DPs (n) is less than the number of FRs (m), the design is coupled (n < m). In this case, the design is restructured so that the number of FRs equals the number of DPs.
- If the number of FRs is less than the defined number of DPs, the design is called redundant (m < n). Such designs should be restructured by reducing the number of DPs.
  - If the number of DPs is equal to the number of FRs, the design is classified as follows;
  - If the relationship matrix is in diagonal form, the design is called uncoupled design and shows the ideal situation. The formulation of ideal design is given in Eq. 4

$$A = \begin{bmatrix} A_{11} & 0 & \dots & 0 \\ 0 & A_{22} & \dots & 0 \\ \dots & \dots & \dots & \dots \\ 0 & 0 & \dots & A_{mm} \end{bmatrix} \tag{4}$$

In other words, ideal design is expressed as follows; if $A_{ij} = 1$ for each $i = j$ in the relationship matrix and $A_{ij} = 0$ for other cases, the design is uncoupled design (Eq. 5).

$$A_{ij} = 1 \;\; for \;\; i = j \;\; ve \;\; A_{ij} = 0 \;\; for \; i \neq j \tag{5}$$

  - If the relationship matrix is in the form of a lower triangle or upper triangular, the design is called decoupled design and it is an acceptable design although it is undesirable. In such designs, the independence among the functional requirements is not fully met, but independence between functional requirements can be achieved by following a certain order. The lower triangular matrix is defined by Eq. 6.

$$A = \begin{bmatrix} A_{11} & 0 & \dots & 0 \\ A_{21} & A_{22} & \dots & 0 \\ \dots & \dots & \dots & \dots \\ A_{m1} & A_{m2} & A_{m3} & A_{mn} \end{bmatrix} \tag{6}$$

If the relationship matrix is $A_{ij} = 1$ for $i \leq j$, the design is called decoupled design (Eq. 7).

$$\forall i, j \;\; için \;\; i > j \;\; ve \;\; A_{ij} = 0 \tag{7}$$

- The designs other than those defined above are called coupled design.

As a result, uncoupled or decoupled designs are called the "*working design*", and coupled designs are called the "*non-working design*". Accordingly, Eqs. 8 and 9 are used to define the working and non-working design, respectively [18–20].

$$i = j \text{ için } a_{ij} = 1 \quad ve \quad \sum_{i=1}^{m}\sum_{j=i+1}^{m} a_{ij} = 0 \tag{8}$$

$$i = j \quad \text{için} \quad a_{ij} \neq 1 \text{ or} \sum_{i=1}^{m}\sum_{j=i+1}^{m} a_{ij} > 0 \tag{9}$$

3.2 *Information Axiom*

In the application, there can be more than one design that satisfies the independence axiom. However, one of these designs is only superior to others. Therefore, the second axiom of the axiomatic design method, *the information axiom*, presents an analytical method that evaluates the developed design alternatives. Thus, it ensures that the best design is chosen among the designs that satisfy the independence axiom. The information axiom is represented by the information content value (I_j) which is expressed by the probability (p_j) of satisfying a defined functional requirement (FR_j). Information content value is expressed by Eq. 10 and in the literature, this equation is known as the Shannon equation [5, 17, 21, 22].

$$I_j = \log_2 \frac{1}{p_j} \tag{10}$$

In Eq. 10, the information content is defined with a logarithmic expression in two bases. When a large number of functional requirements need to be met at the same time, a logarithmic function has been chosen in the calculation in order to provide a collectability feature to the information content. In the application, there is generally m number of FRs in design, and the information content for the whole system (I_{system}) is obtained with Eq. 11 [17].

$$I_{system} = -\sum_{j=1}^{m} \log_2 p_j = \sum_{j=1}^{m} \log_2 (1/p_j) \tag{11}$$

The probability of success (p_j) presents the satisfying level of the functional requirement and it is obtained by the ratio of the intersection of the design range (defined by the designer) and the system range (defined by the system) to the system range. A graphical representation of the intervals is given in Fig. 2. The vertical

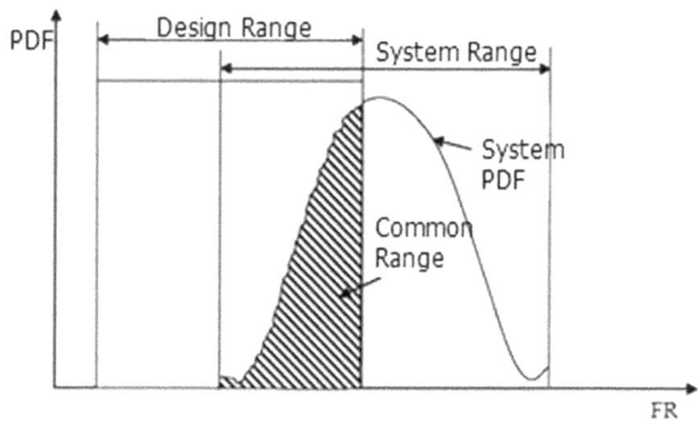

Fig. 2 The definition of design range, system range, and common range

axis of the graph represents the probability density (*PDF*) and the horizontal axis represents FR depending on the mapping [5, 17].

The probability density function (PDF) of the system in Fig. 2 shows the system range boundaries for the defined FR. The common area (A_{cr}) is the intersection area of the design range and the system range (A_{sr}). The intersection area presents the satisfaction level of a FR. As a result, the system range and the intersection range present the probability of success of the defined objective, and the probability of success (p_j) is calculated by Eq. 12. Thus, Eq. 13 is used to obtain information content value.

$$p_j = \frac{A_{cr}}{A_{sr}} \tag{12}$$

$$I_j = \log_2 \frac{A_{sr}}{A_{cr}} \tag{13}$$

In order to achieve the design goal, the information axiom aims at minimizing the information content. Therefore, the design with the smallest I value describes it as the best design. If the probability value is equal to 1, the information content value will be zero. Conversely, if the probability value is 0, the information content value is infinite. This means; if the probability value is small, we need to provide more information to meet the functional requirements [5].

3.3 Fuzzy Independence Axiom

The independence axiom, which is the first axiom, requires the independence of functional requirements with different design parameters. As mentioned before, the relationship between FRs and DPs is shown with a matrix whose elements are 0 and 1's. While 0 indicates that there is no relationship between the functional requirement and the related design parameter, 1 indicates the existence of a relationship between the functional requirement and the design parameter. The degree of relationship

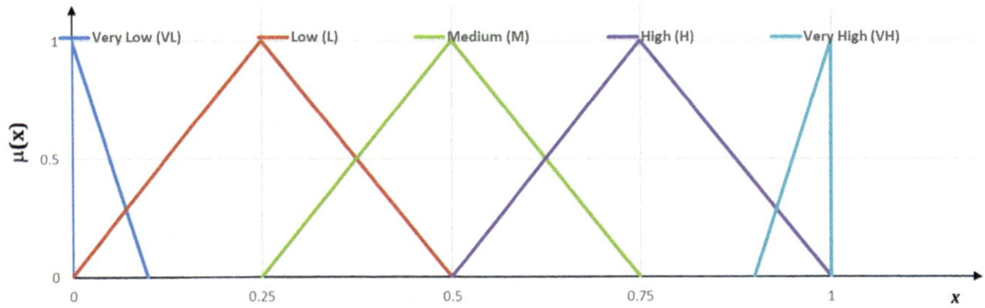

Fig. 3 Linguistic scale for the evaluation of relationships among FRs and DPS

is not specified here. Thus, weak and strong relationships are considered to the same degree in the relationship matrix. Whereas in real problems, relationships can be graded and if the relationships are shown with their degrees in the relationship matrix, it is easier to comment on the uncoupled or coupled of the design. Besides, emphasizing the importance of functional requirements in design matrices will be more useful in determining the best design. A relationship matrix given with Eq. 14 can be established for a design by using the linguistic expressions given in Fig. 3, [19, 20].

$$\begin{Bmatrix} FR_1 \\ FR_2 \end{Bmatrix} = \begin{bmatrix} VH & H \\ H & VH \end{bmatrix} \begin{Bmatrix} DP_1 \\ DP_2 \end{Bmatrix} \tag{14}$$

The independence of the created design is obtained by Eq. 15 in the case of a relative evaluation of the relationships in the design matrix ranging between [0, 1] [20].

$$\tilde{C} = \frac{\sum_{i=1}^{m} \sum_{j=i+1}^{m} \tilde{a}_{ij}}{\sum_{i=1}^{m} \sum_{j=i+1}^{m} 1} \tag{15}$$

where $\tilde{C}\left(C^l,\ C^m,\ C^u\right)$ is the coefficient of design and and $\tilde{a}_{ij}\left(a^l,\ a^m,\ a^u\right)$ is the relationship value between FR_i and DP_j. Equation 15 produces a value in the range of [0, 1] and shows the independence degree of a design. At this point, the significance level of the relationship between the functional requirement and the design parameter requires expert knowledge and the negligible value of the relationship is indicated by γ. In other words, γ indicates the tolerance value defined by Suh [5]. If this coefficient is bigger than 0 or the defined tolerance value ($\tilde{C} > \gamma$), it means that the design is a coupled design. If the coefficient of dependency is equal to zero ($\tilde{C} = (0; 0; 0)$) or less than γ ($\tilde{C} \leq \gamma$), the design is called either decoupled or uncoupled design. Equation 16 is used to define whether the design is uncoupled or decoupled [20].

$$\tilde{c} = \frac{\sum_{i=1}^{m} \sum_{j=1}^{i-1} \tilde{a}_{ij}}{\sum_{i=1}^{m} \sum_{j=1}^{i-1} 1} \tag{16}$$

where $\tilde{c}$ is called the independence coefficient. If we obtain $\tilde{c} = (0; 0; 0)$, the design is uncoupled. Otherwise, the design is decoupled. It is obtained as $\tilde{C} = (0, 0, 0)$ and $\tilde{c} = (0.8, 0.9, 1.0)$ for the design matrix given by Eq. 14 [19, 20]. The comment on whether the design is acceptable or not is based on the selected tolerance value. If there is more than one design that satisfies the first axiom, we can use Eqs. 15 and 16 to select the best one instead of the information axiom. Since these equations will produce different values for each design, it is necessary to compare these outputs to say which design is better. Two different methods can be used for the comparison. The first of them is to convert the fuzzy numbers into crisp numbers by using Eq. 17.

$$w_r = \frac{\tilde{w}_r}{\sum_{j=1}^{n} \tilde{w}_j} = \frac{w_{rl} + w_{rm} + w_{ru}}{\sum_{j=1}^{n} \tilde{w}_j} \tag{17}$$

As the second method, a fuzzy sorting method can be used. For fuzzy sorting, the method suggested by Tran and Duckstein [23] can be used. The method is based on the measurement of the distances between the defined targets (the largest rational number and the smallest rational number) and the fuzzy values. In the method, the fuzzy number with the least distance from the largest rational number and the greatest distance from the smallest rational number is ranked as the first. The largest and smallest values are determined by Eqs. 18 and 19.

$$Max(I) \geq \sup\left(\bigcup_{i=1}^{I} s(\tilde{A}_i)\right) \tag{18}$$

$$Min(I) \leq \inf\left(\bigcup_{i=1}^{I} s\left(\tilde{A}_i\right)\right) \tag{19}$$

where $s(A_i)$ is the fuzzy numbers with a membership degree equal to 1. The distances, D_{maks} and D_{min}, are obtained by Eqs. 20–21. Equation 20 is used when $\tilde{A}$ is a trapezoidal fuzzy number and Eq. 21 is used when $\tilde{A}$ is a triangular fuzzy number.

$$\begin{aligned} D^2\left(\tilde{A}, M\right) = & \left(\frac{a_2 + a_3}{2} - M\right)^2 + \frac{1}{2}\left(\frac{a_2 + a_3}{2} - M\right) \times [(a_4 - a_3) - (a_2 - a_1)] \\ & + \frac{1}{3}\left(\frac{a_3 - a_2}{2}\right)^2 + \frac{1}{6}\left(\frac{a_3 - a_2}{2}\right) \times [(a_4 - a_3) + (a_2 - a_1)] \\ & + \frac{1}{9}\left[(a_4 - a_3)^2 + (a_2 - a_1)^2\right] \\ & - \frac{1}{9}[(a_2 - a_1) \times (a_4 - a_3)] \end{aligned} \tag{20}$$

$$D^2\left(\tilde{A}, M\right) = \Big[(a_2 - M)^2 + \frac{1}{2}(a_2 - M)[(a_3 + a_1) - 2M]$$

$$+\frac{1}{9}\left[(a_3 - a_2)^2 + (a_2 - a_1)^2\right] - \frac{1}{9}\left[(a_2 - a_1)(a_3 - a_2)\right]\Bigg] \quad (21)$$

$$D_{max} = \sqrt{D^2\left(\tilde{A}, Maks\right)} \quad (22)$$

$$D_{min} = \sqrt{D^2\left(\tilde{A}, Min\right)} \quad (23)$$

3.4 Fuzzy Information Axiom

The information axiom which enables the selection of the best design among designs that provide functional independence was first extended to the fuzzy environment by Kulak and Kahraman [24, 25]. In the studies, functional requirements (*FRs*) were used in the solution of decision-making problems by expressing them with triangular fuzzy numbers. In Fig. 4, the triangular fuzzy numbers, $S_i = (a, b, c)$ and $FR_i = (\alpha, \beta, \theta)$ are used to define the system feature and the functional requirement, respectively. Equation 24 is used to calculate the information content.

$$I = \log_2 \frac{\text{TFN of the system range}}{\text{TFN of the common srange}} \quad (24)$$

The fuzzy information axiom has been used in many different applications in the literature. Kahraman and Cebi [26] extended the definition of FRs to be used in different types of multiple criteria decision-making problems. These are (i) exact value problems, (ii) expected value problems, (iii) ranking problems, (iv) threshold value problems, and (v) mixed problems [26].

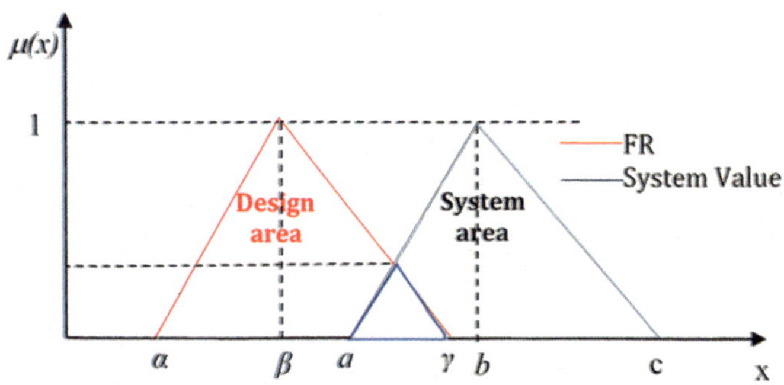

Fig. 4 Definition of the design, system, and common area

4 Design Principles of Smart Containers

The main purpose of cargo transportation is to ensure that the product is delivered from the supply point to the demand point at the planned time, in the desired quantity, in the desired quality, and at low cost. In the event of this activity with air transport, airport handling steps are entered between the supply point warehouse and the demand point warehouse. A smart container that will ensure the transportation of the products should guarantee the accuracy of the product, the process will take place at the planned time, the desired amount, and the desired quality. In this case, the functional requirements and design parameters required for the smart container is created by using the independence axiom as follows;

FR0. Make the container traceable.

FR1. Monitoring of the container in the exit warehouse.

FR2. Monitoring the shipment of the container.

FR3. Report deviations.

FR4. Create the container registry.

DP0. Intelligent container design.

DP1. Container Filling System Design.

DP2. Container Monitoring System Design.

DP3. Deviation warning system design.

DP4. Recording System Design.

Figure 5 presents the first level of the design decomposition, and the relationship matrix between FR and DP is given by;

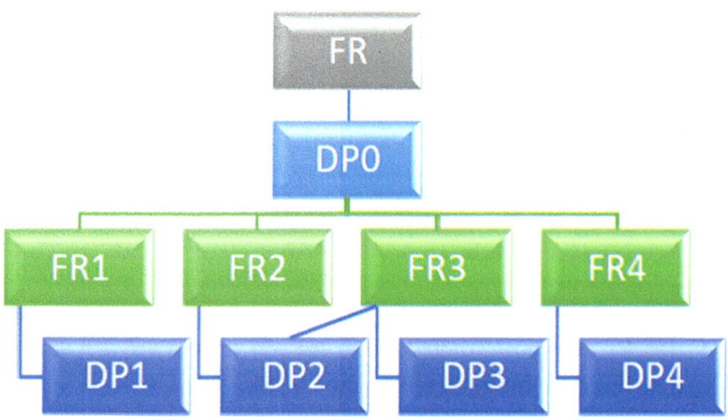

Fig. 5 The first level of the smart container design

$$\begin{Bmatrix} FR1 \\ FR2 \\ FR3 \\ FR4 \end{Bmatrix} = \begin{bmatrix} VH & & & \\ & VH & & \\ & H & VH & \\ & & & VH \end{bmatrix} \begin{Bmatrix} DP1 \\ DP2 \\ DP3 \\ DP4 \end{Bmatrix}$$

By using Eqs. 15–17 given in Sect. 3, the design is accepted as a decoupled design. In this step, design parameters can be explained as follows;

DP1. Container Filling System Design: The container filling system is the system that ensures the correct product, and the desired number of containers are loaded in the manufacturer's warehouse.

DP2. Container Tracking System Design: It is a system that enables the tracking of the shipment of the container and control of the load conditions.

DP3. Deviation warning system design: It is the system that controls the deviations in the process from loading the load in the container to the delivery to the customer and warns or notifies in case of deviation.

DP4. Recording System Design: Notifications and controls defined in the first three stages of the system design are carried out for each shipment and work on a shipment basis. The registry system is a blockchain-based system that enables the recording of all shipments made with the container.

The decomposition process continues to obtain the applicable design parameters. The decomposition of FR1 and DP1 can be explained as follows;

FR1. Monitoring of the container in the exit warehouse.

FR11. Ensure correct product is loaded in container.

FR12. Provide the desired amount of loading.

FR13. Calculate the fill time.

FR14. Calculate the estimated delivery time.

DP1. Container Filling System Design.

DP11. Product verification system.

DP12. Quantity verification system.

DP13. Time control system.

DP14. Delivery time calculation.

Figure 6 presents the decomposition of "FR1. Monitoring of the container in the exit warehouse", and the relationship matrix between FR1 and DP1 is given by;

$$\begin{Bmatrix} FR11 \\ FR12 \\ FR13 \\ FR14 \end{Bmatrix} = \begin{bmatrix} VH & H & & \\ H & VH & & \\ & & VH & \\ & & H & VH \end{bmatrix} \begin{Bmatrix} DP11 \\ DP12 \\ DP13 \\ DP14 \end{Bmatrix}$$

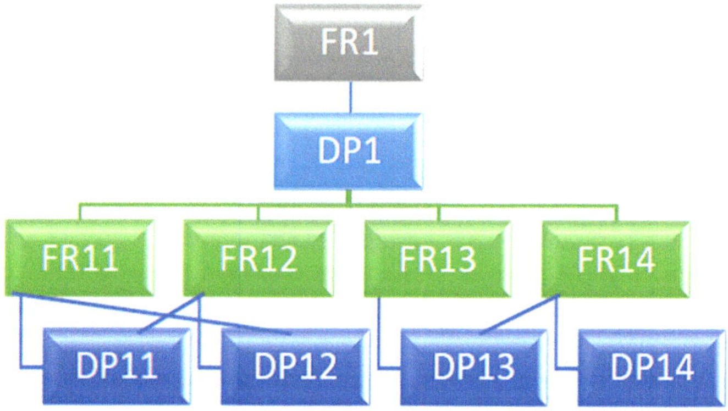

Fig. 6 The decomposition of "FR1. Monitoring of the container in the exit warehouse"

By using Eqs. 15–17 given in Sect. 3, the design is accepted as a decoupled design. In this step, design parameters can be explained as follows;

DP11. Product verification system: It is the system design that will be used in loading the product into the correct container. This system consists of RFID tags, an RFID tag reader, and software. Each container will be given a unique code and the RFID tag on the load loaded on the container and the RFID code reader that will be placed on the container lid opening will be read and transferred to the cloud system and will be matched with the load information to be loaded into the container with the help of a software. Thus, it is planned to ensure that the correct load is loaded into the container.

DP12. Quantity verification system: It is a system design that will define the amount of product to be loaded by calculating the filling weight with the help of a weight sensor simultaneously while filling the container. Here, in the previous step, the product information and product weight received from the RFID tag will be compared with the container weight and the amount of loaded product will be verified.

DP13. Time control system: It is the system that records the time during the container filling process. With the help of a sensor that detects container lid opening and closing in the filling area, the processing time is calculated by the system by recording the start and end time of the process. Calculated times are used for process improvement purposes in subsequent shipments.

DP14. Delivery time calculation: It is the system that calculates the estimated delivery time between the container starting point and the destination point. The duration calculation is calculated based on the distances of the container from the warehouse to the airport, from the airport to the airport and from the airport to the delivery warehouse, the alternative routes in the time period the cargo will move and the historical traffic data on the route to be followed. The purpose of the system is to take precautions and improve the process against delays that may occur during the process.

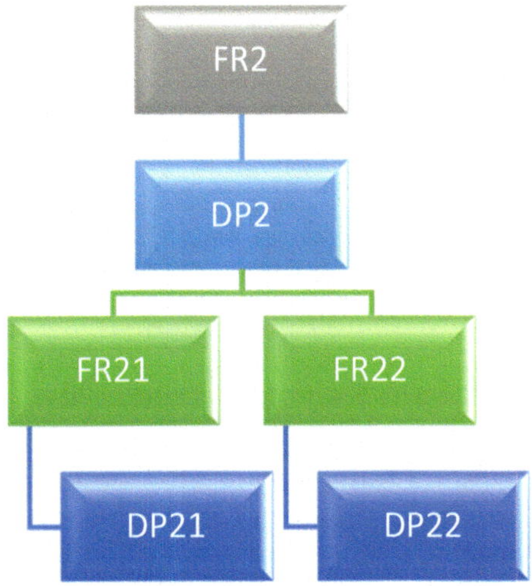

Fig. 7 The decomposition of FR2 "Monitoring the shipment of the container"

FR2 is satisfied by DP2 but DP2 is not applicable yet therefore the decomposition of FR2 is as follows;

FR2. Monitoring the shipment of the container.

FR21. Track container movement in real-time.

FR22 Checking of container internal conditions.

DP2. Container Monitoring System Design.

DP21. Tracking system design.

DP22. Air conditioning and control system.

Figure 7 presents the decomposition of FR2 "Monitoring the shipment of the container" and the relationship matrix between FR1 and DP1 is given by,

$$\begin{Bmatrix} FR21 \\ FR22 \end{Bmatrix} = \begin{bmatrix} VH & \\ & VH \end{bmatrix} \begin{Bmatrix} DP21 \\ DP22 \end{Bmatrix}$$

By using Eqs. 15–17 given in Sect. 3, the design is accepted as an uncoupled design. These parameters are explained as follows;

DP21. Tracking system design: It is the system that allows the movement of the container along the way to be monitored and recorded.

DP22. Air conditioning and control system: It is the system that controls the container's internal conditions and regulates the air conditioning conditions of the cargo.

The design parameters DP21 and DP22 are not applicable yet therefore the decompositions continue. The decomposition of FR21 is as follows.

FR21. Track container movement in real-time.

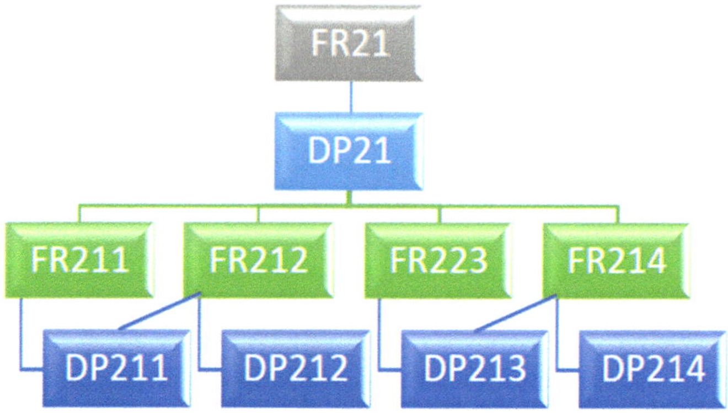

Fig. 8 The decomposition of "FR21. Track container movement in real-time

FR211. Follow the routes passed.

FR212. Provide unexpected container movement detection.

FR213. Make the transfer error-free.

FR214. Prevent container contents from changing.

DP21. Tracking system design.

DP211. Route tracking system.

DP212. Movement tracking system.

DP213. Transfer confirmation system.

DP214. Load control system.

Figure 8 presents the decomposition of "FR21. Track container movement in real-time", and the relationship matrix between FR1 and DP1 is given by;

$$\begin{Bmatrix} FR211 \\ FR212 \\ FR213 \\ FR214 \end{Bmatrix} = \begin{bmatrix} VH & & & \\ M & VH & & \\ & & VH & \\ & & H & VH \end{bmatrix} \begin{Bmatrix} DP211 \\ DP212 \\ DP213 \\ DP214 \end{Bmatrix}$$

By using Eqs. 15–17 given in Sect. 3, the design is accepted as a decoupled design. In this step, design parameters can be explained as follows;

DP211. Route tracking system: It is a GPS based system that provides instant recording and display of the locations where the container passes. This system also records the moment when the container leaves the warehouse and the time it arrives at its destination and verifies the delivery information.

DP212. Movement tracking system: It is a system design that records the speed data during the container road movement by reading it by a microcontroller, transfers the collected data to the cloud system, and produces results such as sudden braking, acceleration, and crash by analyzing the data with software. The acceleration of the vehicle is monitored in order to observe the starting and stopping movements of the vehicle so that the driver's sudden acceleration and braking behavior can be detected. The system is used to detect driver-induced driving errors that have the potential to damage the cargo by tracking the movement of the cargo on the highway.

DP213. Transfer confirmation system: It is the system that verifies that the cargo arriving at the departure airport is loaded on the correct cargo plane, unloaded in the right area, and transferred to the correct vehicle. For this, RFID card readers in the aircraft cabin area are used. The load information read by the system during loading and unloading is transferred to the cloud request and compared with the load content that needs to be transported. After the container is unloaded into the area, the GPS information of the container and the GPS information of the vehicle that will meet the load are matched over the cloud, and the information that the cargo has landed in the correct area and loaded to the correct vehicle is confirmed.

DP214. Load control system: It is a system that prevents the loss (in other words theft) of the cargo in the container and the addition of an unplanned load to the cargo. The system is a system that detects unplanned stops on the route and cargo cover opening and sends a notification to the operator via the cloud system. Here, hardware components such as cover opening-closing sensor, weight sensor, GPS used in other systems will be used.

The decomposition of "FR22. Checking of container internal conditions" is as follows:

FR22. Checking of container internal conditions.

FR221 Prevent deviation in container inner temperature.

FR222 Prevent deviation of container interior moisture.

FR223 Prevent deviation in container internal pressure.

DP22. Air conditioning and control system.

DP221. Temperature control system.

DP222. Humidity control system.

DP223. Pressure control system.

Figure 9 presents the decomposition of "FR22. Checking of container internal conditions", and the relationship matrix between FR1 and DP1 is given by;

$$\begin{Bmatrix} FR221 \\ FR222 \\ FR223 \end{Bmatrix} = \begin{bmatrix} VH & & \\ & VH & \\ & & VH \end{bmatrix} \begin{Bmatrix} \text{DP221} \\ \text{DP222} \\ \text{DP223} \end{Bmatrix}$$

By using Eqs. 15–17 given in Sect. 3, the design is accepted as an uncoupled design. In this step, design parameters can be explained as follows;

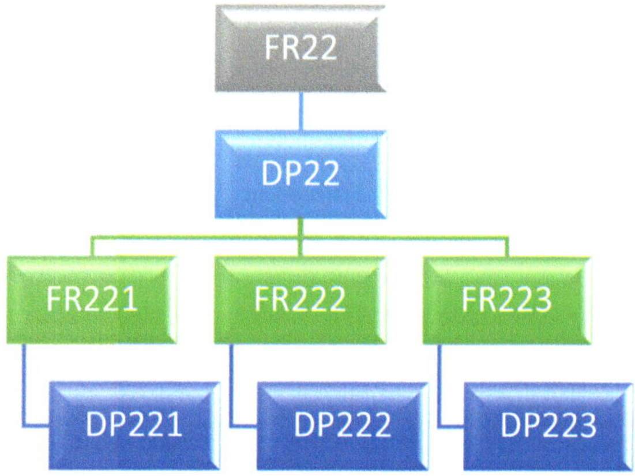

Fig. 9 The decomposition of "FR22. Checking of container internal conditions"

DP221. Temperature control system: It is a system that prevents the material transported from being adversely affected by ensuring that the container's internal temperature is within the prescribed limits.

DP222. Humidity control system: It is a system that prevents the material transported from being adversely affected by ensuring that the internal humidity of the container is within the prescribed limits.

DP223. Pressure control system: It is a system that prevents the material transported from being adversely affected by ensuring that the internal pressure of the container is within the prescribed limits.

The decomposition of FR3 and DP3 can is as follows;

FR3. Report deviations.

FR3.1 Avoid detour.

FR3.2 Avoid deviation from planned arrival time.

FR3.3 Prevent deviation from temperature value.

FR3.4 Prevent deviation from moisture value.

FR3.5 Prevent deviation from pressure value.

FR3.6 Prevent overloading.

FR3.7 Prevent incorrect product loading.

FR3.8 Prevent erroneous transfer operations.

DP3. Deviation warning system design.

DP3.1 Off route warning system.

DP3.2 Time deviation warning system.

DP3.3 Temperature deviation warning system.

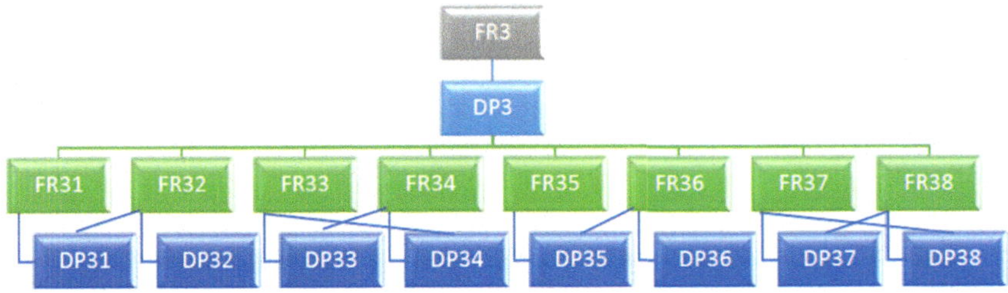

Fig. 10 The decomposition of "FR3. Report deviations"

DP3.4 Humidity deviation warning system.

DP3.5 Pressure deviation warning system.

DP3.6 Overweight warning system.

DP3.7 Faulty product loading warning system.

DP3.8 Faulty transfer warning system.

Figure 10 presents the decomposition of "FR3. Report deviations", and the relationship matrix between FR3 and DP3 is given by;

$$\begin{Bmatrix} FR31 \\ FR32 \\ FR33 \\ FR34 \\ FR35 \\ FR36 \\ FR37 \\ FR38 \end{Bmatrix} = \begin{bmatrix} VH & & & & & & & \\ H & VH & & & & & & \\ & VH & L & & & & & \\ & & L & VH & & & & \\ & & & & VH & M & & \\ & & & & M & VH & & \\ & & & & & & VH & L \\ & & & & & & L & VH \end{bmatrix} \begin{Bmatrix} DP31 \\ DP32 \\ DP33 \\ DP34 \\ DP35 \\ DP36 \\ DP37 \\ DP38 \end{Bmatrix}$$

By using Eqs. 15–17 given in Sect. 3, the design is accepted as a decoupled design. In this step, design parameters can be explained as follows;

DP31 Deviation from the route warning: If it is detected that the container deviates from the route during transportation, the system sends a notification to the operator and triggers the execution of the necessary actions to take the process under control. The operator defines the source of the deviation in question and is the system that ensures the necessary planning updates.

DP32 Time deviation warning: It is the system that triggers the commissioning of alternative plans, if necessary planning for the new process, by sending a notification to the operator and the customer via the cloud system when the planned arrival time is behind.

DP33 Temperature deviation warning: In cases where the temperature inside the container is important for the cargo transported, a notification regarding the temperature change is sent to the operator. It is the system that transmits the data received from the temperature control system to the center through the cloud system.

DP34 Moisture deviation warning: In cases where the humidity in the container is important for the cargo carried, a notification regarding moisture change is sent to the operator. It is the system that transmits data received from the humidity control system to the center via the cloud system.

DP35 Pressure deviation warning: In cases where the pressure in the container is important for the cargo carried, a notification regarding the pressure change is sent to the operator. It is the system that transmits data received from the pressure control system to the center via the cloud system.

DP36. Overweight warning system: It is a system that keeps container weight limits under control. It is a system that notifies the loading control operator via the cloud system when the weight of the load exceeds the container limits with the help of a sensor that detects the weight of the load during loading.

DP37 Faulty product loading warning system: It is the system that warns the container with light/sound and sends a notification to the operator in case the wrong product is loaded into the container during container filling in the warehouse area or the product is loaded into the wrong container.

DP38 Incorrect transfer warning system: It is the system that sends a detection notification to the center via the cloud system in case the loaded container is transferred to the aircraft at the airport, the load is unloaded in the wrong area at the destination and transferred to the wrong vehicle. Thus, the operator is able to prevent the error by making the necessary intervention.

Apart from these features, some physical container features are not considered in the scope of this study since intelligent but are likely to be requested in line with customer demands. These are;

- Frigo feature can be required for foods (especially if frozen or cold air is needed in foods).
- If it is fluid, special insulation may be needed.
- If the material inside is a load that can solidify due to seasonal effects (asphalt, glucose), equipment that will ensure that their fluidity and properties do not deteriorate by heating them with steam or air may be requested.
- Waterproof features that will not be affected by effects such as rain and moisture can be requested.
- In rapid geographic changes, it may be requested to have a ventilation system against humidity/condensation.
- It can be requested to have a structure suitable for loading or unloading the products according to their type (e.g. pallet load can be opened from the front, can be loaded from the top for bulk metal type cargo such as asphalt soda ash).
- A twist stop feature may be requested for healthy operation in vehicles that are shipped with loading and unloading vehicles during loading, unloading, and travel.

Although these features are not necessary for smart container design, they are frequently encountered in container design and these design features can be expanded with this study as the subject of another study.

5 Conclusion

The use of smart containers in the supply chain contributes to the monitoring of many activities and the development of the process, from the empty container arriving at the loading area to the unloading of the container at the destination. The processes contributed by a smart container can be summarized as follows;

- It makes it possible to record the handling times realized in the filling area and to make a performance analysis regarding the handling times realized in the warehouses at the point of supply.
- By ensuring real-time tracking of cargo movement, precautions can be taken in advance for possible delays in the process.
- It ensures the security of the cargo against negative factors such as theft, spoilage, and damage during the journey between the supply point warehouse and the destination point.
- It makes it possible to record the handling times realized in the unloading area and to make a performance analysis regarding the handling times realized in the warehouses at the supply point.
- For each shipment, cargo movements are recorded with a timestamp and a transaction log for the container is created.
- Confidence is given to stakeholders by simultaneously sharing cargo status information with third parties.
- The collected data can be analyzed and used in performance improvement.

As the limitation of the proposed study, the geometrical design parameters of a container are not considered since the current geometric design of a container is found acceptable in the industry. Fur the further study, the design performance of smart container designs which satisfy the fuzzy independence axiom may be evaluated by using the fuzzy information axiom.

References

1. Ceccarelli, M.: Smart containers real-time smart container data for supply chain excellence. UNECE UN/CEFACT (2019)
2. Smart containers, business requirements specification. United Nations Economic Commission for Europe (2019)
3. Schumacher, J., Gschweidl, M., Rieder, M.: EURIDICE—an enabler for intelligent cargo for the logistics sector (2010)

4. Hofman, W., Punter, M., Bastiaansen, H., Cornelisse, E., Dalmolen, S.: Semantic technology for enabling logistics innovations—towards intelligent cargo in the physical internet. Int. J. Adv. Logistics **5**, 58–69 (2016)
5. Suh, N.P.: Design of thinking design machine. Ann. CIRP **39**(1), 145–149 (1990)
6. Büyüközkan, G., Havle, C., Feyzioglu, O.: A new digital service quality model and its strategic analysis in aviation industry using interval-valued intuitionistic fuzzy AHP. J. Air Transp. Manage. **86** (2020)
7. Valdés, R., Comendador, V.F., Sanz, A., Castán, J.A.: Aviation 4.0: more safety through automation and digitization. WIT Trans. Built Environ. **174**, 225–236 (2018)
8. Malygin, I., Komashinsky, V., Tsyganov, V.V.: International experience and multimodal intelligent transportation system of Russia. In: 2017 Tenth International Conference Management of Large-Scale System Development (MLSD), pp. 1–5 (2017)
9. May, A., Anslow, A., Ojiako, U., Wu, Y., Marshall, A. Chipulu, M.: Optimisation of key performance measures in air cargo demand management. J. Transp. Supply Chain Manage. **8**(1) (2014)
10. May, A., Anslow, A., Wu, Y., Ojiako, U., Chipulu, M., Marshall, A.: Prioritisation of performance indicators in air cargo demand management: an insight from industry. Supply Chain Manage.: An Int. J. **19**(1), 108–113 (2012)
11. Zhou, L., Lou, C.X.: Intelligent cargo tracking system based on the internet of things. In: 15th International Conference on Network-Based Information Systems, pp. 489–493 (2012)
12. Dalmolen, S., Cornelisse, E., Hofman, W., Basiaansen, H., Punter, M., Knoors, F.: Improving sustainability through intelligent cargo and adaptive decision making. In: Proceedings of the e-Freight Conference, pp. 1–16 (2012)
13. Forcolin, M., Fracasso, E., Tumanischvili, F., Lupieri, P.: EURIDICE—IoT applied to logistics using the intelligent cargo concept. In: 17th International Conference on Concurrent Enterprising, pp. 1–9 (2011)
14. Sternberg, H., Hagen, A., Paganelli, P., Lumsden, K.: Intelligent cargo—enabling future's sustainable and accountable transportation system. World J. Sci. Technol. Sustain. Dev. **7**, 253–262 (2010)
15. Matthews, M.L., Bryant, D.J., Webb, R.D.G., Harbluk, J.L.: Model for situation awareness and driving: application to analysis and research for intelligent transportation systems. Transp. Res. Rec. **1779**(1), 26–32 (2001)
16. Suh, N.P.: Axiomatic Design: Advance and Applications. Oxford University Press, Oxford (2001)
17. Suh, N.P.: Complexity in engineering. CIRP Ann.—Manuf. Technol. **54**(2), 46–63 (2005)
18. Cebi, S.: Developing a fuzzy decision support system based on axiomatic design principles. Ph.D. Thesis, Istanbul Technical University, Istanbul, Turkey (2010)
19. Cebi, S., Kahraman, C.: Determining design characteristics of automobile seats based on fuzzy axiomatic design. Int. J. Comput. Intell. (forthcoming) (2010)
20. Cebi, S., Kahraman, C.: Extension of axiomatic design principles under fuzzy environment. Expert Syst. Appl. **37**(3), 2682–2689 (2010)
21. Shannon, C.E., Weaver, W.: The mathematical theory of communication. University of Illionis Press, Urbana, IL (1949)
22. Shannon, C.: A mathematical theory of communication. Bell Syst. Tech. J. 379–423 & 623–656 (1948)
23. Tran, L., Duckstein, L.: Comparison of fuzzy numbers using a fuzzy distance measure. Fuzzy Sets Syst. **130**(3), 331–341 (2002)
24. Kulak, O., Kahraman, C.: Fuzzy multi-attribute selection among transportation companies using axiomatic design and analytic hierarchy process. Inf. Sci. **170**, 191–210 (2005a)
25. Kulak, O., Kahraman, C.: Multi-attribute comparison of advanced manufacturing systems using fuzzy versus crisp axiomatic design approach. Int. J. Prod. Econ. **95**, 415–424 (2005b)
26. Kahraman, C., Cebi, S.: A new multi-attribute decision making method: hierarchical fuzzy axiomatic design. Expert Syst. Appl. **36**(3), 4848–4861 (2009)

Intelligent and Fuzzy Applications in Aviation 4.0 Unmanned Aerial Vehicle Technologies

Blockchain Applications in UAV-Towards Aviation 4.0

Reza Vatankhah Barenji and Mazyar Ghadiri Nejad

Abstract Aviation 4.0 is one of the most important contexts that started to attract many researchers in recent years. The main element of Aviation 4.0 is to employ and manage unmanned aircraft vehicles (UAVs). Employing UAVs in the Aviation 4.0 era will minimize the human dependency level of the aviation management systems. Embedding UAVs with Blockchain technology (BCT) might become an increasing phenomenon since the UAVs should be operated in an environment with many obstacles and possibilities of cyber-attacks. Therefore, using BCT, regulators in the whole world will be able to track the identities of UAVs flying in their territories. This chapter provides detailed information about the BCT and explains its implementation architecture, and the potential applications of this technology in the UAV in the Aviation 4.0 era are also stated.

Keywords Aviation 4.0 · Blockchain technology · Cyber-physical system · Unmanned aircraft vehicles (UAVs) · Applications

1 Introduction

Industry 4.0 is a newly merged paradigm for employing a cyber-physical technology in embedded devices (all the devices that hold a processor for decision-making) aiming to boost the smartness level of the device or system. One of the manifestations of Industry 4.0 in the aircraft industry is Aviation 4.0. Generally, aviation is the activity surrounding mechanical flights and the aircrafts industry. The aircraft includes fixed-wing and rotary-wing types, morphable wings, wing-less lifting bodies, as well as and lighter than aircraft such as hot air balloons and unmanned aerial vehicles (UAVs). The main object in the Aviation 4.0 era is boosting

R. V. Barenji
Industrial Engineering Department, Hacettepe University, Ankara, Turkey
e-mail: reza.vatankhah@hacettepe.edu.tr

M. G. Nejad (✉)
Industrial Engineering Department, Girne American University, Kyrenia, TRNC, Turkey
e-mail: mazyarghadirinejad@gau.edu.tr

C. Kahraman and S. Aydın (eds.), *Intelligent and Fuzzy Techniques in Aviation 4.0*, Studies in Systems, Decision and Control 372,
https://doi.org/10.1007/978-3-030-75067-1_18

the smartness level of UAVs and improving the humanness decision-making level of the aircraft [1]. Besides that from a system point of view, Aviation 4.0 attempts to provide a platform and approach to good mange the included UAVs with a minimum level of human dependency by employing a cyber-physical technology. It is expected that in Aviation 4.0, the UAVs will be more intelligent, and therefore the managing system will be smart too. A UAV, the acronym for an unmanned aerial vehicle, generally known as a drone, is an aircraft with no human pilot onboard and a kind of unmanned vehicle. The UAV market is anticipated to rise to around $21 billion by 2022. Nowadays, a huge majority of the market (around 82%) is focused on military applications. Commercial UAVs are predicted to account for around $2.5 billion in revenue by 2021—around a 19% year-over-year progress rate. UAVs are being explored for a diversity of applications like accuracy agriculture (crop spraying), terrain and environmental checking, infrastructure checking (bridges and dams), public security surveillance, commercial freight, border control, and oil and gas pipeline monitoring, among numerous other applications. An unmanned flying object that can be programmed to accomplish tasks that are very dangerous, time-consuming, or difficult for humans is a vast technological leap toward a more automated and productive world. However, UAVs are susceptible to be lost, to be destroyed, or to be physically hijacked.

The next generation of UAVs must be adapted to operate in such an environment with the existence of lots of obstacles and possibilities of cyber-physical attacks. The Federal Aviation Administration (FAA) has severe rules and regulations that bound the usage of UAVs in public air space, because of various issues in UAV networks such as intra-UAV communication, UAV security, air data security, data storage, and management. Containing a consistent communication channel in UAV networks is crucial since they are inclusive of high mobility and sparse deployment [2].

Using BCT, governments in the whole world may trace the identities of UAVs flying in their territories. A BCT, in the humblest of terms, is a time-stamped series of unchallengeable records of data that is managed by a bunch of computers not owned by any solitary entity. Each of these data blocks (i.e. block) is protected and bound to each other using cryptographic principles. Some potential applications of BCT in UAV networks are network safety, decentralized storage, inventory management, and surveillance. In the following, some of the main roles of the BCT in aviation, according to Alladi et al. [3] have been presented.

1.1 Role of BCT in Aviation

- A combination of AirforceNETs and UAV networks by BCT might be used to advance novel models that integrate the facility of exchanging data between an Airforce and end-users, including passengers, goods, post parcels, etc.
- BCT can record the coordinates of all the UAVs to prevent physical crashes, storms, etc.

- A distributed BCT-enabled UAV network may fall the decision-making load of the control tower.
- BCT can solve the trust problem since it is a transparent electronic ledger. Considering the load information of diverse regions, UAVs may be assigned to their regions of action confirming uniform distribution as far away as possible.
- UAVs may interconnect with each other through a shared channel and avoid being hacked by third parties who may acquire access to the channel.
- The BCT-enabled network may offer data source and entity authentication between the Helicopters/UAVs and third party agents. The Source of the data may be traced utilizing this property for disaster relief action and in health care applications, like the I-ward project which uses robot teams to help health laborers.
- Given a common communication channel in the form of a BCT, Helicopters/UAVs may have demands for aid from other Helicopters/UAVs in emergency circumstances like a fuel shortage, system errors, sensor fault, lifeguard, etc.

This chapter provides brief information about the BCT and its implementation architecture and highlights the potential applications of the technology in the UAV industry.

The organization of this chapter is the following. In the second section, the literature review containing the implementation of BCT architecture layers, the state of the art on Aviation 4.0, and UAV applications on Aviation 4.0 are provided. The methodology of the developed framework for BCT-based UAV in Aviation 4.0 is stated in the third section. Then, applications of the developed framework are reported in the fourth section. Finally, the study has been concluded and the other possible applications of the framework are discussed.

2 Literature Review

2.1 Blockchain Technology and Its Implementation Architecture

For the first time, Nakamoto introduced BCT by proposing Bitcoin [4, 5]. This proposal, like the creation of the Internet, became popular very soon and started a revolution in this area, as BCT has been a hot topic during these years. BCT uses a combination of cryptography, distributed systems, peer-to-peer networks, etc. to provide a safe and trustable network safely to allow the information to be verified and recorded in a space with public access without delete permission. Additionally, its process is led to a transparent, time-stamped, and decentralized fashion [6, 7].

In the BCT architecture, there are six different layers as follows:

- Data layer
- Network layer
- Consensus layer

- Incentive layer
- Contract layer
- Application layer.

2.1.1 Data Layer

In BCT architecture, the data layer is the first-bottom layer. It includes the timestamped data blocks. Each of the data blocks, which is encoded by a hash, contains a block header and a block body. This hash connects each block to other blocks, in the way that each block has its hash, which is called the Current hash, and the previous block hash. The moment that the block has been created, is recorded in the Timestamp. To have an appropriate pattern for the block hash, a random number is generated, which is called Nonce. The transaction related to the block is stored in the Merkle root based on a special mechanism. The block body contains the hashes and the transactions of the block. Figure 1 illustrates a block in a BCT [3, 8, 9]. It has been shown that how previous and the next blocks are connected to each current block.

2.1.2 Network Layer

The network layer is the second layer of a BCT to distribute, pushing the transactions forward, and authentication the transactions. There are two main network architectures or strategies such as Client–Server and Peer-to-Peer. BCT uses the latter one by giving the same concession to the blocks, which are next to each other, for verification and validation of the information include accepting or rejecting them. This means that only the information that the blocks ensure it/them is/are recorded. Such an operation looks like gathering digital signatures after verifying the information,

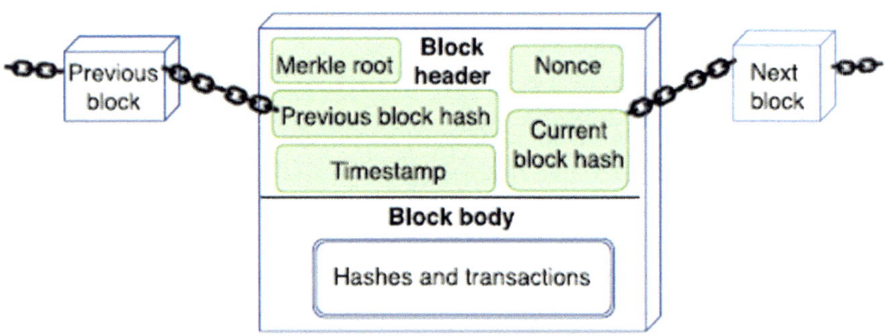

Fig. 1 Structure of a block in a BCT

with its mechanism. In this regard, just some of the blocks have the right to verify and accept or reject the information which has private keys, where the other blocks have public keys [10, 9].

2.1.3 Consensus Layer

This layer is one of the most important layers in the BCT networks, which is created to achieve a consensus among all the participant's opinions in accepting or rejecting a transaction. This is very important in BCT because the consensus among participants is needed to create a decentralized institution. Therefore, having some protocols are necessary to make the participants sure about this consensus [11, 12]. According to Xie et al. [9], the main mechanisms for consensus are listed as the following:

- Practical Byzantine Fault Tolerance [13]
- Proof of Work [5]
- Proof of Stake [14]
- Delegated Proof of Stake [15].

In addition to the above-mentioned consensus algorithms, there are other well-known consensus algorithms as the followings:

- Proof of Bandwidth [16]
- Proof of Reliability [17]
- Tendermint [18]
- Ripple [19]
- Stellar [20].

2.1.4 Incentive Layer

This layer is one of the main elements of BCT and it is the driving power used in BCT. Such a layer combines the technology of BCT with economical parameters and generates a two-pronged plan for miners. The miners finance a huge amount of their powers and efforts for mining the blocks and then, gaining rewards like some incentive points in terms of Bitcoin or other digital currencies [3].

2.1.5 Contract Layer

This is the fifth layer of the technology of BCT. It includes the capability of coding the BCT and performing numerous scripts, computerized methods, and smart contracts to record complicated transactions. The set of rules and regulations that allow the parties of a contract to interact and make transactions we call the smart contract. After establishing the contract between the parties and finalizing the rules, all the provisions of the contract are encrypted and approved by the parties. Then, this contract is sent to all blocks in the BCT to be approved by them [21, 22].

2.1.6 Application Layer

This layer is the last one, or in other words, the uppermost layer of the BCT containing its applications in the considered issue. According to the literature, there are many different areas that scientists have applied the BCT for them, which some of them are listed below [3, 23]:

- Smart cities
- Smart grids and energy management
- Smart communities
- Edge computing
- Cloud computing
- Internet of Things
- Machine learning
- Deep learning
- Brain-Computer Interface
- Healthcare applications
- Industry 4.0
- Tracking and registration of police
- Mobile data offloading.

Figure 2 illustrates the BCT architecture generally.

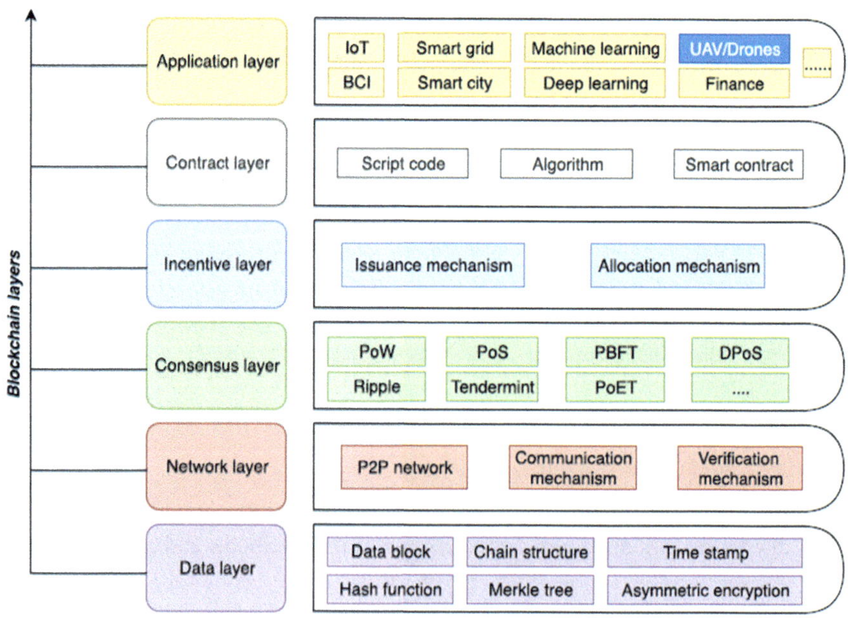

Fig. 2 Common BCT architecture

2.2 BCT Classification

From the Zheng et al.'s point of view [24], BCT is classified into three categories considering the possession and the spectators indorsed to contribute in the procedure of block addition, confirmation, and validation as follows (Fig. 3).

2.2.1 Public BCT

In a public BCT, unauthorized consensus takes place in such a way that everyone can participate in the consensus process, and at the same time, viewing all the records is accessible to all. This group of BCTs has less variability than other BCT categories because the number of people participating in it is very large and more than other categories [24]. It should be noted that public BCTs are less efficient than private and consortium BCTs [25].

2.2.2 Private

In this category of BCTs, unlike the public BCT, only those individuals or nodes can participate that are permitted, which only applies to those who belong to a particular and qualified network. Hence, because this category is completely under the supervision of a specific network, it is considered as a centralized group or network. Due to the smaller number of participants in this category, the ability to manipulate the content in this category is higher than the public category, but on the other hand, such networks have a higher efficiency than the public BCTs [3, 26].

2.2.3 Consortium BCT

A consortium BCT is mainly decentralized because unlike public and private BCTs, only a few predefined organizations can participate in it. On the other hand, such a BCT has a consensus process in each of its missions. Therefore, these BCTs are

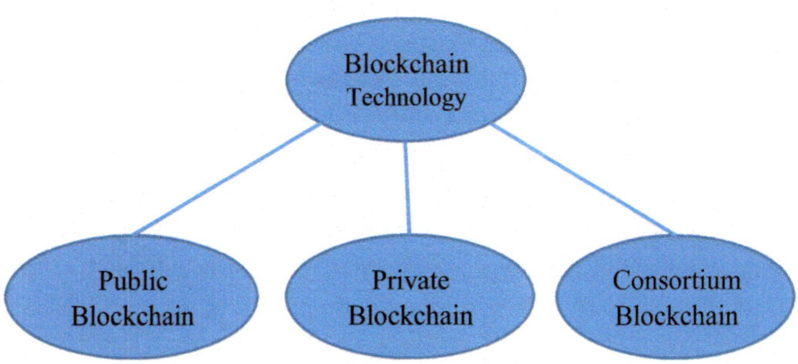

Fig. 3 Classification of BCT-based on ownership

more efficient than public and private BCTs. Moreover, due to the smaller number of participants in this category than the mentioned categories, the possibility of manipulating information in this category is more and easier than the rest [27, 24].

2.3 BCT and Aviation System Applications

Since there should be a direct and unmediated connection between all participating members for UAV applications, a private network is often required for drones based on the BCT. In this regard, García-Magariño et al. [28] analyzed a real-life case study that each UAV in the network of a BCT must get permission or license to connect the monitoring fleet. In this case, waiting to get permission is time-consuming for UAVs and it is needed to have a precise plan for their battery charging. In a recent study, Hassija et al. [29] performed a deep research on utilizing the BCT by considering the smart contracts phenomenon. In the following, some of the main BCT applications in UAV systems are scrutinized.

2.3.1 Supply Chain Management

One of the uses of UAVs is related to inventory control systems and their management in organizations, especially in manufacturing plants. In such cases, RFID tags are affixed to inventory packages to be used by UAVs to read and collect the names and specifications of each inventory. These systems then use BCT to send all the collected data and information into a specific network and verify them through protocols [30].

In general, supply chain management includes all activities related to the supply of raw materials and prefabricated parts, production process, marketing of products and sales, and delivery to the customer. Considering the importance of this issue, the creation and utilization of any strategy and facility including the use of systems automation, use of IoT Industrial technology [31], development of appropriate algorithms for performing complex calculations, and use of drones can because of many improvements. These improvements may be reducing time, reducing costs, increasing profit, the accuracy of operations, increasing the efficiency of available resources and the satisfaction of shareholders can be fruitful [32].

In this regard, UAVs can play an important role in surveying and statistics of warehouses and collecting information about inventories [33–35]. UAVs can also be used to control and check the delivery of products to distributors or customers [36]. BCT can be used to increase the reliability of data collection in such systems. In these systems, a set of data and databases are used and the information is always updated. Since unit personnel performs these activities, there is always the possibility of human error and often these activities are not performed in real-time. In such systems, drones can be used with the help of industry and can be used to automate these systems and to control, validate, and track information [37].

These applications of UAVs are some of the results of the Industrial Revolution 4.0, which can be integrated with BCT and give many activities, including authentication and third-party verification, which may be in the presence of untrustworthy people, to intelligent and automated transactions and contracts. Moreover, UAV systems using BCT can be programmed to act automatically and independently with the customers and even make decisions in this regard [38].

2.3.2 Automation Procedure

Process automation is one of the BCT applications by using drones. In this process, the data and information collected by the UAVs are conveyed to a cyber-system, which is linked to a BCT. In this BCT, each of the blocks contains a section of inventory information that is scanned by UAVs. Besides, the time related to the moment of adding information is stored in the blocks, and then the required information is recorded, stored, and controlled. Since UAVs need to be charged during the work, by creating a predetermined schedule, the operations related to charging the UAVs can be performed. Thus, the whole process of scanning inventory information as well as storing them can be completely automatically done [39].

Another benefit of BCT is the ability to create smart contracts. In this way, special contracts can be created from the beginning of the procedure. Then, after storing and confirming the information related to inventories, the amount of the inventories for parts are checked. If the inventory of any product is less than the specified value for that product, called the safety stock, the contract of requesting the goods are automatically sent to the marketing and finance department. Moreover, this contract may also be sent to the related supplier for the product directly [40].

If UAVs are used in multi-agent systems that smart contracts have been prepared and planned for them, UAVs can be programmed to make decisions on their own [41, 42]. One of the notable cases in this regard is the use of UAVs in BCT systems related to the category of Atrium BCT. Kapitonov et al., [38], have carefully discussed this issue. According to them, in these systems, smart intelligent contracts are used to transfer the necessary commands during operations between active UAVs as well as between UAVs and robotic and automated systems. Such robotic agents can be used in many daily business and economic activities [43–47]. Additionally, such techniques and technologies may be useful to increase the automatization of production lines, and manufacturing floors [48, 49]. In such activities, each agent can physically and independently intervene with other agents to carry out the operations. It is obvious that after completing all the activities, the agent related to the programmed UAV can deposit a certain and predetermined amount, according to the contract, to the account of the agent of another UAV.

It should be noted that the currency of the deposit is one of the digital currencies such as bitcoin, etc. which is common and payable in BCT networks. Additionally, all of these UAVs must be connected to a specific network with predetermined rules and regulations, to be able to create such smart and intelligent contracts, as well as make such deposits, automatically. By having such systems, and relying on BCT, it is possible to connect a large number of UAVs in a network and make difficult decisions that cannot be made by a single UAV.

2.3.3 Creating New Business Models

To create and develop a new business model, a network of UAVs and using BCT may be connected to each other. As mentioned above, in this regard, it should be possible to exchange information communication between the end-user and UAVs. One of the most famous and reputable business models that are used frequently today, is the commercial model in the field of Internet of Things, called "Sensing as a service". In the last decade, many scientists and researchers studied this business model [50–52]. For instance, using this type of model and its operational implementation, users can purchase sensor information by obtaining access and considering the necessary protocols, and then use them in their businesses. It is obvious that here the network and the possibility of communication through the BCT can perform all these operations automatically, without the intervention of operators. For more information, the readers may check the research performed by Ferrer [53] and Nofer [54].

2.3.4 Collision Free Movement of UAVs

Non-collision of UAVs with each other and no disruption in the information system and data transmission by UAVs is one of the most essential points [55]. These types of issues seriously and carefully should be examined, and the existence of these requirements in any network of UAVs should be ensured. In this regard, the use of BCT can help. In this way, information containing coordinates, including latitude and longitude is provided to the database and stored instantly. Using this information, as well as the creation of sophisticated computer algorithms, UAVs can accurately calculate and determine the optimal route from origin to destination and complete their operations without encountering or contacting any obstacles. To get more information about such an algorithm utilized in this issue, the research that has been done by [56] may be useful.

2.3.5 Uniform Load Sharing

Shipping and distribution operations, including products, packages, cargo, postal items, etc., may result in improper load balancing between service providers and contractors. On the other hand, there may not be enough confidence in the freight forwarders. Therefore, this problem can be solved by using BCT and with the help of electronic and automatic registration of information. Different approaches can be adopted for this purpose. For example, in addition to using drones to send, it is possible to distribute the load evenly. Depending on the coordinates and location of customers, it is possible to specific geographical areas to each UAV, in a way that the load is the same in each area. Then each area may be assigned to a deliverer. This innovation can be done by using IoT systems in UAVs [57].

2.3.6 Data and Entity Authentication

As mentioned earlier, UAVs can communicate with each other by connecting to an internal network. Based on the defined parameters regarding network access, the information in a network can be available to everyone (public key) or certain individuals (private key). This helps UAVs to keep their information and data from being stolen, altered, or manipulated by unknown individuals. People who have access to the private network can also be helped to validate and verify the information so that unverified information is not available to everyone. Then, after confirmation, the information can be spread throughout the network and all people can be given free access. In this regard, the use of BCT can assist in the operations of obtaining digital approval and signature. This technique can be used in operations related to helping people in the event of accidents, including accidents, illness, etc. [58].

2.3.7 Cooperation and Fast Synchronization

If UAVs are used in various operations, there is always a possibility of technical defects, sensor failures, UAV battery recharging problems, operating system errors, and other emergency issues. In such cases, a UAV can communicate with other UAVs by using a network connection and BCT, and request for help. The UAV(s) can then make the right decision at that moment, which is made with the help of other UAVs. These decisions are also stored in the network memory for later review and analysis. Using the information stored in the memory about such problems and decisions, UAVs can execute the command in the event of similar problems with shorter time and less processing power. With this technology, UAVs do not need pre-operation training.

In the case of military operations, which are vital, this technique can be used when many UAVs are used and the probability of their damages or being targeted by their sedans are high. Additionally, the use of this technology is not limited to UAVs. The same can be considered for the use of ground robots and using BCT-based networks, to help the performing such operations as well as possible. For instance, when robots try to cross a river, which has been studied by Ferrer [53], or when a robot tries to serve the CNC machine to produce different types of parts in parallel [59–61]. In the first one, UAVs and robots can use the information and decisions of previous operations, by using BCT, in operations with similar environmental and climatic conditions. Similar to previous BCT applications, in this approach, UAVs can share data and information, especially confidential information, by creating communication networks through BCT.

2.3.8 Distributed Decision-Making Mechanism

If UAVs are operating on an interconnected network, they may be confused to make quick decisions due to an event, an incident, or a crash. In the event of such confusion, new addresses are created, in which each address contains one of the decision alternatives. In this regard, UAVs can make distributed decisions using BCT, as well as using special mechanisms [53]. Since the decision-making process, which is the selection of an alternative among some alternatives, is a binary variable with a value of zero or one, UAVs assign values of zero or one to each of these alternatives after making each decision. To choose the best alternative among several alternatives, the one with the highest number of points is selected and implemented. In this system, the drone's network can make decisions in-group, with the participation of all members. Similarly, all the made decisions are stored in a memory, and provided to the UAVs to use the previous decisions in the face of similar events and happenings.

2.3.9 Secure Communication Channel

As mentioned before, BCT uses a peer-to-peer connection for its agents to create reliable communication and save the system from probable attacks and threats. There are two main elements in the BCT encryption such as signatures, which must be digitally, and public-key cryptography, to enable all the members of the network to have access to information. To provide services and create the necessary capabilities in the network, two special keys with the names of public and private keys can be defined for each member.

For more explanations, in real-world cases, these keys may be defined in such a way that public keys are like facilities that are publicly and often freely available to everyone, and private keys provide special facilities only for some special people who have the code of decryption. For the cases of networks of UAVs that are connected to BCT, the public key means that all UAVs within the network have access to each other and can be identified and accessed throughout the system. These UAVs can exchange information with each other. Only those UAVs that have private keys can access certain information and decrypt the system. Because it is not possible to decrypt using the public key, UAVs with this feature can only receive and send messages, data, and information.

2.3.10 Inter-Service Operability of UAVs

When it comes to providing UAV services on a network, we may encounter a dense network, including signal interference, the possibility of UAVs colliding, and information traffic. In this case, instead of using many UAVs in dense areas that each of which serves only a few users, it is better to consider a smaller number of UAVs to serve that area. Therefore, each of them may provide more users. Naturally, this will reduce signal interference, reduce accidents, and so on. In such situations, a small number of UAVs can connect to form a small network and serve all users through the use of BCT. On the other hand, considering that the service systems of competing companies always try to perform the same operation, these networks can also bring the full trust of users to the service level.

Another application of BCT in servicing UAVs is related to UAVs' load balancing, which according to Sharma et al [62], in the literature, researchers used the concept of weighted BCT in such cases. To balance the loads, a consensus must be performed between the UAVs in the network. Such loads that are transferred based on the customers' demand and with the supply by the sellers, cause a charge, which must be transferred from the customer to the seller's account for each operation. For this purpose, an area is assigned to each vendor. It should be noted that if there are cooperation networks, these areas can be wider. In this category of BCT applications, UAVs can communicate and interact with each other without the need for a private key and identification, and it is enough to have access to the public key. When in a specific area, all vendors and suppliers subscribe to each other for services, UAVs with a private key can send information and data to other UAVs. In this regard, and the event of unforeseen events, sellers can turn the private key into the public key, or vice versa, to prevent further losses. Besides, having a private key is necessary to send and distribute information to sources outside the network. All necessary protocols and policies in this area must be established and approved between customers and service providers.

3 Methodology

As mentioned before, from two different perspectives Aviation 4.0 can be brought up to the table. From the first point of view, the aim is to boost the smartness of an aircraft, while in another view the aim is employing Industry 4.0 merging technologies to autonomy control and manage a group of aircraft. There are many instances of technologies that might be considered such as the internet of things technology for aircrafts identification and connection to the internet, cyber-physical system application for most powerful decision-making in device and system level. In this section, the BCT is considered to embed into Aviation 4.0 aiming to track and trace a group of UAVs in the system.

As shown in Fig. 4, two parts are considered in the framework. In the right part of the framework, the UAVs of the system are placed. On the left side of the framework,

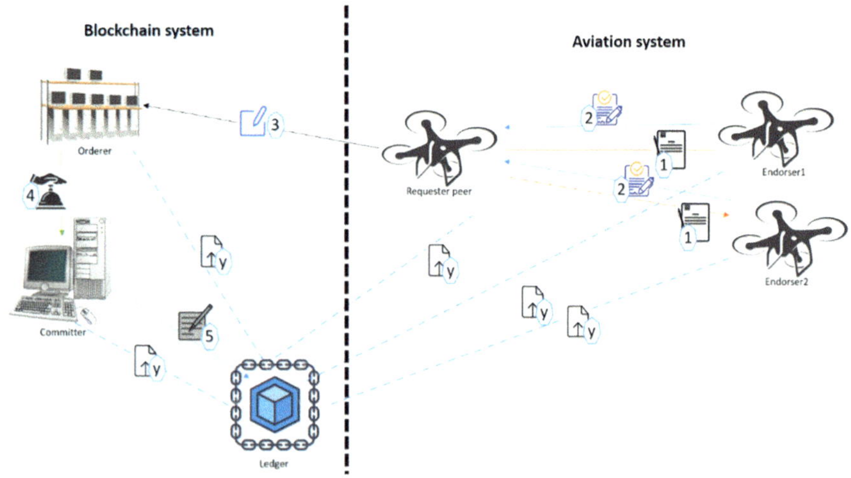

Fig. 4 The BCT-based aviation system framework

the considered BCT system has been presented. A private ledger system is considered in this framework since the data generated in the system is highly sensitive important data and its misuse might harm the human. Each of these UAVs holds an IoT device for sharing the required information with the data collectors, and the IoT devices act like a normal peer in our framework where they can broadcast a transaction proposal to read and or write on the private ledger. As soon as a transaction proposal is broadcasted by a normal peer to the other peers, the proposal will be forwarded to the endorses for the endorsing process. It is expected that some involved UAVs of the aviation system will be in charge of the endorsing process.

The number of the endorsing node can be changed from two to many, depending on system architecture and requirements. The endorsed transactions will be forward to the normal peer and the normal peer after refinement will forward the endorsed transaction into the client (the sender of the request) node for the mining process. The client will collect the endorsed transaction and will create a suitable mined hash function for the transaction and send the mined transaction to the committer. Committer is in charge of writing the transition into the ledger.

As soon as the transaction is written to the ledger, a completion message will be sent to the requester peer. Depends on the type of transactions, different types of information can be recorded in the ledger using this framework. For each type of information, a particular type of ledger should be considered. In the case that it is expected more than one type of information should be recorded, more ledgers and their channels should be designed and added, and considered in the framework. In Fig. 4, you may see the details of the proposed BCT for an aviation-based framework.

4 Applications of BCT-Based Aviation System

As mentioned in previous sections, BCT can be used in Aviation 4.0 and UAV, generally for many purposes. The developed framework will be used to illustrate and record coordination information of the UAVs of the system. UML sequence diagram is used to represent the recorded coordinates in the ledger. Since only one type of information may be recorded at each moment, only a ledger and its corresponding channel are considered for each recording. As shown in Fig. 5, the procedure is like the following:

- A UAV in the system requests to write its coordinate information on the ledger. Since the blocks of the ledger are time-sensitive, the time of recording the information in the ledger will be recorded. This information will avoid double climes.
- The requester UAV will provide transactions containing the coordination information of the UAV, and broadcast this transaction to all the other UAVs that are active in the system. Two UAVs are required in the system to endorse the transaction. Here, it should be mentioned that endorsing is not meaning as proofing the correctness of the broadcasted transaction since it is requester UAV's responsibility that shares correct information. Endorsing here means that two other UAVs will prove that this transaction belongs to the requester using private and public key and digital signature technologies.
- As soon as the transaction is endorsed, the endorsed transaction will go to the order and committer peers for further miming and writing process.
- At the end of the procedure, a completion message will be sent to the requester UAV and the transaction will be written into the ledger.

Third parties for decision-making processes can use the information of the ledgers. Just as an example, a third party controller of the system that has access to the ledger

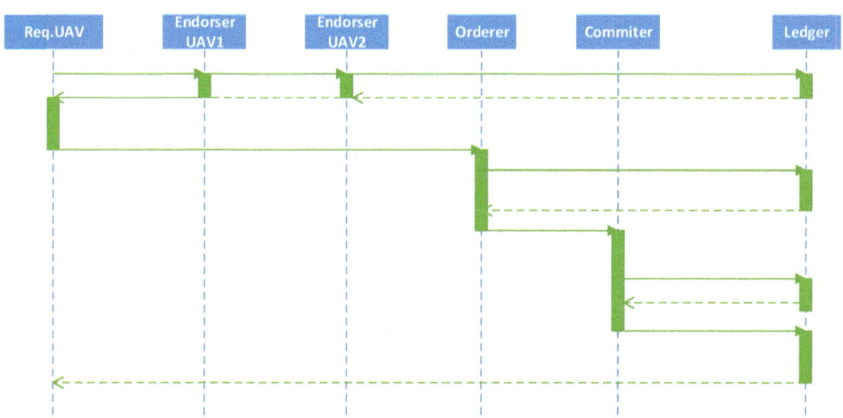

Fig. 5 Coordination recording procedure in the ledger

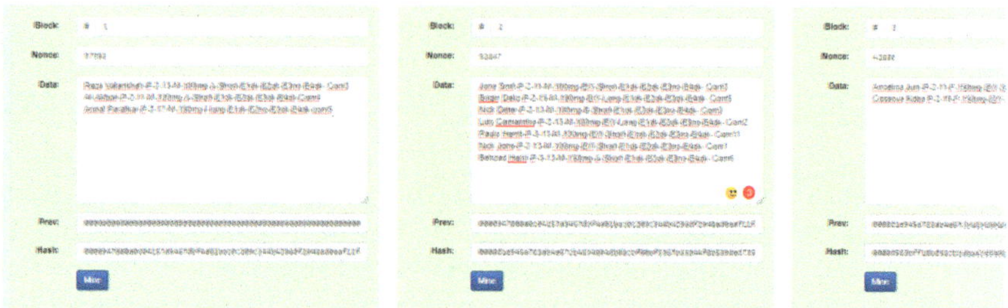

Fig. 6 Representation of a sample BCT ledger for some coordination transactions

coordination information of this system can avoid the UAV from possible accidents. Depends on this application, if the UAVs are using to cover a vast space, this third part control and management system can assign the UAVs to the most appropriate locations and thanks to the information available in the ledger can run the system with a minimum number of resources.

Figure 6, represents the block structure of the ledger recording coordination information of the UAV in the Aviation 4.0 era. As shown in the figure, each of the blocks holds a block number, a nonce, a space to record the transitions, a hash of the previous block, and a hash of the block. As soon as a transaction is added to the block, the time index of the transaction will also be recorded in the block. The previous hash function is used to improve the safety and security of the data recording, since in the case that when a transaction in the previous block is changed, the hash function of the block will be changed, and then the hash value of the next blocks will be changed. Such a procedure will avoid the system from any unauthorized and fraudulent information. The client, the peer in the framework, is in charge of finding a hash value to the block that started with four "Zero" digits. To do so the order should change the nonce value in the block until a suitable hash value is extracted. It is noted that such a trial and error approach is both times consuming and case sensitive.

5 Conclusion and Discussion

Industry 4.0 indicates the new stage of the modernized industrial revolution with a high concentration on interconnectivity, flexible automation, artificial intelligence, and real-time data exchange and sharing using advanced technologies such as cyber-physical systems, IoT, cloud computing, big data, and advanced robotics and virtual reality. Aviation 4.0 is a manifestation of Industry 4.0 in the UAV industry. Aviation 4.0 can be defined as the digitalization of aviation industries from the device and flight, and operations points of view by networked systems, and uses intensively digital models and ontologies. The main missions of using this paradigm are reducing the service costs and also improve the quality of the services by; (a) improving the smartness of the contributing flights (b) connecting and integrating all the contributing flights, (c) providing real-time status and awareness information to the regulating authorities. Aviation 4.0 is powered by the cyber-physical platform which enhances the processing of big data, maximizes the interconnectivity and collaboration, optimizes the artificial intelligence, and distributed cloud architectures. This modern digitalization allows aviation industries and organizations to change their network policies with higher efficiency and performance in different aspects. Networked embedded systems control and monitor the UAVs usually with feedback circles from cloud-based computation units. The economic and societal potential of the paradigm is vastly greater than what has been realized, and major investments should be made worldwide to develop the technology. The challenges are still considerable because aviation systems should follow the tough instructions of the regulating organizations to grasp safety and reliability requirements. Moreover, the used tools for the design, monitor, and control of the UAV are qualitatively different from other industries.

This chapter provides the basic technical details of the BCT and explains its implementation procedures in general applications. Then it highlights some applications of BCT in UAVs available in the literature and explain the steps of achieving such utilizations, considering the Aviation 4.0 phenomenon. Finlay, a real-case utilization of UAVs in the industry is explained in detail. Systematically explained procedures and explaining BCT in the Aviation 4.0 era may lead the researchers to make new ideas in near future and facilitate their related understandings.

References

1. Valdés, R.A., Comendador, V.F.G., Sanz, A.R., Castán, J.P.: Aviation 4.0: more safety through automation and digitization. In: Aircraft Technology. IntechOpen (2018)
2. Friedberg, W., Copeland, K., Duke, F., O'Brien, K., III., Darden, E., Jr.: Guidelines and technical information provided by the US Federal Aviation Administration to promote radiation safety for air carrier crew members. Radiat. Prot. Dosim. **86**(4), 323–327 (1999)
3. Alladi, T., Chamola, V., Sahu, N., Guizani, M.: Applications of blockchain in unmanned aerial vehicles: a review. Veh. Commun.100249 (2020)
4. Nakamoto, S.: Re: Bitcoin P2P e-cash paper. The Cryptography Mailing List (2008)
5. Nakamoto, S.: Bitcoin: A peer-to-peer electronic cash system. Manubot (2019)
6. Crosby, M., Pattanayak, P., Verma, S., Kalyanaraman, V.: Blockchain technology: beyond bitcoin. Appl. Innov. **2**(6–10), 71 (2016)
7. Ferrer-Gomila, J.-L., Hinarejos, M.F., Isern-Deyà, A.-P.: A fair contract signing protocol with blockchain support. Electron. Commer. Res. Appl. **36**, 100869 (2019)
8. Cui, G., Shi, K., Qin, Y., Liu, L., Qi, B., Li, B.: Application of block chain in multi-level demand response reliable mechanism. In: 2017 3rd International Conference on Information Management (ICIM), pp. 337–341. IEEE (2017)
9. Xie, J., Tang, H., Huang, T., Yu, F.R., Xie, R., Liu, J., Liu, Y.: A survey of blockchain technology applied to smart cities: research issues and challenges. IEEE Commun. Surv. Tutorials **21**(3), 2794–2830 (2019)
10. Neudecker, T., Hartenstein, H.: Network layer aspects of permissionless blockchains. IEEE Commun. Surv. Tutorials **21**(1), 838–857 (2018)
11. Abraham, I., Malkhi, D.: The blockchain consensus layer and BFT. Bull. EATCS **3**(123) (2017)

12. Lakshman, T., Agrawala, A.K.: Efficient decentralized consensus protocols. IEEE Trans. Software Eng. **5**, 600–607 (1986)
13. Castro. M, Liskov. B.: Practical Byzantine fault tolerance. In: OSDI, vol. 1999, pp. 173–186 (1999)
14. King, S., Nadal, S.: Ppcoin: Peer-to-Peer Crypto-Currency with Proof-of-Stake, p. 1. Self-published paper, August 19 (2012)
15. Larimer, D.: Delegated proof-of-stake (dpos). Bitshare Whitepaper (2014)
16. Ghosh, M., Richardson, M., Ford, B., Jansen, R.: A TorPath to TorCoin: proof-of-bandwidth altcoins for compensating relays. Naval Research Lab, Washington, DC (2014)
17. Miller, A., Juels, A., Shi, E., Parno, B., Katz, J.: Permacoin: Repurposing bitcoin work for data preservation. In: 2014 IEEE Symposium on Security and Privacy, pp. 475–490. IEEE (2014)
18. Kwon, J.: Tendermint: consensus without mining. Draft v 06, fall **1**(11) (2014)
19. Schwartz, D., Youngs, N., Britto, A.: The ripple protocol consensus algorithm. Ripple Labs Inc White Paper **5**(8) (2014)
20. Mazieres, D.: The stellar consensus protocol: a federated model for internet-level consensus. Stellar Dev. Found. **32** (2015)
21. Chen, S., Shi, R., Ren, Z., Yan, J., Shi, Y., Zhang, J.: A blockchain-based supply chain quality management framework. In: 2017 IEEE 14th International Conference on e-Business Engineering (ICEBE), pp. 172–176. IEEE (2017)
22. Kosba, A., Miller, A., Shi, E., Wen, Z., Papamanthou, C.: Hawk: The blockchain model of cryptography and privacy-preserving smart contracts. In: 2016 IEEE Symposium on Security and Privacy (SP), pp. 839–858. IEEE (2016)
23. Qian, Y., Jiang, Y., Chen, J., Zhang, Y., Song, J., Zhou, M., Pustišek, M.: Towards decentralized IoT security enhancement: a blockchain approach. Comput. Electr. Eng. **72**, 266–273 (2018)
24. Zheng, Z., Xie, S., Dai, H., Chen, X., Wang, H.: An overview of blockchain technology: architecture, consensus, and future trends. In: 2017 IEEE international congress on big data (BigData congress), pp. 557–564. IEEE (2017)
25. Tang, H., Shi, Y., Dong, P.: Public blockchain evaluation using entropy and TOPSIS. Expert Syst. Appl. **117**, 204–210 (2019)
26. Rouhani, S., Deters, R.: Performance analysis of ethereum transactions in private blockchain. In: 2017 8th IEEE International Conference on Software Engineering and Service Science (ICSESS), pp. 70–74. IEEE (2017)
27. Li, Z., Kang, J., Yu, R., Ye, D., Deng, Q., Zhang, Y.: Consortium blockchain for secure energy trading in industrial internet of things. IEEE Trans. Industr. Inf. **14**(8), 3690–3700 (2017)
28. García-Magariño, I., Lacuesta, R., Rajarajan, M., Lloret, J.: Security in networks of unmanned aerial vehicles for surveillance with an agent-based approach inspired by the principles of blockchain. Ad Hoc Netw. **86**, 72–82 (2019)
29. Hassija, V., Saxena, V., Chamola, V.: Scheduling drone charging for multi-drone network based on consensus time-stamp and game theory. Comput. Commun. **149**, 51–61 (2020)
30. Fernández-Caramés, T.M., Blanco-Novoa, O., Suárez-Albela, M., Fraga-Lamas, P.: A UAV and blockchain-based system for industry 4.0 inventory and traceability applications. In: Multidisciplinary Digital Publishing Institute Proceedings, vol. 1, p. 26 (2018)
31. Fraga-Lamas, P., Fernández-Caramés, T.M., Castedo, L.: Towards the internet of smart trains: a review on industrial IoT-connected railways. Sensors **17**(6), 1457 (2017)
32. Fernández-Caramés, T.M., Fraga-Lamas, P., Suárez-Albela, M., Vilar-Montesinos, M.: A fog computing and cloudlet based augmented reality system for the industry 4.0 shipyard. Sensors **18**(6), 1798 (2018)
33. Dasdemir, E., Köksalan, M., Öztürk, D.T.: A flexible reference point-based multi-objective evolutionary algorithm: an application to the UAV route planning problem. Comput. Oper. Res. **114**, 104811 (2020)
34. Ghadiri Nejad, M., Banar, M.: Emergency response time minimization by incorporating ground and aerial transportation. Ann. Optim. Theor. Pract. **1**(1), 43–57 (2018)
35. Shavarani, S.M., Nejad, M.G., Rismanchian, F., Izbirak, G.: Application of hierarchical facility location problem for optimization of a drone delivery system: a case study of amazon prime air in the city of San Francisco. Int. J. Adv. Manuf. Technol. **95**(9–12), 3141–3153 (2018)

36. Beul, M., Droeschel, D., Nieuwenhuisen, M., Quenzel, J., Houben, S., Behnke, S.: Fast autonomous flight in warehouses for inventory applications. IEEE Rob. Autom. Lett. **3**(4), 3121–3128 (2018)
37. Barenji, R.V., Barenji, A.V., Hashemipour, M.: A multi-agent RFID-enabled distributed control system for a flexible manufacturing shop. Int. J. Adv. Manuf. Technol. **71**(9–12), 1773–1791 (2014)
38. Kapitonov, A., Lonshakov, S., Krupenkin, A., Berman, I.: Blockchain-based protocol of autonomous business activity for multi-agent systems consisting of UAVs. In: 2017 Workshop on Research, Education and Development of Unmanned Aerial Systems (RED-UAS), pp. 84–89. IEEE (2017)
39. Sobreira Leite, G., Bessa Albuquerque, A., Rogerio Pinheiro, P.: Process automation and blockchain in intelligence and investigation units: an approach. Appl. Sci. **10**(11), 3677 (2020)
40. Mohanta, B.K., Panda, S.S., Jena, D.: An overview of smart contract and use cases in blockchain technology. In: 2018 9th International Conference on Computing, Communication and Networking Technologies (ICCCNT), pp. 1–4. IEEE (2018)
41. Barenji, A.V., Barenji, R.V., Hashemipour, M.: Flexible testing platform for employment of RFID-enabled multi-agent system on flexible assembly line. Adv. Eng. Softw. **91**, 1–11 (2016)
42. Barenji, R.V.: A blockchain technology based trust system for cloud manufacturing. J. Intell. Manuf.1–15 (2021)
43. Nejad, M.G., Kashan, A.H.: An effective grouping evolution strategy algorithm enhanced with heuristic methods for assembly line balancing problem. J. Adv. Manuf. Syst. **18**(03), 487–509 (2019)
44. Nejad, M.G., Kashan, A.H., Shavarani, S.M.: A novel competitive hybrid approach based on grouping evolution strategy algorithm for solving U-shaped assembly line balancing problems. Prod. Eng. **12**(5), 555–566 (2018)
45. Nejad, M.G., Kovács, G., Vizvári, B., Barenji, R.V.: An optimization model for cyclic scheduling problem in flexible robotic cells. Int. J. Adv. Manuf. Technol. **95**(9–12), 3863–3873 (2018)
46. Nejad, M.G., Shavarani, S.M., Vizvári, B., Barenji, R.V.: Trade-off between process scheduling and production cost in cyclic flexible robotic cells. Int. J. Adv. Manuf. Technol. **96**(1–4), 1081–1091 (2018)
47. Vizvári, B., Guden, H., Nejad, M.G.: Local search based meta-heuristic algorithms for optimizing the cyclic flexible manufacturing cell problem. Ann. Optim. Theory Pract. **1**(3), 15–32 (2018)
48. Mosallaeipour, S., Nazerian, R., Ghadirinejad, M.: A two-phase optimization approach for reducing the size of the cutting problem in the box-production industry: a case study. In: Industrial Engineering in the Industry 4.0 Era, pp. 63–81. Springer (2018)
49. Mosallaeipour, S., Nejad, M.G., Shavarani, S.M., Nazerian, R.: Mobile robot scheduling for cycle time optimization in flow-shop cells, a case study. Prod. Eng. Res. Devel. **12**(1), 83–94 (2018)
50. Mizouni, R., El Barachi, M.: Mobile phone sensing as a service: Business model and use cases. In: 2013 Seventh International Conference on Next Generation Mobile Apps, Services and Technologies, pp. 116–121. IEEE (2013)
51. Noyen, K., Volland, D., Wörner, D., Fleisch, E.: When money learns to fly: towards sensing as a service applications using bitcoin. arXiv preprint arXiv:14095841 (2014)
52. Perera, C., Zaslavsky, A., Christen, P., Georgakopoulos, D.: Sensing as a service model for smart cities supported by internet of things. Trans. Emerg. Telecommun. Technol. **25**(1), 81–93 (2014)
53. Ferrer, E.C.: The blockchain: a new framework for robotic swarm systems. In: Proceedings of the future technologies conference, pp. 1037–1058. Springer (2018)
54. Nofer, M., Gomber, P., Hinz, O., Schiereck, D.: Blockchain. Business & information. Syst. Eng. **59**(3), 183–187 (2017)
55. Han, S.-C., Bang, H., Yoo, C.-S.: Proportional navigation-based collision avoidance for UAVs. Int. J. Control Autom. Syst. **7**(4), 553–565 (2009)

56. Sharma, V., Sabatini, R., Ramasamy, S.: UAVs assisted delay optimization in heterogeneous wireless networks. IEEE Commun. Lett. **20**(12), 2526–2529 (2016)
57. Fan, Q., Ansari, N.: Towards traffic load balancing in drone-assisted communications for IoT. IEEE Internet Things J **6**(2), 3633–3640 (2018)
58. Thiel, S., Häbe D, Block, M.: Co-operative robot teams in a hospital environment. In: 2009 IEEE International Conference on Intelligent Computing and Intelligent Systems, pp. 843–847. IEEE (2009)
59. Nejad, M.G., Güden, H., Vizvári, B., Barenji, R.V.: A mathematical model and simulated annealing algorithm for solving the cyclic scheduling problem of a flexible robotic cell. Adv. Mech. Eng. **10**(1), 1687814017753912 (2018)
60. Nejad, M.G., Güden, H., Vizvári, B.: Time minimization in flexible robotic cells considering intermediate input buffers: a comparative study of three well-known problems. Int. J. Comput. Integr. Manuf. **32**(8), 809–819 (2019)
61. Nejad, M.G., Shavarani, S.M., Güden, H., Barenji, R.V.: Process sequencing for a pick-and-place robot in a real-life flexible robotic cell. Int. J. Adv. Manuf. Technol. **103**(9–12), 3613–3627 (2019)
62. Sharma, V., You, I., Kul, G.: Socializing drones for inter-service operability in ultra-dense wireless networks using blockchain. In: Proceedings of the 2017 International Workshop on Managing Insider Security Threats, pp. 81–84 (2017)

Intelligent and Fuzzy UAV Transportation Applications in Aviation 4.0

Mahmoud Golabi and Mazyar Ghadiri Nejad

Abstract In the last decade, Aviation 4.0 has attracted lots of researchers' attention and with huge scientific progress, it has become one of the most important issues that researchers have focused on. According to the literature, different applications have been proposed for Aviation 4.0, especially in the transportation area. The intelligent and fuzzy UAV transportation applications are of the most important applications in Aviation 4.0. In this study, various related researches from the literature are visited under different categories of delivery operation problems by using UAVs such as facility location problems, vehicle routing problems, path-planning problems, and their applications. Moreover, a case study related to the uncertainty condition of UAV applications is considered. Fuzzy mathematical modelling is developed to address the problem, and a fuzzy possibilistic-based method is utilized to deffuzify the model. Then, a genetic algorithm is proposed to solve the problem and the results of some randomly generated cases are obtained and discussed. It is shown that the proposed method is a suitable approach for decision-makers to make an aerial delivery system.

Keywords Aviation 4.0 · Blockchain technology · Cyber-physical system · Unmanned aircraft vehicles (UAVs) · Applications

1 Introduction

The logistics industry faces multiple tasks like high operation cost, high environmental impact, and complexity in the performance of the supply chain. To tackle the challenges, the inclusion of new technologies such as Unmanned Aircraft Systems (UASs), can help develop new transportation and delivery systems to progress from old-style systems and approaches. However, using such technologies takes

M. Golabi
Université de Haute-Alsace, IRIMAS UR 7499, F-68100 Mulhouse, France
e-mail: mahmoud.golabi@uha.fr

M. G. Nejad (✉)
Department of Industrial Engineering, Girne American University, 99428 Kyrenia, TRNC, Turkey
e-mail: mazyarghadirinejad@gau.edu.tr

C. Kahraman and S. Aydın (eds.), *Intelligent and Fuzzy Techniques in Aviation 4.0*, Studies in Systems, Decision and Control 372,
https://doi.org/10.1007/978-3-030-75067-1_19

unique operational challenges that add new characteristics to the transportation and distribution network configurations.

UASs are mainly used for military and strike programs and a diversity of civilian tasks such as infrastructure checking, accuracy agriculture, package distribution services, photography, exploration, and saving operations. Transportation agencies are in a sole situation to influence the emerging technology, due to the nature of this occupation, the massive demand, and the countless advantages in terms of falling accidents, justifying congestion, and cost-saving. Considering a study by the American Association of State Highway and Transportation Officials (AASHTO), 33 state departments of transportation have performed or prospected applications of unmanned aerial vehicles (UAVs) in numerous features of transportation, including checking bridges, gathering traffic data, and serving crash clear-up [1].

In transportation-related applications, small UAVs, frequently referred to drones, are progressively affordable, easy to transport and launch, and can be prepared with cameras that gather information for transportation agencies. However, using a sole UAV may be problematic to complete a task with enormous computational loads and extensive flight range with the curb of load capacity and power supply. To break through individual restrictions, the multi-UAV system is defined as a multi-agent system that can be monitored independently and remotely [2]. Possible usages of UAVs consist of accident sight photography, surveying, security examinations, building data collection, and checking the condition of roadways. However, there are presently restrictions that delay the use of UAVs by state departments of transportation. The restrictions are connected to the requirement to gain authorization from the Federal Aviation Administration (FAA), which has stated anxiety about the capability of UAVs to see and avoid manned aircraft. Several tests that have been led by the Washington State Department of Transportation (WSDT) to assess the technical abilities and institutional anxieties related to the use of UAVs designated that the UAVs can achieve effective and autonomous aerial surveillance, but anxieties remain about the reliability of UAVs and the consequences of a crash [3].

Aviation 4.0, which is utilizing and applying concepts of Industry 4.0 in the aviation industry, is related to cyber-physical systems' designing to help humanities for doing their jobs and activities automatically, through integrating them to components of cyber-physical systems [4]. The use of industry 4.0 technology, which mainly includes automation, the internet of things, artificial intelligence, big data analytics, and digitalization, can make a huge difference in the aviation industry. Using these changes, new mechanisms can be created and used in the aviation industry, that as their benefits, it should be implied to increasing the efficiency and increasing the operational reliability. Some practical studies in the field of Aviation 4.0 that lead to increased flight safety include instantaneous information updates, cockpit safety calculations, aircraft condition forecasting, emergency rescue operations, and more [5].

In the last decade, intelligent systems have been considered vastly by scientists [6]. The fuzzy logic system as one of the intelligent systems is a powerful nonlinear method in solving optimization problems. This system, which was invented by Zadeh

[7, 8], is a flexible method and can be integrated with other optimization algorithms such as a genetic algorithm.

In recent years, many studies have been conducted on the basis of such combinations and have led to the development of powerful tools for solving uncertain problems, especially in the field of aviation 4.0 [9].

In the next section of this study, delivery operations by using UAVs are discussed in detail. Different types of problems related to this issue such as facility location problems, vehicle routing problems, path planning problems, and their applications are scrutinized, based on the literature. The methodology of this study includes the fuzzy approach of modelling, and a crisp equivalent of the fuzzy model is proposed in the third section. In section four, the case study and the results, obtained by the solution approach are presented, and finally, the study is concluded in the last section.

2 Literature Review

As the use of the internet has grown dramatically during the last decade, it provided limitless occasions for businesses resulting in adopting electronic commerce (e-commerce) in business transactions. According to Nisar and Prabhakar [10], the growth of internet users purchasing online is increasing annually. Business to customer contacts raised the demand for home distributions, subsequent in increasing the need for the supply chain to solve the problem of Last-Mile Delivery (LMD). It refers to the process of transmission goods from transportation hubs to a final destination in the supply chain [11, 12]. The problem with the LMD is that it signifies an inconsistency between speed and cost; clients expect fast same-day or on-demand distribution yet they endure highly worth sensitive and incline to prefer the cheapest available distribution choices. The key challenge for businesses is to find ways to minimize shipping expenditures while providing consumers with rational distribution times and reasonable distribution costs. Additionally, there are many other circumstances that increase LMD challenges such as incorrect customers' addresses, cramped customers' locations, drivers' shortages, and some adverse economic conditions such as rising fuel prices. In addition to these circumstances, delivery windows, customers' density, missed deliveries or environmental impacts are other factors that can adversely affect the efficiency of the last-mile delivery.

Novel technologies like UAVs, robots, and artificial intelligence are being used to replace the traditional delivery methods with improved logistics systems to tackle the challenges of LMD such as reducing the operational cost and the complexity of the supply chain.

The application of drones in delivery operations is known to be significantly efficient. Less maintenance and operator costs along with faster delivery are the main advantages of drones in comparison with traditional delivery vehicles. This delivery option does not also need huge investment costs in road infrastructures. As a result, the main organizations in the delivery industry have been interested in applying drones in their operations. The Amazon Prime Air project [13], the Parcelcopter

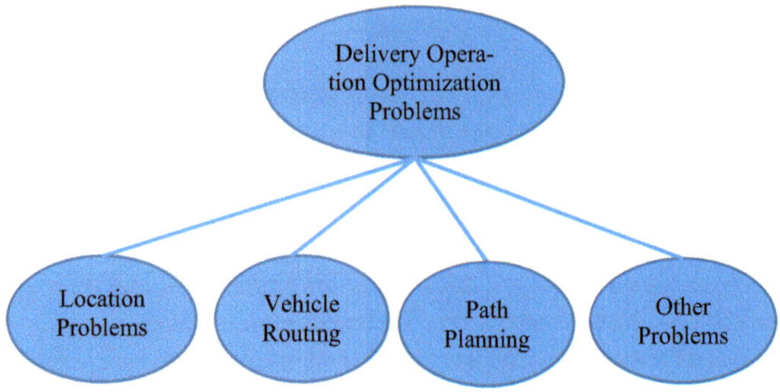

Fig. 1 Classification of delivery operation optimization problems

project of DHL [14], the Google Project Wing [15], and the Swiss Post's project for the delivery of light parcels [16] are some examples of the application of drones in real-life delivery operation.

Optimizing the delivery operation from different perspectives is thoroughly studied in the literature. The optimization problems in this area could be classified as vehicle routing problems to determine the delivery routes for the applied vehicles, path planning problems to determine the optimal path to reach the destination, and location problems for launching and/or refuelling stations. It is noteworthy to mention that the considered optimization problems are not limited to delivery operations and could be studied for other applications of drones in real-life operations such as monitoring, mapping, and surveillance. Figure 1 illustrates the main classification of delivery operation optimization problems. However, there are other categories besides the mentioned main classes, which belong to the class of "Other Problems" in the figure.

2.1 Vehicle Routing Problem

One of the main issues in the delivery operation is to find the delivery routes for available delivery vehicles. This problem, which is known as Vehicle Routing Problem (VRP) aims to determine the best routes of applied vehicles to deliver the items to the set of customers' locations. There are several types of VRP studied in the literature. Single-depot [17] and multiple-depot [18] VRPs account for the number of available depots in the network. Capacitated Vehicle Routing Problem (CVRP) is a prominent type of routing problems to account for the maximum possible traveling distance of vehicles which is also called distance-constraint vehicle routing problem [19], the payload capacity of vehicles [20], and the depot capacity [21].

Dynamic VRP is another class of routing problems to incorporate the new online unforeseen orders of customers [22]. Pillac et al., [23] published a review article related to the DVRP and scrutinized this issue. According to the literature, the difference between DVRP and static VRP is that in DVRP new information about the problem is introduced in real-time, as the vehicles execute the current routes. This new information can be new service requests from a known set of customers, in which the customers can submit their requests up to a certain cut-off time before the execution of routes start. Another approach is to start with a known set of customers and plan the routes for these customers by solving them as a static VRP, then deal with the new requests by either accepting them and modifying the future routes or rejecting the requests. The new requests can also be from customers who are not known in advance. Furthermore, the new information can belong to multiple customers or to a single new customer arriving at a uniformly chosen random location. Moreover, this new information can be handled as soon as it arrives or it can be aggregated and processed at the start or the end of specific time periods. In the proposed research, we follow the latter approach where new information from customers is handled at the end of each time period. Open VRP is another ilk of routing problems in which the vehicles are not required to return to the starting depot [24]. Figure 2 shows the classification of VRPs.

The vehicle routing problem for UAV drones is reviewed by Rojas Viloria [25]. The authors classified the literature based on the developed objective functions, applied solution algorithms, considered constraints and restrictions, various applications in real-life problems, and the usage of any complementary vehicle. The restricted battery and payload capacity is the main concern of Drones' Routing Problem (DRP). The time window, within which the delivery should be accomplished, is another constraint that can be added to the routing problem. The mixed-integer linear programming (MILP) model provided by Guerriero [26] formulates a multi-objective DRP to optimize the travelled distance, the number of visited customers, and the number of applied drones. The authors developed a set of heuristics to solve a generated instance related to video capturing of a football game using filming drones. They used the travelled distance and coverage objectives as the main parameters of evaluating the performance of solution methods in comparison with

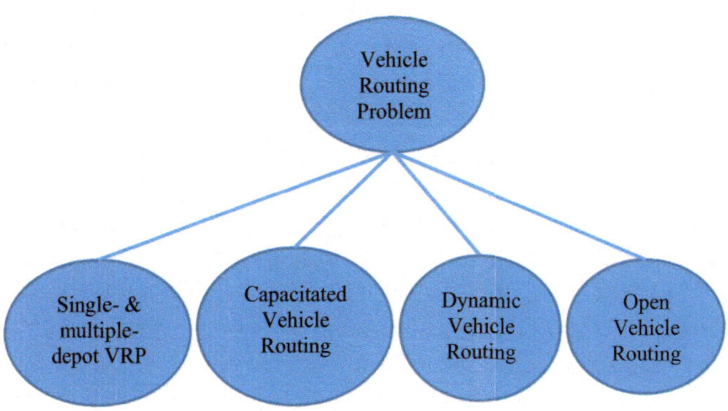

Fig. 2 Classification of vehicle routing problems

the results obtained from solving the concomitant mathematical model. The VRP developed by Zhen et al. [27] to be applied to monitoring drones is an example of VRP in non-delivery operations. The authors tried to determine the sequence of visited vertices along with the height at which the visits are taken place to have a more reliable image of service time. Thus, the drones' routes are optimized in a manner the total monitoring time is minimized. While the authors solved the small instance using CPLEX, they developed a Tabu search method to solve medium and large size instances. It should be mentioned that the CPLEX software is one of the well-known application to solve linear problems, which has been used in many studies [28]. Dhein et al. [29] studied a collaborative system requiring communication and coordination and situation awareness to find trajectories of multiple-drones such that the dispersion between the trajectories is minimized. A local search genetic algorithm is applied to solve the problem as it is a very effective method according to the literature [30]. Alvear et al. [31], developed a metaheuristic algorithm, called Pollution-driven UAV Control (PdUC) algorithm, which is based on a chemotaxis metaheuristic and a local particle swarm optimization strategy, to perform air pollution monitoring tasks utilizing UAVs equipped with off-the-shelf sensors. Mersheeva and Friedrich [32] proposed an algorithm based on Variable Neighbourhood Search to solve the problem of UAV routing in a disaster situation and for the rescue mission planning. Peng et al. [33] developed a hybrid genetic algorithm, which supported the collaboration of a ground vehicle and multiple UAVs for effective package distribution. Their algorithm for routing and scheduling allowed to carry multiple UAVs by the vehicle and multiple parcels concurrently to customers in diverse sites. Khan et al. [34], proposed a bio-inspired clustering scheme for flying ad hoc networks (FANETs) by using the hybrid mechanism of glow-worm swarm optimization (GSO) and krill herd (KH) to minimize the battery resource for UAVs routing. Their proposed scheme was using energy-aware cluster formation and cluster head election based on the GSO algorithm.

The Green Vehicle Routing Problem (G-VRP) introduced by Erdigan & Miller-Hooks [35] accounts for the on-route refuelling or recharging possibility of delivery vehicles to visit or serve a higher number of customers or demand points [36]. A capacitated version of G-VRP for electric vehicles considering time windows is studied by Schneider [37]. Considering a fleet of different electric vehicles with various battery and payload capacities, Hiermann et al. [38] studied a G-VRP to determine the delivery routes and the location of charging stations. As it is mentioned by Dorling et al. [39], locating recharging or refuelling stations in the network may not be economically reasonable for delivery companies. They also mentioned that although G-VRP accounts for several visits of charging or refuelling stations, considering single-visit of depots for each vehicle will definitely lead to a huge fleet size. The Multi-Trip Vehicle Routing Problem (MTVRP) introduced by Azi et al. [40] could decrease the fleet size and vehicle procurement costs by considering multiple trips for each delivery vehicle. Using a linear formulation to calculate the consumed energy based on the payload and flying altitude, Dorling et al. [39] studied two MTVRPs to account for minimizing the delivery cost within a delivery time limitation and minimizing the total delivery time subject to budget restrictions. They proposed a

Simulated Annealing (SA) algorithm to resolve the problems. There are other algorithms applied to solve MTVRP in the literature are branch-and-price algorithm [41], large neighbourhood search algorithm [40], variable neighbourhood algorithm [42], and memetic algorithm [43]. The multi-objective real-time G-VRP developed by Coelho et al. [44] allows multiple-deliveries per each drone's flight respecting the capacity restrictions. The travelled distance, the total delivering time, the number of applied drones, the recorded drones' maximum speed, the time for collecting each delivery package, the time for collecting the last package, and the total consumed energy are the considered objective function of this study. The authors used a black-box dynamic search algorithm to find the non-dominated solutions obtained by a branch-and-bound optimization technique.

Due to the technical restrictions such as the endurance and payload limits, covering more than one customer in each drone flight seems to be impossible [45]. Additionally, flying regulations [46] and safety and privacy issues raise doubts about long drone flights in delivery operations [47]. The emergence of the synchronized truck-drone delivery system (TDDS) consisting of a truck that is complemented by a delivery drone is a solution to the restrictions of the traditional last-mile drone delivery systems. While the truck is traversing its designated routes to serve the assigned customers, the drone that launches and lands on the truck covers the remaining demands for whom the truck delivery operation is not reasonable due to different factors such as land conditions or road congestion. This new concept of last-mile delivery operation is now considered by Mercedes-Benz [48] and UPS [49]. The time saved using a truck-drone system in comparison with just trucks in last-mile delivery operation is scrutinized by Wang et al. [50] to account for worst-case analysis depending on the number of allocated UAVs to each truck and the drone's speed compared with the hosting truck's speed. The one-truck one-drone TDDS studied by Agatz et al. [51] highlights the effect of drone's flying speed on both the truck's and drone's traveling and waiting time. Two heuristics and two exact algorithms solved the developed MILP for this study.

Yoon [52] studied a one-truck multiple-drone TDDS to determine the number of assigned drones to each truck along with the best truck and drone routes to minimize the total cost of a system comprising of the traveling costs and the drone procurement costs. They formulated the problem as a MILP model and solved it on the case study of New York.

Collecting the drones at the same location as launching them before the truck's movement to the next destination is the assumption done by Moshref-Javadi et al. [53]. After applying the bound analysis to highlight the improvements of the system in customers' waiting time, the authors suggested a hybrid Tabu Search-Simulated Annealing algorithm as the solution method.

Madani and Ndiaye [54] developed a mathematical model to get the best configuration of a set of UAV paths and to find the best locations for the truck to release and gather the UAV in order to minimize the traveling cost obligatory to visit all clients on a single tour. The proposed model is referred to as the Travelling Salesman Problem with a Moving Depot (TSP-MD). Moshref-Javadi et al. [55] developed a MILP model to determine the assignment of demands to truck and drones, the routing of the truck, and the locations of drones' launching and returning along the designated truck's route such that the aggregate customers' waiting time is minimized. They used a bound analysis to compare the performance of the truck-drone delivery system with a just-truck system. The findings reveal a reduction of 46.8% in the customers' waiting time having only two drones synchronized to the delivering truck. Inspired by the Adaptive Large Neighbourhood Search Algorithm (ALNS), they proposed a Truck and Drone Routing Algorithm (TDRA) to solve the developed model. The effect of parcel weight and prohibited flying zones is reflected by Jeong et al. [56]. On the premise that drones could also be launched or collected on some discrete points along the network edges traversed by the truck, Schermer et al. [57] formulated a novel MILP and solved it using an algorithm derived from variable neighborhood search and Tabu search methods. Schermer et al. [58] developed a MILP model for a multi-truck multi-drone delivery system such that the makespan, i.e. the total time required to visit all the customers and collect all the trucks and drones at the depot, is minimized. The multi-truck TDDS studied by Sacramento et al. [59] considers time-limit constraints in a manner that the whole expenditure of the system is minimized. The authors used an ALNS metaheuristic algorithm as the solution method. The two-phase and single-phase heuristic methods developed by Schermer et al. [60] to account for inserting the drone and exchanging the role delivery vehicle is reported to be effective in solving large-scale TDDS.

Using drones as a means of providing supplies to trucks is studied by Mccunney and Van Cauwenberghe [61]. In this study, the delivery items are shipped to the predetermined collecting points to be loaded by the delivery truck.

A specific version of this study called a vehicle-assisted multi-drone delivery system in which the vehicle does not serve the customers and is just used to carry multiple-drones to deliver different orders to a set of different customers is studied by Carlsson and Song [62]. While determining the routes of the vehicle using a continuous approximation technique, they used a heuristic to find the launching time of drones. The same idea of using the vehicle as a moving depot is considered by Chang and Lee [63] to minimize the vehicle's and drones' return time to the main depot, where a three-phase algorithm is applied as the solution method. The three phases are to use the k-means clustering algorithm to group the customers, determine the vehicle's route, and relocate clusters to increase the covered area respectively. Luo et al. [64] studied a two-echelon location routing problem for a moving-depot one-vehicle one-drone delivery system in which launching and retaking the drone is performed at determined locations. They developed a MILP model and solved it by using two heuristics to determine the vehicle's tour and then decomposing it into sub-tours assigned to the drone.

Peng et al. [33] studied the same problem by categorizing it into three sub-problems including an anchor location problem to find the location of vehicle stop points, the routing of drones considering the endurance constraints, and the route distribution delivered to drones that are called the Bin Packing Problem (BPP). Further information about BPP could be found in Leung et al. study [65].

Considering both delivery and pick-up options at the customers' locations, Karak and Abdelghany [66] studied multiple drones mounted on a single moving vehicle in which each drone can visit more than one customer at each flight. They assumed that multiple drones can be launched at any stopping station of the vehicle and drones can return to any station along the vehicle route. If a drone arrives at the designated station earlier than the vehicle, then it has to wait in an idle mood until the vehicle arrives. The aim of this study is to determine the vehicle's and the drones' routes such that the total cost of the system is minimized. The authors represented the problem as a MILP model and solved it using an extended Clarke and Wright algorithm [67] to account for both the truck's and drones' routes. The drop-pickup problem is also studied by Ham [68] considering an integrated system of multiple-trucks, multiple-drones, time-window constraints, and multiple-visits of customer nodes where a constraint-programming approach proposed as the solution technique.

2.2 *Path Planning Problem*

The path planning is a prominent optimization problem to find the best collision-free path from the origin or starting point to the destination in terms of different criteria such as the path length, the path smoothness, the path risk, the consumed energy, etc. The path planning problems could be static or dynamic. If all the obstacles are static and pre-known, the problem is categorized as a static path planning problem, otherwise, dealing with moving obstacles changes the problem to a dynamic one [69]. The flight path of a static pre-known environment could be offline [70], otherwise the drone needs to determine the path in an online manner [71]. See Fig. 3 about the categories of path planning problems.

The solution methods related to path planning problems could be categorized as (a) classical methods including graph-based search techniques, and sampling-based approaches; and (b) computational intelligence techniques (See Fig. 4). A* and Dijkstra are among the most prominent classical graph-based solution algorithms for the path planning problem. These methods use a greedy logic to determine the shortest path between any two vertices of a connected graph. They are generally easy

Fig. 3 Classification of path planning problems

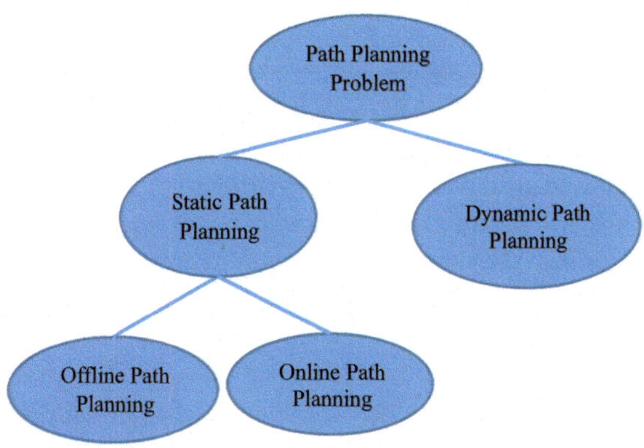

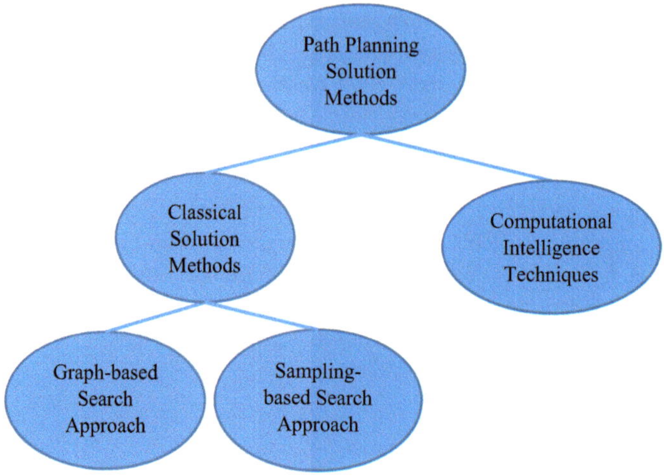

Fig. 4 Path planning solution approaches

to solve algorithms to find the path based on the single criteria of travelled distance. Probabilistic Roadmaps [72] and Rapidly-exploring Random Trees [73] are among the most prominent sampling-based approaches for finding feasible solutions in high-dimensional without guarantying the optimality. Fuzzy Systems, Neural Networks, and Metaheuristics are among the computational intelligent techniques for solving different types of problems in various fields including path-planning problems [74–79]. Pandey et al. [80] made a survey to review the applied metaheuristic algorithms in the literature. Tuning the algorithm parameters is another subject studied in the literature as Kok and Rajendran [81] proposed an optimization technique to tune the parameters of the Differential Evolution algorithm to solve a 3-D path planning problem focusing on the path length and computational time as the cost functions.

To determine the set of visiting points while considering the battery level and having the highest possible resolution quality of taken images in a monitoring mission using flying drones, Bandeira et al.[82] proposed two algorithms for the concomitant drone path planning problem. The first algorithm is based on a traveling salesman problem and the second one is a variant of TSP called Close-enough TSP algorithm. The first algorithm gives high priority to the quality of taken images by visiting the centre of each point in the trajectory, while the second one aims to focus on the path length. The coverage path planning problem (a variant of path planning problems to cover all the points of a specific area) studied by Di Franco and Buttazzo [83] aims to minimize the consumed energy while assuring the quality of taken images by drone. They applied two energy-controlling techniques: the first technique is the offline check to assure that the required energy is sufficient to perform the path; the second one is an online controlling level to guarantee the actual battery level is enough to arrive at the designated point. On the premise of working in an obstacle-free discrete environment, Ahmed et al. [84] studied a multi-drone path planning problem by proposing two greedy-based heuristics called Greedy Least Cost and First detect First Reserved. The obtained results revealed a significant trade-off between complexity and energy consumption. Gautam and Verma [85] used a combination of the Genetic Algorithm and Artificial Neural Network to determine the drone path based on the path length and distance from the radars that are considered as the obstacles. The authors used the results obtained from the Genetic Algorithms to train Neural Networks. Mardani et al. [86] studied a 2-D offline path planning problem to maximize the communication quality given a network coverage to have the highest quality of video streaming by flying drones. They account for wind effect, path post-smoothing, and available battery restrictions. They used two variants of the classical A* algorithm to solve the problem.

3-D A* algorithm is a novel approach for finding the exact solutions of drone path planning problems [87]. The algorithm considers 27 possible movements to account for all the possible moving directions of each network cell. The algorithm accounts for the possibility of flying above the obstacles and limits the prohibited airspace just up to the height of existing obstacles. Considering only one objective function, which is generally the path length, is the main drawback of this algorithm. Golabi et al. [88] developed a novel MILP model for a multi-objective 3-D path planning problem. They guaranteed the collision freeness considering both the possibilities of bypassing the static obstacles to find an alternative way, or adjusting the altitude and flying above them. The objective functions are to minimize the path length, consumed energy, and path risk simultaneously. The consumed energy is a linear function of traveled distance, fluid density of air at different altitude levels, and the energy required to increase the altitude. They used Strength Pareto Evolutionary Algorithm II (SPEA-II), Non-dominated Sorting Genetic Algorithm II, and III (NSGA-II and NSGA-III) algorithms to solve the problem.

2.3 Facility Location Problem

The Facility Location Problem (FLP) is a prominent branch of Operation Research (OR) applied to determine the best locations for establishing facilities [89, 90]. As it is studied by Klose and Drexl [91], Hale and Moberg [92], and Boloori Arabani and Farahani [93], FLP might be categorized to the median, covering, and centre problems. The objective function of median problems is to minimize the aggregate traveling times or costs [94–96]. The covering problems intend to maximize the total number of covered demands [97, 98]. The centre problem, which is mainly used in emergency logistics, is to minimize the maximum traveling or service times [99, 100]. Stochastic approach is another method that is utilized for different types of problem [101]. Therefore, stochastic FLP is another branch of location problems in which at least one of the applied parameters, such as the demand or service time, is uncertain [102, 103]. The congested location problem is a new branch of FLP in which the clients have to wait in a queuing system to get the service [89]. Capacitated FLP accounts for the capacity restrictions in serving the customers [104]. Mobile

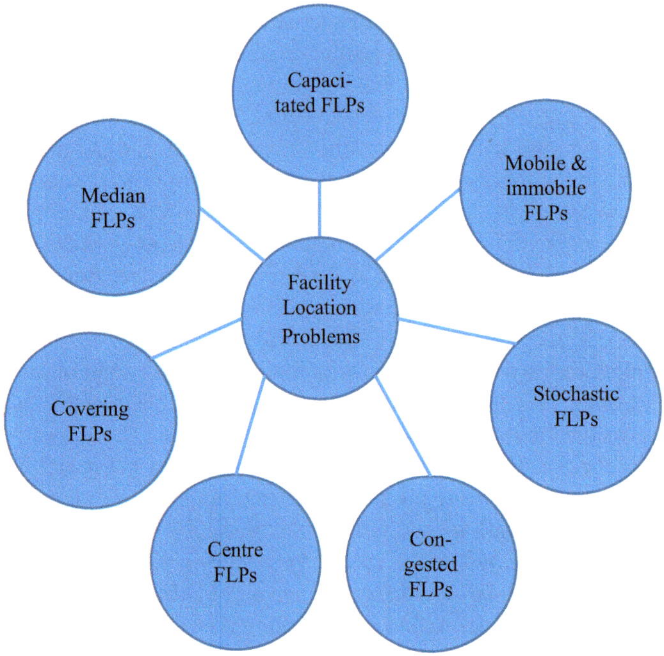

Fig. 5 Classification of facility location problems

and immobile FLPs consider a system with just the moving servers or the moving customers, respectively [105, 106]. Figure 5 shows the classifications of FLPs.

The FLP is broadly used to determine the location of launching and/or refuelling places in drone-based transportation operations. Hong et al. [107] studied a covering location problem to determine the locations of refuelling stations used for extending the flying range of commercial delivery drones. They developed a MILP model and solved it using an efficient heuristic incorporating minimum spanning trees, greedy subtraction, spatial interchange, and simulated annealing. In order to determine the best locations of fast charging stations under driving range uncertainty, Kchaou Boujelben and Gicquel [108] proposed two stochastic programming models where the first one is to maximize the expected flow coverage and the second one is based on chance constraints. They also proposed two solution algorithms that the first one uses a strong MILP for a location-allocation model and the second one is based on a Tabu search algorithm. The authors proved the efficiency of the developed Tabu search algorithm in providing good results with less computational time.

Chauhan et al. [109] proposed an integer linear programming for a covering location problem to determine the location of launching stations considering the energy consumption and range constraints. They developed a new greedy algorithm along with a 3-stage heuristic based on decomposition and local exchange principles comprising a facility location and allocation problem, multiple knapsack sub-problems, and a final local random search stage. Considering the medical drone equipped with an on-board automated external defibrillator and continuous demands, Pulver and Wei [110] developed a new location model to locate drones and launching stations such that the preliminary and back-up coverage is maximized. The authors highlighted the role of vital drones in reducing the traveling times of life-saving equipment. The stochastic facility location problem studied by Kim et al. [111] to determine the location of launching stations, the time for opening the facilities, and the number of delivery drones allocated to each facility such that the total cost of the system is minimized. The problem is designed for effectively coping with a large-scale disaster. The authors used the Benders Decomposition equipped with linear rounding methods to solve the problem in a reasonable time. Golabi et al. [112] presented a new model for combined mobile and immobile edge-based facility location problems with stochastic continuous demands. They assumed a post-disaster situation in which the demand of those located on collapsed or inaccessible network edges is covered using delivery drones equipped with a set of relief items dropped in specific distances along the edges. The mixed-integer non-linear programming model proposed in this research account for drones' payload and endurance retractions. They developed Genetic, Memetic, and Simulated Annealing algorithms to solve the problem on the real case study of Tehran. Considering the facilities capacity restrictions, Shavarani et al. [113] extended the previous study in a multi-objective context to optimize both the number of uncovered demands as well as the total costs of the distribution system. They used NSGA-II and NSGA-III methods as the solution approaches.

Shavarani et al. [114] studied a bi-level edge-based FLP to determine the location of launching and recharging places in the network. They used a Memetic algorithm to solve the problem in a real case study. The node-based congested bi-level fuzzy location problem studied by Shavarani et al. [115] is another study to determine the locations of launching and recharging stations simultaneously. Using the M/G/K queuing framework, the authors considered the maximum waiting time threshold to visit the customers. The uncertain nature of applied parameters is tackled using fuzzy variables. The authors used the fuzzy probabilistic approach to defuzzification of the problem. Finally, they solved the model using a Memetic algorithm.

2.4 Other Related UAV Problems

Besides the applications of drones in delivery operations, there are some well-known problems related to UAVs that are explained in this section.

Vertical take-off and landing are one of the problems that have been addressed in the literature, as a precise and safe take-off and landing is vital for UAVs to prevent an accident. In this regard, a study by Kelner and Ziołkowski [116] presented the structure of an independent system and a Doppler-based navigation process that permits automatic landing approaches and the precision assessment of their proposed solution made by simulation. Recently, Yang & Sun [117] developed a new data fusion algorithm based on the fuzzy complementary Kalman filter to estimate the conditions and states of the UAVs' landings and verified the solution method by simulations and experiments.

Urban intersections control problems are another type of problem including the outstanding impacts of UAVs. For instance, UAVs were successfully used at the study of the urban intersection, which was called the bird's eye view, as an accurate and cost-effective method to analyze the safety of pedestrian-vehicle conflicts [118]. Moreover, Xu et al. [119] applied UAVs at low-altitude to detect vehicles at signalized intersections and illustrated that their approach is robust to illumination changes and cars' in-plane rotation.

3 Methodology

In order to deliver a clearer image to the readers, one of the aforementioned works is selected and studied in detail. Here, the multi-level congested capacitated fuzzy facility location problem in the UAV-based delivery operation studied by Shavarani et al. [115] is selected for scrutinizing. A new version of this study which is easier to understand is proposed and the concomitant solution approach is developed.

The study aims to determine the best locations for launching and refuelling stations in a manner that the total cost of the system is minimized and the customers' waiting time to receive the ordered items does not exceed a known threshold. Since the delivery operation is done through moving UAVs, the authors used an M/G/K queuing framework to calculate the waiting time of customers in a system possessing K servers, or in other words K drones in this study. In this queuing system, the demand generation follows the Poisson distribution and the service time, which is equivalent to the flying time to reach the customers in this study, has a general distribution. The authors considered the demand generation rate of each network node, the cost of opening the facilities, the price of drones, and the distance capacity of drones as the uncertain variables. The distance capacity which could be translated to the maximum distance could be travelled by a drone in a working shift is a function of the endurance of the drone's speed and endurance. The objective function is to minimize the total cost including the cost of opening the facilities, the costs for purchasing drones, and the drones' usage cost, which is a function of the total distance travelled by a drone in the planning horizon. Here a simplified version of this work is studied in which there is only one type of facilities, called the launching stations. The queuing formulations to calculate the expected waiting time of customers along the related constraints are also neglected.

The authors deployed the possibilistic programming approach for the defuzzification of the proposed fuzzy model. To clarify the reason for selecting this approach it is useful to have a short review of the related approaches proposed in the literature. Generally, when dealing with uncertainty there is a need to construct models that are immune and insensitive to the uncertainty in the data. To do so, Soyster proposed a linear optimization approach in 1973 [120]. He tried to take the supreme values of the uncertain factors into account which could be translated to reduce the feasible region and consequently, losing good results to guarantee the feasibility of the possible solution. The other methods proposed to overcome the over-conservatism of Soyster's method increased the complexity of the problem due to their non-linearity. The robust optimization method proposed by Bertsimas and Sim [121] provides control on conservatism. Although this method keeps the linearity of the problem, it increases the number of decision variables and constraints. The possibilistic approach, as proposed by Jiménez [122] first transforms the fuzzy trapezoidal numbers into equivalent intervals. In this approach, the feasibility level of the relations is defined by the decision-maker who controls the level of reliability and conservatism using a coefficient termed α. The higher the value of α, the more conservative the model. By considering $\alpha = 1$, the model provided by the possibilistic approach would be the same as that of Soyster's method. Furthermore, the possibilistic approach does not add to the complexity of the problem.

3.1 Fuzzy Mathematical Model

The following sets, scalars, parameters, and variables are used in developing the mathematical model:
Sets:
V: The set of network nodes $(v, v' \in V)$
J: The set of candidate locations $(j, j' \in J, J \subseteq V)$
Scalars:
M: A large positive number
Parameters:
$\tilde{cl}_j$: The cost of opening a facility at candidate location j
$\tilde{c}_d$: The price of each drone
c_g : The usage cost of drone vehicles per distance unit
$t_{jj'}$: The Euclidean distance between facility j and facility j'
t_{jv}: The Euclidean distance between facility j and node v
$\tilde{\lambda}_v$: The demand located on node v
$\tilde{\vartheta}$: The speed of each drone
$\tilde{\rho}$: The distance capacity of each drone during the planning period
E : The endurance of each drone
M_t: The maximum allowed length of the route from the demand point to its assigned launch station
Variables:
$y_j = 1$ if a facility is established at location j; 0 otherwise
$x_{jv} = 1$ if node v is assigned to facility j; 0 otherwise
n_j: The total number of required drones in facility j
n_d: The total number of required drones
$\tilde{n}_g$: The aggregate utilization of drones

$\widetilde{\gamma}_j$: The request for service at facility j

Using the mentioned notations, the mathematical model would be as follows:

$$Min \sum_j \widetilde{cl}_j y_j + n_d \widetilde{c}_d + \widetilde{n}_g c_g \quad (1)$$

Subject to:

$$\sum_j x_{jv} = 1 \quad \forall v \in V \quad (2)$$

$$x_{jv} \le y_j \quad \forall j \in J, v \in V \quad (3)$$

$$\sum_{j' \in J} t_{j'v} x_{j'v} \le t_{jv} + M(1 - y_j) \quad \forall j \in J, v \in V \quad (4)$$

$$n_j \ge \sum_v \widetilde{\lambda}_v t_{jv} x_{jv} / \tilde{\rho} \quad \forall j \in J \quad (5)$$

$$n_g = \sum_j \sum_v \widetilde{\lambda}_v t_{jv} x_{jv} \quad (6)$$

$$n_d = \sum_j n_j \quad (7)$$

$$y_j, x_{jv} \in \{0, 1\} \quad \forall j \in J, v \in V \quad (8)$$

$$n_j, n_g, n_d \ge 0 \quad \forall j \in J, v \in V \quad (9)$$

The objective function of the model (1) minimizes the total costs of the system. Constraint (2) assigns each demand node to only one opened facility. Constraint (3) assures that demand is only assigned to opened facilities. The assignment of demands to the closest opened facility is guaranteed by constraint (4). Constraint (5) determines the minimum number of required drones for each facility. Drone utilization is reflected in constraint (6). The total number of drones is specified in Eq. (7). Binary and positive variables are defined in constraints (8) and (9) respectively.

3.2 The Crisp Equivalent of the Fuzzy Model

Due to dealing with fuzzy variables, the model should be converted to its crisp equivalent format. To do so, the concepts and definitions of the applied possibilistic approach are described first.

Definition 1 Let $\tilde{\ell} = (\ell^1, \ell^2, \ell^3, \ell^4)$ be a trapezoidal fuzzy number with the following membership function:

$$\mu_{\tilde{\ell}}(x) = \begin{cases} \frac{x-\ell^1}{\ell^2-\ell^1} & \ell^1 \le x \le \ell^2 \\ 1 & \ell^2 \le x \le \ell^3 \\ \frac{\ell^4-x}{\ell^4-\ell^3} & \ell^3 \le x \le \ell^4 \\ 0 & \text{Otherwise} \end{cases} \tag{10}$$

The expected interval and expected value (EI & EV respectively) of the trapezoidal fuzzy number $\tilde{\ell} = (\ell^1, \ell^2, \ell^3, \ell^4)$ are defined as follows Jiménez et al. [123]:

$$EI\left(\tilde{\ell}\right) = \left[E_1^{\ell}, E_2^{\ell}\right] = \left[\frac{\ell^1 + \ell^2}{2}, \frac{\ell^3 + \ell^4}{2}\right] \tag{11}$$

$$V\left(\tilde{\ell}\right) = \frac{E_1^{\ell} + E_2^{\ell}}{2} = \frac{\ell^1 + \ell^2 + \ell^3 + \ell^4}{4} \tag{12}$$

Definition 2 According to Jiménez [122], for any pair of fuzzy numbers ℓ and Υ, the degree to which ℓ is bigger than Υ is defined as:

$$\mu_M\left(\tilde{\ell}, \tilde{\Upsilon}\right) = \begin{cases} 0 & E_2^{\ell} - E_1^{\Upsilon} < 0 \\ \frac{E_2^{\ell} - E_1^{\Upsilon}}{E_2^{\ell} - E_1^{\Upsilon} - \left(E_1^{\ell} - E_2^{\Upsilon}\right)} & 0 \in \left[E_1^{\ell} - E_2^{\Upsilon}, E_2^{\ell} - E_1^{\Upsilon}\right] \\ 1 & E_1^{\ell} - E_2^{\Upsilon} > 0 \end{cases} \tag{13}$$

For the cases $\mu_M\left(\tilde{\ell}, \mathrm{s}\right) \ge \alpha$, it is said that $\tilde{\ell}$ is bigger than or equal to $\tilde{\Upsilon}$ at least in the degree of α. This relation is represented by $\tilde{\ell} \ge_{\alpha} \tilde{\Upsilon}$.

Definition 3 For a pair of fuzzy numbers like ℓ and Υ, the numbers are equal in degree of α (the feasibility level which is set by the decision-maker) if the following relationship holds Parra et al. [124]:

$$\frac{\alpha}{2} \le \mu_M\left(\tilde{\ell}, \tilde{\Upsilon}\right) \le 1 - \frac{\alpha}{2} \tag{14}$$

Definition 4 If $\updownarrow$ is defined as a linear combination of $\tilde{\Upsilon}^1$ and $\tilde{\Upsilon}^2$, then mathematically it is true that EV (ℓ) = EV (linear combination of $\tilde{\Upsilon}^1$ and $\tilde{\Upsilon}^2$).

In order to illustrate how the possibilistic method converts a fuzzy model to its crisp equivalent, consider the following general mathematical formulation:

$$Min\ \tilde{c}^T x \tag{15}$$

Subject to

$$\widetilde{a}_i x \geq \widetilde{b}_i \quad i = 1, 2, \ldots, l \tag{16}$$

$$\widetilde{a}_i x = \widetilde{b}_i \quad i = l + 1, \ldots, m \tag{17}$$

$$\ell = p_1 \sum \widetilde{\Upsilon}^1 / \widetilde{\Upsilon}^2 \tag{18}$$

$$\mathrm{x} \geq 0 \tag{19}$$

According to Jiménez (1996), vector $x \in \mathbb{R}^n$ will be feasible with α degree if $min\{\mu_M(a_i^{\sim}\,\mathrm{x}, b_i^{\sim}) = \alpha,\ i = 1, \ldots, m\}$. Hence, α is considered as the degree of feasibility for the model which is decided by the decision-maker (DM). A higher value of α corresponds to smaller feasible solution space, and consequently, the optimal solution becomes more conservative but less interesting.

Based on (13) and (14), the relation $a_i^{\sim} x \geq b_i^{\sim}$ and $a_i^{\sim}\mathrm{x} = b_i^{\sim}$ can be expressed as the following:

$$\frac{E_2^{a_i x} - E_1^{b_i}}{E_2^{a_i x} - E_1^{b_i} - \left(E_1^{a_i x} - E_2^{b_i}\right)} \geq \alpha \quad i = 1, 2, \ldots, l \tag{20}$$

$$\frac{\alpha}{2} \leq \frac{E_2^{a_i x} - E_1^{b_i}}{E_2^{a_i x} - E_1^{b_i} - \left(E_1^{a_i x} - E_2^{b_i}\right)} \leq 1 - \frac{\alpha}{2} \quad i = l + 1, \ldots, m \tag{21}$$

Substituting the equivalent equations into the original model implies the following key relations:

$$\mathrm{Min}\left[EV\left(c^{\sim T}\right)\right] \tag{22}$$

Subject to:

$$\left((1-\alpha)E_2^{a_i} + \alpha E_1^{a_i}\right)x \geq \alpha E_2^{b_i} + (1-\alpha)E_1^{b_i} \quad i = 1, 2, \ldots, l \tag{23}$$

$$\left((1 - \frac{\alpha}{2})E_2^{a_i} + \frac{\alpha}{2}E_1^{a_i}\right)x \geq \frac{\alpha}{2}E_2^{b_i} + \left(1 - \frac{\alpha}{2}\right)E_1^{b_i} \quad i = l + 1, \ldots, m \tag{24}$$

$$\left(\frac{\alpha}{2}E_2^{a_i} + \left(1 - \frac{\alpha}{2}\right)E_1^{a_i}\right)x \leq \left(1 - \frac{\alpha}{2}\right)\alpha E_2^{b_i} + \left(\frac{\alpha}{2}\right)E_1^{b_i} \quad i = l + 1, \ldots, m \tag{25}$$

$$EV(\ell) = EV(p_1 \sum \widetilde{\Upsilon}^1 / \widetilde{\Upsilon}^2) \tag{26}$$

$$x \geq 0 \tag{27}$$

Using the mentioned formulations, the crisp equivalent of the fuzzy objective function can be written as the following:

$$Min\sum_{j}\left(\frac{cl_j^1+cl_j^2+cl_j^3+cl_j^4}{4}y_j\right)+\frac{c_d^1+c_d^2+c_d^3+c_d^4}{4}n_d + \frac{n_g^1+n_g^2+n_g^3+n_g^4}{4}c_g \tag{28}$$

Constraints (5) and (6) can be written as:

$$n_j \geq \sum_{(v)}\left(\alpha\left(\frac{\lambda_{vv'}^3+\lambda_{vv'}^4}{2}\right)+(1-\alpha)\left(\frac{\lambda_{vv'}^1+\lambda_{vv'}^2}{2}\right)\right)t_{jv}x_{jv}/\left(\alpha\left(\frac{\rho^3+\rho^4}{2}\right)+(1-\alpha)\left(\frac{\rho^1+\rho^2}{2}\right)\right) \quad \forall j \in J \tag{29}$$

$$n_g \geq \sum_{j}\sum_{(v)}\left(\alpha\left(\frac{\lambda_{vv'}^3+\lambda_{vv'}^4}{2}\right)+(1-\alpha)\left(\frac{\lambda_{vv'}^1+\lambda_{vv'}^2}{2}\right)\right)t_{jv}x_{jv} \tag{30}$$

$$n_g \leq \sum_{j}\sum_{(v)}\left(\alpha\left(\frac{\lambda_{vv'}^1+\lambda_{vv'}^2}{2}\right)+(1-\alpha)\left(\frac{\lambda_{vv'}^3+\lambda_{vv'}^4}{2}\right)\right)t_{jv}x_{jv} \tag{31}$$

4 Solution Method and Results

As Shavarani et al. [115] applied the Genetic Algorithm (GA) to solve their developed model, the same solution method is also used in this study. Inspired by the evolution of species in nature, Holland [125] introduced the GA for the first time and up to now, there are many researches that used GA to solve their Np-hard problems in different fields [126–128]. The process of natural selection starts with the selection of fittest individuals from a population. They produce offspring, which inherit the characteristics of the parents and will be added to the next generation. If parents have better fitness, their offspring will be better than parents and have a better chance at surviving. This process keeps on iterating and at the end, a generation with the fittest individuals will be found. The process begins with a set of individuals which is called a population. Each individual is a solution of the studied problem. An individual is characterized by a set of parameters (variables) known as genes. Genes are joined into a string to form a chromosome (solution). The fitness function determines how fit an individual is. It gives a fitness score to each individual. The probability that an individual will be selected for reproduction is based on its fitness score. The idea of selection phase is to select the fittest individuals and let them pass their genes to the next generation. The next generation is reproduced using the crossover operation in which offspring are created by exchanging the genes of selected parents. Each generated offspring will have all of its properties from its parents. To add new features to such offspring, mutation operation is applied to maintain diversity within the population and prevent premature convergence. In this operations, some genes along the generated offspring are randomly selected and their value is randomly changed. The algorithm terminates if the population has converged such that producing offspring which are significantly different from the previous generation is not possible. See Eiben and Smith [129] for a fuller description.

The steps of the proposed GA in this study, could be summarized as follows:

Step 1. Initialization

The parameters of GA comprising the size of the initial population, the crossover probability, the mutation probability, and the number of iterations that are initialized in this step.

Step 2. Initial population

GA initiates with a population of randomly generated solutions. Each solution is called a chromosome and each item along the length of the chromosome is called a gene. The length of the chromosome in this study is set to be equal to the number of candidate locations. The gene related to each candidate location is a randomly generated binary value. Zero values indicate that the corresponding candidate location is not chosen for establishing the facilities.

Step 3. Evaluation of the fitness value

In order to calculate the fitness value, the total demand assigned to each facility based on the proximity rule mentioned in the mathematical model is calculated. Once the aggregated demand of each facility is determined, the aggregate length of required flying tours to cover all the customers will be determined and used as a basis for calculating the number of drones assigned to each location.

Step 4. Crossover operation

Here, a one-point-cut crossover strategy is applied to mate two randomly selected parents and generated two offspring chromosomes. A randomly selected crossover point is selected along the length of parental chromosomes. Then, by exchanging the second part of each parent, two offspring chromosomes are generated.

Step 5. Mutation operation

In this step, a random number of genes are selected and their related status is changed such that an opened facility turns to be closed and vice versa.

Step 6. Updating the population

After adding the generated offspring, the initial population is sorted based on the fitness value. Then the size of the population is kept at its initial value by keeping the best solutions and discarding the rest.

Step 7. Termination

The algorithm continues up to reaching the pre-determined number of iterations. The pseudo-code of the applied GA is illustrated in Fig. 6.

The parameters of applied GA is set to be equal to the values proposed by Shavarani et al. [115]. The algorithm is applied to randomly generate connected networks of different sizes and specifications. The candidate locations are selected randomly among the network nodes. The fuzzy values are taken directly from the

```
Initialize GA parameters (population size: Npop, crossover probability: Pc, mutation
probability: Pm, number of iterations: MaxIt)
t ← 0
P(t) ← Randomly generated individuals
Eval[P(t)] : Evaluate the fitness value of the individuals
Best ← Best individual
while t < MaxIt do
    S ← Select individuals ∈ [P(t)]
    Q ← Crossover(S) ∪ Mutation (S)
    P(t) ← [P(t) + Q]
    Eval[P(t)]
    Update Best
    P(t + 1) ← Select best individuals ∈ [P(t)]
return Best
```

Fig. 6 The pseudo-code of applied GA

main study. The GA is coded in MATLAB and implemented on an Intel Core i5-3320 M@2.6 GHz laptop with 4 GB RAM. The obtained results are summarized in Table 1.

To show the trends of the objective function values and the execution (CPU) times, in seconds, of the four generated cases that have been reported in Table 1, Fig. 7 and Fig. 8 have been presented. It should be mentioned that cases have been varied based on the two elements of the number of the nodes and number of the candidate locations.

Table 1 The obtained results

#	Number of nodes	Number of candidate locations	Objective	Time
1	1000	20	5268.14	920
2	1500	25	5954.29	1162
3	2000	30	6647.71	1443
4	3000	40	6912.34	1972

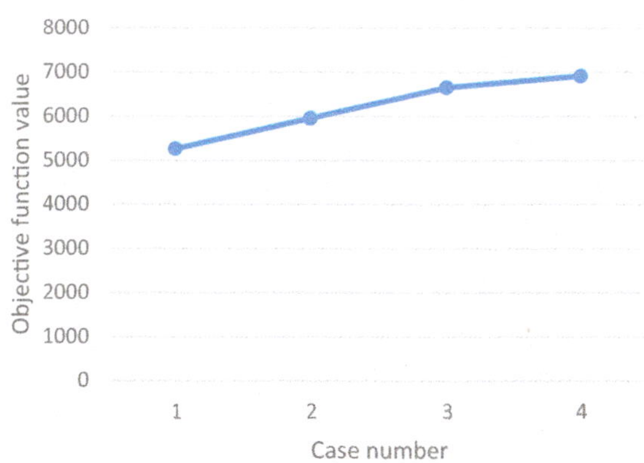

Fig. 7 The objective values of the cases

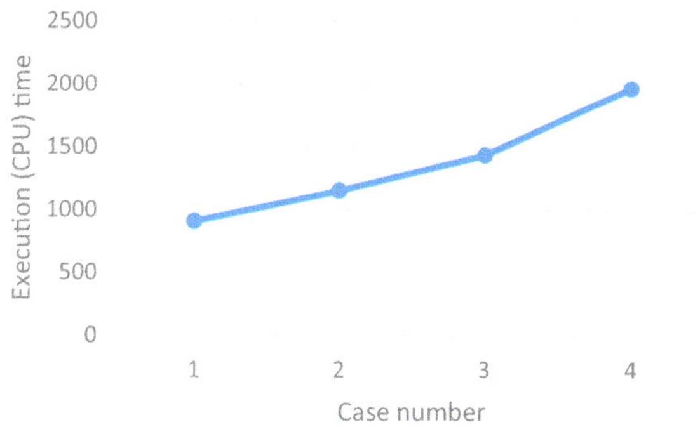

Fig. 8 The CPU time to solve the cases

5 Conclusion

In this study, the authors focused on intelligent and fuzzy UAV transportation applications in aviation 4.0. Various studies from the literature have been visited under different categories of related problem types of delivery operations by using UAVs such as facility location problems, vehicle routing problems, path-planning problems, and their applications. Among many different types of UAV-based delivery system problems and considering the dimensions of this study, a case study under uncertainty condition has been considered. At first, fuzzy mathematical modeling has been developed to address the problem, then, a fuzzy possibilistic-based method has been used to deffuzify the model and be changed to a crisp equivalent model. To solve the problem, a genetic algorithm has been developed and all of its seven steps including Initialization, Initial population, Evaluation of the fitness value, Crossover operation, Mutation operation, Updating the population, and Termination, have been explained in details. The results of some randomly generated cases have been proposed. According to the presented results, for a case of 40 candidate locations and 3000 nodes, the presented solution method can solve such a problem in less than 2000s or let say around half an hour. According to this CPU time for these kinds of Np-hard problems, such a solution method is considered as a very fast one to solve the problem, which Therefore, the presented method seems to be a suitable tool for decision-makers in this field to make an aerial delivery system.

For further studies, there are many topics in this area to be visited by researchers. One of the important ones may be focused on vehicle routing problems for some cased that the payload of UAVs may have different demands, say more than one and limited, to be satisfied on every UAVs' delivery travel.

References

1. Ni, D., Yu, G., Rathinam, S.: Unmanned aircraft system and its applications in transportation. J. Advan. Transp. **2017** (2017)
2. Li, X., et al.: An Aggregate Flow Based Scheduler in Multi-task Cooperated UAVs Network. Chinese Journal of Aeronautics, (2020)
3. McCormack, E.: Exploring transportation applications of small unmanned aircraft. Inst. Transp. Eng. ITE J. **79**(12), 32 (2009)
4. Mosterman, P.J., Zander, J.: Cyber-physical systems challenges: a needs analysis for collaborating embedded software systems. Softw. Syst. Model. **15**(1), 5–16 (2016)
5. Valdés, R.A., et al.: Aviation 4.0: more safety through automation and digitization. In: Aircraft Technology. IntechOpen, (2018)
6. Ernest, N., Sathyan, A., Cohen, K.: Genetic fuzzy single and collaborative tasking for UAV operations. In: Multi-Rotor Platform-based UAV Systems, pp. 217–242. Elsevier, (2020)
7. Zadeh, L.A.: Fuzzy sets. Inf. Control **8**(3), 338–353 (1965)
8. Zadeh, L.A.: Fuzzy sets versus probability. Proc. IEEE **68**(3), 421–421 (1980)
9. Sathyan, A., Ernest, N.D., Cohen, K.: An efficient genetic fuzzy approach to UAV swarm routing. Unmanned Syst. **4**(02), 117–127 (2016)
10. Nisar, T.M., Prabhakar, G.: What factors determine e-satisfaction and consumer spending in e-commerce retailing? J. Retail. Consum. Serv. **39**, 135–144 (2017)
11. James, J., Lam, A.Y.: Autonomous vehicle logistic system: joint routing and charging strategy. IEEE Trans. Intell. Transp. Syst. **19**(7), 2175–2187 (2017)
12. Bidgoli, H.: The Handbook of Technology Management, Supply Chain Management, Marketing and Advertising, and Global Management, vol. 2. Wiley (2010)
13. Whitwam, R.: Amazon Reveals how Prime Air Drone Delivery will Work, still doesn't know How much it will Cost. (2016)
14. Deutsche Post, D.: DHL Parcelcopter Launches Initial Operations for Research Purposes. Press release (2014)
15. Stewart, J.: Google Tests Drone Deliveries in Project Wing Trials. BBC World Service Radio, (2014)
16. Kornatowski, P.M., et al.: Last-centimeter personal drone delivery: field deployment and user interaction. IEEE Robot Autom. Lett. **3**(4), 3813–3820 (2018)
17. Li, S., Chen, L.-H.: Optimization of the VRP with single depot based on vehicle coordination strategy. In: 2010 International Conference on Intelligent Computation Technology and Automation. IEEE, (2010)
18. Lim, A., Wang, F.: Multi-depot vehicle routing problem: a one-stage approach. IEEE Trans. Autom. Sci. Eng. **2**(4), 397–402 (2005)
19. Mester, D., Bräysy, O.: Active-guided evolution strategies for large-scale capacitated vehicle routing problems. Comput. Oper. Res. **34**(10), 2964–2975 (2007)
20. Dethloff, J.: Vehicle routing and reverse logistics: the vehicle routing problem with simultaneous delivery and pick-up. OR-Spektrum **23**(1), 79–96 (2001)
21. Nunkaew, W., Phruksaphanrat, B.: A fuzzy goal programming for a multi-depot distribution problem. In: AIP Conference Proceedings. American Institute of Physics, (2010)
22. Montemanni, R., et al.: Ant colony system for a dynamic vehicle routing problem. J. Comb. Optim. **10**(4), 327–343 (2005)
23. Pillac, V., et al.: A review of dynamic vehicle routing problems. Eur. J. Oper. Res. **225**(1), 1–11 (2013)
24. Sariklis, D., Powell, S.: A heuristic method for the open vehicle routing problem. J. Oper. Res. Soc. **51**(5), 564–573 (2000)
25. Rojas Viloria, D., et al.: Unmanned aerial vehicles/drones in vehicle routing problems: a literature review. Int. Trans. Oper. Res. (2020)
26. Guerriero, F., et al.: A multi-objective approach for unmanned aerial vehicle routing problem with soft time windows constraints. Appl. Math. Model. **38**(3), 839–852 (2014)

27. Zhen, L., et al.: A vehicle routing problem arising in unmanned aerial monitoring. Comput. Oper. Res. **105**, 1–11 (2019)
28. Nejad, M.G., et al.: An optimization model for cyclic scheduling problem in flexible robotic cells. Int. J. Advan. Manufact. Technol. **95**(9–12), 3863–3873 (2018)
29. Dhein, G., et al.: Minimizing dispersion in multiple drone routing. Comput. Oper. Res. **109**, 28–42 (2019)
30. Vizvári, B., Guden, H., Nejad, M.G.: Local search based meta-heuristic algorithms for optimizing the cyclic flexible manufacturing cell problem. Ann. Optim. Theory Pract. **1**(3), 15–32 (2018)
31. Alvear, O., et al.: Using UAV-based systems to monitor air pollution in areas with poor accessibility. J. Advan. Transp. **2017** (2017)
32. Mersheeva, V., Friedrich, G.: Routing for continuous monitoring by multiple micro UAVs in disaster scenarios. In: Proceedings of the 20th European Conference on Artificial Intelligence. IOS Press (2012)
33. Peng, K., et al.: A hybrid genetic algorithm on routing and scheduling for vehicle-assisted multi-drone parcel delivery. IEEE Access **7**, 49191–49200 (2019)
34. Khan, A., Aftab, F., Zhang, Z.: BICSF: bio-inspired clustering scheme for FANETs. IEEE Access **7**, 31446–31456 (2019)
35. Erdoğan, S., Miller-Hooks, E.: A green vehicle routing problem. Transp. Res. Part E: Logistics Transp. Rev. **48**(1), 100–114 (2012)
36. Lin, C., et al.: Survey of green vehicle routing problem: past and future trends. Expert Syst. Appl. **41**(4), 1118–1138 (2014)
37. Schneider, M., Stenger, A., Goeke, D.: The electric vehicle-routing problem with time windows and recharging stations. Transp. Sci. **48**(4), 500–520 (2014)
38. Hiermann, G., et al.: The electric fleet size and mix vehicle routing problem with time windows and recharging stations. Eur. J. Oper. Res. **252**(3), 995–1018 (2016)
39. Dorling, K., et al.: Vehicle routing problems for drone delivery. IEEE Trans. Syst. Man, Cybern. Syst. **47**(1), 70–85 (2016)
40. Azi, N., Gendreau, M., Potvin, J.-Y.: An adaptive large neighborhood search for a vehicle routing problem with multiple routes. Comput. Oper. Res. **41**, 167–173 (2014)
41. Hernandez, F., et al.: Branch-and-price algorithms for the solution of the multi-trip vehicle routing problem with time windows. Eur. J. Oper. Res. **249**(2), 551–559 (2016)
42. Cheikh, M., et al.: A variable neighborhood search algorithm for the vehicle routing problem with multiple trips. Electron. Notes Discret. Math. **47**, 277–284 (2015)
43. Cattaruzza, D., et al.: A memetic algorithm for the multi trip vehicle routing problem. Eur. J. Oper. Res. **236**(3), 833–848 (2014)
44. Coelho, B.N., et al.: A multi-objective green UAV routing problem. Comput. Oper. Res. **88**, 306–315 (2017)
45. French, S.: 6 Myths about Amazon Prime Air and Drone Delivery, Debunked. MarketWatch, December **2**
46. Welch, A.: A cost-benefit analysis of Amazon prime air. (2015)
47. Wohlsen, M.: The next big thing you missed: Amazon? s delivery drones could work–they just need trucks. Wired.com (2014)
48. Yurek, E.E., Ozmutlu, H.C.: A decomposition-based iterative optimization algorithm for traveling salesman problem with drone. Transp. Res. Part C: Emerg. Technol. **91**, 249–262 (2018)
49. Campbell, J.F., Sweeney, D., Zhang, J.: Strategic design for delivery with trucks and drones. Supply Chain Analytics Report SCMA (04 2017), (2017)
50. Wang, X., Poikonen, S., Golden, B.: The vehicle routing problem with drones: several worst-case results. Optim. Lett. **11**(4), 679–697 (2017)
51. Agatz, N., Bouman, P., Schmidt, M.: Optimization approaches for the traveling salesman problem with drone. Transp. Sci. **52**(4), 965–981 (2018)
52. Yoon, J.J.: The traveling salesman problem with multiple drones: an optimization model for last-mile delivery. Massachusetts Ins. Technol. (2018)

53. Moshref-Javadi, M., Hemmati, A., Winkenbach, M.: A truck and drones model for last-mile delivery: a mathematical model and heuristic approach. Appl. Math. Model. **80**, 290–318 (2020)
54. Madani, B., Ndiaye, M.: Autonomous vehicles delivery systems classification: introducing a TSP with a moving depot. In 2019 8th International Conference on Modeling Simulation and Applied Optimization (ICMSAO). IEEE, (2019)
55. Moshref-Javadi, M., Lee, S., Winkenbach, M.: Design and evaluation of a multi-trip delivery model with truck and drones. Transp. Res. Part E: Logistics Transp. Rev. **136**, 101887 (2020)
56. Jeong, H.Y., Song, B.D., Lee, S.: Truck-drone hybrid delivery routing: payload-energy dependency and no-fly zones. Int. J. Prod. Econ. **214**, 220–233 (2019)
57. Schermer, D., Moeini, M., Wendt, O.: A hybrid VNS/Tabu search algorithm for solving the vehicle routing problem with drones and en route operations. Comput. Oper. Res. **109**, 134–158 (2019)
58. Schermer, D., Moeini, M., Wendt, O.: A matheuristic for the vehicle routing problem with drones and its variants. Transp. Res. Part C: Emer. Technol. **106**, 166–204 (2019)
59. Sacramento, D., Pisinger, D., Ropke, S.: An adaptive large neighborhood search metaheuristic for the vehicle routing problem with drones. Transp. Res. Part C: Emer. Technol. **102**, 289–315 (2019)
60. Schermer, D., Moeini, M., Wendt, O.: Algorithms for solving the vehicle routing problem with drones. In: Asian Conference on Intelligent Information and Database Systems. Springer, (2018)
61. McCunney, B., Cauwenberghe, K.: Simulation Test Bed for Drone-Supported Logistics Systems. (2019)
62. Carlsson, J.G., Song, S.: Coordinated logistics with a truck and a drone. Manage. Sci. **64**(9), 4052–4069 (2018)
63. Chang, Y.S., Lee, H.J.: Optimal delivery routing with wider drone-delivery areas along a shorter truck-route. Expert Syst. Appl. **104**, 307–317 (2018)
64. Luo, Z., Liu, Z., Shi, J.: A two-echelon cooperated routing problem for a ground vehicle and its carried unmanned aerial vehicle. Sensors **17**(5), 1144 (2017)
65. Leung, S.C., et al.: Extended guided tabu search and a new packing algorithm for the two-dimensional loading vehicle routing problem. Comput. Oper. Res. **38**(1), 205–215 (2011)
66. Karak, A., Abdelghany, K.: The hybrid vehicle-drone routing problem for pick-up and delivery services. Transp. Res. Part C: Emer. Technol. **102**, 427–449 (2019)
67. Clarke, G., Wright, J.W.: Scheduling of vehicles from a central depot to a number of delivery points. Oper. Res. **12**(4), 568–581 (1964)
68. Ham, A.M.: Integrated scheduling of m-truck, m-drone, and m-depot constrained by time-window, drop-pickup, and m-visit using constraint programming. Transp. Res. Part C: Emer. Technol. **91**, 1–14 (2018)
69. Bekhti, M., et al.: Path planning of unmanned aerial vehicles with terrestrial wireless network tracking. In: 2016 Wireless Days (WD). IEEE, (2016)
70. Nikolos, I.K., et al.: Evolutionary algorithm based offline/online path planner for UAV navigation. IEEE Trans. Syst., Man, Cybern. Part B (Cybern) **33**(6), 898–912 (2003)
71. Peng, X., Xu, D.: Intelligent online path planning for UAVs in adversarial environments. Int. J. Adv. Rob. Syst. **9**(1), 3 (2012)
72. Kavraki, L.E., et al.: Probabilistic roadmaps for path planning in high-dimensional configuration spaces. IEEE Trans. Rob. Autom. **12**(4), 566–580 (1996)
73. LaValle, S.M., Kuffner, J.J., Jr.: Randomized kinodynamic planning. Int. J. Rob. Res. **20**(5), 378–400 (2001)
74. Zhao, Y., Zheng, Z., Liu, Y.: Survey on computational-intelligence-based UAV path planning. Knowl.-Based Syst. **158**, 54–64 (2018)
75. Nejad, M.G., Kashan, A.H., Shavarani, S.M.: A novel competitive hybrid approach based on grouping evolution strategy algorithm for solving U-shaped assembly line balancing problems. Prod. Eng. Res. Devel. **12**(5), 555–566 (2018)

76. Nejad, M.G., Kashan, A.H.: An effective grouping evolution strategy algorithm enhanced with heuristic methods for assembly line balancing problem. J. Advan. Manuf. Syst. **18**(03), 487–509 (2019)
77. Vatankhah Barenji, R., Ghadiri Nejad, M., Asghari, I. Optimally sized design of a wind/photovoltaic/fuel cell off-grid hybrid energy system by modified-gray wolf optimization algorithm. Energy Environ. **29**(6), 1053–1070 (2018)
78. Mosallaeipour, S., et al.: Mobile robot scheduling for cycle time optimization in flow-shop cells, a case study. Prod. Eng. Res. Devel. **12**(1), 83–94 (2018)
79. Ghadiri Nejad, M., et al.: A mathematical model and simulated annealing algorithm for solving the cyclic scheduling problem of a flexible robotic cell. Advan. Mech. Eng. **10**(1), 1687814017753912 (2018)
80. Pandey, P., Shukla, A., Tiwari, R.: Aerial path planning using meta-heuristics: a survey. In: 2017 Second International Conference on Electrical, Computer and Communication Technologies (ICECCT). IEEE, (2017)
81. Kok, K.Y., Rajendran, P.: Differential-evolution control parameter optimization for unmanned aerial vehicle path planning. PLoS ONE **11**(3), e0150558 (2016)
82. Bandeira, T.W., et al.: Analysis of path planning algorithms based on travelling salesman problem embedded in UAVs. In: 2015 Brazilian Symposium on Computing Systems Engineering (SBESC). IEEE, (2015)
83. Di Franco, C., Buttazzo, G.: Coverage path planning for UAVs photogrammetry with energy and resolution constraints. J. Intell. Rob. Syst. **83**(3–4), 445–462 (2016)
84. Ahmed, S., et al.: Energy efficient path planning techniques for UAV-based systems with space discretization. In: 2016 IEEE Wireless Communications and Networking Conference. IEEE. (2016)
85. Gautam, S.A., Verma, N.: Path planning for unmanned aerial vehicle based on genetic algorithm & artificial neural network in 3D. In: 2014 International Conference on Data Mining and Intelligent Computing (ICDMIC). IEEE, (2014)
86. Mardani, A., Chiaberge, M., Giaccone, P.: Communication-aware UAV path planning . IEEE Access **7**, 52609–52621 (2019)
87. Niu, L., Zhuo, G.: An improved real 3D A* algorithm for difficult path finding situation. Proceeding of the International Archives of the Photogrammetry, Remote Sensing and Spatial Information Sciences, p. 37. (2008)
88. Golabi, M., et al.: Bypassing or flying above the obstacles? A novel multi-objective UAV path planning problem. arXiv preprint arXiv:2004.08279, (2020)
89. Golabi, M., Izbirak, G., Arkat, J.**:** Multiple-server facility location problem with stochastic demands along the network edges. J. Eng. Res. **6**(4) (2018)
90. Ghadiri Nejad, M., Banar, M.: Emergency response time minimization by incorporating ground and aerial transportation. Ann. Optim. Theor. Pract. **1**(1), 43–57 (2018)
91. Klose, A., Drexl, A.: Facility location models for distribution system design. Eur. J. Oper. Res. **162**(1), 4–29 (2005)
92. Hale, T.S., Moberg, C.R.: Location science research: a review. Ann. Oper. Res. **123**(1–4), 21–35 (2003)
93. Arabani, A.B., Farahani, R.Z.: Facility location dynamics: an overview of classifications and applications. Comput. Ind. Eng. **62**(1), 408–420 (2012)
94. Daskin, M.S., Maass, K.L.: The p-median problem. In: Location Science. pp. 21–45, Springer, (2015)
95. Jamshidi, M., Median location problem. In: Facility Location. pp. 177–191. Springer, (2009)
96. Mladenović, N., et al.: The p-median problem: a survey of metaheuristic approaches. Eur. J. Oper. Res. **179**(3), 927–939 (2007)
97. Berman, O., Krass, D.: The generalized maximal covering location problem. Comput. Oper. Res. **29**(6), 563–581 (2002)
98. Church, R., ReVelle, C.: The maximal covering location problem. In: Papers of the Regional Science Association. Springer (1974)

99. Garfinkel, R., Neebe, A., Rao, M.: The m-center problem: minimax facility location. Manage. Sci. **23**(10), 1133–1142 (1977)
100. Tamir, A.: The k-centrum multi-facility location problem. Discrete Appl. Math. **109**(3), 293–307 (2001)
101. Ghadirinejad, M., et al.: A stochastic model for the ethanol pharmacokinetics. Iran. J. Public Health **45**(9), 1170 (2016)
102. Bieniek, M.: A note on the facility location problem with stochastic demands. Omega **55**, 53–60 (2015)
103. Snyder, L.V.: Facility location under uncertainty: a review. IIE Trans. **38**(7), 547–564 (2006)
104. Melkote, S., Daskin, M.S.: Capacitated facility location/network design problems. Eur. J. Oper. Res. **129**(3), 481–495 (2001)
105. Bespamyatnikh, S., et al.: Mobile facility location. In: Proceedings of the 4th International Workshop on Discrete Algorithms and Methods for Mobile Computing and Communications. (2000)
106. Raghavan, S., Sahin, M., Salman, F.S.: The capacitated mobile facility location problem. Eur. J. Oper. Res. **277**(2), 507–520 (2019)
107. Hong, I., Kuby, M., Murray, A.T.: A range-restricted recharging station coverage model for drone delivery service planning. Transp. Res. Part C: Emerg. Technol. **90**, 198–212 (2018)
108. Boujelben, M.K., Gicquel, C.: Efficient solution approaches for locating electric vehicle fast charging stations under driving range uncertainty. Comput. Oper. Res. **109**, 288–299 (2019)
109. Chauhan, D., Unnikrishnan, A., Figliozzi, M.: Maximum coverage capacitated facility location problem with range constrained drones. Transp. Res. Part C: Emerg. Technol. **99**, 1–18 (2019)
110. Pulver, A., Wei, R.: Optimizing the spatial location of medical drones. Appl. Geogr. **90**, 9–16 (2018)
111. Kim, D., Lee, K., Moon, I.: Stochastic facility location model for drones considering uncertain flight distance. Ann. Oper. Res. **283**(1), 1283–1302 (2019)
112. Golabi, M., Shavarani, S.M., Izbirak, G.: An edge-based stochastic facility location problem in UAV-supported humanitarian relief logistics: a case study of Tehran earthquake. Nat. Hazards **87**(3), 1545–1565 (2017)
113. Shavarani, S.M., Golabi, M., Izbirak, G.: A capacitated biobjective location problem with uniformly distributed demands in the UAV—supported delivery operation. Int. Trans. Oper. Res. (2019)
114. Shavarani, S.M., et al.: Application of hierarchical facility location problem for optimization of a drone delivery system: a case study of Amazon prime air in the city of San Francisco. Int. J. Advan. Manufact. Technol. **95**(9–12), 3141–3153 (2018)
115. Shavarani, S.M., et al.: A congested capacitated multi-level fuzzy facility location problem: an efficient drone delivery system. Comput. Oper. Res. **108**, 57–68 (2019)
116. Kelner, J.M., Ziółkowski, C.: Doppler effect-based automatic landing procedure for UAV in difficult access environments. J. Advan. Transp. **2017** (2017)
117. Yang, Y., et al.: Robust optimization for integrated scrap steel charge considering uncertain metal elements concentrations and production scheduling under time-of-use electricity tariff. J. Cleaner Prod. **176**, 800–812 (2018)
118. Chen, P., et al.: Surrogate safety analysis of pedestrian-vehicle conflict at intersections using unmanned aerial vehicle videos. J. Advan. Transp. **2017** (2017)
119. Xu, Y., et al.: Car detection from low-altitude UAV imagery with the faster R-CNN. J. Advan. Transp. **2017** (2017)
120. Soyster, A.L.: Convex programming with set-inclusive constraints and applications to inexact linear programming. Oper. Res. **21**(5), 1154–1157 (1973)
121. Bertsimas, D., Sim, M.: The price of robustness. Oper. Res. **52**(1), 35–53 (2004)
122. Jiménez, M.: Ranking fuzzy numbers through the comparison of its expected intervals. Int. J. Uncertainty, Fuzziness Knowl.-Based Syst. **4**(04), 379–388 (1996)
123. Jiménez, M., et al.: Linear programming with fuzzy parameters: an interactive method resolution. Eur. J. Oper. Res. **177**(3), 1599–1609 (2007)

124. Parra, M.A., et al.: Solving a multiobjective possibilistic problem through compromise programming. Eur. J. Oper. Res. **164**(3), 748–759 (2005)
125. Holland, J.H.: Genetic algorithms and the optimal allocation of trials. SIAM J. Comput. **2**(2), 88–105 (1973)
126. Nejad, M.G., et al.: Process sequencing for a pick-and-place robot in a real-life flexible robotic cell. Int. J. Advan. Manuf. Technol. **103**(9–12), 3613–3627 (2019)
127. Nejad, M.G., Güden, H., Vizvári, B.: Time minimization in flexible robotic cells considering intermediate input buffers: a comparative study of three well-known problems. Int. J. Comput. Integr. Manuf. **32**(8), 809–819 (2019)
128. Nejad, M.G., et al.: Trade-off between process scheduling and production cost in cyclic flexible robotic cells. Int. J. Advan. Manuf. Technol. **96**(1–4), 1081–1091 (2018)
129. Eiben, A.E., Smith, J.E.: Introduction to Evolutionary Computing. Springer, (2015)

Spherical Fuzzy Inference Systems (S-FIS) to Control UAVs' Communication Technologies

Yaser Donyatalab and Fariba Farid

Abstract Unmanned aerial vehicles (UAVs) have significant applications and will undoubtedly cause many business developments and flourishing in various fields. This chapter considers the UAVs communication problem and provides a trustworthy solution for selecting the appropriate communication technology and controlling it. So, we introduce the novel spherical fuzzy inference system (S-FIS) with S-FIS properties. The communication technology of UAVs contains numerous parameters, and these parameters include uncertainty information. This chapter consider three types of parameters related to UAVs' communication networks and performance. These parameters are viewed under a spherical fuzzy environment and, we evaluate the most critical detectors to indicate which communication technology is the most appropriate. The proposed S-FIS is utilized to select the optimal communication technology among four types of communication technologies by using MATLAB Fuzzy Logic ToolboxTM.

Keywords Spherical fuzzy sets · Spherical fuzzy inference system (S-FIS) · UAVs communication technology

1 Introduction

Unmanned aerial vehicles (UAVs), also known as Unmanned aircraft systems (UASs) or drones, can be piloted or controlled from the ground. UAVs have significant attention in the last decade due to the incredible ability to control without a pilot on board. It also leads to catch high maneuverability and low cost with high performance [1]. Many applications of UAVs emerged, such as surveillance, package

Y. Donyatalab (✉) · F. Farid
Industrial Engineering Department, Istanbul Technical University, Macka, Istanbul, Turkey
e-mail: donyatalab18@itu.edu.tr

F. Farid
e-mail: farid20@itu.edu.tr

C. Kahraman and S. Aydın (eds.), *Intelligent and Fuzzy Techniques in Aviation 4.0*,
Studies in Systems, Decision and Control 372,
https://doi.org/10.1007/978-3-030-75067-1_20

delivery, personal business aerial photography, video streaming, prehospital emergency care (PEC), search and rescue operations, communication and telecommunication enhancement, law enforcement, traffic control, and precision agriculture scientific explorations [2–6]. UAVs' applications have increased by assisting the wireless network [7] due to the improved flexibility, efficiency, and reliability of drones. Many wireless network connections are proposed for UAVs based on wireless communications, which each proposed technology of communications for UAVs has its features [8]. There will be appeared a list of advantages and disadvantages for each kind of technical communication.

Moreover, it is possible to have a hybrid or multi communication technology UAV, which means that UAV is capable of two or more communication technologies together. However, it does not even solve the problem; in contrast, some other issues lie with this type of UAV. The most appropriate communications technology for different situations is continuously the main topic of the appeared issue here. Nevertheless, it is considered a real scenario-based problem for such UAVs, so the uncertainty behind such cases is undoubtedly a significant parameter in determining the proper and optimal answer.

Fuzzy set theory, which Zadeh proposed in 1965 [9], has appeared as one of the most persuasive techniques to demonstrate the vagueness of the real scenario-based cases. Defining the linguistic variables in fuzzy logic [10] extended the applicability of fuzzy in different problems that could directly carry the human brains' ability to consider real-world cases' uncertainty to those cases' mathematical nature [11]. Fuzzy inference system (FIS) is one of those linguistic-based applications of fuzzy logic that gained lots of attention from its first year proposed by Mamdani [12] to last years. This much attention in various applications shows the FIS models' high ability. Since imprecise parameters that are undeniable in all problems are treated as approximate values instead of definite ones, the results are much trustable [13]. In addition, it is considered to mention that the revolution of fuzzy sets and introducing many extensions of fuzzy sets increased the applicability of fuzzy logic theory [14]. The content of the fuzzy extensions will be discussed in detail in the literature review part. Some of those essential extensions are ordinary fuzzy sets, type-2 fuzzy sets, interval-valued fuzzy sets, intuitionistic fuzzy sets, hesitant fuzzy sets, Pythagorean fuzzy sets, picture fuzzy sets, q-rung ortho pair fuzzy sets, and spherical fuzzy sets.

Motivated by the above discussion, we proposed a fuzzy inference system based on the Mamdani method in a spherical fuzzy environment to solve communication technology selection problems with UAVs. In this problem, a limited time issue lies in many parameters that include uncertainty of the real scenarios. So, a dynamic decision-making model is needed in a moment that can handle the complexities of UAVs. So, we proposed a novel Spherical Fuzzy Inference System (S-FIS) based on the Mamdani model to handle the complexities of the problem and make the appropriate decision based on IF–THEN rules taken from the expert system. At the same time, our proposed model is dynamically selecting UAVs' communication technology and controlling it continuously during the mission period. The rest of the chapter is designed as follows: In Sect. 2, a critical review of fuzzy logic and inference systems is given and discussed different aspects of UAVs. The methodology is divided into two main parts: preliminaries of FIS and spherical fuzzy sets (SFSs) in Sect. 3. In Sect. 3, the proposed novel spherical fuzzy inference system (S-FIS) is introduced. In Sect. 4, the case study about the UAVs is given in detail. The proposed S-FIS is then developed by MATLAB Fuzzy Logic ToolboxTM to control UAV's communication technology dynamically. In Sect. 5, the results are summarized with discussions, and finally, the chapter is concluded in Sect. 5.

2 Literature Review

2.1 Fuzzy Logic and Fuzzy Inference Systems (FIS)

Fuzzy logic, introduced by Zadeh in 1965 [9], originally is a type of multivalued logic, a branch of propositional calculus replacing "true or false" with a broader range of values. Since fuzzy sets can state propositions in more than two values, they directly state uncertainty about elements. Fuzzy sets are trendy in different branches of science due to the appreciated advantages. An apparent uncertainty in human life stems from humanity's brain in analyzing activities in a linguistic base. This approach triggers a complicated system for affirmation. The scientists suggest that the best solution for problems is to simplify them. But simplifying a problem by splitting it into smaller components causes denying some parts of the fact that the problem lies in [15]. This kind of simplification was the pivotal issue of the first generation of fuzzy sets, i.e., ordinary fuzzy sets [9]. Applying fuzzy logic and using its advantages in modeling complex systems growths, first by the increased complexity of the systems, together with a lack of enough numerical data for complex systems [16]. However, the ordinary fuzzy sets could state some complexities of uncertainties but not practically [17]. To solve this inadequacy, Zadeh introduced type 2 fuzzy sets with intervals [10] in 1975. Type-2 fuzzy sets have been investigated by Dubois and Prade [18] and Mizumoto and Tanaka [19, 20]. Because of the intervals defined for the memberships, these new sets are called interval-valued fuzzy sets (IVFS). In parallel with Zadeh, Sambuc [21], Jahn [22], and Grattan-Guinness [23] have introduced IVFS in different papers. Atanassov extended the concept of uncertainty by introducing intuitionistic fuzzy sets (IFSs) and by adding a non-membership degree [24]. IFSs not only describe this fact that how much it is possible for an element to be in a fuzzy set by a membership degree, but also declare how much it is not possible to have that element in the fuzzy set by a non-membership degree.

The inference is defined as a process or mechanism that makes us capable of deriving logical results and achievements from premise knowledge and assumptions [25] or, in other words, a process of reaching new knowledge from existing knowledge. A Fuzzy inference system (FIS) is a reasoning framework-based fuzzy logic that can handle the real cases and problems that lie in the uncertainty of the real world [26]. FIS is very practical and, at the same time, efficient to be used in fields that there exists no exact model for the system's behavior [26]. For instance, we could mention to expert systems [27, 28], control systems [29, 30] and decision support systems [31]. Meanwhile, decision support system (DSS) are generally progressing in usual pace, but it is flourished by different artificial intelligence (AI) methodologies together with distributed information system technology [32, 33]. Also, we could refer to various applications of FIS based on different authors as [34–37].

2.2 *Fuzzy Sets Evolution and Spherical Fuzzy Sets (SFSs)*

Fuzzy set is one of the most fundamental and influential tools in modeling human perceptions. The fuzzy set theory, developed by Zadeh [9], has an important impact on the progress of the decision-making problems in an uncertain environment [38]. Zadeh [9] developed fuzzy sets with notions of inclusion, union, intersection, complement, relation, convexity, and linguistic hedges as a class of objects with a continuum of grades of membership. The ordinary fuzzy sets have been extended to many new different types of fuzzy sets [9, 39]. The last and the most variety of fuzzy sets is Spherical Fuzzy Sets (SFSs), which have a realistic base of data type [38]. The other extensions of ordinary fuzzy sets are as follows: Type-2 fuzzy sets [10], Interval-valued fuzzy sets [10, 40], Intuitionistic fuzzy sets [24], fuzzy multisets [41], non-stationary fuzzy sets [17], hesitant fuzzy sets [42], Pythagorean fuzzy sets [43], picture fuzzy sets [44], q-rung Ortho pair fuzzy sets [45]. Kahraman and Kutlu Gündoğdu in 2018 introduced spherical fuzzy sets as an extension of picture fuzzy sets and Pythagorean fuzzy sets [46, 47]. The basic idea behind SFSs is to allow evaluators to generalize other extensions of fuzzy sets by defining a membership function on a spherical surface and independently assign the parameters of that function with a larger domain. Sequentially, Ashraf and Abdullah [81] proposed spherical fuzzy sets with some operational rules and aggregation operations based on Archimedean t-norm and t-conorms [48]. In Fig. 1, the gradual evolution of the fuzzy sets is illustrated.

2.3 *Unmanned Aerial Vehicles (UAVs)*

In recent years, the evolution of a promising technology as the Unmanned aerial vehicles (UAVs) has changed the global UAV industry. UAVs, commonly known as drones, are aircraft piloted by remote control or embedded computer programs without human onboard [49]. Historically, UAVs were mainly used in military applications deployed in hostile territory for remote surveillance and armed attack to reduce the pilot losses [49]. In November 2017, FAA further launched a national program in Washington, namely the "Drone Integration Pilot Program," to explore the expanded use of drones, including beyond-visual-line-of sight (BVLoS) flights, night-time operations, and flights above people [50]. Additionally, it is anticipated that UAVs offer highly productive business opportunities in the future. With the

Fig. 1 Extensions of fuzzy sets

emergence of new opportunities in a broad UAV range, many different types of valuable research have appeared recently. Several magazines, such as IEEE Wireless Community [1, 51–53], some surveys as [54] and some articles as [55, 56] on wireless communications and networks with UAVs have studied. Among various survey articles, Khawaja et al. [55], Khuwaja et al. [57] and Matolak and Sun [58] focused on air-ground channel models and experimental measurement results of UAV communications. In some survey-based articles like [54, 59, 60], researchers mainly address ad hoc networks for UAV communications by focusing on UAV-UAV communications. Hayat et al. [61] and Shakhatreh et al. [62] summarized the unmanned aerial vehicle networks for civil applications regarding communications viewpoint applications. The UAV-enabled IoT services are overviewed, focusing on data collection, delivery, and processing on Hossein Motlagh et al. [63]. Shakeri et al. in [64] summarized the design challenges of Multi-UAV Systems in cyber-physical applications. Recent studies [6, 56] provide UAV application surveys in cellular networks, focusing on academic literature and industry activities, respectively.

In modern technologies, the applications of UAV communication appeared as 5G. Zhang et al. [65] modeled a hybrid beamforming scheme for 5G and beyond cellular mobile communications, which anticipated an increment impact on UAV communication. The advent of new UAV communication technologies such as 5G creates more resilient, reliable, and robust communication networks [66]. Nowadays, the revolution of fifth-generation and beyond fifth-generation (5G/B5G) bring about the accomplishment of wireless networks dream of achieving a large number of goals in terms of higher coverage and connectivity, ultra-reliable low latency communication (URLLC), support for a massive number of devices via machine type communication (MTC), greater bandwidth and throughput [67, 68].

Practically, there are different types of UAVs based on broad applications of them. UAV classification is used in helping differentiate kinds of existing systems, and regulatory has importance. Further requirements may be imposed on different UAV categories based on their characteristics. According to other criteria, there is no single standard for UAV classification, and UAVs can be assorted into different types [49]. There has a large number of metrics for UAVs classification such as mean take-off weight (MTOW), size, operating conditions, functionality, capabilities, endurance, wing configuration, control methods, cruising range, flying altitude, maximum speed, and energy supplying methods or any combination of these and other characteristics. It should be pointed out that while some of these criteria have the least impact on the requirements of the systems' safety performance, they still take great interest in operational, commercial, legal, and possibly other points of view [2, 69].

Nowadays, the enthusiasm for using UAVs in civilian and commercial applications has increased due to UAVs' manufacturing technologies' advancement and their reducing cost, making them more easily accessible to the public. UAVs have found numerous applications in a proliferation of fields, such as aerial inspection, photography, precision agriculture, traffic control, search and rescue, package delivery, and telecommunications [49].

The most critical technology of UAS (unmanned aircraft systems) is wireless communication. On the one hand, UAVs need to exchange safety–critical information with various parties, such as remote pilots, nearby aerial vehicles, and air traffic controllers, to ensure safe, reliable, and efficient flight operation. This technology is commonly known as control and non-payload communication (CNPC) [7]. On the other hand, depending on their missions, UAVs may need to timely transmit and/or receive mission-related data, such as aerial image, high-speed video, and data packets for relaying, to/from various ground entities, such as UAV operators, end-users, or ground gateways. This technology is known as payload communication. Enabling reliable and secure CNPC links is necessary for the large-scale deployment and wide usage of UAVs [70]. Four candidate communication technologies are communication technology, satellite communication technology, ad hoc network communication technology, and cellular network communication technology (5G network) compared to meet both the CNPC and payload communication requirements in various UAV applications. It provides a need to achieve seamless connectivity and high reliability/throughput for both air-to-air and air-to-ground wireless communications in the 3D space [49].

1. Direct link communication technology:

The characteristics of its simplicity and low-cost cause the direct-link communication between UAVs and its associated ground node over the unlicensed band was most commonly used for commercial UAVs in the past, where the ground node can be a joystick, remote controller, or ground station. However, due to the limitations such as vulnerable to interference, limited range, and low data rate, simple direct-link communication cannot be a scalable solution for supporting the large-scale deployment of UAVs in the future [49].

2. Satellite communication technology:

This technology underpins UAV communications due to its global coverage. Immensely, satellites can help transfer data transmitted between widely separated UAVs and ground gateways, particularly useful for UAVs above the ocean and remote areas where the terrestrial network (Wi-Fi or cellular) coverage is unavailable. Furthermore, satellite signals can also be used for navigation and localization of UAVs. However, several disadvantages of satellite-enabled UAV communications are costly, heavy/bulky/energy consuming communication equipment, high latency, and large-signal attenuation [49].

3. Ad Hoc network communication technology:

Mobile ad hoc network (MANET) is an infrastructure-free and dynamically self-organizing network for enabling peer-to-peer communications among mobile devices, such as laptops, cellphones, and walkie-talkies. This technology has some limitations: extra cost, low spectrum efficiency, intermittent connectivity, and complex routing protocol, to name but a few [49].

4. Cellular network communication technology (5G network):

It is an undeniable fact that the three technologies mentioned above generally cannot cost-effectively support large-scale UAV communications. For this reason, there has been significantly growing interested recently in leveraging the existing as well as future generation cellular networks for enabling UAV-ground communications [51]. On account of the almost ubiquitous coverage of the cellular network worldwide and its high-speed optical backhaul and advanced communication technologies, CNPC and payload communication requirements for UAVs can be potentially met, regardless of the density of UAVs considering their distances from the corresponding ground nodes [49].

3 Methodology

In this section, we will first explain the computational aspects of our study and give a definition of SFSs and mathematical expressions of SFSs operations; then, we will talk about the fuzzy inference system (FIS). In the following of this section, we will show the technical aspects of our proposed methodology and demonstrate the given theorems. The concept of SFSs, Which is introduced by Gündoğdu and Kahraman in [71], is the extended and upper generation of all fuzzy sets which introduced before in the literature part. The SFSs significant advantage is that it will prepare a decision-maker to have a larger domain expressing their ideas about uncertainty [72]. Another privilege of SFSs is that can have an independently stated hesitancy degree [73].

3.1 Preliminaries

In this section, we will give the main definitions of FIS, such as Mamdani, Takagi–Sugeno (TSK), and Tsukamoto fuzzy inference systems.

Fuzzy Inference Systems

A Fuzzy Inference System (FIS) is a framework based on fuzzy logic and fuzzy set theory, mainly concentrating on fuzzy reasoning to prepare computation bases for solving most real case problems [74]. Therefore, a FIS is a reasoning mechanism that uses Fuzzy Production Rules (FRP), which are known as if–then rules to handle the knowledge of experts [75]. So, every FIS could be concluded using the experts' knowledge and experience to prepare a rule base and then obtain appropriate controlling actions [76]. Linguistic variables define the rule base, and controlling actions are based on output variables of FIS. Fuzzy logic also causes approximate interpolation among those input and out variables in a FIS [16]. There are three types of fuzzy inference systems (FIS) as following: 1. Mamdani model [12], 2. Takagi–Sugeno (TSK) model [30, 77] and 3. Tsukamoto model [78].

Mamdani fuzzy inference system:

The Mamdani model [12] first proposed a model as an inference system in controlling the steam engine and boiler combination, which consists of linguistic rules to control the whole procedure by getting human experiences. In this first proposed, two FIS were used to control the entrance heat to the boiler, and the other one was opening the accelerator pedal, which all inside the engine cylinder, respectively [74]. Figure 2 shows the Mamdani fuzzy inference system to have a better understanding of this model.

The classical Mamdani FIS includes the following steps [9]:

1. *Fuzzy Inputs*: In this step, crisp or linguistic inputs will be transferred to fuzzy values by using transformation matrices. So, the input variables are represented by fuzzy values, and the output of this step is also fuzzy sets
2. *Fuzzy Rules*: A FIS with p inputs and l linguistic terms for each of inputs leads to $r = p^l$ rules. Straightforwardly, the size of the rule base increases exponentially by increasing the dimension and complexity of the system.

$$\boldsymbol{IF}\ \left(x_1\ is\ A_1^k\right) \boldsymbol{AND}\ \left(x_2\ is\ A_2^k\right) \boldsymbol{OR}\ \left(x_3\ is\ A_3^k\right)\ \ldots\ \left(x_n\ is\ A_n^k\right) \boldsymbol{THEN}\ \left(y\ is\ B^k\right) \quad (1)$$

where $x_i;\ \forall i = 1, 2, \ldots n$ are input variables and y is an output variable of a fuzzy inference system with $k = 1, 2, \ldots, r$ number of rules. Also, $A = \left\{A_1^k, A_2^k, \ldots, A_n^k\right\}$

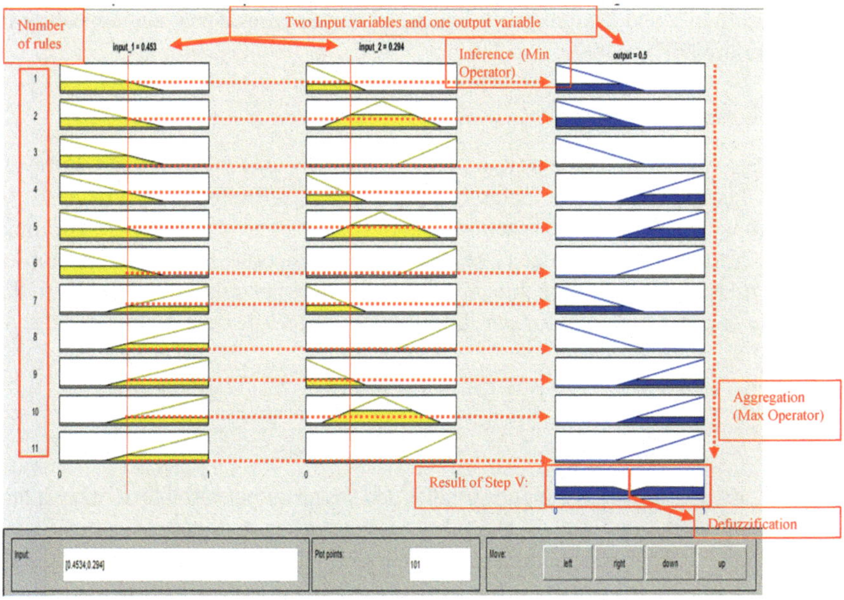

Fig. 2 An example of Mamdani type of FIS

and $B = \{B^1, B^2, \ldots, B^r\}$ are set of linguistic terms of input and output variables for kth rule, respectively. We could also have different fuzzy set operations among different inputs in the antecedent part of each rule. For example, in Eq. (1), both AND (minimum) and OR (maximum) operators are applied.

3. *Apply Fuzzy Aggregation Operator*: In this step, the first step's fuzzified input, which is the antecedent part of the rules, should be combined.
4. *Implication*: The defined rules should be applied one by one, and the antecedent part of each rule determines the cut point membership value of the corresponding membership function.
5. *Apply Fuzzy Aggregation Operator to Fuzzy Outputs*: The results of each rule are aggregated to get final aggregated membership values, which consists of the aggregated fuzzy region.

The result of Mamdani inference method through steps 1, 2, 3, 4, and 5 is formulated as follow:

$$\mu_B^k(y) = \max_k \left[\min_i \left\{ \mu_{A_1}^k(x_1), \mu_{A_2}^k(x_2), \ldots, \mu_{A_n}^k(x_n) \right\} \right], \tag{2}$$

6. *Defuzzification*: A defuzzification formula is applied to determine the representative crisp value of the aggregated fuzzy region. In this study, the centroid is applied to defuzzify the fuzzy outputs. The reason to select the centroid method

is continuity and disambiguate, caused by a more prevalence and appealing in literature [34].

Takagi–Sugeno fuzzy inference system:

The Takagi–Sugeno model (also knonwn as TSK fuzzy inference system) was developed by Takagi, Sugeno, and Kang in [30] and [77]. TSK model is a data-driven method in which membership functions and if–then rules are developed by a training set [16]. Although the parameters of membership functions improved, and subsequently, rules are optimized in this process to decrease the training error [25]. In the following, a usual form of a fuzzy rule in TSK model's rule base is shown:

$$\mathbf{IF}\left(\mathrm{x}_1 \text{ is } \mathrm{A}_1^{\mathrm{k}}\right)\mathbf{AND}\left(\mathrm{x}_2 \text{ is } \mathrm{A}_2^{\mathrm{k}}\right)\mathbf{OR}\left(\mathrm{x}_3 \text{ is } \mathrm{A}_3^{\mathrm{k}}\right) \ldots \left(\mathrm{x}_\mathrm{n} \text{ is } \mathrm{A}_\mathrm{n}^{\mathrm{k}}\right)\mathbf{THEN}(\mathrm{y} = \mathrm{f}(\mathrm{x}_1, \mathrm{x}_2)), \tag{3}$$

where x_i; $\forall i = 1, 2, \ldots n$ are input variables and y is an output variable of fuzzy inference system with $k = 1, 2, \ldots, r$ number of rules. Also, $A = \left\{A_1^k, A_2^k, \ldots, A_n^k\right\}$ are set of linguistic terms of input variables for kth rule. Also, it is clear that in antecedent part of each rule, we could have different fuzzy set operations among different inputs; for example, above in Eq. (3), both AND (minimum) and OR (maximum) operators are applied. While $y = f(x_1, x_2)$ is the consequent part of the rule and a polynomial of input variables, function (f) could be any appropriate function in describing the fuzzy model in defined domain and codomain of the problem.

TSK models can mimic some complicated real case problems easily, which is focused by different researchers in their brilliant studies [30, 79, 80]. However, unlike Mamdani fuzzy models, the Sugeno fuzzy models cannot handle the compositional rule of inference strictly in their fuzzy reasoning mechanism. So, the most view problem with TSK models are referring to the feature mentioned above. It is because Sugeno fuzzy models could not propagate the fuzziness from inputs to outputs in a proper way.

Tsukamoto fuzzy inference system:

Tsukamoto fuzzy models, proposed by Yahachiro Tsukamoto in [78], consider a monotonical membership function for the consequent part of the fuzzy rules. So, the Tsukamoto fuzzy model will result in a crisp output value based on the rule's firing strength. In this way, the overall output value of Tsukamoto FIS will be weighted average outputs of rules [74].

Spherical Fuzzy Sets

Gündoğdu and Kahraman in 2019 introduced SFSs with their operations [71]. In the same year, they developed interval-valued SFSs with some aggregation and basic operators [47]. Additionally, the Analytic Hierarchy Process (AHP) method in the SFS environment is developed by the same authors [73]. Additionally, at the same time, Ashraf and Abdullah [81] proposed some new aggregation operators for SFS. The idea behind SFS is to let decision-makers have more flexibility, as illustrated

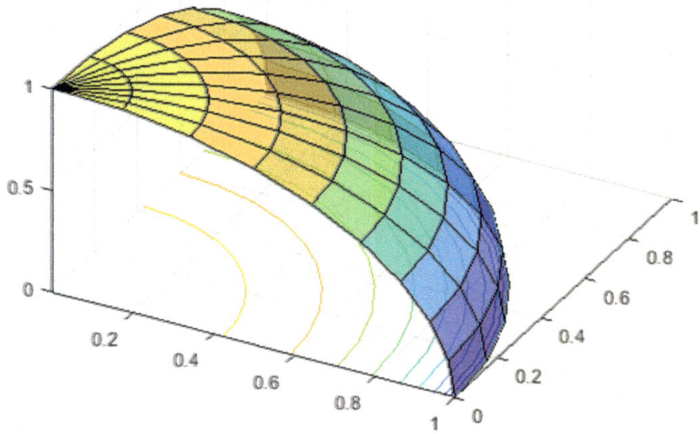

Fig. 3 Operating domain of SFSs

in Fig. 3. Real life cannot be explained by one or two uncertainty degrees; it makes SFSs more applicable for business [38].

Definition 1 [47] Let X be the universal set and $x_i \in X; \forall i = 1, 2, \ldots n$, the SFS $\tilde{A}_s$ of the universe discourse X is defined as x_i as an element of $\tilde{A}_s$ with membership, non-membership and hesitancy degree values, then $\tilde{A}_s$ will be expressed mathematically in the form of:

$$\tilde{A}_s = \left\{ \langle x_i; \mu_{\tilde{A}_s}(x_i), \vartheta_{\tilde{A}_s}(x_i), h_{\tilde{A}_s}(x_i) \rangle | x_i \in X \right\}, \tag{4}$$

where $\mu_{\tilde{A}_s}(x_i), \vartheta_{\tilde{A}_s}(x_i), h_{\tilde{A}_s}(x_i)$ stands for membership, non-membership and hesitancy degrees, respectively and:

$$\begin{cases} \mu_{\tilde{A}_s}: X \rightarrow [0, 1], x_i \in X \rightarrow \mu_{\tilde{A}_s}(x_i) \in [0, 1] \\ \vartheta_{\tilde{A}_s}: X \rightarrow [0, 1], x_i \in X \rightarrow \vartheta_{\tilde{A}_s}(x_i) \in [0, 1] \\ h_{\tilde{A}_s}: X \rightarrow [0, 1], x_i \in X \rightarrow h_{\tilde{A}_s}(x_i) \in [0, 1], \end{cases} \tag{5}$$

which satisfies the condition that the sum square of these values to be between 0 and 1:

$$S_{\tilde{A}_s}(x_i) = \mu^2_{\tilde{A}_s}(x_i) + \vartheta^2_{\tilde{A}_s}(x_i) + h^2_{\tilde{A}_s}(x_i) \rightarrow 0 \leq S_{\tilde{A}_s}(x_i) \leq 1, \tag{6}$$

then refusal degree $R_{\tilde{A}_s}$ of x_i in the spherical fuzzy set $\tilde{A}_s$ will be as follows:

$$R_{\tilde{A}_s}(x_i) = \sqrt{1 - S_{\tilde{A}_s}(x_i)}, \tag{7}$$

Definition 2 [47] Let X be the universal set and $x_i \in X; \forall i = 1, 2, \ldots n$ then $\tilde{A} = \{\langle x_i; \mu_{\tilde{A}}(x_i), \vartheta_{\tilde{A}}(x_i), h_{\tilde{A}}(x_i)\rangle | x_i \in X\}$ and $\tilde{B} = \{\langle x_i; \mu_{\tilde{B}}(x_i), \vartheta_{\tilde{B}}(x_i), h_{\tilde{B}}(x_i)\rangle | x_i \in X\}$ be two spherical fuzzy sets (SFSs) and then we have:

$$\tilde{A}^c = \{\langle x_i; \vartheta_{\tilde{A}}(x_i), \mu_{\tilde{A}}(x_i), h_{\tilde{A}}(x_i)\rangle | x_i \in X\}, \tag{8}$$

$$\tilde{A} \subseteq \tilde{B} = \begin{cases} \mu^2_{\tilde{A}}(x_i) \leq \mu^2_{\tilde{B}}(x_i) \\ \vartheta^2_{\tilde{A}}(x_i) \geq \vartheta^2_{\tilde{B}}(x_i) \\ h^2_{\tilde{A}}(x_i) \leq h^2_{\tilde{B}}(x_i) \end{cases}, \tag{9}$$

$$\tilde{A} = \tilde{B} iff : \tilde{A} \subseteq \tilde{B} \wedge \tilde{B} \subseteq \tilde{A}, \tag{10}$$

Definition 3 [47] Let X be the universal set and $x_i \in X; \forall i = 1, 2, \ldots n$ then $\tilde{A} = \{\langle x_i; \mu_{\tilde{A}}(x_i), \vartheta_{\tilde{A}}(x_i), h_{\tilde{A}}(x_i)\rangle | x_i \in X\}$ and $\tilde{B} = \{\langle x_i; \mu_{\tilde{B}}(x_i), \vartheta_{\tilde{B}}(x_i), h_{\tilde{B}}(x_i)\rangle | x_i \in X\}$ be two spherical fuzzy sets (SFSs) and $k \geq 0$. Therefore, some basic operators are defined as below:

$$\begin{aligned} \tilde{A} \cup \tilde{B} = \{\langle x_i; &max\{\mu_{\tilde{A}}(x_i), \mu_{\tilde{B}}(x_i)\}, min\{\vartheta_{\tilde{A}}(x_i), \vartheta_{\tilde{B}}(x_i)\}, \\ &min\left\{\left(1 - \left(\max\{\mu_{\tilde{A}}(x_i), \mu_{\tilde{B}}(x_i)\}\right)^2 - \left(min\{\vartheta_{\tilde{A}}(x_i), \vartheta_{\tilde{B}}(x_i)\}\right)^2\right)^{0.5}, \right. \\ &max\{h_{\tilde{A}}(x_i), h_{\tilde{B}}(x_i)\}\}\rangle | x_i \in X\} \end{aligned} \tag{11}$$

$$\begin{aligned} \tilde{A} \cap \tilde{B}_s = \{\langle x_i; &min\{\mu_{\tilde{A}}(x_i), \mu_{\tilde{B}}(x_i)\}, max\{\vartheta_{\tilde{A}}(x_i), \vartheta_{\tilde{B}}(x_i)\}, \\ &max\left\{\left(1 - \left(\max\{\mu_{\tilde{A}}(x_i), \mu_{\tilde{B}}(x_i)\}\right)^2 - \left(min\{\vartheta_{\tilde{A}}(x_i), \vartheta_{\tilde{B}}(x_i)\}\right)^2\right)^{0.5}, \right. \\ &min\{h_{\tilde{A}}(x_i), h_{\tilde{B}}(x_i)\}\}\rangle | x_i \in X\}, \end{aligned} \tag{12}$$

$$\begin{aligned} \tilde{A} \oplus \tilde{B} = \Bigg\{\Bigg\langle &x_i; \left(\mu^2_{\tilde{A}}(x_i) + \mu^2_{\tilde{B}}(x_i) - \mu^2_{\tilde{A}}(x_i) \cdot \mu^2_{\tilde{B}}(x_i)\right)^{0.5}, \\ &\vartheta_{\tilde{A}}(x_i) \cdot \vartheta_{\tilde{B}}(x_i), \Big(\Big(1 - \mu^2_{\tilde{B}}(x_i)\Big) h^2_{\tilde{A}}(x_i) \\ &+ \Big(1 - \mu^2_{\tilde{A}}(x_i)\Big) h^2_{\tilde{B}}(x_i) - h^2_{\tilde{A}}(x_i) \cdot h^2_{\tilde{B}}(x_i)\Big)^{0.5}\Bigg\rangle | x_i \in X\Bigg\}, \end{aligned} \tag{13}$$

$$\begin{aligned} \tilde{A} \otimes \tilde{B}_s = \Bigg\{\Bigg\langle &x_i; \mu_{\tilde{A}}(x_i) \cdot \mu_{\tilde{B}}(x_i), \left(\vartheta^2_{\tilde{A}}(x_i) + \vartheta^2_{\tilde{B}}(x_i) - \vartheta^2_{\tilde{A}}(x_i) \cdot \vartheta^2_{\tilde{B}}(x_i)\right)^{0.5}, \\ &- \vartheta^2_{\tilde{A}}(x_i) \cdot \vartheta^2_{\tilde{B}}(x_i)\Big)^{0.5}, \Big(\Big(1 - \vartheta^2_{\tilde{B}}(x_i)\Big) h^2_{\tilde{A}}(x_i) \end{aligned}$$

$$+\left(1-\vartheta_{\tilde{A}}^{2}(x_i)\right)h_{\tilde{B}}^{2}(x_i)-h_{\tilde{A}}^{2}(x_i).h_{\tilde{B}}^{2}(x_i)\right)^{0.5}\Big\rangle|x_i\in X\},\tag{14}$$

$$k\tilde{A}=\left\{\left\langle x_i;\left(1-\left(1-\mu_{\tilde{A}}^{2}(x_i)\right)^{k}\right)^{0.5},\left(\vartheta_{\tilde{A}}(x_i)\right)^{k},\right.\right.$$
$$\left.\left.\left(\left(1-\mu_{\tilde{A}}^{2}(x_i)\right)^{k}-\left(1-\mu_{\tilde{A}}^{2}(x_i)-h_{\tilde{A}}^{2}(x_i)\right)^{k}\right)^{0.5}\right\rangle|x_i\in X\right\},\tag{15}$$

$$\tilde{A}^{k}=\left\{\left\langle x_i;\left(\mu_{\tilde{A}}(x_i)\right)^{k},\left(1-\left(1-\vartheta_{\tilde{A}}^{2}(x_i)\right)^{k}\right)^{0.5},\right.\right.$$
$$\left.\left.\left(\left(1-\vartheta_{\tilde{A}}^{2}(x_i)\right)^{k}-\left(1-\vartheta_{\tilde{A}}^{2}(x_i)-h_{\tilde{A}}^{2}(x_i)\right)^{k}\right)^{0.5}\right\rangle|x_i\in X\right\},\tag{16}$$

Remark 1 Assume $\widetilde{\mathrm{A}}_{\mathrm{j}}=\langle\mu_{\widetilde{\mathrm{A}}_{\mathrm{j}}},\vartheta_{\widetilde{\mathrm{A}}_{\mathrm{j}}},\mathrm{h}_{\widetilde{\mathrm{A}}_{\mathrm{j}}}\rangle;\forall\mathrm{j}=1,2,\ldots\mathrm{n}$ be a collection of spherical fuzzy numbers (SFNs), the generalized addition will be as follows:

$$\underset{j=1}{\overset{n}{\oplus}}\tilde{A}_j=\left(\tilde{A}_1\oplus\tilde{A}_2\oplus\cdots\oplus\tilde{A}_n\right)$$
$$=\left\langle\begin{array}{c}\left(1-\prod_{j=1}^{n}\left(1-\mu_{\tilde{A}_j}^{2}\right)\right)^{0.5},\ \prod_{j=1}^{n}\vartheta_{\tilde{A}_j},\\ \left(\prod_{j=1}^{n}\left(1-\mu_{\tilde{A}_j}^{2}\right)-\prod_{j=1}^{n}\left(1-\mu_{\tilde{A}_j}^{2}-h_{\tilde{A}_j}^{2}\right)\right)^{0.5}\end{array}\right\rangle,\tag{17}$$

Remark 2 Assume $\tilde{A}_j=\langle\mu_{\tilde{A}_j},\vartheta_{\tilde{A}_j},h_{\tilde{A}_j}\rangle;\forall j=1,2,\ldots n$ be a collection of spherical fuzzy numbers (SFNs), the generalized multiplication will be as follows:

$$\underset{j=1}{\overset{n}{\otimes}}\tilde{A}_j=\left(\tilde{A}_1\otimes\tilde{A}_2\otimes\cdots\otimes\tilde{A}_n\right)$$
$$=\left\langle\begin{array}{c}\prod_{j=1}^{n}\mu_{\tilde{A}_j},\ \left(1-\prod_{j=1}^{n}\left(1-\vartheta_{\tilde{A}_j}^{2}\right)\right)^{0.5},\\ \left(\prod_{j=1}^{n}\left(1-\vartheta_{\tilde{A}_j}^{2}\right)-\prod_{j=1}^{n}\left(1-\vartheta_{\tilde{A}_j}^{2}-h_{\tilde{A}_j}^{2}\right)\right)^{0.5}\end{array}\right\rangle,\tag{18}$$

Definition 4 [47] Score (Sc) and accuracy (Ac) functions of sorting a SFN are defined respectively as follows:

$$Sc\left(\tilde{A}\right) = \left(\mu_{\tilde{A}} - \frac{h_{\tilde{A}}}{2}\right)^2 - \left(\vartheta_{\tilde{A}} - \frac{h_{\tilde{A}}}{2}\right)^2, \tag{19}$$

$$Ac\left(\tilde{A}\right) = \mu_{\tilde{A}}^2 + \vartheta_{\tilde{A}}^2 + h_{\tilde{A}}^2, \tag{20}$$

Remark 3 Note if $\tilde{A}$ and $\tilde{B}$ are SFNs then $\tilde{A} < \tilde{B}$ if and only if:

i. $Sc\left(\tilde{A}\right) < Sc\left(\tilde{B}\right)$i,

or

ii. $Sc\left(\tilde{A}\right) = Sc\left(\tilde{B}\right)$ii and $Ac\left(\tilde{A}\right) < Ac\left(\tilde{B}\right)$ and $Ac\left(\tilde{A}\right) < Ac\left(\tilde{B}\right)$.

Definition 5 [47] Let X be the universal set and $x_i \in X; \forall i = 1, 2, \ldots n$ then $\tilde{A} = \left\{\langle x_i;\, \mu_{\tilde{A}}(x_i), \vartheta_{\tilde{A}}(x_i), h_{\tilde{A}}(x_i)\rangle | x_i \in X\right\}$ be a spherical fuzzy set (SFS) with corresponding weight vector $w = \{w_1, w_2, \ldots, w_n\}; w_i \in [0, 1]; \sum_{i=1}^n w_i = 1$. Spherical Fuzzy Weighted Arithmetic Mean ($SFWAM$) is defined as:

$$\begin{aligned} SFWAM_w\left(\tilde{A}\right) &= \underset{i=1}{\overset{n}{\otimes}} w_i\tilde{A}(x_i) = w_1\tilde{A}(x_1) + w_2\tilde{A}(x_2) + \cdots + w_n\tilde{A}(x_n) \\ &= \left\{\left(1 - \prod_{i=1}^n\left(1 - \mu_{\tilde{A}}^2(x_i)\right)^{w_i}\right)^{0.5}, \prod_{i=1}^n\left(\vartheta_{\tilde{A}}(x_i)\right)^{w_i}, \right. \\ &\left.\left(\prod_{i=1}^n\left(1 - \mu_{\tilde{A}}^2(x_i)\right)^{w_i} - \prod_{i=1}^n\left(1 - \mu_{\tilde{A}}^2(x_i) - h_{\tilde{A}}^2(x_i)\right)^{w_i}\right)^{0.5}\right\} \end{aligned} \tag{21}$$

Definition 6 [47] Let X be the universal set and $x_i \in X; \forall i = 1, 2, \ldots n$ then $\tilde{A} = \left\{\langle x_i;\, \mu_{\tilde{A}}(x_i), \vartheta_{\tilde{A}}(x_i), h_{\tilde{A}}(x_i)\rangle | x_i \in X\right\}$ be a spherical fuzzy set (SFS) with corresponding weight vector $w = \{w_1, w_2, \ldots, w_n\}; w_i \in [0, 1]; \sum_{i=1}^n w_i = 1$. Spherical Fuzzy Weighted Geometric Mean ($SFWGM$) is defined as:

$$\begin{aligned} \mathrm{SFWGM_w}\left(\tilde{A}\right) &= \underset{i=1}{\overset{n}{\otimes}} \left(\tilde{A}(x_i)\right)^{w_i} = \left(\tilde{A}(x_1)\right)^{w_1} \times \left(\tilde{A}(x_2)\right)^{w_2} \times \cdots \times \left(\tilde{A}(x_n)\right)^{w_n} \\ &= \left\{\prod_{i=1}^n\left(\mu_{\tilde{A}}(x_i)\right)^{w_i}, \left(1 - \prod_{i=1}^n\left(1 - \vartheta_{\tilde{A}}^2(x_i)\right)^{w_i}\right)^{0.5}, \right. \\ &\left.\left(\prod_{i=1}^n\left(1 - \vartheta_{\tilde{A}}^2(x_i)\right)^{w_i} - \prod_{i=1}^n\left(1 - \vartheta_{\tilde{A}}^2(x_i) - h_{\tilde{A}}^2(x_i)\right)^{w_i}\right)^{0.5}\right\}, \end{aligned} \tag{22}$$

3.2 Proposed Methodology

This section first introduces the novel spherical fuzzy transformation function and then proved and studied different properties for spherical fuzzy transformation functions.

Spherical Fuzzy Transformation Function

Transformation functions proposed for different fuzzy sets transform a fuzzy set, usually a higher generation, to a bellower generation, almost to an ordinary fuzzy set. For example, we could refer to the transformation function proposed by Atanassov in [82] to transform any intuitionistic fuzzy set to an ordinary fuzzy set. It is defined that any transformation function is a mapping function between two types of fuzzy sets. Also, any mapping function is not acceptable without preserving original fuzzy sets' features to the end of the transformation process. It is defined as some properties to determine an appropriate mapping function as a transformation function. We will define a transformation function among spherical fuzzy sets onto ordinary fuzzy sets in the following. first. We will then introduce the appropriate properties for spherical fuzzy transformation function and show how those properties are satisfied for our introduced transformation function.

Definition 7 Let X is the universal set and $x_i \in X; \forall i = 1, 2, \ldots n$ then $\tilde{A} = \left\{ \langle x_i;\ \mu_{\tilde{A}}(x_i), \vartheta_{\tilde{A}}(x_i), h_{\tilde{A}}(x_i) \rangle | x_i \in X \right\}$be a SFS, so that $\mu(x_i)$,$\vartheta(x_i)$, $h(x_i)$ and $R(x_i)$ are membership, non-membership, hesitancy and refusal degrees, respectively. Every $\Theta_\lambda \in [0, 1]^{[0,1]\times[0,1]\times[0,1]}$ is a mapping function which: $\Theta_\lambda : SFS(X) \rightarrow FS(X)$. So, we will have Transformation function (Θ_λ) as follows:

$$\begin{cases} \Theta_\lambda : \tilde{A}_{SFS} \rightarrow \tilde{A}_{FS} \\ \tilde{A} = \left\{ \langle x_i;\ \mu_{\tilde{A}}(x_i), \vartheta_{\tilde{A}}(x_i), h_{\tilde{A}}(x_i) | x_i \in X \right\} \rightarrow \Theta_\lambda \\ \left(\tilde{A}\right) = \left\{ x_i;\ \mu_{\Theta_\lambda\left(\tilde{A}\right)}(x_i),\ \vartheta_{\Theta_\lambda\left(\tilde{A}\right)}(x_i) \right\rangle | x_i \in X \right\}, \end{cases} \tag{23}$$

$$\mu_{\Theta_\lambda\left(\tilde{A}\right)}(x_i) = \sqrt{\mu^2_{\tilde{A}}(x_i) + \lambda^2\left(h^2_{\tilde{A}}(x_i) + R^2_{\tilde{A}}(x_i)\right)} \text{ and}$$
$$\vartheta_{\Theta_\lambda\left(\tilde{A}\right)}(x_i) = 1 - \mu_{\Theta_\lambda\left(\tilde{A}\right)}(x_i), \tag{24}$$

where $\lambda \in [0, 1]$.

Definition 8 Let X be the universal set and $x_i \in X; \forall i = 1, 2, \ldots n$ then $\tilde{A} = \left\{ \langle x_i;\ \mu_{\tilde{A}}(x_i), \vartheta_{\tilde{A}}(x_i), h_{\tilde{A}}(x_i) \rangle | x_i \in X \right\}$ and $\tilde{B} = \left\{ \langle x_i;\ \mu_{\tilde{B}}(x_i), \vartheta_{\tilde{B}}(x_i), h_{\tilde{B}}(x_i) \rangle | x_i \in X \right\}$be two SFSs, so that$\mu(x_i)$,$\vartheta(x_i)$, $h(x_i)$ and $R(x_i)$ are membership, non-membership, hesitancy and refusal degrees, respectively and λ, $\lambda_1 and \lambda_2 \in [0, 1]$. Transformation function (Θ_λ) which is defined above will satisfy the following properties:

(T1): $\mu_{\Theta_\lambda(\tilde{A})}(x_i) \geq \mu_{\tilde{A}}(x_i)$

(T2): $\Theta_{\lambda_2}\left(\Theta_{\lambda_1}\left(\tilde{A}\right)\right) = \Theta_{\lambda_1}\left(\tilde{A}\right)$

(T3): $\Theta_{\lambda_1}\left(\tilde{A}\right) \subseteq \Theta_{\lambda_2}\left(\tilde{A}\right)$, *if* $\lambda_1 \leq \lambda_2$

(T4): $\Theta_\lambda\left(\tilde{A}\right) \subseteq \Theta_\lambda\left(\tilde{B}\right)$, *if* $\tilde{A} \subseteq \tilde{B}$

(T5): $\Theta_\lambda\left(\tilde{A}^c\right) = \Theta_\lambda\left(\tilde{A}\right)$, *if* : $\mu_{\tilde{A}}(x_i) = \vartheta_{\tilde{A}}(x_i)$

Theorem 1 *The function* (Θ_λ) *defined in Eq.* (23) *is a transformation function of SFSs.*

Proof To prove Theorem 1, we have shown that Eq. (23) satisfies properties in Definition 8 as follows:

(T1): $\mu_{\Theta_\lambda(\tilde{A})}(x_i) \geq \mu_{\tilde{A}}(x_i)$

Based on Definition 7, we have: $\mu_{\Theta_\lambda(\tilde{A})}(x_i) = \sqrt{\mu^2_{\tilde{A}}(x_i) + \lambda^2\left(h^2_{\tilde{A}}(x_i) + R^2_{\tilde{A}}(x_i)\right)}$ and we know that $0 \leq h^2_{\tilde{A}}(x_i) + R^2_{\tilde{A}}(x_i) \leq 1$ also $0 \leq \lambda \leq 1$, so $0 \leq \lambda^2\left(h^2_{\tilde{A}}(x_i) + R^2_{\tilde{A}}(x_i)\right) \leq 1$:

$$\mu_{\tilde{A}}(x_i) \leq \sqrt{\mu^2_{\tilde{A}}(x_i) + \lambda^2\left(h^2_{\tilde{A}}(x_i) + R^2_{\tilde{A}}(x_i)\right)} = \mu_{\Theta_\lambda(\tilde{A})}(x_i)$$

(T2): $\Theta_{\lambda_2}\left(\Theta_{\lambda_1}\left(\tilde{A}\right)\right) = \Theta_{\lambda_1}\left(\tilde{A}\right)$

Based on Definition 8, we have:

$\Theta_{\lambda_1}\left(\tilde{A}\right)$: $\mu_{\Theta_{\lambda_1}(\tilde{A})}(x_i) = \sqrt{\mu^2_{\tilde{A}}(x_i) + \lambda_1^2\left(h^2_{\tilde{A}}(x_i) + R^2_{\tilde{A}}(x_i)\right)}$ and $\vartheta_{\Theta_{\lambda_1}(\tilde{A})}(x_i) = 1 - \mu_{\Theta_{\lambda_1}(\tilde{A})}(x_i)$

We have: $h_{\Theta_{\lambda_1}(\tilde{A})}(x_i) = R_{\Theta_{\lambda_1}(\tilde{A})}(x_i) = 0$, also we know:

$$\Theta_{\lambda_2}\left(\Theta_{\lambda_1}\left(\tilde{A}\right)\right): \mu_{\Theta_{\lambda_2}(\Theta_{\lambda_1}(\tilde{A}))}(x_i) = \sqrt{\mu^2_{\Theta_{\lambda_1}(\tilde{A})}(x_i) + \lambda_2^2\left(h^2_{\Theta_{\lambda_1}(\tilde{A})}(x_i) + R^2_{\Theta_{\lambda_1}(\tilde{A})}(x_i)\right)}$$
$$= \mu_{\Theta_{\lambda_1}(\tilde{A})}(x_i) \text{ and } \vartheta_{\Theta_{\lambda_2}(\Theta_{\lambda_1}(\tilde{A}))}(x_i) = 1 - \mu_{\Theta_{\lambda_2}(\Theta_{\lambda_1}(\tilde{A}))}(x_i)$$
$$= 1 - \mu_{\Theta_{\lambda_1}(\tilde{A})}(x_i) = \vartheta_{\Theta_{\lambda_1}(\tilde{A})}(x_i).$$

So, we show that $\Theta_{\lambda_2}\left(\Theta_{\lambda_1}\left(\tilde{A}\right)\right) = \Theta_{\lambda_1}\left(\tilde{A}\right)$.

(T3): $\Theta_{\lambda_1}\left(\tilde{A}\right) \subseteq \Theta_{\lambda_2}\left(\tilde{A}\right), if\, \lambda_1 \le \lambda_2$

We know that: $\Theta_{\lambda_1}\left(\tilde{A}\right) \subseteq \Theta_{\lambda_2}\left(\tilde{A}\right) iff: \begin{cases} \mu_{\Theta_{\lambda_1}\left(\tilde{A}\right)}(x_i) \le \mu_{\Theta_{\lambda_2}\left(\tilde{A}\right)}(x_i) \\ \vartheta_{\Theta_{\lambda_1}\left(\tilde{A}\right)}(x_i) \ge \vartheta_{\Theta_{\lambda_2}\left(\tilde{A}\right)}(x_i) \end{cases}; \forall i$

Based on Definition 8, we have:

$$\Theta_{\lambda_1}\left(\tilde{A}\right): \mu_{\Theta_{\lambda_1}\left(\tilde{A}\right)}(x_i) = \sqrt{\mu_{\tilde{A}}^2(x_i) + \lambda_1^2\left(h_{\tilde{A}}^2(x_i) + R_{\tilde{A}}^2(x_i)\right)} \text{ and}$$
$$\vartheta_{\Theta_{\lambda_1}\left(\tilde{A}\right)}(x_i) = 1 - \mu_{\Theta_{\lambda_1}\left(\tilde{A}\right)}(x_i)$$

$$\Theta_{\lambda_2}\left(\tilde{A}\right): \mu_{\Theta_{\lambda_2}\left(\tilde{A}\right)}(x_i) = \sqrt{\mu_{\tilde{A}}^2(x_i) + \lambda_2^2\left(h_{\tilde{A}}^2(x_i) + R_{\tilde{A}}^2(x_i)\right)} \text{ and}$$
$$\vartheta_{\Theta_{\lambda_2}\left(\tilde{A}\right)}(x_i) = 1 - \mu_{\Theta_{\lambda_2}\left(\tilde{A}\right)}(x_i)$$

$$\begin{aligned} \mu_{\Theta_{\lambda_1}\left(\tilde{A}\right)}(x_i) &= \sqrt{\mu_{\tilde{A}}^2(x_i) + \lambda_1^2\left(h_{\tilde{A}}^2(x_i) + R_{\tilde{A}}^2(x_i)\right)} \\ &\le \sqrt{\mu_{\tilde{A}}^2(x_i) + \lambda_2^2\left(h_{\tilde{A}}^2(x_i) + R_{\tilde{A}}^2(x_i)\right)}. \\ &= \mu_{\Theta_{\lambda_2}\left(\tilde{A}\right)}(x_i) \text{ if:} \lambda_1 \le \lambda_2 \end{aligned}$$

So, $\mu_{\Theta_{\lambda_1}\left(\tilde{A}\right)}(x_i) \le \mu_{\Theta_{\lambda_2}\left(\tilde{A}\right)}(x_i)\, if: \lambda_1 \le \lambda_2$. Also we have: $\mu_{\Theta_{\lambda_1}\left(\tilde{A}\right)}(x_i) \le \mu_{\Theta_{\lambda_2}\left(\tilde{A}\right)}(x_i) \to \vartheta_{\Theta_{\lambda_1}\left(\tilde{A}\right)}(x_i) = 1 - \mu_{\Theta_{\lambda_1}\left(\tilde{A}\right)}(x_i) \ge 1 - \mu_{\Theta_{\lambda_2}\left(\tilde{A}\right)}(x_i) = \vartheta_{\Theta_{\lambda_2}\left(\tilde{A}\right)}(x_i)$.

So, we show that $\Theta_{\lambda_1}\left(\tilde{A}\right) \subseteq \Theta_{\lambda_2}\left(\tilde{A}\right), if\, \lambda_1 \le \lambda_2$.

(T4): $\Theta_{\lambda}\left(\tilde{A}\right) \subseteq \Theta_{\lambda}\left(\tilde{B}\right), if\, \tilde{A} \subseteq \tilde{B}$

We know that: $\Theta_{\lambda}\left(\tilde{A}\right) \subseteq \Theta_{\lambda}\left(\tilde{B}\right): \begin{cases} \mu_{\Theta_{\lambda}\left(\tilde{A}\right)}(x_i) \le \mu_{\Theta_{\lambda}(\tilde{B})}(x_i) \\ \vartheta_{\Theta_{\lambda}\left(\tilde{A}\right)}(x_i) \ge \vartheta_{\Theta_{\lambda}(\tilde{B})}(x_i) \end{cases}; if\, \tilde{B} \subseteq \tilde{A}$

Based on Definition 2, we have: $\tilde{A} \subseteq \tilde{B}, if: \begin{cases} \mu_{\tilde{A}}(x_i) \le \mu_{\tilde{B}}(x_i) \\ \vartheta_{\tilde{A}}(x_i) \ge \vartheta_{\tilde{B}}(x_i)\ \forall x_i \in X \\ h_{\tilde{A}}(x_i) \le h_{\tilde{B}}(x_i) \end{cases}$. Then based on Definition 8, we have:

$$\Theta_{\lambda}\left(\tilde{A}\right): \mu_{\Theta_{\lambda}\left(\tilde{A}\right)}(x_i) = \sqrt{\mu_{\tilde{A}}^2(x_i) + \lambda^2\left(h_{\tilde{A}}^2(x_i) + R_{\tilde{A}}^2(x_i)\right)} \text{ and}$$
$$\vartheta_{\Theta_{\lambda}\left(\tilde{A}\right)}(x_i) = 1 - \mu_{\Theta_{\lambda}\left(\tilde{A}\right)}(x_i)$$

$$\Theta_{\lambda}\left(\tilde{B}\right): \mu_{\Theta_{\lambda}(\tilde{B})}(x_i) = \sqrt{\mu_{\tilde{B}}^2(x_i) + \lambda^2\left(h_{\tilde{B}}^2(x_i) + R_{\tilde{B}}^2(x_i)\right)} \text{ and}$$

$$\vartheta_{\Theta_\lambda(\tilde{B})}(x_i) = 1 - \mu_{\Theta_\lambda(\tilde{B})}(x_i)$$

$\sqrt{\mu_{\tilde{A}}^2(x_i) + \lambda^2\left(h_{\tilde{A}}^2(x_i) + R_{\tilde{A}}^2(x_i)\right)} \geq \sqrt{\mu_{\tilde{B}}^2(x_i) + \lambda^2\left(h_{\tilde{B}}^2(x_i) + R_{\tilde{B}}^2(x_i)\right)} \rightarrow$ $\mu_{\Theta_\lambda\left(\tilde{A}\right)}(x_i) \geq \mu_{\Theta_\lambda(\tilde{B})}(x_i)$, also we have: $\mu_{\Theta_{\lambda_1}\left(\tilde{A}\right)}(x_i) \geq \mu_{\Theta_{\lambda_2}\left(\tilde{A}\right)}(x_i) \rightarrow \vartheta_{\Theta_{\lambda_1}\left(\tilde{A}\right)}(x_i) = 1 - \mu_{\Theta_{\lambda_1}\left(\tilde{A}\right)}(x_i) \leq 1 - \mu_{\Theta_{\lambda_2}\left(\tilde{A}\right)}(x_i) = \vartheta_{\Theta_{\lambda_2}\left(\tilde{A}\right)}(x_i)$.

So, we show that $\Theta_\lambda\left(\tilde{B}\right) \subseteq \Theta_\lambda\left(\tilde{A}\right)$, *if* $\tilde{B} \subseteq \tilde{A}$.

(T5): $\Theta_\lambda\left(\tilde{A}^c\right) = \Theta_\lambda\left(\tilde{A}\right)$, *if* : $\mu_{\tilde{A}}(x_i) = \vartheta_{\tilde{A}}(x_i)$

We know that $\tilde{A} = \left\{\langle x_i; \mu_{\tilde{A}}(x_i), \vartheta_{\tilde{A}}(x_i), h_{\tilde{A}}(x_i)\rangle | x_i \in X\right\}$ then $\tilde{A}^c = \left\{\langle x_i; \vartheta_{\tilde{A}}(x_i), \mu_{\tilde{A}}(x_i), h_{\tilde{A}}(x_i)\rangle | x_i \in X\right\}$ then based on Definition 8, we have:

$$\Theta_\lambda\left(\tilde{A}\right) : \mu_{\Theta_\lambda\left(\tilde{A}\right)}(x_i) = \sqrt{\mu_{\tilde{A}}^2(x_i) + \lambda^2\left(h_{\tilde{A}}^2(x_i) + R_{\tilde{A}}^2(x_i)\right)} \text{ and}$$
$$\vartheta_{\Theta_\lambda\left(\tilde{A}\right)}(x_i) = 1 - \mu_{\Theta_\lambda\left(\tilde{A}\right)}(x_i)$$

$$\Theta_\lambda\left(\tilde{A}^c\right) : \mu_{\Theta_\lambda\left(\tilde{A}^c\right)}(x_i) = \sqrt{\vartheta_{\tilde{A}}^2(x_i) + \lambda^2\left(h_{\tilde{A}}^2(x_i) + R_{\tilde{A}}^2(x_i)\right)} \text{ and}$$
$$\vartheta_{\Theta_\lambda\left(\tilde{A}^c\right)}(x_i) = 1 - \mu_{\Theta_\lambda\left(\tilde{A}\right)}(x_i)$$

We have: $\mu_{\tilde{A}}(x_i) = \vartheta_{\tilde{A}}(x_i)$, So:
$\sqrt{\mu_{\tilde{A}}^2(x_i) + \lambda^2\left(h_{\tilde{A}}^2(x_i) + R_{\tilde{A}}^2(x_i)\right)} = \sqrt{\vartheta_{\tilde{A}}^2(x_i) + \lambda^2\left(h_{\tilde{A}}^2(x_i) + R_{\tilde{A}}^2(x_i)\right)}$, so we will get $\mu_{\Theta_\lambda\left(\tilde{A}\right)}(x_i) = \mu_{\Theta_\lambda\left(\tilde{A}^c\right)}(x_i)$, also we get: $\vartheta_{\Theta_\lambda\left(\tilde{A}\right)}(x_i) = 1 - \mu_{\Theta_\lambda\left(\tilde{A}\right)}(x_i) = 1 - \mu_{\Theta_\lambda\left(\tilde{A}^c\right)}(x_i) = \vartheta_{\Theta_\lambda\left(\tilde{A}^c\right)}(x_i)$

So, we show that $\Theta_\lambda\left(\tilde{A}\right) = \Theta_\lambda\left(\tilde{A}^c\right)$.

Based on (T1), (T2), (T3), (T4) and (T5) we have shown that the function (Θ_λ) defined in Eq. (23) is a transformation function of SFSs.

Describing in detail the above-defined and proved transformation function (Θ_λ) is that we do not consider the hesitancy degree in FS. A non-membership degree is considered dependent on membership, so in an equal position, membership of an SFS is much more realistic. The other explanation that could mention is that a part of the membership degree related to hesitancy is out of our hand during considering uncertainty and fuzzifying in FS and will not be considered, so it is added to the membership. Nevertheless, in an SFS, we could independently consider it trivial, which spherical fuzzy provided it.

Spherical Fuzzy Inference System (S-FIS)

As discussed before, the Fuzzy Inference System (FIS) is a procedure based on the fuzzy logic theory that maps a set of input variables of the problem's domain onto a set of possible appropriate outputs codomain [34]. Set of input and output variables are defined as fuzzy sets, as we know it is available that every fuzzy set has just one or several elements in both fuzzy or crisp states. Altogether, inputs and outputs of the problem can be both fuzzy or crisp values in the problem's domain and codomain, so the problem's inputs and outputs can be represented by fuzzy membership functions, then it may be used a fuzzy mapping function through FIS. We define Spherical Fuzzy Inference System (S-FIS) based on Mamdani Model.

Definition 9 A Spherical Fuzzy Inference system (S-FIS) based on the Mamdani model is a fuzzy inference system based on spherical fuzzy sets (SFSs). S-FIS consists of Spherical Fuzzy Rules (SFR) including IF–THEN rules and the following components:

1. *Spherical Fuzzification:* In this step, linguistic scales for input and output variables are determined and used to transfer linguistic values to spherical fuzzy values. So, the variables are represented by spherical fuzzy values. Assume that we have L linguistic scale for our variables, $(l = 1, 2, 3 \ldots L)$ so that we will have:

$$\widetilde{LS}_s = \langle LS_l; \mu_{\tilde{LS}_s}(LS_l), \vartheta_{\tilde{LS}_s}(LS_l), h_{\tilde{LS}_s}(LS_l)\rangle, \tag{25}$$

where, LS_l is the crisp part of each spherical fuzzy linguistic scale, and $\mu_{\tilde{LS}_s}(LS_l), \vartheta_{\tilde{LS}_s}(LS_l) and h_{\tilde{LS}_s}(LS_l)$ are the spherical fuzzy part of LS_l in the spherical fuzzy set. $\tilde{LS}_s$ shows the uncertainty around that by spherical fuzzy degrees: membership, non-membership, and hesitancy degrees, respectively. Table 1 shows that linguistic scales and corresponding spherical fuzzy numbers.

2. *Spherical Fuzzy Rule (SFR) Base:* A S-FIS with p inputs and l linguistic terms for each input consists of $r = p^l$ rules. To determine spherical fuzzy rules, the methodology has the following sub-steps:

   a. *Spherical Fuzzy Transformation Function*: the spherical fuzzy matrices for input and output variables will be transformed into fuzzy input and output variables. The result of this step will be transformed spherical fuzzy factor for variables based on transformation function (Θ_λ), which is defined in Definition 7.
   b. *Evaluate Transformed Gaussian Fuzzy numbers for linguistic scales*: Gaussian membership function [74] with a center parameter of LS_l and width parameter of σ is used for each linguistic scale as $Gauss\,(LS_l, \sigma)$ [83].

   Then, the Transformed Gaussian Spherical Fuzzy Linguistic Scales will be as follows:

$$\begin{aligned}\widetilde{LS}_s a &= \left\langle LS_l; \mu_{\widetilde{LS}_s}(LS_l), \vartheta_{\widetilde{LS}_s}(LS_l), h_{\widetilde{LS}_s}(LS_l) \rightarrow \widetilde{TLS}\right\rangle \\ &= \left\langle Gauss\,(LS_l, \sigma); \mu_{\Theta_\lambda(\widetilde{LS}_s)}(LS_l), \vartheta_{\Theta_\lambda(\widetilde{LS}_s)}(LS_l)\right\rangle, \end{aligned} \tag{26}$$

Table 1 Spherical fuzzy linguistic scales for input $\left(\widetilde{LSI}_s\right)$ and output $\left(\widetilde{LSO}_s\right)$ variables

Spherical fuzzy linguistic scales for input $\left(\widetilde{LSI}_s\right)$ and output $\left(\widetilde{LSO}_s\right)$ variables		$\widetilde{LS}_s = \langle \mathrm{LS}_l; \mu_{\widetilde{LS}_s}(\mathrm{LS}_l), \vartheta_{\widetilde{LS}_s}(\mathrm{LS}_l), h_{\widetilde{LS}_s}(\mathrm{LS}_l) \rangle$		
Scale ($l = 1, 2, \ldots, 5$)		LSI_l	LSO_l	$\left(\mu_{\widetilde{LS}_s}(\mathrm{LS}_l), \vartheta_{\widetilde{LS}_s}(\mathrm{LS}_l), h_{\widetilde{LS}_s}(\mathrm{LS}_l)\right)$
$l = 1$	Absolutely High Possible (AHP)	10	1	(0.9, 0.1, 0.1)
$l = 2$	High Possible (HP)	7.5	0.75	(0.7, 0.3, 0.3)
$l = 3$	Equally Possible (EP)	5	0.5	(0.5, 0.5, 0.5)
$l = 4$	Low Possible (LP)	2.5	0.25	(0.3, 0.7, 0.3)
$l = 5$	Absolutely Low Possible (ALP)	0	0	(0.1, 0.9, 0.1)

c. *Rules in SFR*: The rules are defined in SFR base of S-IFS as follows:

$$\begin{aligned} &\textbf{\textit{IF}}\left(x_1 \ is\ LSI_1^k\right) \textbf{\textit{AND}}\left(x_2 \ is\ LSI_2^k\right) \ldots \left(x_n \ is\ LSI_n^k\right) \\ &\textbf{\textit{THEN}}\left(y_1 \ is\ LSO_1^k\right) \textbf{\textit{AND}}\left(y_2 \ is\ LSO_2^k\right) \ldots \left(y_m \ is\ LSO_m^k\right) \end{aligned} \quad (27)$$

where $x_i; \forall i = 1, 2, \ldots n$ are the input variables and $y_j; \forall j = 1, 2, \ldots m$ are the output variables of S-IFS with $k = 1, 2, \ldots, r$ number of rules. Also, $\tilde{LSI}_s = \{LSI_1^k, LSI_2^k, \ldots, LSI_n^k\}$ and $\tilde{LSO}_s = \{LSO_1^k, LSO_2^k, \ldots, LSO_m^k\}$ are set of linguistic scales of input and output variables for kth rule, respectively. It is clear that in antecedent part of each rule we could have different fuzzy set operations among other inputs.

Based on sub-steps *a* and *b*, we could rewrite the spherical fuzzy IF–THEN rule in Eq. (27) as follows:

$$\begin{aligned} &\textbf{\textit{IF}}\left(x_1 \ is\ \mu_{\Theta_\lambda(\widetilde{LSI}_s)}(LSI_1) * Gauss\left(\frac{LSI_1}{\mu_{\Theta_\lambda(\widetilde{LSI}_s)}(LSI_1)}, \sigma\right)\right) \\ &\textbf{\textit{AND}}\left(x_2 \ is\ \mu_{\Theta_\lambda(\widetilde{LSI}_s)}(LSI_2) * Gauss\left(\frac{LSI_2}{\mu_{\Theta_\lambda(\widetilde{LSI}_s)}(LSI_2)}, \sigma\right)\right) \\ &\ldots\left(x_n \ is\ \mu_{\Theta_\lambda(\widetilde{LSI}_s)}(LSI_n) * Gauss\left(\frac{LSI_n}{\mu_{\Theta_\lambda(\widetilde{LSI}_s)}(LSI_n)}, \sigma\right)\right) \\ &\textbf{\textit{THEN}}\left(y_1 \ is\ \mu_{\Theta_\lambda\left(\widetilde{LSO}_s\right)}(LSO_1) * Gauss\left(\frac{LSO_1}{\mu_{\Theta_\lambda\left(\widetilde{LSO}_s\right)}(LSO_1)}, \sigma\right)\right) \\ &\textbf{\textit{AND}}\left(y_2 \ is\ \mu_{\Theta_\lambda\left(\widetilde{LSO}_s\right)}(LSO_2) * Gauss\left(\frac{LSO_2}{\mu_{\Theta_\lambda\left(\widetilde{LSO}_s\right)}(LSO_2)}, \sigma\right)\right) \\ &\ldots\left(y_m \ is\ \mu_{\Theta_\lambda\left(\widetilde{LSO}_s\right)}(LSO_m) * Gauss\left(\frac{LSO_m}{\mu_{\Theta_\lambda\left(\widetilde{LSO}_s\right)}(LSO_m)}, \sigma\right)\right), \end{aligned} \quad (28)$$

Assume that $c_i = \frac{LSI_i}{\mu_{\Theta_\lambda(\tilde{LSI}_s)}(LSI_i)}$ and $c_j = \frac{LSO_j}{\mu_{\Theta_\lambda(\tilde{LSO}_s)}(LSO_j)}$ are the center parameters of normalized Gaussian membership function of input and output variables, respectively. Then we could rewrite IF–THEN formula for S-FIS in Eq. (29) as follows:

$$\begin{aligned} &\textbf{\textit{IF}}\left(x_1 \ is\ \mu_{\Theta_\lambda(\widetilde{LSI}_s)}(LSI_1) * Gauss\ (c_1, \sigma)\right) \\ &\textbf{\textit{AND}}\left(x_2 \ is\ \mu_{\Theta_\lambda(\widetilde{LSI}_s)}(LSI_2) * Gauss\ (c_2, \sigma)\right) \end{aligned}$$

$$\ldots \left(x_n \; is \; \mu_{\Theta_\lambda(\widetilde{LSI}_s)}(LSI_n) * Gauss\,(c_n, \sigma)\right)$$

$$\boldsymbol{THEN} \left(y_1 \; is \; \mu_{\Theta_\lambda\left(\widetilde{LSO}_s\right)}(LSO_1) * Gauss\,(c_1, \sigma)\right)$$

$$\boldsymbol{AND} \left(y_2 \; is \; \mu_{\Theta_\lambda\left(\widetilde{LSO}_s\right)}(LSO_2) * Gauss\,(c_2, \sigma)\right)$$

$$\ldots \left(y_m \; is \; \mu_{\Theta_\lambda\left(\widetilde{LSO}_s\right)}(LSO_m) * Gauss\,(c_m, \sigma)\right), \quad (29)$$

3. *Apply Fuzzy Aggregation Operator:* In this step, the first step's fuzzified input, which is the antecedent part of the rules, should be aggregated.
4. *Implication:* The defined rules should be applied one by one, and the antecedent part of each rule determines the cut point membership value of the corresponding membership function.
5. *Apply Fuzzy Aggregation Operator to Fuzzy Outputs:* The results of each rule are aggregated to get final aggregated membership values which consists the aggregated fuzzy region.

The result of Mamdani inference method through steps 1, 2, 3, 4 and 5 is formulated as follow:

$$\mu^k_{LSO_j}(y_j) = \max_k \left[\min_i \left\{ \mu^k_{LSI_1}(x_1), \mu^k_{LSI_2}(x_2), \ldots, \mu^k_{LSI_n}(x_n) \right\} \right], \quad (30)$$

6. *Defuzzification:* A defuzzification formula is applied to determine the representative crisp value of the aggregated fuzzy region. In this chapter, the centroid method is applied to defuzzify the outputs [34].

4 An Application of S-FIS: To Control and Select UAVs' Communication Technologies

As mentioned before, wireless communication is one of the significant steps ahead for UAVs controlling and navigation systems. It is referred that there are two types of communication and information exchange need for UAVs [49]:

1. Control and non-payload communication (CNPC)
   a. Communication for UAV command and control
   b. Communication for Air Traffic Control (ATC) relay
   c. Communication supporting the "Sense and Avoid" approach.
2. Payload communication

The first group of communication is to guarantee a safe, trustworthy, and efficient flight operation. The second group of communication is directly related to the UAVs' mission and its different applications. The International Telecommunication Union (ITU) is determined and defined the needed CNPC to ensure the safety and security in UAV operations in three sub-categories. All three subcategories of CNPC encourage the UAVs to have a safe and trustworthy flight operation simultaneously.

Some technologically different alternatives for the construction of wireless communication's bases for UAVs are listed below:

1. Direct link communication technology
2. Satellite communication technology
3. Ad-hoc network communication technology
4. Cellular network communication technology (5G network).

Beforehand in the literature review part, these communications technologies and the advantages are discussed. As derived from literature, cellular networks, especially 5G and mm-Wave frequencies, definitely catch ultra-high-speed and real-time transmissions in the future.

Similarly, we know that the technologies, as mentioned earlier, except cellular networks cannot support large-scale communications cost-effectively or, if they are, cause a considerable amount of costs. It is also not economically sensible to build the ground bases for these new and dedicated networks to catch this goal. It is an intense interest in the cellular network, especially 5Gs in UAV-ground communications [51]. So, 5G cellular networks, which is almost ubiquity coverage all over the world as well as its high-speed and advanced communication technologies, which is at the same time so cost-effective, can fulfill both CNPC and payload communication requirements of UAVs. However, despite referred significant advantages for UAVs with 5G cellular network technology, some disadvantages and some inabilities for 5G cellular services can make problems in that way for UAVs. One of the considerable inabilities of 5G cellular-based UAVs is remote areas far from urbanistic and city areas like deserts, oceans, seas, and forests. In facing this kind of scenario, other technologies mentioned above have to be used and support UAV communications. It could also be derived from the above that the future of wireless communication for preparing the supportive bases of large-scale UAV communications will have a higher architecture consisting of UAV-ground, UAV-UAV, and UAV-Satellite communications. Also, each UAV has two or multi communication technologies. Hence, such a hybrid or multi-connectivity increases the capabilities of UAVs. The other side of the coin is that UAVs have different applications based on their qualifications and designs for professional and special missions. So used technology to communicate with UAVs will be determined based on their applications and missions. Therefore, input and output variables are determined based on their applications. Simultaneously, using a high-performance and very UpToDate technology to a UAV with a simple mission is not reasonable due to high costs. Some applications and missions for UAVs are listed as follows [61, 63, 84–87]:

1. Geological and geographical aerial inspections
2. Personal business aerial photography and video streaming by using drones
3. Prehospital Emergency Care (PER)

4. Search and rescue operations
5. communication and telecommunication enhancement
6. Law enforcement
7. Traffic control
8. Package delivery
9. Precision agriculture
10. Scientific explorations.

The UAV could have a hybrid or multi-way connectivity to apply for different applications or missions based on the above mentioned.

The question is which communication technology is most appropriate for a specific application or mission of UAV. Nevertheless, each application and mission have its conditions besides UAVs' features and determine UAVs' appropriate communication technology in those different conditions. The essential parameters for input and output variables as follows:

The most important input variables or parameters which are considered in this model are as shown below:

1. Operational Costs:
    a. Physical Characteristics
        Size
        Weight
        Max speed or power
        Maintenance costs
    b. Energy Consumption
        Max energy capacity
        Energy usage rate
2. Communication Network's Coverage (Availability and Accessibility):
    a. Data rate for command and control
    b. Payload data rate
    c. Payload traffic latency
    d. Packet error rate
3. Performance Rate:
    a. Cruising range
    b. Altitude coverage
    c. Latency

The most critical output variables or indicator factors in this chapter are as follows:

1. Safety and security
2. Reliability
3. Efficiency.

Another crucial issue is that there is a time limitation in the real scenarios' nature. To select within different network technologies that apply appropriate communication networks for special conditions to catch the peak point of satisfaction for goals needs time. This problem is a very complicated decision-making process; besides that, it is a dynamic problem with time limitations. Since we face such a dynamic and complex problem, we offer an S-FIS solution developed using MATLAB Fuzzy Logic ToolboxTM to control and decide UAV's communication technology dynamically. We try to determine the proper communication technology based on the defined parameters and indicator factors. Thus, we proposed a novel Spherical Fuzzy Inference System (S-FIS) based on the Mamdani model to handle the complexities of the problem and make the appropriate decision based on IF–THEN rules taken from the expert system. Our proposed model dynamically selects the communication technology for UAVs and controls it continuously during the mission period.

The spherical fuzzification of input and output variables in our proposed model is based on the linguistic scales are shown in the following Tables 1, 2 and 3.

The input variables $I_i; \forall i = 1, 2, 3$ are defined to cover both control and non-payload communications and payload communications for S-FIS and consist of 1. Operational Costs (OC), 2. Communication Network's coverage and 3. Performance

Table 2 Linguistic scales of input variables $(\widetilde{LSI}_s)$

Input variables (I_i) $(i = 1, 2, 3)$	Spherical fuzzy linguistic scales of input variables $(\widetilde{LSI}_s)$	
	Scale $(l = 1, 2, \ldots, 5)$	$\widetilde{LSI}_s = \langle LSI_l; \mu_{\widetilde{LSI}_s}(LSI_l), \vartheta_{\widetilde{LSI}_s}(LSI_l), h_{\widetilde{LSI}_s}(LSI_l)\rangle$
• Operational Costs (OC) • Communication Network's Coverage (CNC) • Performance Rate (PR)	AHP	$\langle 10; (0.9, 0.1, 0.1)\rangle$
	HP	$\langle 7.5; (0.7, 0.3, 0.3)\rangle$
	EP	$\langle 5; (0.5, 0.5, 0.5)\rangle$
	LP	$\langle 2.5; (0.3, 0.7, 0.3)\rangle$
	ALP	$\langle 0; (0.1, 0.9, 0.1)\rangle$

Table 3 Linguistic scales for output variables $(\widetilde{LSO}_s)$

Output variables (O_j) $(j = 1, 2, 3)$	Spherical fuzzy linguistic scales for output variables $(\widetilde{LSO}_s)$	
	Scale $(l = 1, 2, \ldots, 5)$	$\widetilde{LSO}_s = \langle \mathrm{LSO}_l; \mu_{\widetilde{LSO}_s}(\mathrm{LSO}_l), \vartheta_{\widetilde{LSO}_s}(\mathrm{LSO}_l), h_{\widetilde{LSO}_s}(\mathrm{LSO}_l)\rangle$
• Safety and Security (S&S) • Reliability (R) • Efficiency (E)	AHP	$\langle 1; (0.9, 0.1, 0.1)\rangle$
	HP	$\langle 0.75; (0.7, 0.3, 0.3)\rangle$
	EP	$\langle 0.5; (0.5, 0.5, 0.5)\rangle$
	LP	$\langle 0.25; (0.3, 0.7, 0.3)\rangle$
	ALP	$\langle 0; (0.1, 0.9, 0.1)\rangle$

Table 4 Transformed spherical fuzzy linguistic scales $\widetilde{LS}_s \rightarrow \Theta_\lambda(\widetilde{LS}_s)$

Spherical fuzzy linguistic scales $(\widetilde{LS}_s)$		Transformed spherical fuzzy linguistic scales ($\lambda = 0.5$) $\widetilde{LS}_s \rightarrow \Theta_\lambda(\widetilde{LS}_s)$	
Scale ($l = 1, 2, \ldots, 5$)		$(\mu_{\widetilde{LS}_s}(\mathrm{LS}_l), \vartheta_{\widetilde{LS}_s}(\mathrm{LS}_l), h_{\widetilde{LS}_s}(\mathrm{LS}_l))$	$\left(\mu_{\Theta_\lambda(\widetilde{LS}_s)}(LS_l), \vartheta_{\Theta_\lambda(\widetilde{LS}_s)}(LS_l)\right)$
$l = 1$	(AHP)	(0.9, 0.1, 0.1)	(0.924662, 0.075338)
$l = 2$	(HP)	(0.7, 0.3, 0.3)	(0.771362, 0.228638)
$l = 3$	(EP)	(0.5, 0.5, 0.5)	(0.612372, 0.387628)
$l = 4$	(LP)	(0.3, 0.7, 0.3)	(0.441588, 0.558412)
$l = 5$	(ALP)	(0.1, 0.9, 0.1)	(0.234521, 0.765479)

Rate classes. These classes of inputs contain some sub-classes and some sub-sub-classes to encounter all aspects and parameters that are related to both types of UAVs communications. So, the referred input variables are designed to measure the indicator factors like security and safety and security (S&S), Reliability (R), and Efficiency (E), which are our S-FIS model output variablesO_j; $\forall j = 1, 2, 3$.

In this chapter, we entirely consider 42 rules for S-FIS, and bellow, you could see two examples of those rules which are defined based on Eq. (27):

$$\textbf{\textit{IF}}\ (''OC''\ is\ AHP)\ \textbf{\textit{AND}}\ (''CNC''\ is\ HP)\ \textbf{\textit{AND}}\ (''PR''\ is\ HP)$$
$$\textbf{\textit{THEN}}\ (''S\&S''\ is\ HP)\ \textbf{\textit{AND}}\ (''R''\ is\ AHP)\ \textbf{\textit{AND}}\ (''E''\ is\ LP)$$

After constructing the S-FIS rules we will use the spherical fuzzy transformation function (Θ_λ), which is defined in Definition 7, for linguistic scales, and then we will Evaluate Transformed Gaussian fuzzy numbers for those linguistic scales (Table 5).

Then, the Transformed Gaussian Spherical Fuzzy Linguistic Scales for input $(\widetilde{LSI}_s)$ and output $\left(\widetilde{LSO}_s\right)$ variables by using Eq. (26) is given in (Table 5).

Subsequently, based on Transformed Gaussian Spherical Fuzzy Linguistic Scales for input $(\widetilde{LSI}_s)$ and output $\left(\widetilde{LSO}_s\right)$ variables, which are shown in Table 5 and using Eq. (28), we could rewrite the spherical fuzzy IF–THEN example rule for our proposed S-FIS model:

$$\textbf{\textit{IF}}\left(''OC''\ \ is\ \mu_{\Theta_\lambda(\widetilde{LSI}_s)}(AHP) * Gauss\,(c_1, \sigma)\right)$$
$$\textbf{\textit{AND}}\left(''CNC''\ is\ \mu_{\Theta_\lambda(\widetilde{LSI}_s)}(HP) * Gauss\,(c_2, \sigma)\right)$$
$$\textbf{\textit{AND}}\left(''PR''\ is\ \mu_{\Theta_\lambda(\widetilde{LSI}_s)}(HP) * Gauss\,(c_3, \sigma)\right)$$
$$\textbf{\textit{THEN}}\left(''S\&S''\ \ is\ \mu_{\Theta_\lambda\left(\widetilde{LSO}_s\right)}(AHP) * Gauss\,(c_1, \sigma)\right)$$
$$\textbf{\textit{AND}}\left(''R''\ is\ \mu_{\Theta_\lambda\left(\widetilde{LSO}_s\right)}(AHP) * Gauss\,(c_2, \sigma)\right)$$

Table 5 Transformed Gaussian spherical fuzzy linguistic scales for input $\left(\widetilde{\mathrm{LSI}}_{\mathrm{s}}\right)$ and output $\left(\widetilde{\mathrm{LSO}}_{\mathrm{s}}\right)$ variables

Spherical fuzzy linguistic scales for input $(\widetilde{LSI}_s)$ and output $\left(\widetilde{LSO}_s\right)$ variables		Transformed Gaussian spherical fuzzy linguistic scales for input $(\widetilde{LSI}_s)$ and output $\left(\widetilde{LSO}_s\right)$ variables ($\lambda = 0.5$) $\widetilde{TLSI} = \langle Gauss(c_i, \sigma);\ \mu_{\Theta_\lambda(\widetilde{LSI}_s)}(LSI_l), \vartheta_{\Theta_\lambda(\widetilde{LSI}_s)}(LSI_l)\rangle$ $\widetilde{TLSO} = \left\langle Gauss\left(c_j, \sigma\right);\ \mu_{\Theta_\lambda\left(\widetilde{LSO}_s\right)}(LSO_l), \vartheta_{\Theta_\lambda\left(\widetilde{LSO}_s\right)}(LSO_l)\right\rangle$		
Scale ($l = 1, 2, \ldots, 5$)		$Gauss(c_i, \sigma)$	$Gauss\left(c_j, \sigma\right)$	$\left(\mu_{\Theta_\lambda(\widetilde{LS}_s)}(LS_l), \vartheta_{\Theta_\lambda(\widetilde{LS}_s)}(LS_l)\right)$
$l = 1$	(AHP)	$Gauss(10, \sigma)$	$Gauss(1, \sigma)$	(0.924662, 0.075338)
$l = 2$	(HP)	$Gauss(7.5, \sigma)$	$Gauss(0.75, \sigma)$	(0.771362, 0.228638)
$l = 3$	(EP)	$Gauss(5, \sigma)$	$Gauss(0.5, \sigma)$	(0.612372, 0.387628)
$l = 4$	(LP)	$Gauss(2.5, \sigma)$	$Gauss(0.25, \sigma)$	(0.441588, 0.558412)
$l = 5$	(ALP)	$Gauss(0, \sigma)$	$Gauss(0, \sigma)$	(0.234521, 0.765479)

$$\boldsymbol{AND}\left(''E'' \text{ is } \mu_{\Theta_\lambda\left(\widetilde{LSO_s}\right)}(LP) * Gauss\,(c_{3i}, \sigma)\right), \tag{31}$$

After constructing rules, the S-FIS procedure will be performed step by step to get the final result based on defined IF–THEN rules for our S-FIS model, as given in Figs. 4 and 5.

As an example, let's assume a situation to see the result of the S-FIS for selecting and controlling the communication technology of UAVs (Table 6).

Communication technology of a UAV is strongly affected by the determined parameters, and each input variable affects the indicator factors. Indicator factors of the problem can direct the appropriate communication technology of UAV in its specific conditions. In the designed S-FIS model, the Reference Indicator Factors (RIF) are defined for indicator factors, shows the relation of communication technology with each indicator factor. So, we have output variables or indicator factors $O_j; \forall j = 1, 2, 3.$ we consider four different communication technologies for UAVs, which are referred to above as well. Next, we will have $CT_k; \forall k = 1, 2, \ldots, 4$ as a set of considered communication technologies of UAVs.

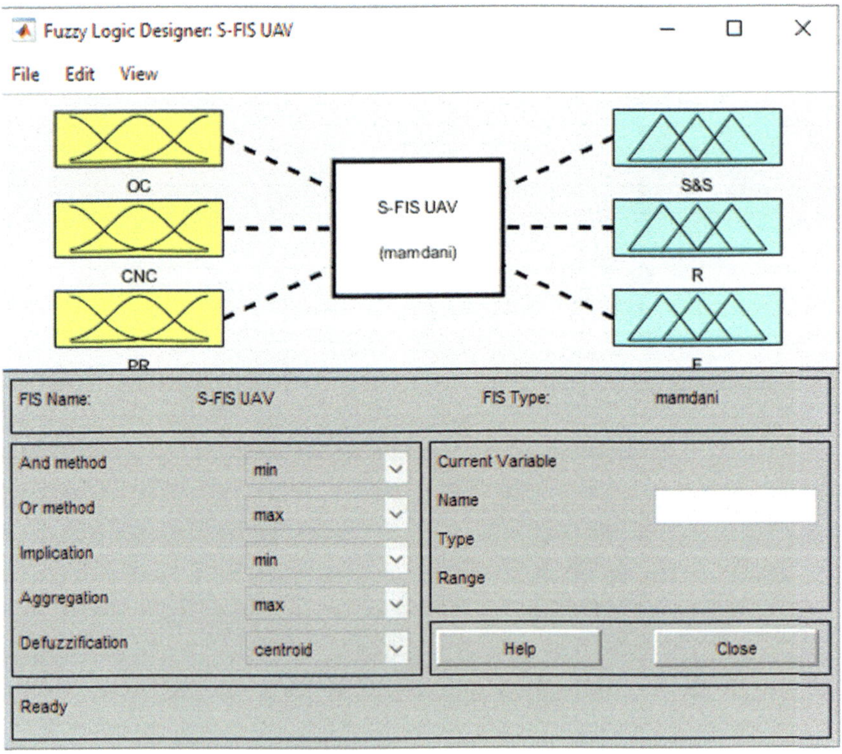

Fig. 4 S-FIS for UAV communication technology selecting and controlling

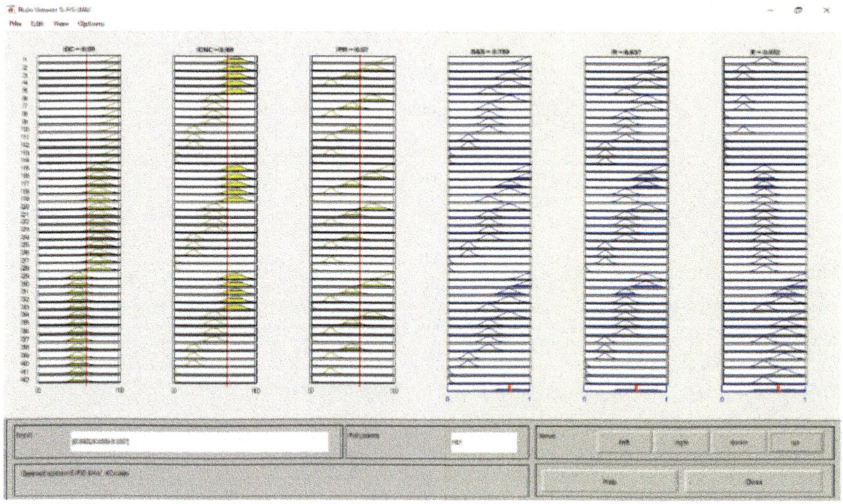

Fig. 5 S-FIS rules view for UAV communication technology controlling

Table 6 Evaluated output variables based on given inputs in proposed S-FIS model

Input variables (I_i) ($i = 1, 2, 3$)	Evaluated values for inputs based on S-FIS	Output variables (O_j) ($j = 1, 2, 3$)	Obtained values for output based on S-FIS
OC	5	S&S	0.75
CNC	8	R	0.507
PR	4.5	E	0.746

1. Direct link communication technology: CT_1
2. Satellite communication technology: CT_2
3. Ad-hoc network communication technology: CT_3
4. Cellular network communication technology (5G network): CT_4.

$$S(A, B) = 1 - \left(\frac{1}{q}\sum_{z=1}^{q}(a_z - b_z)^2\right)^{0.5}$$
$$= 1 - \left(\frac{1}{q}\left[(a_1 - b_1)^2 + (a_2 - b_2)^2 + \ldots + \left(a_q - b_q\right)^2\right]\right)^{0.5}, \quad (32)$$

where, A and B are supposed to be set of numbers as $A = \{a_z; z = 1, 2, \ldots, q\}$ and $B = \{b_z; z = 1, 2, \ldots, q\}$.

After that, Reference Indicator Factors (RIF) will be as $RIF_{jk} = RIF\left(O_j, CT_k\right); \forall j, k$. You could see RIF_{jk} in Table 7.

Table 7 Reference *Indicator Factors* $\left(RIF_{jk}\right)$

RIF_{jk}	$O_1 = S\&S$	$O_2 = R$	$O_3 = E$
CT_1	0.2	0.2	0.4
CT_2	0.9	0.7	0.2
CT_3	0.5	0.6	0.4
CT_4	0.7	0.9	0.6

Table 8 Similarity measure $S\left(RIF_{jk}, IF\right); \forall j, k$ among indicator factor (IF) and reference indicator factors $\left(RIF_{jk}\right)$

Similarity values	$IF = \{O_1 = 0.75, O_1 = 0.507\ and\ O_1 = 0.746\}$
$S\left(RIF_{j1}, IF\right)$	0.585084
$S\left(RIF_{j2}, IF\right)$	0.654619
$S\left(RIF_{j3}, IF\right)$	0.747767
$S\left(RIF_{j4}, IF\right)$	0.756234

After evaluating a set of indicator factors $IF = \left\{O_j\right\}$ based on reference indicator factors, the similarity values can be calculated using Eq. (32). The similarity values have obtained, as shown in Table 8.

According to the obtained similarity measures in Table 8, it can be said that the most appropriate alternative within UAV' communication networks is Cellular network communication technology (CT_4). Also, Ad-hoc network communication technology (CT_3) stands in the second place. The conclusion results of S-FIS based on input and output variables are given in Figs. 6, 7, 8, 9, 10, 11, 12, 13 and 14.

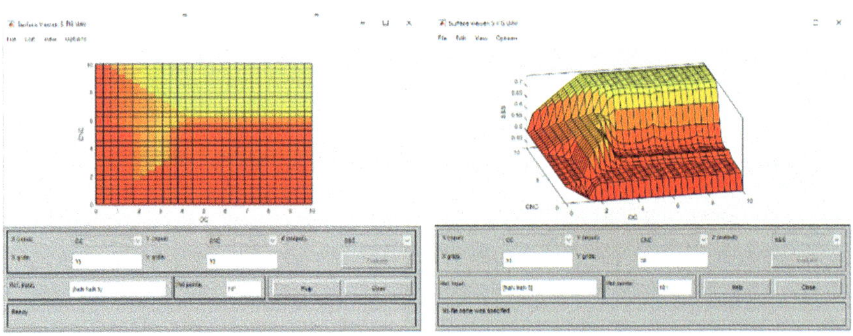

(a) (b)

Fig. 6 S-FIS diagram results by considering input variables (OC) and (CNC) with output variable (S&S) for UAV communication technology controlling **a**: 2D surface diagram, **b**: 3D surface diagram

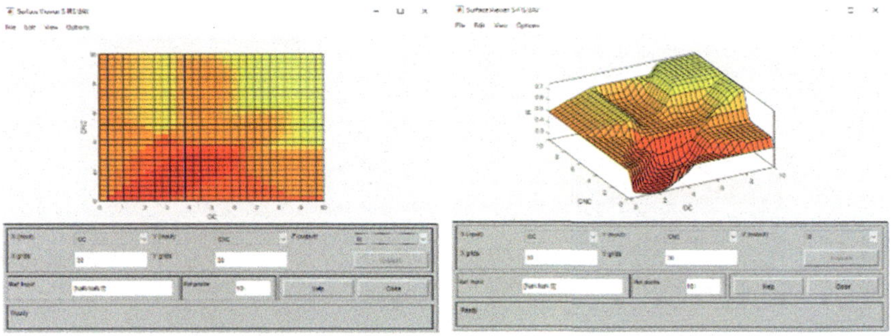

(a) (b)

Fig. 7 S-FIS diagram results by considering input variables (OC) and (CNC) with output variable (R) for UAV communication technology controlling **a**: 2D surface diagram, **b**: 3D surface diagram

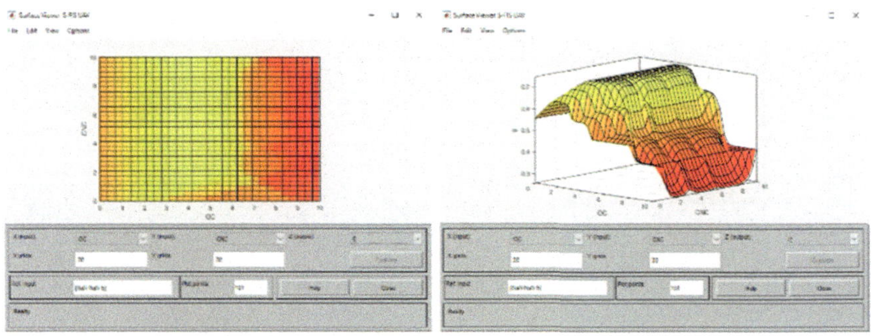

(a) (b)

Fig. 8 S-FIS diagram results by considering input variables (OC) and (CNC) with output variable (E) for UAV communication technology controlling **a**: 2D surface diagram, **b**: 3D surface diagram

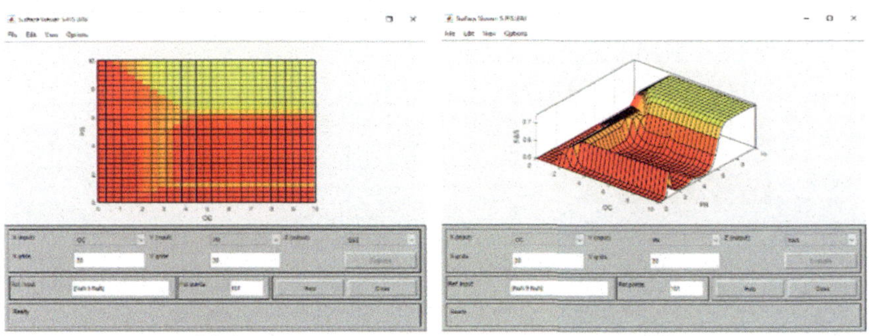

(a) (b)

Fig. 9 S-FIS diagram results by considering input variables (OC) and (PR) with output variable (S&S) for UAV communication technology controlling **a**: 2D surface diagram, **b**: 3D surface diagram

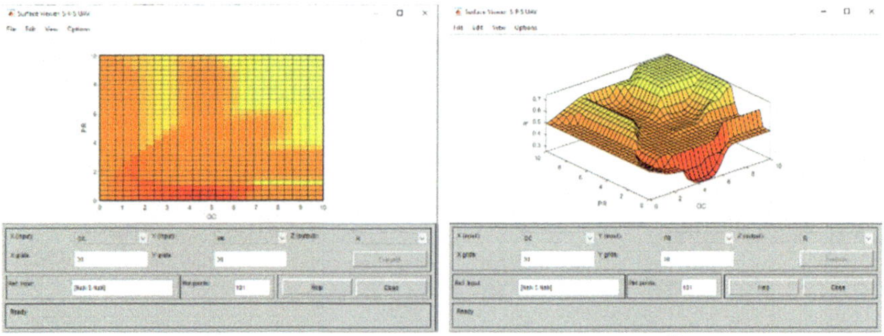

(a) (b)

Fig. 10 S-FIS diagram results by considering input variables (OC) and (PR) with output variable (R) for UAV communication technology controlling **a**: 2D surface diagram, **b**: 3D surface diagram

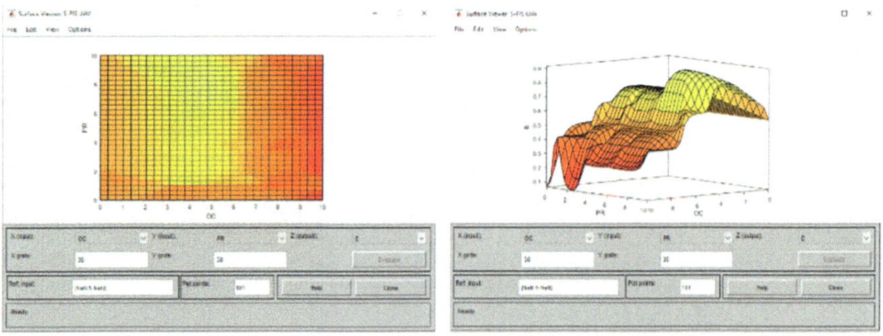

(a) (b)

Fig. 11 S-FIS diagram results by considering input variables (OC) and (PR) with output variable (E) for UAV communication technology controlling **a**: 2D surface diagram, **b**: 3D surface diagram

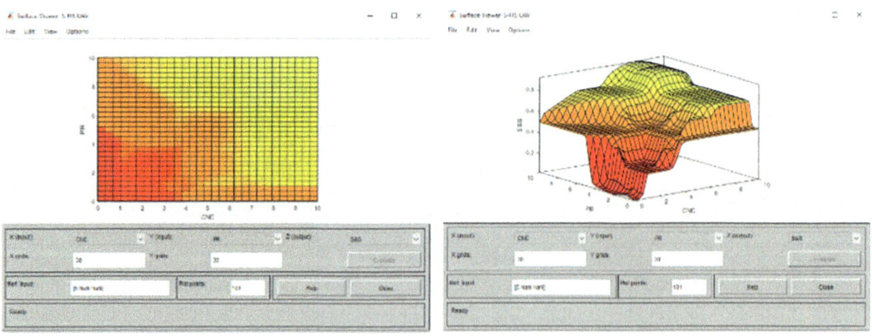

(a) (b)

Fig. 12 S-FIS diagram results by considering input variables (CNC) and (PR) with output variable (S&S) for UAV communication technology controlling **a**: 2D surface diagram, **b**: 3D surface diagram

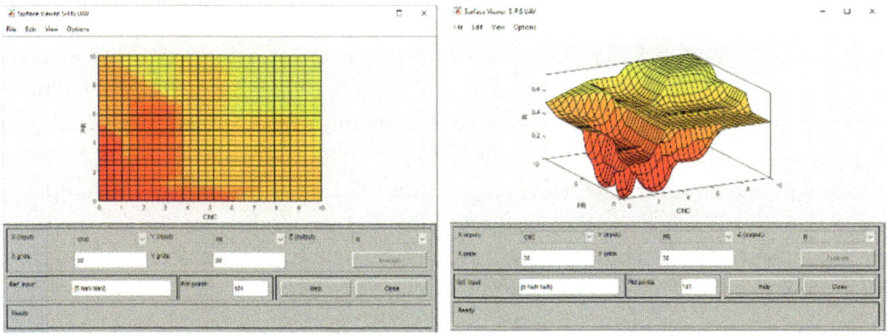

(a) (b)

Fig. 13 S-FIS diagram results by considering input variables (CNC) and (PR) with output variable (R) for UAV communication technology controlling **a**: 2D surface diagram, **b**: 3D surface diagram

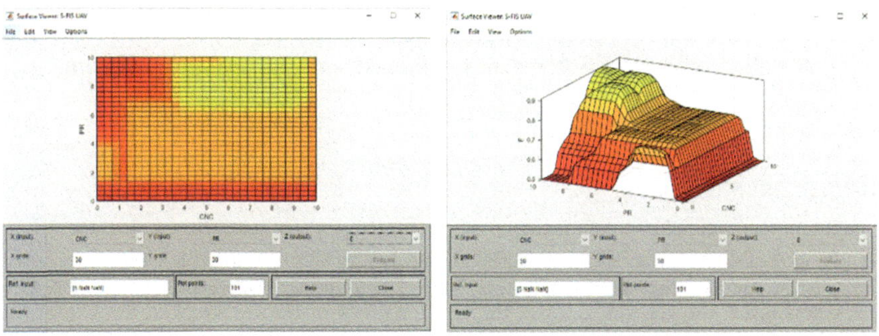

(a) (b)

Fig. 14 S-FIS diagram results by considering input variables (CNC) and (PR) with output variable (E) for UAV communication technology controlling **a**: 2D surface diagram, **b**: 3D surface diagram

5 Conclusion

Fuzzy inference systems are one of the prevalent implementations of fuzzy logic. Spherical fuzzy sets are one of the latest and significant extensions of fuzzy sets. However, we could not find any study about spherical fuzzy inference systems when we covered the literature. Then we introduced spherical fuzzy inference systems based on operations of spherical fuzzy sets. Therefore, our study can be considered an initial attempt to introduce S-FIS, which uses both SFFs characteristics and the technical power of the fuzzy inference system to solve the most complex problems.

The proposed S-FIS model based on the Mamdani model comprises spherical fuzzy rule (SFR) with IF–THEN rules. The proposed novel S-FIS consists of the Mamdani model's steps based on spherical fuzzy: 1. Spherical fuzzification, 2. Spherical fuzzy rule (SFR), 3. Aggregation of different input variables,4. Implication, 5. Aggregation of the output of rules, and 6. Defuzzification.

The UAVs communication technology with all complexities in its nature is the best choice to apply S-FIS and show the high reliability of S-FIS's optimal solution. The proposed S-FIS was illustrated and discussed in detail. Also, UAVs' communication technology was evaluated under the uncertainty of the defined parameters. The three types of parameters were considered in this study: Operational cost, communication networks' coverage, and performance rate. These main parameters include sub-parameters. The important leading indicators of communication technologies are safety and security, reliability, and efficiency. Hence, it is considered 42 rules for S-FIS to solve the UAVs communication technology problem and find the indicator parameters based on input variables, which all are defined based on SFSs. Then it is solved by using the MATLAB Fuzzy Logic ToolboxTM, and all results for output variables (indicators) are given above, which then evaluated by using similarity measures to find the most appropriate network for UAV in that moment and in that outstanding condition. The proposed methodology could dynamically select and control UAVs' communication by continuously considering the time limitation.

For future studies, we propose various applications for S-FIS in UAVs such as supplier selection, control systems, navigation, social, and healthcare systems.

References

1. Zeng, Y., Zhang, R., Lim, T.J.: Wireless communications with unmanned aerial vehicles: opportunities and challenges. IEEE Commun. Mag. (2016). https://doi.org/10.1109/MCOM.2016.7470933
2. Valavanis, K.P., Vachtsevanos, G.J.: Handbook of Unmanned Aerial Vehicles (2015)
3. Mu, X., Liu, Y., Guo, L., Lin, J.: Non-orthogonal multiple access for air-to-ground communication. IEEE Trans. Commun. (2020). https://doi.org/10.1109/TCOMM.2020.2973264
4. Beard, R.W., McLain, T.W., Nelson, D.B., Kingston, D., Johanson, D.: Decentralized cooperative aerial surveillance using fixed-wing miniature UAVs. Proc. IEEE (2006). https://doi.org/10.1109/JPROC.2006.876930
5. Mozaffari, M., Kasgari, A.T.Z., Saad, W., Bennis, M., Debbah, M.: Beyond 5G with UAVs: foundations of a 3D wireless cellular network. IEEE Trans. Wirel. Commun. (2019). https://doi.org/10.1109/TWC.2018.2879940
6. Mozaffari, M., Saad, W., Bennis, M., Nam, Y.H., Debbah, M.: A tutorial on UAVs for wireless networks: applications, challenges, and open problems. IEEE Commun. Surv. Tutorials (2019). https://doi.org/10.1109/COMST.2019.2902862
7. Saad, W., Bennis, M., Chen, M.: A vision of 6G wireless systems: applications, trends, technologies, and open research problems. IEEE Netw. (2020). https://doi.org/10.1109/MNET.001.1900287
8. Li, B., Fei, Z., Zhang, Y.: UAV communications for 5G and beyond: recent advances and future trends. IEEE Internet Things J. (2019). https://doi.org/10.1109/JIOT.2018.2887086
9. Zadeh, L.A.: Fuzzy sets. Inf. Control (1965). https://doi.org/10.1016/S0019-9958(65)90241-X
10. Zadeh, L.A.: The concept of a linguistic variable and its application to approximate reasoning-I. Inf. Sci. (Ny) (1975). https://doi.org/10.1016/0020-0255(75)90036-5
11. Zimmermann, H.-J.: Fuzzy Set Theory—and Its Applications (2001)
12. Mamdani, E.H., Assilian, S.: An experiment in linguistic synthesis with a fuzzy logic controller. Int. J. Man. Mach. Stud. (1975). https://doi.org/10.1016/S0020-7373(75)80002-2

13. Kahraman, C., Gülbay, M., Kabak, Ö.: Applications of fuzzy sets in industrial engineering: a topical classification. Stud. Fuzziness Soft Comput. (2006). https://doi.org/10.1007/3-540-33517-x_1
14. Kutlu Gündoğdu, F., Kahraman, C.: A novel spherical fuzzy QFD method and its application to the linear delta robot technology development. Eng. Appl. Artif. Intell. https://doi.org/10.1016/j.engappai.2019.103348 (2020)
15. Beer, C.: Fuzzy thinking: the new science of fuzzy logic. Bart Kosko, Q. Rev. Biol. https://doi.org/10.1086/418985 (1995)
16. Keshwani, D.R., Jones, D.D., Meyer, G.E., Brand, R.M.: Rule-based Mamdani-type fuzzy modeling of skin permeability. Appl. Soft Comput. J. (2008). https://doi.org/10.1016/j.asoc.2007.01.007
17. Garibaldi, J.M., Ozen, T.: Uncertain fuzzy reasoning: a case study in modelling expert decision making. IEEE Trans. Fuzzy Syst. (2007). https://doi.org/10.1109/TFUZZ.2006.889755
18. Carter, H., Dubois, D., Prade, H.: Fuzzy sets and systems—theory and applications. J. Oper. Res. Soc. (1982). https://doi.org/10.2307/2581310
19. Mizumoto, M., Tanaka, K.: Some properties of fuzzy sets of type 2. Inf. Control (1976). https://doi.org/10.1016/S0019-9958(76)80011-3
20. Mizumoto, M., Tanaka, K.: Fuzzy sets and type 2 under algebraic product and algebraic sum. Fuzzy Sets Syst. (1981). https://doi.org/10.1016/0165-0114(81)90056-7
21. Sambuc, R.: Fonctions and floues : application a l'aide au diagnostic en pathologie thyroidienne. Marseille : Faculté de Médecine de Marseille (1975)
22. Jahn, K.-U.: Intervall-wertige Mengen. Math. Nachrichten **68**(1), 115–132 (1975). https://doi.org/10.1002/mana.19750680109
23. Grattan-Guinness, I.: Fuzzy membership mapped onto intervals and many-valued quantities. Zeitschrift für Math Logik und Grundlagen der Math **22**(1), 149–160 (1976). https://doi.org/10.1002/malq.19760220120
24. Atanassov, K.T.: Intuitionistic fuzzy sets. Fuzzy Sets Syst. (1986). https://doi.org/10.1016/S0165-0114(86)80034-3
25. Behret, H., Öztayşi, B., Kahraman, C.: A fuzzy inference system for supply chain risk management. Adv. Intell. Soft Comput. (2011). https://doi.org/10.1007/978-3-642-25658-5_52
26. Khan, M.J., Shamail, S., Awais, M.M.: Decision making in autonomic managers using fuzzy inference system. In: Proceedings of the 5th International Conference on Autonomic and Autonomous Systems, ICAS 2009, (2009). https://doi.org/10.1109/ICAS.2009.37
27. Abraham, A.: Rule-based expert systems. In: Handbook of Measuring System Design (2005)
28. Liao, S.H.: Expert system methodologies and applications-a decade review from 1995 to 2004. Expert Syst. Appl. (2005). https://doi.org/10.1016/j.eswa.2004.08.003
29. Palm, R.: Designs of fuzzy controllers. In: Fuzzy Systems (1998)
30. Takagi, T., Sugeno, M.: Fuzzy identification of systems and its applications to modeling and control. IEEE Trans. Syst. Man Cybern. (1985). https://doi.org/10.1109/TSMC.1985.6313399
31. Dai, J., Wang, S., Yang, X.: Computerized support systems for emergency decision making. Ann. Oper. Res. (1994). https://doi.org/10.1007/BF02048553
32. Jacob, V.S., Pirkul, H.: A framework for networked knowledge-based systems. IEEE Trans. Syst. Man Cybern. (1990). https://doi.org/10.1109/21.47814
33. Sutherland, J.W.: Assessing the artificial intelligence contribution to decision technology. IEEE Trans. Syst. Man Cybern. (1986). https://doi.org/10.1109/TSMC.1986.289277
34. Ross, T.J.: Fuzzy Logic with Engineering Applications, 3rd edn (2010)
35. Nguyen, H.T., Sugeno, M. (Eds.): Fuzzy Systems, vol. 2, 1st edn. Springer, US (1998)
36. Tang, K.L., Mulholland, R.J.: Comparing fuzzy logic with classical controller designs. IEEE Trans. Syst. Man Cybern. SMC-**17**(6), 1085–1087 (1987). https://doi.org/10.1109/tsmc.1987.6499321
37. Diao, Y., Hellerstein, J.L., Parekh, S.: Using fuzzy control to maximize profits in service level management. IBM Syst. J. (2002). https://doi.org/10.1147/sj.413.0403
38. Donyatalab, Y., Farrokhizadeh, E., Garmroodi, S.D.S., Shishavan, S.A.S.: Harmonic mean aggregation operators in spherical fuzzy environment and their group decision making applications. J. Mult. Log. Soft Comput. (2019)

39. Shishavan, S.A.S., Kutlu Gündoğdu, F., Farrokhizadeh, E., Donyatalab, Y., Kahraman, C.: Novel similarity measures in spherical fuzzy environment and their applications. Eng. Appl. Artif. Intell. (2020). https://doi.org/10.1016/j.engappai.2020.103837
40. Jahn, K.-U.: Intervall-wertige Mengen. Math. Nachrichten (1975). https://doi.org/10.1002/mana.19750680109
41. Yager, R.R.: On the theory of bags. Int. J. Gen. Syst. (1986). https://doi.org/10.1080/03081078608934952
42. Torra, V.: Hesitant fuzzy sets. Int. J. Intell. Syst. (2010). https://doi.org/10.1002/int.20418
43. Yager, R.R.: Pythagorean Fuzzy Subsets. 2013 Jt. IFSA World Congr. NAFIPS Annu. Meet. **2**(x), 57–61 (2013). https://doi.org/10.1109/IFSA-NAFIPS.2013.6608375
44. Cuong, B.C., Kreinovich, V.: Picture fuzzy sets—a new concept for computational intelligence problems. In: 2013 3rd World Congress on Information and Communication Technologies, WICT 2013 (2014). https://doi.org/10.1109/WICT.2013.7113099
45. Yager, R.R.: Generalized orthopair fuzzy sets **25**(5), 1222–1230 (2017)
46. Kutlu Gündoğdu, F., Kahraman, C.: From 1D to 3D Membership: Sphericalfuzzy Sets
47. Kutlu Gündoğdu, F., Kahraman, C.: A novel fuzzy TOPSIS method using emerging interval-valued spherical fuzzy sets. Eng. Appl. Artif. Intell. (2019). https://doi.org/10.1016/j.engappai.2019.06.003
48. Ashraf, S., Abdullah, S., Aslam, M., Qiyas, M., Kutbi, M.A.: Spherical fuzzy sets and its representation of spherical fuzzy t-norms and t-conorms. J. Intell. Fuzzy Syst. **36**(6), 6089–6102 (2019). https://doi.org/10.3233/JIFS-181941
49. Zeng, Y., Wu, Q., Zhang, R.: Accessing from the sky: a tutorial on UAV communications for 5G and beyond. Proc. IEEE (2019). https://doi.org/10.1109/JPROC.2019.2952892
50. "UAS Integration Pilot Program." .
51. Zeng, Y., Lyu, J., Zhang, R.: Cellular-connected UAV: potential, challenges, and promising technologies. IEEE Wirel. Commun. (2019). https://doi.org/10.1109/MWC.2018.1800023
52. Van Der Bergh, B., Chiumento, A., Pollin, S.: LTE in the sky: trading off propagation benefits with interference costs for aerial nodes. IEEE Commun. Mag. (2016). https://doi.org/10.1109/MCOM.2016.7470934
53. Lin, X., et al.: The sky is not the limit: LTE for unmanned aerial vehicles. IEEE Commun. Mag. (2018). https://doi.org/10.1109/MCOM.2018.1700643
54. Bekmezci, İ., Sahingoz, O.K., Temel, Ş.: Flying ad-hoc networks (FANETs): a survey. Ad Hoc Netw. **11**(3), 1254–1270. https://doi.org/10.1016/j.adhoc.2012.12.004
55. Khawaja, W., Guvenc, I., Matolak, D.W., Fiebig, U.C., Schneckenburger, N.: A survey of air-to-ground propagation channel modeling for unmanned aerial vehicles. IEEE Commun. Surv. Tutor. (2019). https://doi.org/10.1109/COMST.2019.2915069
56. Fotouhi, A., et al.: Survey on UAV cellular communications: practical aspects, standardization advancements, regulation, and security challenges. IEEE Commun. Surv. Tutor. (2019). https://doi.org/10.1109/COMST.2019.2906228
57. Khuwaja, A.A., Chen, Y., Zhao, N., Alouini, M.S., Dobbins, P.: A survey of channel modeling for UAV communications. IEEE Commun. Surv. Tutorials (2018). https://doi.org/10.1109/COMST.2018.2856587
58. Matolak, D.W., Sun, R.: Unmanned aircraft systems: air-ground channel characterization for future applications. IEEE Veh. Technol. Mag. (2015). https://doi.org/10.1109/MVT.2015.2411191
59. Frew, E.W., Brown, T.X.: Airborne communication networks for small unmanned aircraft systems. Proc. IEEE (2008). https://doi.org/10.1109/JPROC.2008.2006127
60. Gupta, L., Jain, R., Vaszkun, G.: Survey of Important Issues in UAV communication networks. IEEE Commun. Surv. Tutor. (2016). https://doi.org/10.1109/COMST.2015.2495297
61. Hayat, S., Yanmaz, E., Muzaffar, R.: Survey on unmanned aerial vehicle networks for civil applications: a communications viewpoint. IEEE Commun. Surv. Tutor. (2016). https://doi.org/10.1109/COMST.2016.2560343
62. Shakhatreh, H., et al.: Unmanned aerial vehicles (UAVs): a survey on civil applications and key research challenges. IEEE Access **7**, 48572–48634. Institute of Electrical and Electronics Engineers Inc. (2019). https://doi.org/10.1109/ACCESS.2019.2909530.

63. Hossein Motlagh, N., Taleb, T., Arouk, O.: Low-altitude unmanned aerial vehicles-based internet of things services: comprehensive survey and future perspectives. IEEE Internet Things J. (2016). https://doi.org/10.1109/JIOT.2016.2612119
64. Shakeri, R., et al.: Design challenges of multi-UAV systems in cyber-physical applications: a comprehensive survey and future directions. IEEE Commun. Surv. Tutorials (2019). https://doi.org/10.1109/COMST.2019.2924143
65. Zhang, J., Yu, X., Letaief, K.B.: Hybrid beamforming for 5G and beyond millimeter-wave systems: a holistic view. IEEE Open J. Commun. Soc. (2019). https://doi.org/10.1109/ojcoms.2019.2959595
66. Sharma, A., et al.: Communication and networking technologies for UAVs: a survey. J. Netw. Comput. Appl. (2020). https://doi.org/10.1016/j.jnca.2020.102739
67. Marsch, P., et al.: 5G radio access network architecture: design guidelines and key considerations. IEEE Commun. Mag. (2016). https://doi.org/10.1109/MCOM.2016.1600147CM
68. Marsch, P., Da Silva, I., Bulakci, Ö., Tesanovic, M., El Ayoubi, S.E., Säily, M.: Emerging network architecture and functional design considerations for 5G radio access. Trans. Emerg. Telecommun. Technol. **27**(9), 1168–1177 (2016). https://doi.org/10.1002/ett.3073
69. Zeitlin, A.D.: Progress on Requirements and Standards for Sense & Avoid. *Wwwsrv2.Mitre.Org*, (2010)
70. Characteristics of unmanned aircraft systems and spectrum requirements to support their safe operation in non-segregated airspace (2009)
71. Gündoğdu, F.K., Kahraman, C.: Spherical fuzzy sets and spherical fuzzy TOPSIS method. J. Intell. Fuzzy Syst. (2019). https://doi.org/10.3233/JIFS-181401
72. Kahraman, C., Kutlu Gundogdu, F., Cevik Onar, S., Oztaysi, B.: Hospital location selection using spherical fuzzy TOPSIS. In: Proceedings of the 2019 Conference of the International Fuzzy Systems Association and the European Society for Fuzzy Logic and Technology (EUSFLAT 2019) (2019). https://doi.org/10.2991/eusflat-19.2019.12
73. Kutlu Gündoğdu, F., Kahraman, C.: A novel spherical fuzzy analytic hierarchy process and its renewable energy application. Soft Comput. (2019). https://doi.org/10.1007/s00500-019-04222-w
74. Jang, J.S.R., Sun, C.T., Mizutani, E.: Neuro-fuzzy and soft computing—a computational approach to learning and machine intelligence, 1482 B. Rev. IEEE Trans. Autom. Control (1997)
75. Tay, K.M., Lim, C.P.: On the use of fuzzy inference systems for assessment and decision making problems. Intell. Syst. Ref. Libr. **4**, 233–246 (2010). https://doi.org/10.1007/978-3-642-13639-9_10
76. Mamdani, E.H.: Application of fuzzy algorithms for control of simple dynamic plant. Proc. Inst. Electr. Eng. (1974). https://doi.org/10.1049/piee.1974.0328
77. Sugeno, M., Kang, G.T.: Structure identification of fuzzy model. Fuzzy Sets Syst. (1988). https://doi.org/10.1016/0165-0114(88)90113-3
78. Tsukamoto, Y.: An approach to fuzzy reasoning method. In: Readings in Fuzzy Sets for Intelligent Systems (1993)
79. Kasabov, N.K., Song, Q.: DENFIS: dynamic evolving neural-fuzzy inference system and its application for time-series prediction. IEEE Trans. Fuzzy Syst. (2002). https://doi.org/10.1109/91.995117
80. Lohani, A.K., Goel, N.K., Bhatia, K.K.S.: Takagi-Sugeno fuzzy inference system for modeling stage-discharge relationship. J. Hydrol. (2006). https://doi.org/10.1016/j.jhydrol.2006.05.007
81. Ashraf, S., Abdullah, S.: Spherical aggregation operators and their application in multiattribute group decision-making. Int. J. Intell. Syst. (2019). https://doi.org/10.1002/int.22062
82. Atanassov, K.T.: Two theorems for intuitionistic fuzzy sets. Fuzzy Sets Syst. (2000). https://doi.org/10.1016/S0165-0114(99)00112-8
83. Gulley, N., Jang, J.-S.R.: MATLAB fuzzy logic toolbox user's guide. The MathWorks (1997)
84. Erdelj, M., Natalizio, E., Chowdhury, K.R., Akyildiz, I.F.: Help from the sky: leveraging UAVs for disaster management. IEEE Pervasive Comput. (2017). https://doi.org/10.1109/MPRV.2017.11

85. Zhang, L., et al.: A survey on 5G millimeter wave communications for UAV-assisted wireless networks. IEEE Access (2019). https://doi.org/10.1109/access.2019.2929241
86. Chao, H., Cao, Y., Chen, Y.: Autopilots for small unmanned aerial vehicles: a survey. Int. J. Control. Autom. Syst. (2010). https://doi.org/10.1007/s12555-010-0105-z
87. Naqvi, S.A.R., Hassan, S.A., Pervaiz, H., Ni, Q.: Drone-aided communication as a key enabler for 5G and resilient public safety networks. IEEE Commun. Mag. (2018). https://doi.org/10.1109/MCOM.2017.1700451

Technology Analysis for Logistics 4.0 Applications: Criteria Affecting UAV Performances

Aylin Adem, Burcu Yilmaz Kaya, and Metin Dağdeviren

Abstract Industry 4.0 revolution and elements related to it such as IoT technologies, artificial intelligence, cognitive computing, big data analytics, and digitization of systems initiated a transformation in almost every industry and application area. This big revolution step generated a paradigm shift in the aviation industry too, addressing a new concept called Aviation 4.0. As IoT aided logistics application devices, which are designed to be remotely administered, UAVs attract more attention and get popular every day, and employment of this technology in delivery or transportation processes becomes the new trend, due to its competency on operating in difficult or dangerous geographical areas, lower resource consumption, higher mobility level, faster response feature, plus contactless and sustainable delivery potential. 10 different performance indicators related to UAVs are investigated with two different fuzzy number styles, named hesitant and Pythagorean fuzzy numbers, respectively, and AHP MCDM methodology, in this study. The results implied that the first two important key factors were determined as privacy and economic life of UAVs in logistic activities with respect to Aviation 4.0. In order for the process of using UAVs in logistics applications within the scope of aviation 4.0 to be successful, attention must be paid to the features related to the privacy and economic life of the utilized UAVs.

Keywords Aviation 4.0 · Logistics 4.0 · Hesitant fuzzy numbers · Pythagorean fuzzy numbers · Analytical hierarchy process · Unmanned aerial vehicles

1 Introduction

As Industry 4.0 revolution is no longer seen as a new trend but a mandatory technological and organizational re-arrangement stage to systematically fulfill customer needs, its applications on specific fields of variant operations are attracting attention. The digitalization of all system elements and transformation of ongoing processes into

A. Adem (✉) · B. Yilmaz Kaya · M. Dağdeviren
Industrial Engineering Department, Gazi University, 06570 Ankara, Turkey
e-mail: aylinadem@gazi.edu.tr

C. Kahraman and S. Aydın (eds.), *Intelligent and Fuzzy Techniques in Aviation 4.0*, Studies in Systems, Decision and Control 372,
https://doi.org/10.1007/978-3-030-75067-1_21

cyber-physical systems through IoT technology, had led today's broader apprehension on Industry 4.0 application fields. Today, practitioners seek new opportunities for bigger utilization of Industry 4.0 concept, yet, new concepts such as the Aviation 4.0 began to be addressed.

The airline companies undergone many changes during the last three decades, where thanks to Aviation 4.0 concept and the innovations brought along with it, next three decades are expected to be more thrilling in terms of technological improvements and execution reforms. Even today, airline companies are facing with new descriptions of mobility, connectivity, personalized service, and quality experience [1]. With more than 5000 sensors, which generate up to 10 GB of data per second, modern aircraft engines alone are proofs that what digitalization and the IoT technology could furnish for aviation, as part of the upcoming Aviation 4.0 revolution. Besides these, this new era has the potential to make improvements on air transport key performance points, particularly, in an industry where safety levels are so high and the margins for improvement are extremely tight [2].

Unmanned Aerial Vehicles (UAVs) constitute an important IoT-aided device group, which is especially employed in the field of logistics services. As an outstanding application field of Aviation 4.0 concept, this device group sometimes can be referred as Flying Robots (FRs), or, with another more general and familiar appellation "the drone technologies".

These aerial devices are designed to be self-programmed and are able to be remotely administered through a mobile device thanks to the sensors, processors, and wireless connectivity devices. By means of the advantages brought by this novel technology, UAVs were mostly adopted by military forces and defense industries at first. Later on, after civil and commercial sectors also began to recognize their benefits such as ease of deployment, dynamic configuration convenience, comparatively lower investment and maintenance cost levels, higher mobility proposing features, and facilities as faster and accurate response; this technology started to be adopted by a diverse range of application areas and sectors. The main UAV employment areas other than military and aviation based field can be listed as; healthcare sector, arts and entertainment events, public administration, transportation, education services, agriculture, insurance sector, public safety, environmental applications, and logistics. Interested readers can find more detailed information related to these diverse application areas in the studies of Choudhary et al. [3] and Mehta et al. [4]. While UAVs are being introduced to a new application field each day, the economic value of demands in economic markets is constantly and exponentially increasing, too. In 2013, the demand of UAVs for commercial and industrial use was approximately \$0.25 billion, where it was \$19.3 billion in 2019 and projected to be \$45.8 billion by 2025 [4].

Among mentioned many application fields, UAVs are considerably proper to be employed in the field of logistics, where competency of operating in different geographies and physical infrastructures is considered very difficult, dangerous or expensive. Moreover, the UAV technology promises other advantages on delivery circumstances in terms of logistics applications; as instances, in case of limited human capital or energy resources, or may be, some unforeseen constraints, such as the current case of COVID-19 pandemic, UAVs makes contactless and uninterrupted delivery possible and sustainable by reaching the target market or delivery points where and when it is needed. Although many advantageous and valuable contribution presented by UAV technologies, the application framework and required specifications still lacks in standardization and unification, mostly due to the challenges arising from its novelty. As Torens et al. [5] mentioned, there is currently no finished regulation for certification and accreditation for the use of UAVs. Hence, to be able to integrate UAVs into the segregated or non-segregated airspace, some critical issues and related specifications have to be identified and regulated in the first place. The specifications of UAVs are expected to be vary depending on the purpose of use, naturally, but a set of several broadened criteria will remain fundamental for each and every type of UAVs for any purpose of use. These mentioned UAV specifications, or UAV selection criteria, are expected to be numerous, conflicting and mainly expressed linguistically. In this context, multi-criteria decision making (MCDM) methodologies and fuzzy logic will be particularly suitable to handle this type of problem and provide reliable and accurate solutions. Some recent studies related to UAV applications employing MCDM methodologies and fuzzy logic can be listed as [1, 6–18]. Fuzzy scales have an advantage over crisp numbers because they make it easier to handle the uncertainty structure of the data. At the same time, especially Hesitant Fuzzy numbers have another advantage of fuzzy numbers in that they allow decision-makers to convey their thoughts to the decision-making process more easily [19]. Similarly, since the Pythagorean numbers are also linguistically evaluated, the comfort of decision-makers in the decision-making process was provided. The main advantages of using fuzzy scales are that decision-makers can express their thoughts more easily with the logic of computing with words and they handle the uncertainty of the data structure appropriately.

In this chapter, implementation process and some performance related issues about UAVs are enlightened, by means of determining the most important UAV specifications and their effects on overall application performance. In this manner, a set of broadened key performance indicators consisting of ten criteria named "*Privacy*", "*Security*", "*Repair time*", "*Energy efficiency*", "*Audio and video surveillance*", "*Economic life*", "*Flight distance*", "*Communication efficiency*", "*Fault tolerance*" and "*Maneuverability*" are investigated. Analytical Hierarchy Process (AHP), a very well-known and robust MCDM methodology, is employed to analyze, compare and weight the criteria belonging to the key performance indicators set of UAVs. Furthermore, two different fuzzy number presentation types are employed to represent the linguistic expressions and humanitarian aspects on the results and underline their effects; named Hesitant fuzzy number representation and Pythagorean fuzzy number representation. The hierarchical model for the key performance indicators of UAVs is developed based on the literature research and decision makers' experience and opinions. Many of the existing studies in UAV literature generally considered solely the system vulnerabilities of UAVs, and are mainly focused on single or few but generally security and privacy related specifications about UAV technologies for this problem type. In this study, apart from those, different aspects related to performance, implementation and utilization issues related to UAVs are underlined and investigated in the perspective of Aviation 4.0, to be able to introduce a road-map for researchers and practitioners. This study is mainly focused on examining the UAVs' performance criteria. The proposed approach is to evaluate these performance criteria using fuzzy scales and the AHP technique. It is aimed to evaluate the performance criteria of UAVs and to clearly demonstrate which criteria will be prioritized in a possible selection. Especially evaluating and prioritizing critical success factors is sort of a guide for people who will deal with this business. For this evaluation, two different linguistic scales were employed and the obtained results were compared. Thus, it was aimed to reveal the differences between the two linguistic scales.

The remaining of this chapter is organized as follows; the concepts of Aviation 4.0, Logistics 4.0 and UAVs are detailed and related terms and terminology are presented in Sect. 2, employed methodologies are described in Sect. 3, and, mathematical problem and proposed approach are illustrated in Sect. 4. Results are indicated and discussed in Sect. 5.

2 Unmanned Aerial Vehicle and Aviation 4.0 Concepts

Aviation 4.0 can be understood as a concept where Industry 4.0 revolution is implemented and embraced in aviation operations. Aviation 4.0 could be seen as the fourth era of the technology, hence, before addressing the fourth stage of revolution, former stages of automation integration on have to be identified.

Aviation 1.0 was defined as the beginning of the commercial aviation in the beginning of 1900s, where flights were performed under visual flight rules, and there was hardly any instrumental aid to help pilots to fly. This era was dominated by the technological difficulties on how to build and fly an aircraft, where very simple instruments constituted the so-called first steps toward "virtualization of the environment" to provide basic indications required for the flight, such as, anemometer, altimeters, pneumatic and electric gyroscopes, basic mechanical autopilots and servos, or Mach trim compensators [2]. Despite these, the advancements in engine technology and construction could be considered as such rapid innovation and was outstanding even by today's standards, considering the uncertainty in aircraft technology and the diversity of tools and equipment available to engineers at that time. As an instance, in a matter of four years between 1909 and 1913, the speed, altitude and distance records were increased from 47.82 mph to 126.67 mph, 1486 ft. to 20,079 ft., and 145.53 mi to 634.54 mi, respectively, which was swift enough to propel the performance of aircraft and set the stage for the next era.

Aviation 2.0 begins with the first scheduled commercial air travel across borders, which was an international operation by Aircraft Transport and Travel Limited in 1919. This era specifies the time period when aircraft sizes and performances were increased, new civil aviation industries were formed, and innovative designs were being introduced to the aeronautical space, which enabled the replacement of old mechanism by electrical devices. New flight indicator instruments such as the VOR (very high-frequency omnidirectional range) and ILS (instrument landing system) were introduced in this era, which allowed the pilots to follow safely tracks and approach route paths. As board innovations, such as electrical autopilots, auto-throttle, flight directors, airborne weather radars, navigation instruments, inertial platforms belong to this era, too. Aviation 2.0 evolution came with a rise of information to be managed by the pilot, who might be confronted with more than 600 devices and indicators to be monitored and controlled in the cockpit [2]. This era ends with the introduction of the world's first commercial jet airliner in 1952.

Aviation 3.0, the third stage of aero-revolution, started in 1952, which led to true, fast, reliable and affordable transcontinental travel through the first commercial jet airliner. This era involves the massive incorporation of electronics in the cockpit, which was driven by the availability of reliable and usable digital data processing and communication technologies. At the beginning of this era, electronics significantly helped to diminish the clutter of instruments and replace the old indicators with integrated colored displays, cathode ray tube (CRT) and liquid crystal display (LCD). Technological solutions were progressively designed to support the pilots, with the help of aggregated, visualized, understandable information. The amount of information available in the system raised exponentially during this era, as a consequence of higher level digitalization. This third revolution in aviation introduced another new notion, which is the "electronic echo-systems". An A-320 was able to incorporate around 190 computers placed all through the fuselage, which interacts with them without the pilot being aware, can be given as an instance for "electronic echo-systems" [2].

After the introduction of Industry 4.0, and wide-spreading applications of it in various industries; integration of cyber-physical systems into aviation addressed the newest era. Aviation 4.0 includes the concepts of digital and "smart airplane". Today, the amount and diversity of operational data that can be collected onboard and by ground operations raised exponentially, as side effects of IoT applications and digitalization of systems, but are still able to be handled thanks to cloud computing technologies and big data analysis. In Aviation 4.0, the concept and facilities of this new era can be traced in the manufacturing processes of the aircrafts, in-flight operations, maintenance processes, ground operations and tracking systems.

In parallel with the introduction of Aviation 4.0, a new era for transportation, warehousing, and handling activities called Logistics 4.0 was also addressed. Prinz et al. [20] stated that Logistics 4.0 stands in a broader sense of Industry 4.0 including five functional areas valid across organizations, namely; data collection and processing, assistance systems, networking and integration, decentralization and service orientation, and self-organization and autonomy. The key logistics activities of transport, inventory management, material handling, supply chain structure and information flow are affected in Logistics 4.0. In Logistics 4.0 environment, (i) by real-time big data analytics of vehicles, products and facilities optimal routing for material and product transportation could be found, (ii) storage needs are reduced by on-site and on-demand, (iii) rapid manufacturing opportunities, autonomous robots and vehicles along with tracking and decision-making systems are abled to keep control over inventory, (iv) real-time exchange of information among different actors is made possible, and, (v) the information flow is kept intact by smart products and cloud-supported networks [21]. Logistics 4.0 refers to the combination of traditional logistics and supply chain management activities with the innovations added by cyber-physical systems, regarding Industry 4.0 revolution. Logistics 4.0 operates with a different set of component parts, as instances "smart" versions of containers, vehicles, pallets, and transport systems in order to create a fully networked supply stream and to route, transport and perform other logistics tasks in an optimal way.

As a main device available for employment in terms of Logistics 4.0 activities, the UAV technology was born in the early 1920s. UAVs form a fundamental part of an unmanned aircraft system, which generally consists of a drone, a ground control system (GCS), and a communication network between them for exchange of information. There are plenty of employment area of UAVs, where they are considered useful mainly in situations which are too monotonous (i.e., agricultural land surveillance) or too dangerous (i.e., border or nuclear plant surveillance) for human beings. UAVs are complex technological devices which can fly autonomously using preprogrammed flight plans, as well as, are able to create new flight plans during the flights using Simultaneous Location and Mapping [4].

UAVs could be categorized in various ways considering specific attributes like the type of aircraft (rotorcraft or fixed wing), weight, flight altitude (high, medium, low) or relative speed [22]. UAVs also come in distinguish sizes. Most of-the-shelf, commercial "drones" are quad-rotor aircraft, considered as "small UAVs) category, being limited in payload weight and flight altitude and in most cases can be controlled within line-of-sight by the operator through an unlicensed frequency band radio communication link [22]. Large UAVs can be used in defense or missions' critical applications, whereas small UAVs can either to form swarms or perform regular tasks (best for civilian applications) [4]. In addition to these, there is also a concept of multiple UAVs. UAVs are considered to be restricted to limited geographical region due to their energy and visual coverage constraints. A multi-UAV system (more than one UAVs are operated in parallel) can overcome the issues of energy and coverage. Multi-UAV systems also enhance the observation diversity by sensing data from various point of views. Moreover, these systems also increase the UAV data reliability and fault tolerance [4]. When a multi-UAV system is being considered, these systems can be categorized and analyzed according to UAV communication styles, in this case. Communication techniques employed among UAVs can differ in some aspects, too. UAV communication can be of three types; UAV-to-UAV (U2U),) UAV-to-base station, and UAV-to-satellite. In U2U multi-UAV systems, UAVs can pass or share information among themselves for scalability, reliability, and precision results, where in UAV-to-base station systems, a base station can control or direct UAVs through air interface and UAVs can send analyzed data to base station for accurate and effective results. In UAV-to-satellite systems, satellites are used to track UAVs when UAV leaves the line of sight, then base station shifted to a satellite control [4, 22].

As for many other IoT technology applications, security is a big and serious issue for UAVs, and has a vital importance on effectiveness and reliability of the systems. Blockchain technologies with its various features to improve transparency, immutability, and security is considered as a perfect fit to handle UAV security issues. As a main reason, UAVs have to take real-time fast decisions, which can be very risky. Moreover, they have to do proper path planning with obstacle avoidance. In case of a single UAV, it must communicate effectively with the GCS to create a safe path. If the UAVs are working in a swarm, which is generally considered as the civilian application of devices belonging the small UAVs category, then they can communicate among themselves and share information to coordinate [4] safely through blockchain technologies.

Some essential studies belonging to the literature of UAV technology practices related to Aviation 4.0 application can be listed as Choudhary et al. [3], Militaru et al. [22], Petrovsky et al. [23], Torens et al. [5], Valdes et al. [2], Ghafar et al. [24], Büyüközkan et al. [1], Cokorilo [25] Mehta et al. [4]. For more detailed information and beneficial identification of all related terms, interested readers are recommended to review especially these articles. Different problems are held with different methodologies suitable to the problem type in these studies, while, the methodologies employed in this study are presented in details in the next section: Sect. 3.

3 Methods

In this part of the study, the methods employed are presented. For making comparative analysis, two different fuzzy extensions of the AHP technique were employed to weight the key performance indicators related to employment of UAVs in logistic activities in the scope of Aviation 4.0.

AHP technique is based on pairwise comparisons of criteria or alternatives [26] and was developed by Saaty in 1977 [27]. AHP was employed to problems in various areas [28–31]. Saaty proposed this technique with a 1–9 scale, however, as time passes, this scale started to be not suitable in such circumstances which include vagueness, since decision makers are expected to feel more comfortable to express their evaluations in linguistic terms in place of exact values for most of the decision-making processes [32]. In the existing literature, plenty of different MCDM methods have been improved by adding various fuzzy extensions for handled problems covering such situations stated in the works of Ramirez-Atencia et al. [33] Qin et al. [34], Chien et al. [35], Li and Chen [36], Solangi et al. [37], and Sufiyan et al. [38]. In this study, extended AHP methods with Hesitant and Pythagorean numbers were used, respectively. Hence, in this section, the AHP method which is extended with Pythagorean numbers was explained, firstly. Then the AHP method which is extended with Hesitant fuzzy numbers was presented.

3.1 Pythagorean Fuzzy AHP

Yager [39] proposed Pythagorean fuzzy sets and its basic set operations. Karasan et al. [40] specified that the Pythagorean Fuzzy Numbers (PFNs) are an extended form of intuitionistic fuzzy sets. This extension was conducted with the aim of handling the ambiguity of data sets in a more reliable sense [41]. Some basic explanations of PFNs and the computing steps of the Pythagorean fuzzy AHP (PF-AHP) method are given as follows:

Definition Let X be a universe. One Pythagorean fuzzy set P is an element having the following form (Eq. 1) [39];

$$\tilde{P} = \left\{x, P(\mu_p(x), v_p(x)) \middle| x \in X\right\} \tag{1}$$

where

$$\mu_p(x) : X \to [0, 1]\,\text{degree of membership}$$
$$v_p(x) : X \to [0, 1]\,\text{degree of non-membership}$$

of the element $x \in X$ to P. Every element of X satisfies the given inequality below (Eq. 2);

$$0 \le (\mu_P(x))^2 + (v_P(x))^2 \le 1 \tag{2}$$

In PFNs, while, the sum of membership and non-membership degrees can be greater than 1, the sum of squares of membership and non-membership cannot be greater than 1 [42].

For each PFN, where $\tilde{P}\; x \in X,\; \pi_p(x) = \sqrt{1 - \mu_p(x)^2 - v_p(x)^2}$ is the degree of hesitancy of x to $\tilde{P}$. Zhang and Xu [43] defined the Euclidean (d) and Taxicab (T) distance between two PFNs in the given equations below (Eqs. 3–4);

$$d(\tilde{P}_1, \tilde{P}_2) = \frac{1}{2}\left(\left|\mu^2_{P_1} - \mu^2_{P_2}\right| + \left|v^2_{P_1} - v^2_{P_2}\right| + \left|\pi^2_{P_1} - \pi^2_{P_2}\right|\right). \tag{3}$$

$$T(\tilde{P}_1, \tilde{P}_2) = \left(\left|\mu_{P_1} - \mu_{P_2}\right| + \left|v_{P_1} - v_{P_2}\right| + \left|\pi_{P_1} - \pi_{P_2}\right|\right). \tag{4}$$

Given two PFNs $P_1 = P(\mu_{P_1}, v_{P_1}),\; P_2 = P(\mu_{P_2}, v_{P_2})\, and\, \lambda > 0$. Zeng et al. [44] and Zhang and Xu [43] described some mathematical operators for PFNs as follows (Eqs. 5–7);

$$\tilde{P}_1 \oplus \tilde{P}_2 = P\left(\sqrt{\mu^2_{P_1} + \mu^2_{P2}} - \mu^2_{P_1}\mu^2_{P_2}, v_{P_1}, v_{P_2}\right) \tag{5}$$

$$\tilde{P}_1 \otimes \tilde{P}_2 = P\left(\mu_{P_1}, \mu_{P_2}, \sqrt{v^2_{P_1} + v^2_{P2}} - v^2_{P_1}v^2_{P_2}\right) \tag{6}$$

$$\lambda\tilde{P} = P\left(\sqrt{1 - (1 - \mu^2_P)^\lambda}, (v_P)^\lambda\right), \lambda \ge 0\, and\, \lambda \in R. \tag{7}$$

The execution steps of our first methodology PF-AHP are explained from her on. These steps are directly adapted from Gül [45]'s paper.

Step 1. Developing the pairwise comparison matrix (A = $(a_{ik})_{m\times m}$). In this matrix, i.e. while comparing the alternatives or criteria; the linguistic scale which is offered by Ilbahar et al. [46] is utilized. The mentioned scale is presented in Table 1.

Step 2. The elements of the difference matrix (D = $(d_{ik})_{m\times m}$) are computed using Eqs. (8) and (9). This matrix includes the differences between the lower and upper values of the membership and non-membership functions

$$d_{ikL} = \mu_{ikL}^2 - v_{ikU}^2 \tag{8}$$

$$d_{ikU} = \mu_{ikU}^2 - v_{ikL}^2 \tag{9}$$

Step 3. The interval multiplicative matrix ($S = (s_{ik})_{m\times m}$) is computed with the help of Eqs. (10) and (11) respectively:

Table 1 Pythagorean fuzzy scale for decision making process [46]

	Pythagorean fuzzy number			
Linguistic Term	Lower value of membership degree (μ_L)	Upper value of membership degree (μ_U)	Lower value of non-membership degree (v_L)	Upper value of non-membership degree (v_U)
Certainly low important (CLI)	0.00	0.00	0.90	1.00
Very low important (VLI)	0.10	0.20	0.80	0.90
Low important (LI)	0.20	0.35	0.65	0.80
Below average important (BAI)	0.35	0.45	0.55	0.65
Average important (AI)	0.45	0.55	0.45	0.55
Above average important (AAI)	0.55	0.65	0.35	0.45
High important (HI)	0.65	0.80	0.20	0.35
Very high important (VHI)	0.80	0.90	0.10	0.20
Certainly high important (CHI)	0.90	1.00	0.00	0.00
Exactly equal (EE)	0.1965	0.1965	0.1965	0.1965

$$S_{ikL} = \sqrt{1000^{d_{ikL}}} \tag{10}$$

$$S_{ikU} = \sqrt{1000^{d_{ikU}}} \tag{11}$$

Step 4. The hesitancy/determinacy value $\tau = (\tau_{ik})_{mxm}$ is computed using Eq. (12).

$$\tau_{ik} = 1 - (\mu_{ikU}^2 - \mu_{ikL}^2) - (v_{ikU}^2 - v_{ikL}^2) \tag{12}$$

Step 5. The hesitancy/determinacy degrees are multiplied with $S = (s_{ik})_{m\times m}$ matrix of weights $T = (t_{ik})_{m\times m}$ with the help of Eq. (13):

$$t_{ik} = \left(\frac{S_{ikL} + S_{ikU}}{2}\right) * \tau_{ik} \tag{13}$$

Step 6. Last step of this method is normalization process. By applying Eq. (14) priority weights of each element is gained.

$$w_i = \sum_{k=1}^{m} t_{ik} / \sum_{i=1}^{m}\sum_{k=1}^{m} t_{ik} \tag{14}$$

The readers who are interested in AHP and PF-AHP can benefit from the following papers: Dağdeviren et al. [26], Karasan et al. [40], Yücesan and Kahraman [47]. As a digest literature review, generally, PF-AHP method was utilized in the literature with the intention of determining suitable site selection. Kaya et al. [48] selected recycling facility location with the help of PF-AHP. Karasan et al. [40] determined the landfill site with PF-AHP. Yıldız et al. [49] solved the problem of automated teller machine site selection utilizing PF-AHP and TOPSIS integration. Shete et al. [50] utilized PF-AHP method to evaluate sustainable supply chain innovation enablers. Pythagorean numbers are also utilized by combining with other MCDM methods, besides the AHP method. Interested readers could find the following studies selected from the literature as illustrative studies; SWARA [51]; AHP [52]; COPRAS [53]; CODAS [54]; TOPSİS [55]; VIKOR [56]; PROMETHEE [57]; TODIM [58].

3.2 *Hesitant Fuzzy Linguistic Term Set*

The Hesitant Fuzzy Linguistic Term (HFLT) set is a fairly novel issue in fuzzy logic [59]. HFLT sets are explained as sets of linguistic terms that have been developed in a way to take into account decision-makers' indecision/hesitation, which allows decision-makers to transfer their thoughts to the decision-making process more easily [19, 60]. In other words, employment of HFLT sets means trying to make it easier for the decision makers to convey their perspective in their mind to the decision-making process. Rodriguez et al. [61] are the first researchers among all to use these sets for decision making. Since its first days, the HFLT sets were integrated into many different MCDM methods. In order to exemplify this diversity, selected examples from the literature are presented which employed different MCDM methodologies as instances; MULTIMOORA [60], Todim [62], ELECTRE [63], VIKOR [64], TOPSIS [65]; AHP [66]; PROMETHEE [67]; DEMATEL [68]; Copras [69]; ORESTE [70]. The following stages of the application steps of HFLT sets are adapted from the study of Yavuz et al. [19].

Step 1. The semantics and syntax of the used linguistic term set S are defined in this step. S has the following components; "No importance (NI), very low importance (VLI), low importance (LI), medium importance (MI), high importance (HI), very high (VHI), absolute importance (AI)" (Table 2).

Step 2. The context free grammar GH is defined as second step, where GH = {VN, VT, I, P}.

VT = {lower than, greater than, at least, at most, between, and, S_0, S_1, $S_{2,}$, …, S_g}.

I ϵ VN.

For the context free grammar, the production rules are indicated in Fig. 1.

The context-free grammar is actually developed for computer programming languages. Context-Free grammar rules adjust the mechanism among the defined linguistic terms in the general sense and are in short represented by *G* [71]. In context-free grammar rules, the square brackets are used to intent "I" in the form of a straight line to separate each preferred option [72].

Table 2 The scale for linguistic expressions [19]

ni	vli	li	mi	hi	vhi	ai
0	1	2	3	4	5	6

$$P = \begin{pmatrix} \text{I} = \langle\text{primary term}\rangle \mid \langle\text{composite term}\rangle, \\ \langle\text{composite term}\rangle ::= \\ \langle\text{unary relation}\rangle\langle\text{primary term}\rangle \mid \langle\text{binary relation}\rangle \\ \langle\text{primary term}\rangle \mid \langle\text{conjuction}\rangle\langle\text{primary term}\rangle, \\ \text{primary term} ::= S_0 \mid S_1 \mid \ldots. \mid S_g, \langle\text{unary relation}\rangle ::= \\ \text{lower than} \mid \text{greater than} \mid \text{at least} \mid \text{at most}, \\ \langle\text{binary relation}\rangle ::= \text{between}, \langle\text{conjuction}\rangle ::= \text{and} \end{pmatrix}$$

Fig. 1 Production rules

Step 3. The preferences relations p^k provided by experts ({k: 1, 2, …, m}) for both criteria and alternatives, are gathered as the third step. In this step, experts are asked to make evaluations for both alternatives and criteria.

Step 4. The preferences relations into HFLT sets are transformed by using the transformation function E_{GH}. In order to be able to mathematically process the expert's evaluations taken, it is necessary to express verbal statements as a range. This process is called the transformation process and is performed by the E_{GH} function [72].

E_{GH} (high importance) = [high importance]

E_{GH} (at least medium importance) = [medium importance, high importance, very high importance, absolute importance]

Step 5. The envelope of $[p_{ij}k^{-}, p_{ij}k^{+}]$ is obtained for each HFLT set in the fifth step. In this phase, a complete range assignment is conducted to the expressions converted with the E_{GH} function in the previous step. It is desired to obtain a lower limit and upper limit with the range assignment operation just mentioned. The lower limit represents the pessimistic interpretation for the related linguistic expression, whilst the upper limit represents the optimistic interpretation.

Step 6. A linguistic aggregation operator (φ) is selected in this step. If there is more than one expert, it is necessary to combine experts' evaluations in order to express the evaluations of each expert as a single evaluation. This aggregation is conducted with the help of linguistic aggregation operators. Although there are multiple linguistic aggregation operators in the literature, the arithmetic mean method which does not allow for data loss is a frequently used operator. In this study, the arithmetic mean method was employed [73].

Step 7. Optimistic and pessimistic collective preferences for each criterion/alternative are calculated here, by using the linguistic aggregation operator.

Step 8. In this step, the linguistic ranges and midpoints of each criterion/alternative is determined.

Step 9. As the last and ninth step, midpoints are normalized and weights for criteria or alternatives are determined here.

4 Application

In this part of this chapter, different fuzzy extensions of the AHP technique are employed to weight the key performance indicators of UAV employment in logistic activities in the scope of Aviation 4.0. The reasons of employment two different fuzzy extensions to weight the key performance indicators can be explained as; (1) the aim of comparing the results and hence effects of two different fuzzy extension methods, (2) showing the main differences between the two linguistic scales. The mentioned key performance indicators were determined by literature investigations and brainstorming process results of acknowledged experts. Determined indicators are listed as Privacy (C1), Security (C2), Repair Time (C3), Energy Efficiency (C4), Audio and Video Surveillance (C5), Economic Life (C6), Flight Distance (C7), Communication Efficiency (C8), Fault Tolerance (C9), and Maneuverability (C10).

The following explanations of the determined criteria were given with the intention of clarifying them:

- ***Privacy***: What is meant by this criterion is the degree of confidentiality regarding the address where the UAV finally reached and the content it carries.
- ***Security***: This criterion means that if the UAV carries a product, it means the protection of the product without harming its physical integrity.
- ***Repair Time***: This criterion indicates the duration of malfunctions and periodic maintenance caused by a certain period of use. Since the reparability of the UAV at home or abroad will affect this period, the ability to perform repair and maintenance services domestically affects this criterion directly. This period is desired to be short.
- ***Energy Efficiency***: This criterion refers to the battery life of the UAV. How long a UAV can serve with a charged battery matters. This period is desired to outlast.
- ***Audio and video surveillance***: This criterion indicates whether the UAV has this feature or the degree of the feature.
- ***Economic Life***: This criterion stands for the product life cycle of UAV. It is desirable that the period indicated by this criterion is outlast.
- ***Flight distance***: This criterion refers to how far the UAV can fly without interrupting communication with the center. This distance is desired to be long.
- ***Communication* efficiency**: This criterion intends that the communication that the UAV can establish with the center during its flight is healthy.
- ***Fault tolerance***: It can be expressed as a systemic feature. It indicates the ability of the system to continue operating, although some parts of the UAV may lose their functionality.
- ***Maneuverability***: This criterion refers to the maneuverability of the UAV during its movement.

As the first stage of the application, criteria weights which are determined with the help of PF-AHP technique are presented. Table 3 shows the pairwise comparisons of the criteria. The evaluations in Table 3 were conducted using the linguistic scale which was given in Table 1.

Table 4 presents the difference matrix of the PF-AHP calculation process. This matrix is acquired by applying Eqs. (8) and (9) to Table 3.

Table 5 presents the interval multiplicative matrix, which is gained by applying Eqs. (10) and (11) to Table 4 i.e. difference matrix.

Table 6 shows the T matrix of decision making process. This matrix is gained with the help of Eq. (13). Table 7 shows the weights of criteria. These weights are computed using Eq. (14).

The order of the importance of key performance indicators which is determined with respect to PF-AHP method is obtained as follows: Privacy (C1)—Economic Life (C6)—Communication Efficiency (C8)—Security (C2)—Audio and

Table 3 Decision matrix of PF-AHP application

	C1	C2	C3	C4	C5	C6	C7	C8	C9	C10
C1	EE	AI	BAI	HI	AAI	EE	VHI	HI	CHI	AAI
C2		EE	BAI	LI	AI	CLI	AI	LI	EE	CHI
C3			EE	EE	LI	BAI	AI	VLI	AI	BAI
C4				EE	EE	HI	AAI	BAI	EE	EE
C5					EE	LI	EE	LI	VHI	AI
C6						EE	HI	EE	VHI	AAI
C7							EE	AI	HI	EE
C8								EE	HI	EE
C9									EE	AI
C10										EE

Video Surveillance (C5)—Energy Efficiency (C4)—Flight Distance (C7)—Maneuverability (C10)—Repair Time(C3)—Fault Tolerance (C9). After the weights of the key performance indicators were obtained by PF-AHP method, the second scale was utilized to calculate the weights of indicators. Table 8 shows the linguistic evaluations for criteria with respect to HFLT sets.

Table 9 shows the enveloped matrix of the evaluations. The elements of Table 9 were gained with the help of E_{GH} function.

Table 10 gives the interval utilities, midpoints of interval utilities and the weights of indicators. Interval utilities of criteria were computed as follows:

$$\Delta\Bigg(\frac{1}{9}(\Delta^{-1}(mi,3)+\Delta^{-1}(mi,3)+\Delta^{-1}(hi,4)+\Delta^{-1}(mi,3)$$
$$+\Delta^{-1}(mi,3)+\Delta^{-1}(vhi,5)+\Delta^{-1}(mi,3)+\Delta^{-1}(hi,4)+\Delta^{-1}(vli,1))$$
$$=\Delta\left(\frac{1}{9}(3+3+4+3+3+5+3+4+1)\right)=\Delta(3.22)=(mi,+0.22)$$

$$\Delta\Bigg(\frac{1}{9}(\Delta^{-1}(mi,3)+\Delta^{-1}(hi,4)+\Delta^{-1}(hi,4)+\Delta^{-1}(mi,3)$$
$$+\Delta^{-1}(hi,4)+\Delta^{-1}(ai,6)+\Delta^{-1}(hi,4)+\Delta^{-1}(ai,6)+\Delta^{-1}(mi,3)$$
$$=\Delta\left(\frac{1}{9}(3+4+4+3+4+6+4+6+3)\right)=\Delta(4.11)=(hi,+0.11)$$

Midpoints of the interval utilities are found as follows: (3.22 + 4.11)/2 = 3.667. The weights of the key performance indicators are computed by normalizing the midpoints of interval utilities (Table 10).

Table 4 Difference matrix for PF-AHP application

	C1		C2		C3		C4		C5		C6		C7		C8		C9		C10	
C1	0	0	−0.1	0.1	−0.3	−0.1	0.3	0.6	0.1	0.3	0	0	0.6	0.8	0.3	0.6	0.81	1	0.1	0.3
C2	−0.1	0.1	0	0	−0.3	−0.1	−0.6	−0.3	−0.1	0.1	−1	−0.81	−0.1	0.1	−0.6	−0.3	0	0	0.81	1
C3	0.1	0.3	0.1	0.3	0	0	0	0	−0.6	−0.3	−0.3	−0.1	−0.1	0.1	−0.8	−0.6	−0.1	0.1	−0.3	−0.1
C4	−0.6	−0.3	0.3	0.6	0	0	0	0	0	0	0.3	0.6	0.1	0.3	−0.3	−0.1	0	0	0	0
C5	−0.3	−0.1	−0.1	0.1	0.3	0.6	0	0	0	0	−0.6	−0.3	0	0	−0.6	−0.3	0.6	0.8	−0.1	0.1
C6	0	0	0.81	1	0.1	0.3	−0.6	−0.3	0.3	0.6	0	0	0.3	0.6	0	0	0.6	0.8	0.1	0.3
C7	−0.8	−0.6	−0.1	0.1	−0.1	0.1	−0.3	−0.1	0	0	−0.6	−0.3	0	0	−0.1	0.1	0.3	0.6	0	0
C8	−0.6	−0.3	0.3	0.6	0.6	0.8	0.1	0.3	0.3	0.6	0	0	−0.1	0.1	0	0	0.3	0.6	0	0
C9	−1	−0.81	0	0	−0.1	0.1	0	0	−0.8	−0.6	−0.8	−0.6	−0.6	−0.3	−0.6	−0.3	0	0	−0.1	0.1
C10	−0.3	−0.1	−1	−0.81	0.1	0.3	0	0	−0.1	0.1	−0.3	−0.1	0	0	0	0	−0.1	0.1	0	0

Table 5 The interval multiplicative matrix for PF-AHP application

C1		C2		C3		C4		C5		C6		C7		C8		C9		C10	
1.000	1.000	0.708	1.413	0.355	0.708	2.818	7.943	1.413	2.818	1.000	1.000	7.943	15.849	2.818	7.943	16.406	31.623	1.413	2.818
0.708	1.413	1.000	1.000	0.355	0.708	0.126	0.355	0.708	1.413	0.032	0.061	0.708	1.413	0.126	0.355	1.000	1.000	16.406	31.623
1.413	2.818	1.413	2.818	1.000	1.000	1.000	1.000	0.126	0.355	0.355	0.708	0.708	1.413	0.063	0.126	0.708	1.413	0.355	0.708
0.126	0.355	2.818	7.943	1.000	1.000	1.000	1.000	1.000	1.000	2.818	7.943	1.413	2.818	0.355	0.708	1.000	1.000	1.000	1.000
0.355	0.708	0.708	1.413	2.818	7.943	1.000	1.000	1.000	1.000	0.126	0.355	1.000	1.000	0.126	0.355	7.943	15.849	0.708	1.413
1.000	1.000	16.406	31.623	1.413	2.818	0.126	0.355	2.818	7.943	1.000	1.000	2.818	7.943	1.000	1.000	7.943	15.849	1.413	2.818
0.063	0.126	0.708	1.413	0.708	1.413	0.355	0.708	1.000	1.000	0.126	0.355	1.000	1.000	0.708	1.413	2.818	7.943	1.000	1.000
0.126	0.355	2.818	7.943	7.943	15.849	1.413	2.818	2.818	7.943	1.000	1.000	0.708	1.413	1.000	1.000	2.818	7.943	1.000	1.000
0.032	0.061	1.000	1.000	0.708	1.413	1.000	1.000	0.063	0.126	0.063	0.126	0.126	0.355	0.126	0.355	1.000	1.000	0.708	1.413
0.355	0.708	0.032	0.061	1.413	2.818	1.000	1.000	0.708	1.413	0.355	0.708	1.000	1.000	1.000	1.000	0.708	1.413	1.000	1.000

Table 6 T matrix for PF-AHP application

	C1	C2	C3	C4	C5	C6	C7	C8	C9	C10
C1	1.00	0.85	0.43	3.77	1.69	1.00	9.52	3.77	19.45	1.69
C2	0.85	1.00	0.43	0.17	0.85	0.04	0.85	0.17	1.00	19.45
C3	1.69	1.69	1.00	1.00	0.17	0.43	0.85	0.08	0.85	0.43
C4	0.17	3.77	1.00	1.00	1.00	3.77	1.69	0.43	1.00	1.00
C5	0.43	0.85	3.77	1.00	1.00	0.17	1.00	0.17	9.52	0.85
C6	1.00	19.45	1.69	0.17	3.77	1.00	3.77	1.00	9.52	1.69
C7	0.08	0.85	0.85	0.43	1.00	0.17	1.00	0.85	3.77	1.00
C8	0.17	3.77	9.52	1.69	3.77	1.00	0.85	1.00	3.77	1.00
C9	0.04	1.00	0.85	1.00	0.08	0.08	0.17	0.17	1.00	0.85
C10	0.43	0.04	1.69	1.00	0.85	0.43	1.00	1.00	0.85	1.00

Table 7 The weights of criteria calculated with PF-AHP application

Criteria	Weights
C1	0.213
C2	0.122
C3	0.040
C4	0.073
C5	0.092
C6	0.212
C7	0.049
C8	0.131
C9	0.026
C10	0.041

The order of importance of selection criteria which is determined with respect to HFLT sets-AHP method is obtained as follows, Privacy (C1)—Economic Life (C6)—Communication Efficiency (C8)—Security (C2)—Audio and Video Surveillance (C5)—Energy Efficiency (C4)—Flight Distance (C7)—Maneuverability (C10)—Repair Time(C3)—Fault Tolerance (C9).

5 Results and Discussion

The following table is given for the comparison of the weights of the key performance indicators obtained according to both methods, PF-AHP and HFLT-AHP methods (Table 11).

It was observed that the first three and last three sequences of the criteria were the same for both methods. The results of this paper showed that the first two important

Table 8 Linguistic evaluations of criteria for HFLT sets

	C1	C2	C3	C4	C5	C6	C7	C8	C9	C10
C1	–	mi	Between mi and hi	hi	mi	Between mi and hi	Between vhi and ai	Between mi and hi	At least hi	Between vli and mi
C2		–	Between mi and hi	Between vli and li	Between mi and hi	Between vli and li	At least hi	Between vli and li	mi	vhi
C3			–	mi	At most li	vli	mi	Between vli and mi	li	Between li and mi
C4				–	mi	Between mi and hi	Between hi and vhi	Between li and mi	Between mi and vhi	Between li and mi
C5					–	li	Between mi and hi	Between vli and li	Between mi and hi	hi
C6						–	Between hi and vhi	mi	Between hi and vhi	mi
C7							–	li	Between hi and vhi	mi
C8								–	hi	mi
C9									–	Between vli and mi
C10										–

Table 9 Enveloped matrix of the evaluations for HFLT sets

	C1	C2	C3	C4	C5	C6	C7	C8	C9	C10
C1	–	[mi; mi]	[mi; hi]	[hi; hi]	[mi; mi]	[mi; hi]	[vhi; ai]	[mi; hi]	[hi;ai]	[vli; mi]
C2		–	[mi; hi]	[vli; li]	[mi; hi]	[vli; li]	[hi;ai]	[vli; li]	[mi;mi]	[vhi;vhi]
C3			–	[mi; mi]	[ni; li]	[vli; vli]	[mi; mi]	[vli; mi]	[li; li]	[li; mi]
C4				–	[mi; mi]	[mi; hi]	[hi; vhi]	[li; mi]	[mi; vhi]	[li; mi]
C5					–	[li; li]	[mi; hi]	[vli; li]	[mi; hi]	[hi; hi]
C6						–	[hi; vhi]	[mi; mi]	[hi; vhi]	[mi; mi]
C7							–	[li; li]	[hi; vhi]	[mi; mi]
C8								–	[hi; hi]	[mi; mi]
C9									–	[vli; mi]
C10										–

Table 10 Interval utilities, midpoints and the weights of indicators calculated with HFLT sets

	Interval utilities	Midpoints	Weights
C1	[mi, +0.22; hi, +0.11]	3.667	0.122
C2	[mi,−0.33; mi, +0.44]	3.056	0.102
C3	[li,−0.23; mi, −0.45]	2.167	0.072
C4	[mi, −0.11; hi, −0.33]	3.278	0.109
C5	[mi, −0.23; mi, +0.44]	3.111	0.104
C6	[mi, +0.44; hi, 0.00]	3.722	0.124
C7	[li, −0.22; mi,−0.44]	2.167	0.072
C8	[mi, +0.33; hi, +0.00]	3.667	0.122
C9	[li, −0.33; mi,−0.33]	2.167	0.072
C10	[mi, −0.33; mi, +0.33]	3.000	0.100

key factors were determined as privacy and economic life of UAVs. The difference between the weights of indicators that share the first two orders in both methods is very low (such as 0.001). The fact that the difference between the weights of the first two most important indicators is so small can be interpreted as that these two factors are almost equally important. It is especially attractive that the economic life of UAVs is determined as one of the most important critical performance indicators. On the other hand, repair time and fault tolerance was determined as the two least important in both methods. In this paper, a study was conducted to determine the key performance indicators that will affect the success of UAVs in logistics applications within the scope of Aviation 4.0 and to prioritize these indicators. In order to achieve

Table 11 Comparison of the sequences obtained according to both methods

	PF-AHP	HFLT-AHP
C1	1	2
C2	4	6
C3	9	9
C4	6	4
C5	5	5
C6	2	1
C7	7	8
C8	3	3
C9	10	10
C10	8	7

this aim, the AHP method combined with linguistic scales was used. The reason for using linguistic scales during the evaluation of key performance/success factors can be explained by the fact that Pythagorean numbers handle uncertainty in the data more realistically and strongly [46]. In addition, a calculation was conducted with Hesitant Fuzzy linguistic term sets. The reason why Hesitant sets were utilized can be explained as they allow decision-makers to make freer evaluations in making computing with words [19]. In future studies, different linguistic scales and the concept of group decision can be put into use. In this study, the AHP method was used because it was assumed that there was no relationship between the criteria. However, in future studies, there may be situations where the interaction between critical success factors should be taken into account. In such a case, the integration of the DEMATEL and ANP techniques would be a more appropriate solution method for determining the interaction between the criteria and determining the criterion weights calculated according to this interaction.

At the same time, for further studies, the problem of choosing an appropriate/ the best UAV can be addressed by employing the criteria and weights introduced by the current paper. It may be desirable to have a lower number of criteria in selection problems. Therefore, among the indicators introduced here, those showing similar characteristics can be expressed on the basis of some main selection criteria (technical, economic, social, etc.). For example, criteria such as audio and video surveillance, repair time can be represented under the title of technical criteria, and a reduction in the number of criteria can be achieved. In addition, energy efficiency and economic life indicators can be shown under the basic heading of economy criteria. It is clear that UAVs will have a very critical function in the changing and developing world integrated with technology This function will evolve in a way that facilitates the work of human beings, especially in aviation and logistics activities. For this reason, we think that the analysis of the critical success factors or key performance indicators discussed in this study will be beneficial to the literature and private sector employees on this subject.

References

1. Büyüközkan, G., Feyzioğlu, O., Havle, C.A.: Analysis of success factors in aviation 4.0 using integrated intuitionistic fuzzy MCDM methods. https://doi.org/10.1007/978-3-030-23756-1_73. Retrieved from www.scopus.com. (2020)
2. Arnaldo Valdes, R., Gómez Comendador, V.F.: Aviation 4.0: more safety through automation and digitization. Paper Presented WIT Trans. Built Environ. **174**, 225–236 (2018). https://doi.org/10.2495/SAFE170211
3. Choudhary, G., Sharma, V., Gupta, T., Kim, J., You, I.: Internet of Drones (IoD): threats, vulnerability, and security perspectives. In The 3rd International Symposium on Mobile Internet Security (MobiSec'18), Aug 29–Sept 1, 2018, Cebu, Philippines, Article No. 37, pp. 1–13 (2018)
4. Mehta, P., Gupta, R., Tanwar, S.: Blockchain envisioned UAV networks: challenges, solutions, and comparisons. Comput. Commun. **151**, 518–538 (2020). https://doi.org/10.1016/j.comcom.2020.01.023
5. Torens, C., Dauer, J. C., Adolf, F.: Towards autonomy and safety for unmanned aircraft systems. In Advances in Aeronautical Informatics: Technologies Towards Flight 4.0, pp. 105–120. https://doi.org/10.1007/978-3-319-75058-3_8 (2018)
6. Mavris, D.N., Collins, K.B., Schrage, D.P.: A method of qualitative analysis during conceptual design as applied to unmanned aerial vehicles. Paper Presented Ann. Forum Proc. Am. Helicopter Soc. **1**, 642–655 (2004)
7. Fitzgerald, D.L., Mejias, L., Eng, P., Liu, X., Walker, R.: Towards flight trials for an autonomous UAV emergency landing using machine vision. Paper presented at the proceedings of the 2007 Australasian conference on robotics and automation, ACRA 2007 (2007)
8. Fallahi, K., Leung, H., Chandana, S.: An integrated ACO-AHP approach for resource management optimization. Paper presented at the conference proceedings—IEEE international conference on systems, man and cybernetics, pp. 4335–4340.https://doi.org/10.1109/ICSMC.2009.5346794 (2009)
9. Li, X., Ci, L., Yang, M., Cheng, B.: Exploration-exploitation balancing deployment strategy in UAV sensor networks. Information **14**(8), 2701–2710 (2011)
10. Sun, X., Cai, C., Yang, J., Shen, X.: Route evaluation for unmanned aerial vehicle based on type-2 fuzzy sets. Eng. Appl. Artif. Intell. **39**, 132–145 (2015). https://doi.org/10.1016/j.engappai.2014.11.008
11. Sariçiçek, I., Akkuş, Y.: Unmanned aerial vehicle hub-location and routing for monitoring geographic borders. Appl. Math. Model. **39**(14), 3939–3953 (2015). https://doi.org/10.1016/j.apm.2014.12.010
12. Petkovics, I., Simon, J., Petkovics, A., Covic, Z.: Selection of unmanned aerial vehicle for precision agriculture with multi-criteria decision making algorithm. Paper presented at the SISY 2017—IEEE 15th international symposium on intelligent systems and informatics, proceedings, pp. 151–155. https://doi.org/10.1109/SISY.2017.8080543 (2017)
13. Dursun, M., Çuhadar, İ: Risk based multi criteria decision making for secure image transfer between unmanned air vehicle and ground control station. Reliab. Eng. Syst. Saf. **178**, 31–39 (2018). https://doi.org/10.1016/j.ress.2018.05.011
14. Mondal, T., Bhattacharya, I., Pramanik, P., Boral, N., Roy, J., Saha, S., Saha, S.: A multi-criteria evaluation approach in navigation technique for micro-jet for damage & need assessment in disaster response scenarios. Knowl. Based Syst. **162**, 220–237 (2018). https://doi.org/10.1016/j.knosys.2018.09.016
15. Jung, S., Kim, K., Roh, B., Ham, J.: Load balancing algorithm for multiple UAVs relayed tactical ad hoc networks. Paper Presented Proc. Int. Comput. Softw. Appl. Conf. **1**, 944–945 (2019). https://doi.org/10.1109/COMPSAC.2019.00153
16. Raj, A., Sah, B.: Analyzing critical success factors for implementation of drones in the logistics sector using grey-DEMATEL based approach. Comput. Indus. Eng. **138**https://doi.org/10.1016/j.cie.2019.106118 (2019)

17. Aragão, F.V., Cavicchioli Zola, F., Nogueira Marinho, L.H., De Genaro Chiroli, D.M., Braghini Junior, A., Colmenero, J.C.: Choice of unmanned aerial vehicles for identification of mosquito breeding sites. Geospatial Health **15**(1). https://doi.org/10.4081/gh.2020.810 (2020)
18. Karaşan, A., Kaya, İ.: Neutrosophic TOPSIS method for technology evaluation of unmanned aerial vehicles (UAVs).https://doi.org/10.1007/978-3-030-23756-1_80 (2020)
19. Yavuz, M., Oztaysi, B., Cevik Onar, S., Kahraman, C.: Multi-criteria evaluation of alternative-fuel vehicles via a hierarchical hesitant fuzzy linguistic model. Expert Syst. Appl. **42**(5), 2835–2848 (2015)
20. Prinz, C., Morlock, F., Freith, S., Kreggenfeld, N., Kreimeier, D., Kuhlenkötter, B.: Learning factory modules for smart factories in industrie 4.0. Paper Presented Proc. CIRP **54**, 113–118 (2016). https://doi.org/10.1016/j.procir.2016.05.105
21. Strandhagen, J.O., Vallandingham, L.R., Fragapane, G., Strandhagen, J.W., Stangeland, A.B.H., Sharma, N.: Logistics 4.0 and emerging sustainable business models. Adv. Manuf. **5**(4), 359–369. https://doi.org/10.1007/s40436-017-0198-1 (2017)
22. Militaru, G., Popescu, D., Ichim, L.: UAV-to-UAV communication options for civilian applications. Paper presented at the 2018 26th telecommunications forum, TELFOR 2018—proceedings. https://doi.org/10.1109/TELFOR.2018.8612108 (2018)
23. Petrovsky, A., Doole, M., Ellerbroek, J., Hoekstra, J.M., Tomasello, F.: Challenges with obstacle data for manned and unmanned aviation. Paper Presented Int. Arch. Photogrammetry Remote Sens. Spat. Inf. Sci. ISPRS Arch. **42**(4/W10), 143–149. https://doi.org/10.5194/isprs-archives-XLII-4-W10-143-2018 (2018)
24. Ghafar, N.H., Rahman, N.A.A., Mohammad, M.F.N., Shah, M.Z., Hassan, F.: Developing new aviation management postgraduate program in responding to industry 4.0: Key findings from multistages-multilevels market study. Paper Presented IOP Conf. Ser. Mater. Sci. Eng. **645**(1).https://doi.org/10.1088/1757-899X/645/1/012008 (2019)
25. Cokorilo, O.: Urban air mobility: safety challenges. Paper Presented Transport. Res. Proc. **45**, 21–29 (2020). https://doi.org/10.1016/j.trpro.2020.02.058
26. Dağdeviren, M., Yavuz, S., Kılınç, N.: Weapon selection using the AHP and TOPSIS methods under fuzzy environment. Expert Syst. Appl. **36**(8143–8151) (2009)
27. Saaty, T.L.: How to make a decision: the analytic hierarchy process. Euro. J. Oper. Res. **48**, 9–26 (1990)
28. Yedla, S., Shresta, R.M.: Multi-criteria approach for the selection of alternative options for environmentally sustainable transport system in Delhi. Transport. Res. **37**, 717–729 (2003)
29. Aras, H., Erdogmus, S., Koc, E.: Multi-criteria selection for a wind observation station location using analytic hierarchy process. Renew. Energy **29**, 1383–1392 (2004)
30. Tolga, E., Demircan, M.L., Kahraman, C.: Operating system selection using fuzzy replacement analysis andnanalytic hierarchy process. Int. J. Prod. Econ. **97**, 89–117 (2005)
31. Dağdeviren, M.: Decision making in equipment selection: an integrated approach with AHP and PROMETHEE. J. Intell. Manuf. **19**(397–406) (2008)
32. Kim, P.P., Lee, K.J., Lee, B.W.: Selection of an optimal nuclear fuel cycle scenario by goal programming & analytic hierarchy process. Ann. Nucl. Energy **26**, 449–460 (1999)
33. Ramirez-Atencia, C., Rodriguez-Fernandez, V., Camacho, D.: A revision on multi-criteria decision making methods for multi-UAV mission planning support. Expert Syst. Appl. 113708 (2020)
34. Qin, Y., Qi, Q., Scott, P.J., Jiang, X.: An additive manufacturing process selection approach based on fuzzy Archimedean weighted power Bonferroni aggregation operators. Robot. Comput. Integr. Manuf. **64**, 101926 (2020)
35. Chien, F., Wang, C.N., Nguyen, V.T., Nguyen, V.T., Chau, K.Y.: An evaluation model of quantitative and qualitative fuzzy multi-criteria decision-making approach for hydroelectric plant location selection. Energies **13**(11), 2783 (2020)
36. Li, J., Chen, Q.: An outranking method for multicriteria decision making with probabilistic hesitant information. Expert Syst. e12513 (2020)
37. Solangi, Y.A., Tan, Q., Mirjat, N.H., Ali, S.: Evaluating the strategies for sustainable energy planning in Pakistan: an integrated SWOT-AHP and Fuzzy-TOPSIS approach. J. Cleaner Product. **236**, 117655 (2019)

38. Sufiyan, M., Haleem, A., Khan, S., Khan, M.I.: Evaluating food supply chain performance using hybrid fuzzy MCDM technique. Sustain. Product. and Consump. **20**, 40–57 (2019)
39. Yager, R.: Pythagorean membership grades in multi criteria decision making. IEEE Trans. Fuzzy Syst. **22**(4), 958–965 (2014)
40. Karasan, A., Ilbahar, E., Kahraman, C.: A novel pythagorean fuzzy AHP and its application to landfill site selection problem. Soft. Comput. **23**(21), 10953–10968 (2019)
41. Ak, M.F., Gül, M.: AHP–TOPSIS integration extended with Pythagorean fuzzy sets for information security risk analysis. Complex Intell. Syst. 113–126 (2018)
42. Gul, M., Ak, M.F., Guneri, A.F.: Pythagorean fuzzy VIKOR-based approach for safety risk assessment in mine industry. J. Saf. Res. **69**, 135–153 (2019)
43. Zhang, X., Xu, Z.: Extension of topsis to multiple criteria decision making with Pythagorean fuzzy sets. Int. J. Intel. Syst. 1061–1078 (2014)
44. Zeng, S., Chen, J., Li, X.: A hybrid method for Pythagorean fuzzy multiple-criteria decision making. Int. J. Inf. Technol. Decis. Mak. **15**(02), 403–422 (2016)
45. Gul, M.: Application of Pythagorean fuzzy AHP and VIKOR methods in occupational health and safety risk assessment: the case of a gun and rifle barrel external surface oxidation and coloring unit. Int. J. Occup. Saf. Ergon. 1–15 (2018)
46. Ilbahar, E., Karaşan, A., Cebi, S., Kahraman, C.: A novel approach to risk assessment for occupational health and safety using Pythagorean fuzzy AHP & fuzzy inference system. Saf. Sci. **103**, 124–136 (2018)
47. Yücesan, M., Kahraman, G.: Risk evaluation and prevention in hydropower plant operations: a model based on Pythagorean fuzzy AHP. Energy Policy **126**, 343–351 (2019)
48. Kaya, A., Çiçekalan, B., Çebi, F.: Location selection for WEEE recycling plant by using Pythagorean fuzzy AHP. J. Intell. Fuzzy Syst. **38**(1), 1097–1106 (2020)
49. Yildiz, A., Ayyildiz, E., Gumus, A.T., Ozkan, C.: A modified balanced scorecard based hybrid Pythagorean fuzzy AHP-Topsis methodology for ATM site selection problem. Int. J. Inf. Technol. Decis. Mak. (IJITDM) **19**(02), 365–384 (2020)
50. Shete, R.S.P.C., Ansari, Z.N., Kant, R.: A Pythagorean fuzzy AHP approach and its application to evaluate the enablers of sustainable supply chain innovation. Sustain. Product. Consump. (2020)
51. Rani, P., Mishra, A.R., Mardani, A., Cavallaro, F., Štreimikienė, D., Khan, S.A.R.: Pythagorean fuzzy SWARA–VIKOR framework for performance evaluation of solar panel selection. Sustainability **12**(10), 4278 (2020)
52. Tepe, S., Kaya, İ.: A fuzzy-based risk assessment model for evaluations of hazards with a real-case study. Hum. Ecol. Risk Assess. Int. J. (2019)
53. Seker, S.: A novel integrated MCDM approach: an application to selection of the optimal fiber optical access network strategy. J. Intell. Fuzzy Syst. (Preprint), 1–11 (2020)
54. Büyüközkan, G., Göçer, F.: Assessment of additive manufacturing technology by Pythagorean fuzzy CODAS. In *International Conference on Intelligent and Fuzzy System*, pp. 959–968. Springer, Cham (2019)
55. Seker, S., Aydin, N.: Hydrogen production facility location selection for Black Sea using entropy based TOPSIS under IVPF environment. Int. J. Hydrogen Energy (2020)
56. Rani, P., Mishra, A.R., Pardasani, K.R., Mardani, A., Liao, H., Streimikiene, D.: A novel VIKOR approach based on entropy and divergence measures of Pythagorean fuzzy sets to evaluate renewable energy technologies in India. J. Cleaner Product. **238**, 117936 (2019)
57. Zhang, Z.X., Hao, W.N., Yu, X.H., Chen, G., Zhang, S.J., Chen, J.Y.: Pythagorean fuzzy preference ranking organization method of enrichment evaluations. Int. J. Intell. Syst. **34**(7), 1416–1439 (2019)
58. Ren, P., Xu, Z., Gou, X.: Pythagorean fuzzy TODIM approach to multi-criteria decision making. Appl. Soft Comput. **42**, 246–259 (2016)
59. Herrera, F., Martinez, L., Rodríguez, R.M.: Hesitant fuzzy linguistic term sets Adv. Intel. Soft Comput. **122**, 287–295 (2011)
60. Gou, X., Liao, H., Xu, Z., Herrera, F.: Double hierarchy hesitant fuzzy linguistic term set and MULTIMOORA method: a case of study to evaluate the implementation status of haze controlling measures. Inf. Fus. **38**, 22–34 (2017)

61. Rodriguez, R.M., Martinez, L., Herrera, F.: A group decision making model dealing with comparative linguistic expressions based on hesitant fuzzy linguistic term sets. Inf. Sci. **241**, 28–42 (2013)
62. Zhang, Y., Xu, Z.: Efficiency evaluation of sustainable water management using the HF-TODIM method. Int. Trans. Oper. Res. **26**(2), 747–764 (2019)
63. Liao, H., Wu, X., Mi, X., Herrera, F.: An integrated method for cognitive complex multiple experts multiple criteria decision making based on ELECTRE III with weighted Borda rule. Omega (United Kingdom) **93**, 102052 (2020)
64. Narayanamoorthy, S., Geetha, S., Rakkiyappan, R., Joo, Y.H.: Interval-valued intuitionistic hesitant fuzzy entropy based VIKOR method for industrial robots selection. Expert Syst. Appl. **121**, 28–37 (2019)
65. Xian, S., Guo, H.: Novel supplier grading approach based on interval probability hesitant fuzzy linguistic TOPSIS. Eng. Appl. Artif. Intell. **87**, 103299 (2020)
66. Beskese, A., Camci, A., Temur, G.T., Erturk, E.: Wind turbine evaluation using the hesitant fuzzy AHP-TOPSIS method with a case in Turkey. J. Intell. Fuzzy Syst. **38**(1), 997–1011 (2020)
67. Liang, R.-X., Wang, J.-Q., Zhang, H.-Y.: Projection-based PROMETHEE methods based on hesitant fuzzy linguistic term sets. Int. J. Fuzzy Syst. **20**(7), 2161–2174 (2018)
68. Liu, Z., Ming, X., Song, W.: A framework integrating interval-valued hesitant fuzzy DEMATEL method to capture and evaluate co-creative value propositions for smart PSS. J. Cleaner Product. **215**, 611–625 (2019)
69. Darko, A.P., Liang, D.: An extended COPRAS method for multiattribute group decision making based on dual hesitant fuzzy Maclaurin symmetric mean. Int. J. Intell. Syst. **35**(6), 1021–1068 (2020)
70. Wang, X., Gou, X., Xu, Z.: Assessment of traffic congestion with ORESTE method under double hierarchy hesitant fuzzy linguistic environment. Appl. Soft Comput. J. **86**, 105864 (2020)
71. Maiers, J., Sherif, Y.S.: Applications of fuzzy set theory. IEEE Trans. Syst. Man Cybern. **1**, 175–189 (1985)
72. Rodriguez, R.M., Martinez, L., Herrera, F.: Hesitant fuzzy linguistic term sets for decision making. IEEE Trans. Fuzzy Syst. **20**(1), 109–119 (2012)
73. Herrera, F., Martínez, L.: A 2-tuple fuzzy linguistic representation model for computing with words. IEEE Trans. Fuzzy Syst. **8**(6), 746–752 (2000)

A Novel Mathematical Model to Design UAV Trajectory for Search and Rescue Operations in Disaster

Seyed Amin Seyfi-Shishavan, Elmira Farrokhizadeh, and Fatma Kutlu Gündoğdu

Abstract In mass casualty incidents, whether natural or human-made, Search and Rescue Operations (SAR) have a critical role in humanitarian problems. Time and information are crucial in victims' survival. The high success rate in SAR operations depends on achieving reliable data about the number of victims, the severity of the failure, access status, and location status, etc. as soon as possible. Nowadays, the technology of Unmanned Aerial Vehicle (UAV) presents an opportunity to help the rescue teams with avoiding wasting time and accessing areas where searching by rescue teams are costly and impossible to go there. In this study, we design a mathematical model to get an optimal path planning to steer the UAVs based on the potential risk degree (PRD) of the candidate location in the affected area. The proposed model is inspired by the Travel Salesman Problem (TSP) that selects the optimal tour giving priority to districts with high PRDs obtained using the concept of similarity measure in the spherical fuzzy environment, considering the power limitation of UAVs. The priority of candidate locations is evaluated by Jaccard, exponential, and square root cosine similarity measures. The applicability of the proposed model is demonstrated by applying it to an earthquake case. A sensitivity analysis is performed to approve the validity of the method.

Keywords Unmanned Aerial Vehicle (UAV) · Similarity measures · Spherical fuzzy sets · Disaster management · Mathematical model

S. A. Seyfi-Shishavan · E. Farrokhizadeh
Industrial Engineering Department, Istanbul Technical University, Macka, Istanbul, Turkey

F. Kutlu Gündoğdu (✉)
Department of Industrial Engineering, National Defence University, Turkish Air Force Academy, 34149 Istanbul, Turkey
e-mail: kgundogdu@hho.edu.tr

C. Kahraman and S. Aydın (eds.), *Intelligent and Fuzzy Techniques in Aviation 4.0*, Studies in Systems, Decision and Control 372,
https://doi.org/10.1007/978-3-030-75067-1_22

1 Introduction

The number of deaths from natural disasters such as flood, earthquakes, and fire can be highly variable from year-to-year; some years pass with very few deaths before a large disaster event claims many lives. Generally, all of these disasters have an essential aspect that directly or indirectly endangers many people's lives that even seconds are critical in search of survivors and reducing casualties. The most crucial aspect in these cases before any action is Search and Rescue Operation (SAR), performed by specialized teams that even put at risk their lives to save others [1].

Rescue teams often have a limited view of the situation, and given the importance of time in these matters, they need to know where to start their search. Access to intuitive disaster damage assessment (DDE) support information can also simplify the decision-making and rescue process [2]. In [3], Unmanned Aircraft System (UAS) was developed for rapidly obtaining disaster information with image analysis and a photogrammetric workflow. Recently to gather the images of disaster and geographical perceptions of terrestrial targets for providing critical support to search and rescue operations, Unmanned Aerial Vehicles (UAVs) like drones are used because of their low cost and high speed in response [4]. As mentioned, UAVs are quick, rapid, and autonomous equipment that can be controlled remotely. These characteristics allow them to carry out complicated operations with lower operating costs and in the shortest time, while the same operations can be performed by humans with difficulty over a longer time [5]. As presented in Fig. 1, UAVs are deployed by the rescue teams to collect evidence of the victim's presence in the affected area according to the mission, then start to taking the photos, gathering information, and sending the collected data remotely to the ground station or rescue team.

Unmanned Aerial Vehicles (UAVs) widely have been used in dangerous, risky, and unknown environments, such as operations at nuclear power plants, exploring Mars, and monitoring enemy forces on the battlefield [6]. Moreover, UAVs are often used in search and rescue operations for trapped people. The growing recognition

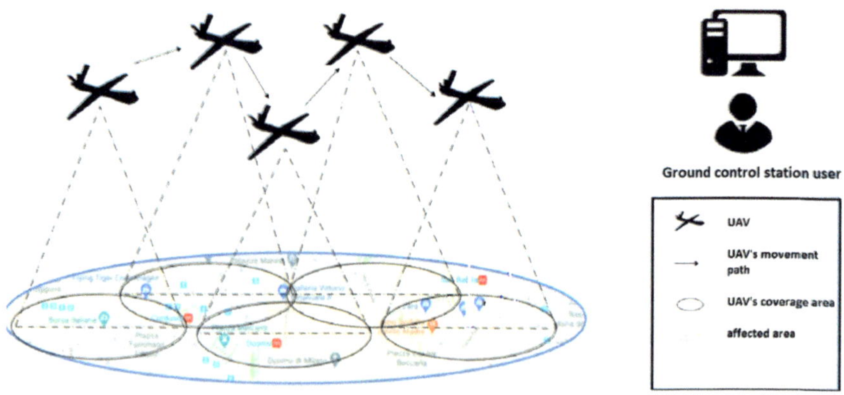

Fig. 1 The conceptual scheme of SAR operation by UAV

of the potential of using UAVs for search and rescue applications is supported by an increasing number of studies in the areas of image recognition for victim detection, path planning, and task allocation. With the expanding use of UAVs for search and rescue programs, researchers have conducted extensive field studies such as imaging to identify the victim, planning the route, and assigning the task [5]. In [7], the authors analyzed a Wilderness Search and Rescue problem with visual-based aerial search as practical aspects. Another study combined both cellular technology and the first Galileo services to localize the smartphones of potential victims [8]. In [9], they utilized camera extrinsic and multi-view geometry techniques to get one-to-one correspondence between infrared and visual images. Silvagni et al. developed a multipurpose UAV for mountain rescue operations [10]. In [11], they described a Fog Computing architecture that deals simultaneously with real-time constraints and synchronization goals for a hybrid human drone collaboration. In [12], the paper presented autonomous UAVs to handle two aspects of the SAR problem: First, the robot needs to be as close as possible to the survivors so they can be detected. Second, the robot must turn around each side because it can only detect the survivors if they are directly in front of it. In other articles, researchers used UAVs with rotary wings to locate survivor locations for maritime SAR operations with an algorithm that divided taken photographs into four quarters to search and repeated the algorithm until a survivor was found [13]. In the other study, Rahmes et al. used automated UAVs based on onboard computers and pre-knowledge data from last seen victims [14]. In [15], Cui et al. proposed algorithms for cooperative MAVs in two phases: aerial imaging and map covering, discovering the search area as the first phase, and then searching with more detail to recognize each building roads as the second phase.

Still, the path planning problem is an essential topic in the planning of the UAV mission. The path planning problem has different UAVs approaches in a risky environment to arrive at the desired area in maximum safety to collect the required information about survivors. Many studies have published about Unmanned Aerial Vehicles (UAVs) path planning problems. In [16], the authors explained the importance of path planning in UAV systems and developed a mathematical model based on the Traveling Salesman Problem (TSP) to find the shortest route. Kallmann and Kapadia [17] reviewed the related articles' hierarchy progress from classical planning and computational geometry techniques and then progressively shifted against more advanced topics. In [18], the algorithms used in UAVs' path planning are listed and compared to each other to help choose the best one. Researchers often select Geometrical methods such as spline interpolation to achieve softer curves [19]. In [20], the authors used a Monte Carlo simulation to find UAVs' best path. In most of the articles to optimization the path between points in UAV path planning, researchers used optimization solution algorithms such as Genetic Algorithms (GA) [21–24], Distribution Algorithms (DA) [25] or A* Algorithms (AA) [26]. In [1], fuzzy logic, PSO algorithm, the original proposal called attraction, and ANFIS methods are developed to calculate the location of the waypoints in UAVs to minimize the distance and the risk degree.

One of the challenging issues in SAR problems by UAVs is having the right and exact information about under review area. It means that if critical data about the potential hazards in the area are known, a safe path for UAVs can obtain by a reliable optimization problem [1]. Logical approaches must be taken to overcome with uncertainty nature of information in such incidents. One of these approaches is fuzzy logic, which is developed by Zadeh in 1965 [27]. Seyfi-Shishavan et al. [28] developed some novel similarity measures for spherical fuzzy sets to analyze the degree of similarity between items. These similarity functions are utilized in this article to determine the potential risk degree of candidate locations.

In hazard situations such as earthquakes, many people can lose or stay under the rubble in uninhabitable places and need to rescue and emergency intervention as soon as possible. Therefore, due to the importance of time and high accuracy of detection, designing an optimal path problem for UAVs is a significant issue. In this model, UAVs are controlled by users in the ground control station and lead toward the affected area. Considering the importance of the time factor in a rescue operation, for a first-time potential risk degree of all candidate locations in the affected area are calculated by similarity measures to obtain the travel priority by optimizing UAVs' path planning inspired by Travel Salesman Problem (TSP). The primary purpose is to get maximum coverage of districts with high potential risk degrees.

The rest of the chapter is structured as follows. First, some basic definitions about spherical fuzzy sets and similarity measures are described in the preliminary section. The definition of the problem is prepared in Sect. 3. The mathematical model and solution methodology are described in Sect. 4. In Sect. 5, an application of the proposed model is prepared to demonstrate the model's applicability and validity. A sensitivity analysis is performed in Sect. 6. Finally, in the last section, the conclusion and results are discussed.

2 Preliminaries and Basic Concepts

This section presents some basic information and mathematical equations that are used in the rest of the paper. The concept of spherical fuzzy sets and some recently developed similarity measures in the spherical fuzzy environment are discussed in this section.

2.1 *Spherical Fuzzy Sets*

Spherical fuzzy sets are the latest extension of picture fuzzy sets and Pythagorean fuzzy sets. Spherical Fuzzy Sets (SFSs) give a larger preference domain for evaluators as well as each decision-maker may also determine his/her hesitancy information independently under the spherical fuzzy environment [29].

Definition 1 A Single valued Spherical Fuzzy Set (SFSs) $\tilde{A}_S$ of the universe of discourse X is given by [29]:

$$\tilde{A}_s = \{\, x, \mu_{\tilde{A}_s}(x), \vartheta_{\tilde{A}_s}(x), I_{\tilde{A}_s}(x) | x \in X \} \tag{1}$$

where $\mu_{\tilde{A}_s}(x), \vartheta_{\tilde{A}_s}(x), I_{\tilde{A}_s}(x)$: $X \rightarrow [0, 1]$ are the degree of membership, non-membership and indeterminacy of x to $\tilde{A}_S$, respectively, and

$$0 \leq \mu^2_{\tilde{A}_s}(x) + \vartheta^2_{\tilde{A}_s}(x) + I^2_{\tilde{A}_s}(x) \leq 1 \tag{2}$$

Then, $\sqrt{1 - \left(\mu^2_{\tilde{A}_s}(x) + \vartheta^2_{\tilde{A}_s}(x) + I^2_{\tilde{A}_s}(x)\right)}$ is defined as refusal degree of x in X.

Definition 2 Assume that $\tilde{A}_s$ and $\tilde{B}_s$ be any two Spherical Fuzzy sets. So the basic operations of SFSs can be defined as follows [29]:

$$\begin{aligned} \tilde{A}_s \oplus \tilde{B}_s = &\left\{\sqrt{\mu^2_{\tilde{A}_s} + \mu^2_{\tilde{B}_s} - \mu^2_{\tilde{A}_s}\mu^2_{\tilde{B}_s}}, \vartheta_{\tilde{A}_s}\vartheta_{\tilde{B}_s}, \right. \\ &\left. \sqrt{\left(1 - \mu^2_{\tilde{B}_s}\right)I^2_{\tilde{A}_s} + \left(1 - \mu^2_{\tilde{A}_s}\right)I^2_{\tilde{B}_s} - I^2_{\tilde{A}_s}I^2_{\tilde{B}_s}}\right\} \end{aligned} \tag{3}$$

$$\begin{aligned} \tilde{A}_s \otimes \tilde{B}_s = &\left\{\mu_{\tilde{A}_s}\mu_{\tilde{B}_s}, \sqrt{\vartheta^2_{\tilde{A}_s} + \vartheta^2_{\tilde{B}_s} - \vartheta^2_{\tilde{A}_s}\vartheta^2_{\tilde{B}_s}}, \right. \\ &\left. \sqrt{\left(1 - \vartheta^2_{\tilde{B}_s}\right)I^2_{\tilde{A}_s} + \left(1 - \vartheta^2_{\tilde{A}_s}\right)I^2_{\tilde{B}_s} - I^2_{\tilde{A}_s}I^2_{\tilde{B}_s}}\right\} \end{aligned} \tag{4}$$

$$\begin{aligned} k\tilde{A}_s = &\left\{\sqrt{1 - \left(1 - \mu^2_{\tilde{A}_s}\right)^k}, \vartheta^k_{\tilde{A}_s}, \sqrt{\left(1 - \mu^2_{\tilde{A}_s}\right)^k - \left(1 - \mu^2_{\tilde{A}_s} - I^2_{\tilde{A}_s}\right)^k}\right\}; \\ &k \geq 0 \end{aligned} \tag{5}$$

$$\begin{aligned} \tilde{A}^k_s = &\left\{\mu^k_{\tilde{A}_s}, \sqrt{1 - \left(1 - \vartheta^2_{\tilde{A}_s}\right)^k}, \sqrt{\left(1 - \vartheta^2_{\tilde{A}_s}\right)^k - \left(1 - \vartheta^2_{\tilde{A}_s} - I^2_{\tilde{A}_s}\right)^k}\right\}; \\ &k \geq 0 \end{aligned} \tag{6}$$

Definition 3 For the SFSs $\tilde{A}_s = \left(\mu_{A_s}, \vartheta_{A_s}, I_{A_s}\right)$ and $\tilde{B}_s = \left(\mu_{B_s}, \vartheta_{B_s}, I_{B_s}\right)$, the followings are valid, provided that k, k_1, and $k_2 \geq 0$ [29].

I. $\tilde{A}_s \oplus \tilde{B}_s = \tilde{B}_s \oplus \tilde{A}_s$ (7)

II. $\tilde{A}_s \otimes \tilde{B}_s = \tilde{B}_s \otimes \tilde{A}_s$ (8)

III. $k(\tilde{A}_s \oplus \tilde{B}_s) = k\tilde{A}_s \oplus k\tilde{B}_s$ (9)

$$\text{IV.}\quad k_1\tilde{A}_s \oplus k_2\tilde{A}_s = (k_1 + k_2)\tilde{A}_s \tag{10}$$

$$\text{V.}\quad \left(\tilde{A}_s \otimes \tilde{B}_s\right)^k = \tilde{A}_s^k \otimes \tilde{B}_s^k \tag{11}$$

$$\text{VI.}\quad \tilde{A}_s^{k_1} \otimes \tilde{A}_s^{k_2} = \tilde{A}_s^{k_1+k_2} \tag{12}$$

Definition 4 Suppose that $\tilde{A}_s$ and $\tilde{B}_s$ are two spherical fuzzy sets. Then to compare these SFSs, the score function (SC) and accuracy function (AC) are defined as follows [29]:

$$SC\left(\tilde{A}_s\right) = \left(\mu_{\tilde{A}_s} - \frac{I_{\tilde{A}_s}}{2}\right)^2 - \left(\vartheta_{\tilde{A}_s} - \frac{I_{\tilde{A}_s}}{2}\right)^2 \tag{13}$$

$$AC\left(\tilde{A}_s\right) = \mu_{\tilde{A}_s}^2 + \vartheta_{\tilde{A}_s}^2 + I_{\tilde{A}_s}^2 \tag{14}$$

After calculating the results of score and accuracy functions, the comparison rules are as follows:

If $SC\left(\tilde{A}_s\right) > SC\left(\tilde{B}_s\right)$, then $\tilde{A}_s > \tilde{B}_s$;

If $SC\left(\tilde{A}_s\right) = SC\left(\tilde{B}_s\right)$ and $AC\left(\tilde{A}_s\right) > AC\left(\tilde{B}_s\right)$, then $\tilde{A}_s > \tilde{B}_s$;

If $SC\left(\tilde{A}_s\right) = SC\left(\tilde{B}_s\right)$, $AC\left(\tilde{A}_s\right) < AC\left(\tilde{B}_s\right)$, then $\tilde{A}_s < \tilde{B}_s$;

If $SC\left(\tilde{A}_s\right) = SC\left(\tilde{B}_s\right)$, $AC\left(\tilde{A}_s\right) = AC\left(\tilde{B}_s\right)$, then $\tilde{A}_s = \tilde{B}_s$;

After presenting the main mathematical operations and comparison rules of spherical fuzzy sets, we present four similarity measures of spherical fuzzy sets in the next sub-section.

2.2 Similarity Measures

Similarity measures and distances are widely used to determine the relationship between two individuals in many domains composed of decision-making, pattern recognition, medical engineering, and network comparison [30]. Here, we present three similarity measures in spherical fuzzy environment developed by [28]. These similarity measures are Jaccard, exponential, and square root cosine similarity functions. The details of these similarity measures will be explained in the following.

2.2.1 Jaccard Similarity Measure

Definition 5 Suppose that there are two sets of SFSs as follows:

$$\tilde{A}_s = \{x_i\langle \mu_A(x_i), \vartheta_A(x_i), I_A(x_i)\rangle | x_i \in X\};\ \forall i = 1, 2, \ldots, n$$

$$\tilde{B}_s = \{x_i\langle \mu_B(x_i), \vartheta_B(x_i), I_B(x_i)\rangle | x_i \in X\};\ \forall i = 1, 2, \ldots, n$$

So that $\mu(x_i)$, $\vartheta(x_i)$ and $I(x_i)$ are membership, non-membership and indeterminacy degrees, respectively. Then, the Jaccard similarity measure of $\tilde{A}_s$ and $\tilde{B}_s$ is defined as Eq. (15) [28]:

$$JS_{SFS}\left(\tilde{A}_s, \tilde{B}_s\right) = \frac{1}{n}\sum_{i=1}^{n}\left[\frac{\left[\mu_A^2(x_i)\mu_B^2(x_i) + \vartheta_A^2(x_i)\vartheta_B^2(x_i) + I_A^2(x_i)I_B^2(x_i)\right]}{\begin{array}{c}\left[\mu_A^4(x_i) + \vartheta_A^4(x_i) + I_A^4(x_i)\right] + \left[\mu_B^4(x_i) + \vartheta_B^4(x_i) + I_B^4(x_i)\right] \\ -\left[\mu_A^2(x_i)\mu_B^2(x_i) + \vartheta_A^2(x_i)\vartheta_B^2(x_i) + I_A^2(x_i)I_B^2(x_i)\right]\end{array}}\right] \quad (15)$$

Definition 6 If each x_i with weight vector $w_i = (w_1, w_2, \ldots, w_n)$ where $w_i \in [0, 1]$ and $\sum_{i=1}^{n} w_i = 1$ be considered, then the weighted Jaccard similarity measure of $\tilde{A}_s$ and $\tilde{B}_s$ is defined as Eq. (16) [28]:

$$WJS_{SFS}\left(\tilde{A}_s, \tilde{B}_s\right) = \sum_{i=1}^{n} w_i \frac{\left[\mu_A^2(x_i)\mu_B^2(x_i) + \vartheta_A^2(x_i)\vartheta_B^2(x_i) + I_A^2(x_i)I_B^2(x_i)\right]}{\begin{array}{c}\left[\mu_A^4(x_i) + \vartheta_A^4(x_i) + I_A^4(x_i)\right] + \left[\mu_B^4(x_i) + \vartheta_B^4(x_i) + I_B^4(x_i)\right] \\ -\left[\mu_A^2(x_i)\mu_B^2(x_i) + \vartheta_A^2(x_i)\vartheta_B^2(x_i) + I_A^2(x_i)I_B^2(x_i)\right]\end{array}} \quad (16)$$

2.2.2 Exponential Similarity Measure

Definition 7 Suppose that there are two sets of SFSs as follows:

$$\tilde{A}_s = \{x_i\langle \mu_A(x_i), \vartheta_A(x_i), I_A(x_i)\rangle | x_i \in X\};\ \forall i = 1, 2, \ldots, n$$

$$\tilde{B}_s = \{x_i\langle \mu_B(x_i), \vartheta_B(x_i), I_B(x_i)\rangle | x_i \in X\};\ \forall i = 1, 2, \ldots, n$$

So that $\mu(x_i)$, $\vartheta(x_i)$ and $I(x_i)$ are membership, non-membership and Indeterminacy degrees, respectively. Then, the exponential similarity measure of $\tilde{A}_s$ and $\tilde{B}_s$ is defined as Eq. (17) [28]:

$$ES_E\left(\tilde{A}_s, \tilde{B}_s\right) = e^{-\left[\sqrt{\frac{1}{n}\sum_{i=1}^{n}\left[(\mu_A(x_i)-\mu_B(x_i))^2+(\vartheta_A(x_i)-\vartheta_B(x_i))^2+(I_A(x_i)-I_B(x_i))^2\right]}\right]} \quad (17)$$

Definition 8 If each x_i with weight vector $w_i = (w_1, w_2, \ldots, w_n)$ were $w_i \in [0, 1]$ and $\sum_{i=1}^{n} w_i = 1$ be considered, then the weighted exponential similarity measure of $\tilde{A}_s$ and $\tilde{B}_s$ is defined as Eq. (18) [28]:

$$WES_E\left(\tilde{A}_s, \tilde{B}_s\right) = \sum_{i=1}^{n} w_i\, e^{-\left[\sqrt{\sum_{i=1}^{n}\left[(\mu_A(x_i)-\mu_B(x_i))^2+(\vartheta_A(x_i)-\vartheta_B(x_i))^2+(I_A(x_i)-I_B(x_i))^2\right]}\right]} \quad (18)$$

2.2.3 Square Root Cosine Similarity Measure

Square root cosine similarity measure is given as follows:

Definition 9 Suppose that there are two sets of SFSs:

$$\tilde{A}_s = \{x_i\langle\mu_A(x_i), \vartheta_A(x_i), I_A(x_i)\rangle | x_i \in X\};\ \forall i = 1, 2, \ldots, n$$

$$\tilde{B}_s = \{x_i\langle\mu_B(x_i), \vartheta_B(x_i), I_B(x_i)\rangle | x_i \in X\};\ \forall i = 1, 2, \ldots, n$$

be two SFSs so that $\mu(x_i)$, $\vartheta(x_i)$ and $I(x_i)$ are membership, non-membership and indeterminacy degrees, respectively. Then the Square Root Cosine Similarity measure of $\tilde{A}_s$ and $\tilde{B}_s$ is given by Eq. (19) [28]:

$$SqrtCS_{SFS}\left(\tilde{A}_s, \tilde{B}_s\right) = \frac{1}{n}\sum_{i=1}^{n}\frac{\left(\sqrt{\mu_A^2(x_i)\mu_B^2(x_i) + \vartheta_A^2(x_i)\vartheta_B^2(x_i) + I_A^2(x_i)I_B^2(x_i)}\right)}{\sqrt[4]{\left(\mu_A^4(x_i) + \vartheta_A^4(x_i) + I_A^4(x_i)\right)}\sqrt[4]{\left(\mu_B^4(x_i) + \vartheta_B^4(x_i) + I_B^4(x_i)\right)}} \quad (19)$$

Definition 10 If each x_i with weight vector $w_i = (w_1, w_2, \ldots, w_n)$ where $w_i \in [0, 1]$ and $\sum_{i=1}^{n} w_i = 1$ be considered, then the weighted square root cosine similarity measure of $\tilde{A}_s$ and $\tilde{B}_s$ is defined as Eq. (20) [28]:

$$WSqrtCS_{SFS}\left(\tilde{A}_s, \tilde{B}_s\right) =$$
$$\sum_{i=1}^{n} w_i \frac{\left(\sqrt{\mu_A^2(x_i)\mu_B^2(x_i) + \vartheta_A^2(x_i)\vartheta_B^2(x_i) + I_A^2(x_i)I_B^2(x_i)}\right)}{\sqrt[4]{\left(\mu_A^4(x_i) + \vartheta_A^4(x_i) + I_A^4(x_i)\right)}\sqrt[4]{\left(\mu_B^4(x_i) + \vartheta_B^4(x_i) + I_B^4(x_i)\right)}} \quad (20)$$

3 Problem Definition

Search and Rescue Operations (SAR) occur in an emergency where people are involved, and their lives are at risk. One of the advanced and useful technologies that nowadays carried out SAR operations are Unmanned Aerial Vehicles (UAVs). These autonomous vehicles have experienced significant development during the last years [1].

In search and rescue operations, two critical factors have a crucial impact on their success rate. These two factors are designing the optimal flight path for UAV and determining the high-risk areas. Researchers used several different methods, such as operation research techniques, simulation, and optimization algorithms, to design the flight path and determine the high potential risk areas. We give some details about these two critical factors.

3.1 Trajectory Planning and Designing

The unmanned aerial vehicle path planning problem is an essential problem in UAV mission planning that requires careful consideration [31]. The autonomous UAVs flight can be based on a preplanned path or can even be calculated online [32]. UAV path planning refers to an optimization problem to generate a set of waypoints from the start point to the destination point with guarantee not to fly out of the designated map [33]. The optimization path's main purpose is to find a flight path with minimum energy consumption, minimum traveled path, and maximize covered area on the premise of completing the UAV mission [34]. Generally, UAVs have finite power capacity to support them in these kinds of missions. Their driving force is usually supplied by a battery that needs to be charged after a while. Therefore, designing an optimal path to cover the vast area with the least energy consumption in a mission is an unavoidable and essential need. In this research, we aim to propose a mathematical model to calculate an optimal flight path for UAVs in search and rescue operations after a disaster, considering the maximization of search and coverage area, taking into account the UAVs' power limitation. Therefore, inspired by Travelling Salesman Problem (TSP), a linear mathematical model is formulated to find out the optimal trajectory to find the injured after an earthquake. During an earthquake, it is possible to destroy some buildings and residential areas. Therefore, some people may be injured or even trapped under rubbles. Due to these areas' inaccessibility, it may be impossible for humans and rescue teams to search for the injured. In this situation, the use of UAVs can help search and rescue operations. Having the best path to fly the UAVs can speed up the surveillance and rescue operations. Planning the optimal trajectory directly depends on determining the areas at the most potential risk so that these areas prioritize the reaching process. In the next section, the concept and detail of areas with high potential risk will be explained.

3.2 The Potential Risk Degree

The first step in trajectory design for UAVs in SAR operations is identifying the target area to estimate the rate of potential damage and the number of casualties. For this purpose, an estimation factor should be calculated to determine the Potential Risk Degree (PRD) of each area. The PRD helps the rescue teams perform search and rescue operations in a targeted manner and prioritize areas with high PRD. This factor is employed to generate the map used for the discrete UAVs path planning. The PRD is determined by analyzing the surface's characteristics, where the search will take place. Different methods have been applied to estimate the potential risk degree considering various factors such as rate of staying, rate of hazard, and accessibility rate of the affected areas after a disaster [1]. In this study, we proposed a novel way to determine the PRD of the target areas using the similarity degree concept will be explained in the next section.

4 Proposed Model and Solution Methodology

As it is noticed before, the present study aims to propose a mathematical model to design an optimal path for UAVs flight in search and rescue operations. As mentioned in Sect. 3.2, first of all, we have to determine the potential risk degree of each area. Then, by maximizing the obtained potential degrees via proposed model, the optimal path will be obtained. The steps of the solution process will be explained step by step in the following sections.

4.1 Potential Risk Degree Calculation

To calculate the PRD of each area, we applied the concept of similarity degree. To take into accounts the uncertainty in the utilized information, we used the concept of spherical fuzzy sets. In the first step, the characteristics of the target area should be specified. We applied the proposed method for a SAR operation after an earthquake.

Therefore, by investigating previous research and literature, four main factors were determined to calculate each area's PRD. These factors are population density (F1), strength of buildings (F2), strength of soil and land (F3), and distance from faults (F4) [35]. All target districts have to be evaluated based on these factors. The evaluation process could be done using different ways such as historical data, geographical maps, expert's comments, simulation techniques and so on. After evaluation process, the similarity degree of each district with reference potential risk degree (RPRD) should be calculated. The RPRD also could be obtained by mentioned ways. The steps and detail of mentioned process is as follows.

Step 1 Constructing the district evaluation matrix as Table 2. Suppose, there are a set of M districts $D = \{D_1, D_2, \ldots, D_m\}$ have to be evaluated based on the set of N factors $F = \{F_1, F_2, ..., F_n\}$ with the corresponding weight vector $w = \{w_1, w_2, \ldots, w_n\}$, where $\sum_{n=1}^{n} w_n = 1$ and $w_n \geq 0$.

Step 2 Evaluate all districts based on all factors using spherical fuzzy scale (see Table 1) as Table 2.

Step 3 Calculate the similarity measure degree between each district D_m and reference potential risk degree (see Table 3) using Eqs. (16), (18), and (20).

Table 1 Linguistic terms of importance

Linguistic terms	Spherical fuzzy number (μ, ϑ, I)
Absolutely low important	(0.1, 0.9, 0.0)
Very low important	(0.2, 0.8, 0.1)
Low important	(0.3, 0.7, 0.2)
Slightly low important	(0.4, 0.6, 0.3)
Equally important	(0.5, 0.5, 0.5)
Slightly much important	(0.6, 0.4, 0.3)
Much important	(0.7, 0.3, 0.2)
Very much important	(0.8, 0.2, 0.1)
Absolutely much important	(0.9, 0.1, 0.0)

Table 2 The district evaluation matrix

Districts / Factors	F_1	F_2	…	F_n
D_1	$(\mu_{11}, \vartheta_{11}, I_{11})$	$(\mu_{12}, \vartheta_{12}, I_{12})$	…	$(\mu_{1n}, \vartheta_{1n}, I_{1n})$
D_1	$(\mu_{21}, \vartheta_{21}, I_{21})$	$(\mu_{22}, \vartheta_{22}, I_{22})$	…	$(\mu_{2n}, \vartheta_{2n}, I_{2n})$
D_1	⋮	⋮	⋮	⋮
D_1	$(\mu_{m1}, \vartheta_{m1}, I_{m1})$	$(\mu_{m2}, \vartheta_{m2}, I_{m2})$	…	$(\mu_{mn}, \vartheta_{mn}, I_{mn})$

Table 3 Reference potential risk degree

Factors	F_1	F_2	...	F_n
RPRD	$(\mu_1, \vartheta_1, \pi_1)$	$(\mu_2, \vartheta_2, \pi_2)$	...	$(\mu_n, \vartheta_n, \pi_n)$

Step 4 Obtain the final PRD of each district.

The obtained PRD of districts will be used in the proposed mathematical model to get the optimal flight path of UAVs. In the next section, the developed mathematical model will be presented and explained.

4.2 Proposed Mathematical Model to Obtain the Flight Trajectory

After calculating the PRD for each district, the optimal flight path could be obtained. To this end, a single objective mathematical model is presented. The aim of this model finds the optimal trajectory for UAVs, considering corresponding constraints, so that the districts with high PRD be covered as soon as possible. To designing the model, some assumptions have to be considered as follows:

- Each district has to be visited just once.
- Each path has to be travelled just once.
- The UAV has flight time limitation due to its power restriction.
- UAV starts from a point, visit districts as much as possible and returns to the starting point.
- UAV has a constant average velocity.

Taking into accounts the above assumptions, the mathematical model can be formulated as follows:

Sets

$Q = \{1, 2, \ldots, n\}$ Set of candidate districts.

Indices

$i \subset Q$ Index of origin districts.
$j \subset Q$ Index of destination districts.

Parameters

t_{ij} Travelling time between origin i and destination j.
a_j Searching time of each district.
T Maximum flight time of UAV.
P_j Potential risk degree of district j.

Decision Variables

y_{ij} 1, If the UAV travel from origin i to destination j; 0, otherwise.

Objective Function

$$MaxZ = \sum_{i \subset Q, i \neq j} \sum_{j \subset Q} P_j * y_{ij} \tag{21}$$

Constraints

$$\sum_{i \subset Q, i \neq j} y_{ij} \leq 1 \quad \forall \mathrm{j} \tag{22}$$

$$\sum_{j \subset Q, i \neq j} y_{ij} \leq 1 \quad \forall \mathrm{i} \tag{23}$$

$$y_{ij} \leq 1 - y_{ji} \quad \forall \mathrm{i, j} \tag{24}$$

$$\sum_{j \subset Q} y_{ij} - \sum_{j \subset Q} y_{ji} = 0 \quad \forall \mathrm{i} \tag{25}$$

$$\sum_{i \subset Q, i \neq j} \sum_{j \subset Q} y_{ij} \leq |Q| - 1 \quad \forall \mathrm{Q} \subseteq \{1, 2, \ldots, n\}, |Q| \geq 2 \tag{26}$$

$$\sum_{i \subset Q, i \neq j} \sum_{j \subset Q} y_{ij} * \left(t_{ij} + a_j\right) \leq T \tag{27}$$

$$y_{ij} \in \{0, 1\} \tag{28}$$

The objective function tries to select districts with high potential risk degrees by maximizing PRD's total amount. Constraints (22) and (23) guarantee that at most one path could be established between two consecutive districts. Constraint (24) ensures that there is no return route between the two areas. Constraints (25) and (26) enforce that there is only a single tour covering districts. Constraint (27) shows the flight time limitation of UAV. Finally, constraint (28) indicates that the decision variable is binary.

Solving the above model gives the optimal path for UAV flight so that the districts with a high degree of the potential risk to be covered as much as possible. In the following section, the application of the proposed methodology will be shown.

5 Application

In a rapid-onset disaster situation, such as an earthquake, a great number of injured people need urgent care, especially in the early post-disaster times [36]. Finding and rescuing casualties in the early hours after a disaster can reduce the number of casualties. Search and rescue operations are carried out for this purpose. This section demonstrates the applicability of the proposed methodology by a numerical example of an earthquake. Suppose, after an earthquake, SAR teams try to search a specified area with UAVs. The target area is divided into ten locations. The aerial travel time among districts is given in Table 4. These amounts are obtained using Eq. (29). We supposed that just one UAV is utilized in operation.

$$t_{ij} = \frac{x_{ij}}{v} \tag{29}$$

The velocity of the UAV is constant and considered 60 km/h. When the UAV reaches any place, the search for that place takes some time as given in Table 5.

Besides, The UAV has a restriction in flight time due to its power limitation. We assume the maximum flight time is one hour. We aim to find the optimal path among these locations to consider all limitations of the problem, and the UAV visit locations are based on their potential risk degree. The model's objective function prioritizes districts with high potential risk degrees by considering all the model's constraints.

As mentioned in Sect. 4.1, the first step of the solution process is determining all districts' PRD. To this end, we will apply the concept of similarity measures.

Table 4 The travel time among locations

District	A	B	C	D	E	F	G	H	I	J
A	0	5	8	6	10	3	5	7	12	2
B	5	0	4	7	9	2	11	14	4	6
C	8	4	0	15	12	4	8	11	2	7
D	6	7	15	0	5	7	10	3	14	6
E	10	9	12	5	0	2	8	7	11	4
F	3	2	4	7	2	0	7	12	3	9
G	5	11	8	10	8	7	0	14	5	9
H	7	14	11	3	7	12	14	0	11	9
I	12	4	2	14	11	3	5	11	0	5
J	2	6	7	6	4	9	9	9	5	0

Table 5 The searching time of each district

District	A	B	C	D	E	F	G	H	I	J
Searching time	5	4	7	9	6	4	3	5	7	6

First of all, the district evaluation matrix has to be established based on Table 1. All locations are evaluated using a spherical fuzzy linguistic scale based on historical information and an expert. The obtained evaluation matrix is shown in Table 6. To determine the potential risk degree, four critical factors are used. The candidate locations are evaluated against these factors. These factors are population density (F1), the strength of buildings (F2), the strength of soil and land (F3), and distance from faults (F4). The corresponding weight of each factor is determined as $w = \{0.25, 0.30, 0.20, 0.25\}$. These essential factors are utilized to estimate the PRD of a location after an earthquake, even though there are many factors in the literature.

To apply the similarity degree functions, the reference potential risk degree for each factor is required. This metric indicates a location with the highest potential risk degree. By comparing locations to this metric, the PRD of each district can be obtained. The RPRD is presented in Table 7.

Table 6 The district evaluation matrix

Factors / Districts	Population density	Strength of buildings	Strength of soil and land	Distance from faults
A	(0.6, 0.4, 0.3)	(0.4, 0.6, 0.3)	(0.4, 0.6, 0.3)	(0.6, 0.4, 0.3)
B	(0.7, 0.3, 0.2)	(0.5, 0.5, 0.5)	(0.6, 0.4, 0.3)	(0.7, 0.3, 0.2)
C	(0.9, 0.1, 0.0)	(0.7, 0.3, 0.2)	(0.3, 0.7, 0.2)	(0.5, 0.5, 0.5)
D	(0.3, 0.7, 0.2)	(0.8, 0.2, 0.1)	(0.8, 0.2, 0.1)	(0.6, 0.4, 0.3)
E	(0.5, 0.5, 0.5)	(0.6, 0.4, 0.3)	(0.4, 0.6, 0.3)	(0.5, 0.5, 0.5)
F	(0.4, 0.6, 0.3)	(0.5, 0.5, 0.5)	(0.4, 0.6, 0.3)	(0.6, 0.4, 0.3)
G	(0.8, 0.2, 0.1)	(0.6, 0.4, 0.3)	(0.3, 0.7, 0.2)	(0.5, 0.5, 0.5)
H	(0.6, 0.4, 0.3)	(0.6, 0.4, 0.3)	(0.5, 0.5, 0.5)	(0.6, 0.4, 0.3)
I	(0.3, 0.7, 0.2)	(0.4, 0.6, 0.3)	(0.7, 0.3, 0.2)	(0.6, 0.4, 0.3)
J	(0.9, 0.1, 0.0)	(0.4, 0.6, 0.3)	(0.4, 0.6, 0.3)	(0.4, 0.6, 0.3)

Table 7 Reference potential risk degree of factors

	Population density	Strength of buildings	Strength of soil and land	Distance from faults
RPRD	(0.9, 0.1, 0.0)	(0.1, 0.9, 0.0)	(0.1, 0.9, 0.0)	(0.1, 0.9, 0.0)

Table 8 The PRD of each district based on three different similarity measures

District	Jaccard similarity measure	Exponential similarity measure	Square root cosine similarity measure
A	0.47	0.56	0.87
B	0.35	0.52	0.71
C	0.51	0.63	0.77
D	0.09	0.41	0.41
E	0.33	0.49	0.76
F	0.30	0.49	0.74
G	0.53	0.61	0.83
H	0.31	0.50	0.74
I	0.26	0.48	0.64
J	0.67	0.70	0.96
Total	3.82	5.39	7.43

After calculating the similarity degrees using three different similarity functions mentioned in Eqs. (16), (18) and (20), each district's final potential risk degree is obtained in Table 8. As can be seen, the PRD of locations changes in different similarity measures due to these similarity measures' properties. For example, the Jaccard similarity measure is the most pessimistic one among them, while the square root cosine similarity measure has more optimistic characteristics. These metrics show the potential risk degree of each district. It means that in locations with higher PRD, the probability of damage, consequently, the higher the number of casualties. Therefore, these districts should have priority in search and rescue operations. Due to the other restrictions and limitations of the SAR operations, all districts may not be searched, but most of the locations with higher potential risk degrees will be searched. By adjusting and modifying the operation's limitations, it is possible to search and rescue more injured. As mentioned before, the obtained PRDs can be utilized in the proposed mathematical model as objective function coefficients. In the following, the results of the model will be presented.

In this step, using obtained PRDs, parameters, and assumptions, the proposed mathematical model is applied to get the optimal flight path. The results of three different similarity measures are shown in Table 9 and compared with them.

The covered districts in exponential and square root similarity measures are the same. Although, in all similarity measures, the districts A, G, F, E, and J are visited by UAV, the Jaccard similarity measure gives the less covered districts than other two methods. As can be seen in Table 8, all visited locations have the highest potential risk degree. In case of objective function, more than 70% of PRD is covered in all similarity measures (2.81 out of 3.82 in Jaccard, 3.85 out of 5.39 in exponential and 5.51 out of 7.43 in square root similarity measure). The coverage rate obtained by different similarity measures are given in Table 10. Also, the graphical representation

Table 9 Results of the proposed model with different similarity measures

Similarity measure	Objective function value	Optimal path and covered locations
Jaccard similarity measure	2.81	A-G-C-F-E-J
Exponential similarity measure	3.85	A-G-I-B-F-E-J
Square root cosine similarity measure	5.51	A-G-I-B-F-E-J

Table 10 The coverage rate obtained by different similarity measures

Similarity measure	Total PRD of districts	Total PRD of covered districts	Coverage rate (%)
Jaccard similarity	3.82	2.81	74
Exponential similarity	5.39	3.85	71
Square root cosine similarity	7.43	5.51	74

of obtained optimal flight path is shown in Figs. 2 and 3. These figures show each district's visiting order by UAV (green arrow) and covered locations (green points). Also, the black points show the unvisited districts.

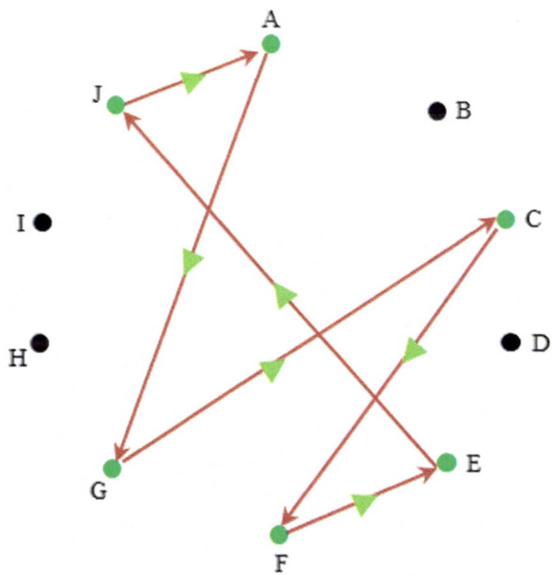

Fig. 2 The flight path of UAV using Jaccard similarity measure

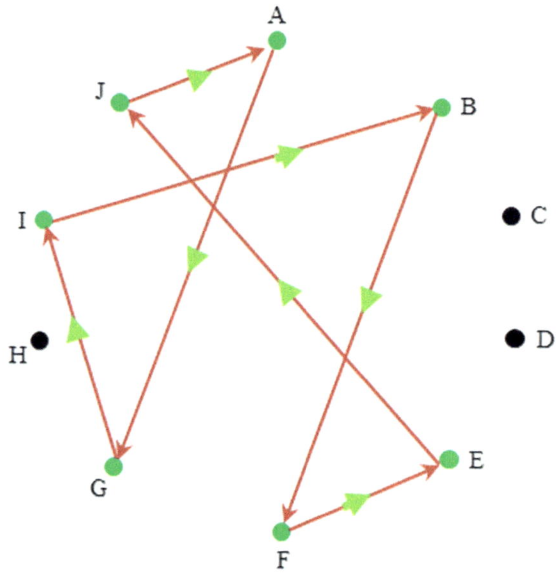

Fig. 3 The flight path of UAV using exponential and square root similarity measures

6 Sensitivity Analyses

To illustrate the validity of the proposed model, a sensitivity analysis is performed. In this problem, the important parameter that can affect the final results and optimal flight path is UAV's flight time limitation. By increasing the flight time, using extra UAVs, or using UAVs with a powerful energy source, further districts can be covered and searched. To investigate the effects of flight time, we performed a sensitivity analysis on this parameter. The flight time is considered one hour in the original model. It is changed as 40, 50, 70 and 80 min. The results are shown in Table 11.

It's clear that by increasing the flight time, the UAV can cover more locations with a high PRD. Hence, more casualties can be saved. The coverage rates after these changes are presented in Table 12. A significant increase in coverage rate is visible by enhancement the flight time duration of UAV. Also, we present the results in Fig. 4, to show the coverage rate changes for more straightforward interpretation and understanding. If it is not possible to increase flight time, using extra UAVs also could be another alternative. In this case, another model can be developed by considering several multi-UAVs for further research.

7 Conclusion

UAVs become increasingly popular as searching and identifying tools in search and rescue operations due to their relatively low cost, ease of operation, and usability in most geographical and climatic conditions. In this research, a mathematical model

Table 11 Results of the sensitivity analysis

Maximum flight time (min)	Jaccard similarity		Exponential similarity		Square root cosine similarity	
	Objective function	Optimal flight path	Objective function	Optimal flight path	Objective function	Optimal flight path
40	2	A-G-E-J	2.76	A-B-F-E-J	4.04	A-B-F-E-J
50	2.53	A-G-C-B-J	3.02	A-J-B-C-G	4.17	A-G-F-C-J
70	3.42	A-J-E-F-B-C-I-G	4.48	A-G-I-C-B-F-E-J	6.28	A-J-E-F-B-C-I-G
80	3.42	A-J-E-F-B-C-I-G	4.48	A-G-I-C-B-F-E-J	6.28	A-J-E-F-B-C-I-G

Table 12 The coverage rates after sensitivity analysis

Maximum flight time (min)	Coverage rate		
	Jaccard similarity (%)	Exponential similarity (%)	Square root cosine similarity (%)
40	52	51	54
50	66	56	56
70	90	83	85
80	90	83	85

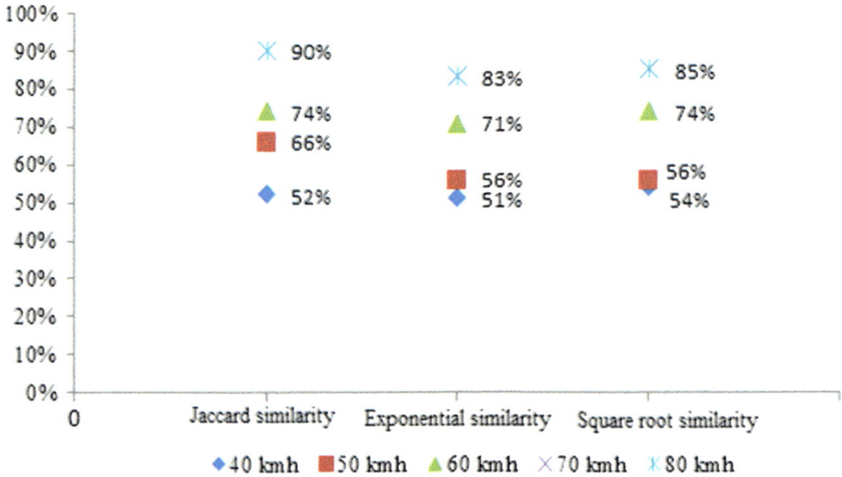

Fig. 4 The graphical representation of coverage rate

is proposed to design an optimal path for UAVs in search and rescue operation in the post-disaster case, taking into accounts the concept of a potential risk degree. To calculate the PRD, different similarity measures of the spherical fuzzy sets were utilized. To demonstrate the applicability of the proposed method, the numerical example of a post-disaster situation was considered. Two similarity measures out of three resulted in the same path. To prove the validity of the model, a sensitivity analysis is performed. The results show that the number of covered districts could be increased by increasing UAVs' flight time.

The study had some assumptions that could be considered as limitations of research. For instance, just one UAV is considered for searching operation, whereas more than one could be utilized to increase the coverage rate. The number of candidate locations was few, while many areas need to be explored in real case problems. In future, these limitations can be handled.

For future studies, a real case study can be also considered to get more realistic results that require an innovative solution methodology based on heuristic or meta-heuristic algorithms to solve it.

References

1. San Juan, V., Santos, M., Andújar, J.M.: Intelligent UAV map generation and discrete path planning for search and rescue operations. Complexity **2018**, 1–17 (2018)
2. Chen, S.W., Li, Y.Z., Xing, S.Q., Wang, X.S., Sato, M.: Urban damage evaluation using polarimetric SAR data. In International Geoscience and Remote Sensing Symposium (IGARSS), pp. 2754–2757 (2014)
3. Xu, Z., et al.: Development of an UAS for post-earthquake disaster surveying and its application in Ms7.0 Lushan Earthquake, Sichuan, China. Comput. Geosci. **68**(2014), 22–30 (2014)
4. Chen, J., et al.: Damage degree evaluation of earthquake area using UAV aerial image. Int. J. Aerosp. Eng. **2016**, 1–10 (2016)
5. Waharte, S., Trigoni, N.: Supporting search and rescue operations with UAVs. In 2010 International Conference on Emerging Security Technologies, pp. 142–147 (2010)
6. Jun, M., Andrea, R.D'.: Path planning for unmanned aerial vehicles in uncertain and adversarial environments. In Butenko, S., Murphey, R., Pardalos, P.M. (eds.) Cooperative Control: Models, Applications and Algorithms, pp. 95–110. Springer, Berlin (2003)
7. Goodrich, M.A., et al.: Supporting wilderness search and rescue using a camera-equipped mini UAV: research articles. J. Field Robot. **25**(1–2), 89–110 (2008)
8. Półka, M., Ptak, S., Kuziora, Ł., Kuczyńska, A.: The use of unmanned aerial vehicles by urban search and rescue groups. In Drones—Applications, pp. 83–95 (2018)
9. Bebis, G., et al. (eds.): Advances in Visual Computing, vol. 9474. Springer International Publishing, Cham (2015)
10. Silvagni, M., et al.: Multipurpose UAV for search and rescue operations in mountain avalanche events avalanche events, vol. 5705 (2017)
11. Avezum, M., Seitz, A., Bruegge, B.: MODCAP : a platform for cooperative search and rescue missions (2019)
12. S. K. B et al.: Towards highly reliable autonomy for urban search and rescue robots. Springer International Publishing Switzerland, pp. 118–129 (2015)
13. Ghazali, S.N.A.M., Anuar, H.A., Alsagoff, S.N.S.Z., Yusoff, Z.: Determining position of target subjects in maritime search and rescue (MSAR) operations using rotary wing unmanned aerial vehicles (UAVs). In 2016 International Conference on Information and Communication Technology (ICICTM), pp. 1–4 (2016)
14. Rahmes, M., Chester, D., Hunt, J., Chiasson, B.: Optimizing cooperative cognitive search and rescue UAVs. In Autonomous Systems: Sensors, Vehicles, Security and the Internet of Everything (2018)
15. Cui, J.Q., et al.: Drones for cooperative search and rescue in post-disaster situation. In 2015 IEEE 7th International Conference on Cybernetics and Intelligent Systems (CIS) and IEEE Conference on Robotics, Automation and Mechatronics (RAM), pp. 167–174 (2015)
16. Phung, M.D., Quach, C.H., Dinh, T.H., Ha, Q.: Enhanced discrete particle swarm optimization path planning for UAV vision-based surface inspection. Autom. Construct. **81**, 25–33 (2017)
17. Kallmann, M., Kapadia, M.: Geometric and discrete path planning for interactive virtual worlds. Synth. Lect. Vis. Comput. **8**(1), 1–201 (2016)
18. Sujit, P.B., Saripalli, S., Sousa, J.B.: An evaluation of UAV path following algorithms. In 2013 European Control Conference (ECC), pp. 3332–3337 (2013)
19. Hernandez-Martinez, E.G., Ferreira-Vazquez, E.D., Fernandez-Anaya, G., Flores-Godoy, J.J.: Formation tracking of heterogeneous mobile agents using distance and area constraints. Complexity **2017**, 1–13 (2017)
20. Kamrani, F., Ayani, R.: UAV path planning in search operations. In Aerial Vehicles, InTech (2009)
21. Besada-Portas, E., de la Torre, L., Moreno, A., Risco-Martín, J.L.: On the performance comparison of multi-objective evolutionary UAV path planners. Inf. Sci. **238**, 111–125 (2013)
22. Auñón, P.G., Peñas, M.S.: Use of genetic algorithms for unmanned aerial systems path planning. In Decision Making and Soft Computing—Proceedings of the 11th International FLINS Conference, FLINS 2014, pp. 430–435 (2014)

23. Davoodi, M., Panahi, F., Mohades, A., Hashemi, S.N.: Multi-objective path planning in discrete space. Appl. Soft Comput. **13**(1), 709–720 (2013)
24. Izzo, D.: Optimization of interplanetary trajectories for impulsive and continuous asteroid deflection. J. Guid. Control Dyn. **30**(2), 401–408 (2007)
25. Yang, P., Tang, K., Lozano, J.A.: Estimation of distribution algorithms based unmanned aerial vehicle path planner using a new coordinate system. In 2014 IEEE Congress on Evolutionary Computation (CEC), pp. 1469–1476 (2014)
26. Ruz, J.J., Pajares, G., Jesus, M., Arevalo, O.: UAV trajectory planning for static and dynamic environments. INTECH Open Access Publisher (2009)
27. Zadeh, L.A.: Fuzzy sets. Inf. Control **8**(3), 338–353 (1965)
28. Seyfi-Shishavan, S.A., Kutlu Gündoğdu, F., Farrokhizadeh, E., Donyatalab, Y., Kahraman, C.: Novel similarity measures in spherical fuzzy environment and their applications. Eng. Appl. Artif. Intell. **94**, 103837 (2020)
29. Gündoğdu, F.K., Kahraman, C.: Spherical fuzzy sets and spherical fuzzy TOPSIS method. J. Intell. Fuzzy Syst. **36**(1), 337–352 (2019)
30. Chiclana, F., Tapia García, J.M., del Moral, M.J., Herrera-Viedma, E.: A statistical comparative study of different similarity measures of consensus in group decision making. Inf. Sci. **221**, 110–123 (2013)
31. Garcia-Aunon, P., Peñas, M.S., de la Cruz García, J.M.: Parameter selection based on fuzzy logic to improve UAV path-following algorithms. J. Appl. Logic **24**, 62–75 (2017)
32. Valavanis, K.P., Vachtsevanos, G.J.: UAV mission and path planning: ıntroduction. In Handbook of Unmanned Aerial Vehicles, pp. 1443–1446. Springer Netherlands, Dordrecht (2015)
33. Huang, C., et al.: A new dynamic path planning approach for unmanned aerial vehicles. Complexity **2018**, 1–17 (2018)
34. Fu, Z., Yu, J., Xie, G., Chen, Y., Mao, Y.: A heuristic evolutionary algorithm of UAV path planning. Wirel. Commun. Mobile Comput. **2018**, 1–11 (2018)
35. Kamranzad, F., Memarian, H., Zare, M.: Earthquake risk assessment for Tehran, Iran. ISPRS Int. J. Geo-Inf. **9**(7), 430 (2020)
36. Salman, F.S., Gül, S.: Deployment of field hospitals in mass casualty incidents. Comput. Ind. Eng. **74**(1), 37–51 (2014)

Index

C. Kahraman and S. Aydın (eds.), *Intelligent and Fuzzy Techniques in Aviation 4.0*, Studies in Systems, Decision and Control 372,
https://doi.org/10.1007/978-3-030-75067-1

Printed by Printforce, the Netherlands